Distributionally Robust Optimization and its Applications in Power System Energy Storage Sizing

Rui Xie · Wei Wei

Distributionally Robust Optimization and its Applications in Power System Energy Storage Sizing

Springer

Rui Xie
Department of Electrical Engineering
Tsinghua University
Beijing, China

Wei Wei
Department of Electrical Engineering
Tsinghua University
Beijing, China

ISBN 978-981-97-2565-6 ISBN 978-981-97-2566-3 (eBook)
https://doi.org/10.1007/978-981-97-2566-3

This Springer imprint is published by the registered company Springer Nature Singapore Pte Ltd.
The registered company address is: 152 Beach Road, #21-01/04 Gateway East, Singapore 189721, Singapore

Preface

The transition of power systems to higher renewable energy penetration is a worldwide trend. However, the high volatility of renewable energy introduces significant uncertainties, creating challenges for efficient power management. Energy storage emerges as a viable solution to enhance the flexibility of the power system when facing these uncertainties, whereas its capacity has essential impacts on the performance and economy. However, in the planning stage, acquiring precise probability distributions of these uncertainties appears challenging. This necessitates the consideration of ambiguous uncertainty in energy storage sizing, leading to the adoption of distributionally robust optimization (DRO), which is an emerging and promising mathematical tool for optimization under ambiguous uncertainty and has advantages in balancing robustness and optimality.

The primary objective of this book is to equip readers with the skills to apply DRO to real-world decision-making problems under ambiguous uncertainties. To this end, this book introduces readers to the mathematical foundations and theories of DRO. It then provides practical implementations of DRO techniques in engineering decision-making scenarios, with a special focus on energy storage sizing in renewable-integrated power systems.

In particular, the main content of this book is divided into two parts: Part I is devoted to the mathematical theory of DRO, which first provides a preliminary as a necessary foundation, then focuses on the DRO modeling and solving methods, including both the general theory and specialized theories under different ambiguity sets established using moments, f-divergence, and Wasserstein distance. The book then transitions from theory to practice. Part II summarizes our research on DRO-based energy storage sizing approaches, focusing on the challenges of handling ambiguous uncertainties in various renewable-integrated power systems such as stand-alone microgrids, large-scale renewable power plants, bulk power grids, and multi-carrier energy networks.

This book intends to bridge the gap between DRO theories and their practical applications. It is designed to appeal to a broad audience including researchers, graduate students, and industry professionals. This book provides clear explanations and accessible guidance for both beginners and experts in the fields of electrical engineering and operations research.

In retrospect, the creation of this book has been an instructive and beneficial journey, and we hope that it serves the purpose of providing an efficient and reliable approach to real-world energy system design under uncertainties.

Beijing, China
February 2024

Rui Xie
Wei Wei

Acknowledgements

The authors sincerely express their appreciation to Prof. Qiang Lu and Prof. Shengwei Mei from Tsinghua University and Prof. Yue Chen from the Chinese University of Hong Kong for their support for the research. We also would like to thank Dr. Yang Cao from Tsinghua University and Dr. Zhongjie Guo from the University of Electronic Science and Technology of China for their contributed materials. We acknowledge Ms. Ella Zhang, Mrs. Sridevi Purushothaman, and other staff from Springer Nature for their help in the book preparation.

The authors would like to acknowledge the financial support from the National Key Research and Development Program of China under Grant No. 2021YFB2400701.

Contents

Acronyms

AC	Alternating current
C&CG	Column-and-constraint generation
CDF	Cumulative distribution function
CVaR	Conditional value-at-risk
DC	Direct current
DDU	Decision-dependent uncertainty
DRO	Distributionally robust optimization
ES	Energy storage
EV	Electric vehicle
KKT	Karush-Kuhn-Tucker
KL	Kullback-Leibler
LP	Linear programming
MCS	Monte Carlo simulation
MILP	Mixed-integer linear programming
PDF	Probability density function
p.u.	Per unit
PV	Photovoltaic
RO	Robust optimization
SAA	Sample average approximation
SDP	Semidefinite programming
SoC	State-of-charge
SOCP	Second-order cone programming
SP	Stochastic programming
VaR	Value-at-risk
WT	Wind turbine

Notations

$\varnothing$	The empty set
$\mathbb{R}$	Set of real numbers
$\mathbb{R}^M$	Set of $M \times 1$-dimensional real vectors
$\mathbb{R}_+$	Set of nonnegative real numbers
$\mathbb{R}_+^M$	Nonnegative orthant in $\mathbb{R}^M$
$\mathbb{C}$	Set of complex numbers
$\mathbb{Z}$	Set of integers
$\mathbb{N}$	Set of nonnegative integers
$\mathbb{N}_+$	Set of positive integers
$\mathbb{S}^M$	Set of $M \times M$-dimensional symmetric real matrices
$\mathbb{S}_+^M$	Set of $M \times M$-dimensional symmetric positive semidefinite real matrices
$\mathbb{S}_{++}^M$	Set of $M \times M$-dimensional symmetric positive definite real matrices
$\mathbb{L}^{M+1}$	Second-order cone in $\mathbb{R}^{M+1}$
C^1	Set of continuously differentiable functions
C^2	Set of twice continuously differentiable functions
$\mathcal{M}(\Omega)$	Set of probability distributions on sample space Ω
$\mathcal{J}(\mathbb{P}_1, \mathbb{P}_2)$	Set of joint probability distributions with marginals $\mathbb{P}_1$ and $\mathbb{P}_2$
$\mathrm{cl}(S)$	Closure of set S
$\mathrm{diam}(S)$	Diameter of set S
$\mathrm{conv}(S)$	Convex hull of set S
$\mathrm{int}(S)$	Interior of set S
∂S	Boundary of set S
$\mathrm{aff}(S)$	Affine hull of set S
$\mathrm{relint}(S)$	Relative interior of set S
K^*	Dual cone of cone K
$1_S(\cdot)$	Indicator function of set S
$\chi_S(\cdot)$	Characteristic function of set S
$h_S(\cdot)$	Support function of set S
$\exp(\cdot)$	Exponential function

$\log(\cdot)$	Natural logarithm function
$(\cdot)_+$	Positive part function
$(\cdot)_-$	Negative part function
$A^\top$	Transpose of matrix A
$\mathrm{tr}(A)$	Trace of matrix A
$\mathrm{diag}(A_1, \ldots, A_n)$	Block diagonal matrix with diagonal blocks $A_1, \ldots, A_n$
$\lvert S \rvert$	The number of elements in set S
$\lVert \cdot \rVert$	A norm
$\lVert \cdot \rVert_*$	Dual norm of $\lVert \cdot \rVert$
$\lVert \cdot \rVert_1$	1-norm
$\lVert \cdot \rVert_2$	2-norm
$\lVert \cdot \rVert_\infty$	∞-norm
$\lVert \cdot \rVert_F$	Frobenius norm
$D_f(\mathbb{P}' \lvert\rvert \mathbb{P})$	f-divergence from $\mathbb{P}'$ to $\mathbb{P}$
$D_{\mathrm{KL}}(\mathbb{P}' \lvert\rvert \mathbb{P})$	Kullback-Leibler divergence from $\mathbb{P}'$ to $\mathbb{P}$
$D_{\chi^2}(\mathbb{P}' \lvert\rvert \mathbb{P})$	Pearson χ^2 divergence from $\mathbb{P}'$ to $\mathbb{P}$
d_{D}	Discrete metric
$d_{\mathrm{TV}}(\mathbb{P}, \mathbb{P}')$	Total variation distance between $\mathbb{P}$ and $\mathbb{P}'$
$d_{\mathrm{H}}(\mathbb{P}, \mathbb{P}')$	Hellinger distance between $\mathbb{P}$ and $\mathbb{P}'$
$d_{\mathrm{W}}(\mathbb{P}_1, \mathbb{P}_2)$	Wasserstein distance between $\mathbb{P}_1$ and $\mathbb{P}_2$
$\langle \cdot, \cdot \rangle$	Inner product
$\mathbb{P}_2 \ll \mathbb{P}_1$	Probability distribution $\mathbb{P}_2$ is absolutely continuous with respect to probability distribution $\mathbb{P}_1$
$A \succeq 0$	Matrix A is positive semidefinite
$A \succ 0$	Matrix A is positive definite
$x \preceq_K y$	$y - x \in K$ with a proper cone K
$x \prec y$	$y - x \in \mathrm{int}(K)$ with a proper cone K
$\mathrm{Lip}(f)$	Lipschitz modulus of function f
$\mathrm{graph}(f)$	Graph of function f
$\mathrm{epi}(f)$	Epigraph of function f
$\partial f(x)$	Subdifferential of function f at point x
f^*	Conjugate function of function f
∇f	Gradient of function f
$\nabla^2 f$	Hessian matrix of function f
$\mathbb{E}^{\mathbb{P}}[X(\omega)]$	Expectation of random variable $X(\omega)$ when the probability distribution is $\mathbb{P}$
$\mathrm{Var}(X)$	Variance of random variable X
$\mathrm{Cov}(X, Y)$	Covariance of random variables X and Y
$\mathcal{N}(\mu, \sigma^2)$	Normal distribution with mean μ and variance σ^2
δ_x	Dirac distribution at point x

Chapter 1
Introduction

Abstract This chapter begins by introducing the emerging energy storage (ES) sizing problem as the power system background of decision-making under uncertainties. Then the key concepts of optimization in the uncertain environment are summarized, including the basic ideas of stochastic programming, robust optimization, and distributionally robust optimization. The chapter then shifts to the critical role of ES in power systems, analyzing the diverse usages of ES and introducing the models for ES operation. Subsequently, a comprehensive review is provided for ES sizing under different scenarios from the perspectives of the generation side, the grid side, and the distribution/demand side. A holistic view of the book's content is offered in the end.

1.1 Background of Energy Storage Sizing Under Uncertainties

Global climate change has drawn significant attention worldwide. To combat global warming, many countries have promulgated plans and policies aimed at reducing carbon emissions. For instance, China announced the goal of carbon peaking in 2030 and carbon neutrality in 2060. The utilization of renewable energy, such as wind and photovoltaic (PV) power, contributes to environmental protection, fossil fuel consumption reduction, and carbon emission decrease. Consequently, the rapid global development of renewable energy is expected to continue, becoming a primary approach towards carbon neutrality.

However, integrating renewable energy into power systems presents challenges in maintaining secure and economic operation, due to the inherent volatility and randomness of renewable power generation. Wind and PV power outputs are limited by real-time weather conditions. In addition, the changing trends of renewable power generation and load demand curves often differ, and so do their peak and valley times. For example, PV power plants can only generate energy during daylight hours, while load demands persist into the night. The volatile and random nature of renewable

R. Xie and W. Wei, *Distributionally Robust Optimization and its Applications in Power System Energy Storage Sizing*,
https://doi.org/10.1007/978-981-97-2566-3_1

power generation puts burdens on power balance and renewable accommodation in power systems.

For systems with a high proportion of renewable power generation, these challenges become even more pronounced. On the one hand, as the share of renewable energy grows, the associated volatility intensifies, increasing the power system's demand for power balancing and ramping capabilities. On the other hand, the declining proportion of traditional generators results in a reduction of flexible resources. Therefore, as power systems transition towards low-carbon goals through renewable energy development, the proportion of renewable energy grows and the power balancing and renewable energy accommodation problems become more critical to address.

The equipment of energy storage (ES) in power systems can help with these problems. ES can store energy in mechanical or chemical forms and achieve temporal electrical energy shifting through charging and discharging processes. Although the use of ES in power systems has grown significantly in recent years, electrochemical ES remains costly. The ES sizing problem is a practical and challenging issue faced by low-carbon-oriented and renewable-integrated power systems, aiming to strike a balance between the economic feasibility and the performance of ES equipment. The main challenges stem from the volatile and stochastic nature of renewable power generation and the temporal correlation characteristic of ES. Therefore, in the remaining part of this chapter, we first introduce the optimization methods considering uncertainties. Then summarize the ES techniques and operation models in power systems. A review of the storage sizing methods under various scenarios is also provided.

1.2 Optimization in the Uncertain Environment

Optimization theory has wide-ranging applications in science, engineering, and various other fields. In real-world scenarios, uncertain factors often arise as parameters in optimization problems. To address this, optimization theories for uncertain environments are required, enabling the modeling, analysis, and solving of optimization problems with random parameters.

Optimization methods that take uncertainties into account include stochastic programming (SP), robust optimization (RO) (Ben-Tal et al. 2009), and distributionally robust optimization (DRO) (Goh and Sim 2010). SP assumes that the accurate probability distribution is known, but it may have drawbacks under biased or inaccurate distribution information. RO considers the worst-case scenario of uncertainty realization, yielding more robust solutions with feasibility guarantees. DRO accounts for the inexactness of empirical probability distributions through the use of ambiguity sets, i.e., pre-defined sets of probability distributions that contain the real probability distribution. This approach also offers theoretical guarantees, while potentially reducing the conservativeness of RO. An appropriate optimization method should be selected based on the specific characteristics of the problem. In the following, we will introduce the basic ideas and main attributes of the three methods one by one.

1.2.1 Stochastic Programming

A typical SP model is given by:

$$\min_{x \in \mathcal{X}} \{f(x) := \mathbb{E}[g(x, \omega)]\} \ . \tag{1.1}$$

In this model, x is the decision variable taking values in the set $\mathcal{X}$. ω represents the uncertain parameters, with the probability distribution $\mathbb{P}$ defined over the sample space Ω. The function $g(x, \omega)$ is a function of both x and ω. For any fixed pair of values $x_0 \in \mathcal{X}$ and $\omega_0 \in \Omega$, the function value $g(x_0, \omega_0)$ can be calculated. If we fix x_0 while treating ω as a random variable, $g(x_0, \omega)$ is also a random variable. Since we cannot know the uncertainty realization, i.e., the value of ω, beforehand, we instead measure the "average" performance, which can be rigorously formulated as the expectation. The notation $\mathbb{E}[\cdot]$ denotes the operation of taking the expectation, which yields $\mathbb{E}[g(x_0, \omega)] = \int_\Omega g(x_0, \omega) \mathrm{d}\mathbb{P}(\omega)$, assuming the expectation exists. Allowing x to vary within $\mathcal{X}$, $\mathbb{E}[g(x, \omega)]$ is a function of x and we denote it by $f(x)$. The objective of the optimization model is to minimize $f(x)$, which may represent the expected loss. The expression $\min_{x \in \mathcal{X}} f(x)$ is a general mathematical programming model that does not explicitly account for uncertainty. However, uncertainty is incorporated into the model through the expectation operation, which defines the objective function $f(x)$.

Another prevalent SP model contains *chance constraints*:

$$\begin{aligned} &\min_{x \in \mathcal{X}} \ f(x) \\ &\text{s.t. } \mathbb{P}(\omega \in S(x)) \geq 1 - \alpha \ . \end{aligned} \tag{1.2}$$

In this model, $S(x)$ is a subset of Ω that depends on the decision variable x. Therefore, $\omega \in S(x)$ is a random event, with $\mathbb{P}(\omega \in S(x))$ representing the event's probability. The chance constraint $\mathbb{P}(\omega \in S(x)) \geq 1 - \alpha$ stipulates that the probability of the random event $\omega \in S(x)$ must be no less than $1 - \alpha$, where the probability threshold $1 - \alpha \in [0, 1]$. In practice, the event $\omega \in S(x)$ may represent security constraints jointly influenced by the decision variable x and the random variable ω. The goal is to ensure security with a probability of at least $1 - \alpha$ with a small number α. Under this chance constraint, the objective function $f(x)$ is minimized, which may represent the cost.

In SP problems such as (1.1) and (1.2), the underlying probability distribution $\mathbb{P}$ of the random variable ω is assumed to be perfectly known. The probability distribution $\mathbb{P}$ significantly influences the optimal solution and its performance. Information about $\mathbb{P}$ is usually obtained through theoretical analysis and historical data, using either parametric or non-parametric approaches. With the parametric method, $\mathbb{P}$ is assumed to follow a specific type of probability distribution, such as normal distribution or Weibull distribution. Within this category, the probability distribution is uniquely determined by the values of certain parameters. For example, a normal distribution can be defined by its mean and variance. These parameters are typically estimated using historical data.

In the non-parametric method, $\mathbb{P}$ is constructed directly from the historical data without assuming a specific probability distribution type or requiring parameter estimation. Suppose there are N samples of ω in the historical data, denoted by $\omega_1, \omega_2, \ldots, \omega_N$. A basic non-parametric probability distribution estimation method involves assigning a probability of $1/N$ to each sample, forming the probability distribution $\mathbb{P}$. Formally,

$$\mathbb{P} := \frac{1}{N} \sum_{n=1}^{N} \delta_{\omega_n} , \tag{1.3}$$

where δ_{ω_n} represents the Dirac distribution with a probability of 1 at ω_n. Consequently, $\mathbb{P}$ constructed in (1.3) is a discrete probability distribution, which is the average of these Dirac distributions at sample data points. For any measurable set $S \subset \Omega$,

$$\mathbb{P}(S) = \frac{1}{N} \sum_{n=1}^{N} \delta_{\omega_n}(S) = \frac{\text{The number of sample data points in } S}{N} .$$

The expected cost can then be expressed as:

$$f(x) = \mathbb{E}^{\mathbb{P}}[g(x, \omega)] = \frac{1}{N} \sum_{n=1}^{N} g(x, \omega_n) ,$$

where $\mathbb{E}^{\mathbb{P}}[\cdot]$ denotes taking an expectation under the probability distribution $\mathbb{P}$. This approach for estimating the expectation is known as the *sample average approximation* (SAA) method.

We can see that the probability distribution $\mathbb{P}$ is usually not perfectly known in real-world scenarios, and this inexactness can significantly undermine the feasibility and optimality of the optimal solutions obtained from SP problems. For instance, if $\mathbb{P}$ is determined using the approximation in (1.3), then $\mathbb{P}$ is only supported on the set consisting of the N sample points. As a result, the optimal solution of the decision variable is merely guaranteed to be feasible for these samples. If the uncertainty realization falls outside of these samples, the optimal solution may be unable to deal with this realization. Another example is related to the problem in (1.2). When $\mathbb{P}$ deviates from the real probability distribution of ω, the probability threshold may become inaccurate, potentially causing the probability guarantee to no longer hold, which endangers the feasibility of the optimal solution. Even when the optimal solution remains feasible, discrepancies between $\mathbb{P}$ and the real probability distribution can make the expectation and event probability imprecise, resulting in suboptimal solutions in most cases.

1.2.2 Robust Optimization

Although estimating the probability distribution accurately is challenging, the sample space of the possible uncertainty realizations is often easier to obtain. Instead of optimizing the decision variable based on the probability distribution, RO seeks the best strategy concerning the worst-case scenario of the uncertainty realizations, where the sample space of the random variable is represented by a set called the uncertainty set. A typical RO problem is as follows:

$$\min_{x \in \mathcal{X}} \left\{ f(x) := \sup_{\omega \in \Omega} g(x, \omega) \right\} . \tag{1.4}$$

Similar to the problem in (1.1), $g(x, \omega)$ in (1.4) is a function of the decision variable x and random variable ω, representing the loss. The uncertainty set is denoted by Ω. For a fixed $x_0 \in \mathcal{X}$, $\sup_{\omega \in \Omega} g(x_0, \omega)$ computes the supremum of the possible losses, i.e., the loss under the worst-case scenario within the uncertainty set. Hence, $f(x)$ is the worst-case loss under the decision variable x. The optimization objective is to find the best choice of x so that the worst-case loss $f(x)$ is minimized.

RO can guarantee feasibility so long as the uncertainty set indeed encompasses all possible uncertainty realization values. In practice, this set can be derived from theoretical bounds and historical data analysis. For example, consider the uncertain power output ω of a wind farm. On the one hand, the power output should be non-negative and no larger than the installed capacity P. In this case, the uncertainty set can be chosen as $\Omega := [0, P]$ to ensure that $\omega \in \Omega$. On the other hand, suppose we have last year's power output data of this wind farm, with the minimum and maximum values being $\underline{P}$ and $\overline{P}$, respectively. It would then be reasonable to define the uncertainty set as $\Omega := [\underline{P}, \overline{P}]$.

The primary drawback of RO is its tendency to produce overly conservative solutions, which may compromise optimality. First, RO focuses on minimizing worst-case losses without considering average losses, while the latter serves as a statistical performance indicator that better reflects optimality. Second, in some instances, the uncertainty may present a highly unfavorable worst-case scenario that significantly inflates costs. However, such a worst-case scenario might rarely occur in practice. Third, RO relies solely on the information regarding the random variable sample space, without considering the probability distribution. Consequently, it lacks sufficient information to achieve satisfactory optimality. Therefore, RO sometimes gives overly conservative solutions.

Some studies have aimed at reducing the conservativeness of RO by selecting a more refined uncertainty set (Bertsimas and Sim 2004). Instead of making the uncertainty set Ω include all possible uncertainty realizations, a smaller subset can be chosen, so long as ensuring that the uncertainty realizations fall within the subset with a high enough probability. In other words, the uncertainty set Ω satisfies:

$$\mathbb{P}(\omega \in \Omega) \geq 1 - \alpha , \tag{1.5}$$

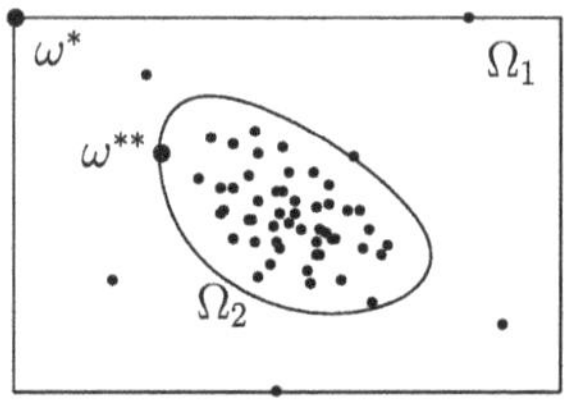

Fig. 1.1 Consider the case where ω is a 2-dimensional random variable with sample space Ω_1. Suppose the probability of ω being in the central area of Ω_1 is high, while the probability of ω being near the boundary of Ω_1 is low. In this scenario, an uncertainty set Ω_2 can be constructed, which is a proper subset of Ω_1. The worst-case scenarios in Ω_1 and Ω_2 are ω^* and ω^{**}, respectively. Then ω^{**} is no worse than ω^* and the conservativeness can be reduced if Ω_1 is replaced by Ω_2

which means the probability of ω lying within Ω is at least $1-\alpha$. An example is illustrated in Fig. 1.1. The RO problem (1.4) is then solved using this uncertainty set Ω. This approach allows for the exclusion of some extreme scenarios that would lead to a significant increase in cost, thereby enhancing the optimality of the solution. At the same time, there is a probability guarantee for unfavorable scenarios, ensuring that the solution is feasible and the realized cost does not exceed the obtained cost value of the optimization problem with a probability of at least $1-\alpha$. Formally, suppose the optimal value of (1.4) is O^* and the optimal solution is x^*, then

$$\mathbb{P}(g(x^*, \omega) \leq O^*) \geq 1 - \alpha \ .$$

Therefore, the uncertainty set improvement leads to solutions with enhanced optimality and probability-guaranteed robustness.

Despite its advantages, this method still exhibits some shortcomings. First, since (1.5) involves taking probability, constructing the uncertainty set relies on the information of the underlying probability distribution $\mathbb{P}$. As a consequence, this type of RO may also be affected by the inexactness of the estimated probability distribution, just like SP. When $\mathbb{P}$ significantly deviates from the real probability distribution, (1.5) may no longer provide an effective probability guarantee. Second, the optimization problem in (1.4) remains focused on the worst-case scenario, albeit within a smaller uncertainty set. The statistical performance, i.e., the expectation, is still not taken into consideration during the optimization, limiting the achievement of better optimality.

1.2.3 Distributionally Robust Optimization

We have seen that SP sometimes cannot guarantee feasibility and may produce suboptimal solutions due to inexact probability distributions, while RO can ensure feasibility but may result in overly conservative solutions because it only relies on the range of uncertainty realizations. DRO combines the ideas from both SP and RO, which is inspired by the following observation: Although the real probability distribution is unknown, it likely belongs to a set of probability distributions that can be determined based on available information. This set is called the ambiguity set. The optimization process considers the worst-case probability distribution within the ambiguity set. Even if the empirical probability distribution is not precise enough, as long as the ambiguity set contains the real probability distribution, the optimal solution is assured to exhibit the desired statistical performance under the real probability distribution.

A wide range of DRO problems shares the following form:

$$\min_{x \in \mathcal{X}} \left\{ f(x) := \sup_{\mathbb{P} \in \mathcal{P}} \mathbb{E}^{\mathbb{P}}[g(x, \omega)] \right\} . \tag{1.6}$$

Compared to the formulation of the SP problem in (1.1), there is an additional term $\sup_{\mathbb{P} \in \mathcal{P}}$ in the definition of $f(x)$ in problem (1.6). The set $\mathcal{P}$ represents the ambiguity set. For a fixed probability distribution $\mathbb{P} \in \mathcal{P}$, the expectation of loss under the decision variable x is $\mathbb{E}^{\mathbb{P}}[g(x, \omega)]$, with the superscript $\mathbb{P}$ emphasizing that the expectation is taken with respect to $\mathbb{P}$. The term $\sup_{\mathbb{P} \in \mathcal{P}} \mathbb{E}^{\mathbb{P}}[g(x, \omega)]$ represents the supremum of loss expectation as $\mathbb{P}$ varies within the ambiguity set $\mathcal{P}$. Therefore, $f(x)$ denotes the worst-case loss expectation under the decision variable x, with the objective of problem (1.6) being to minimize the worst-case loss expectation. Assume that x^* and O^* are the optimal solution and the optimal value of (1.6), respectively, and $\mathbb{P}^*$ is the real probability distribution. As long as $\mathbb{P}^* \in \mathcal{P}$, we have $\mathbb{E}^{\mathbb{P}^*}[g(x^*, \omega)] \leq O^*$, indicating that the real loss expectation is no larger than the obtained optimal value.

Expanding on the idea of chance constraints in problem (1.2), another type of DRO problem includes *robust chance constraints* as follows:

$$\begin{aligned} &\min_{x \in \mathcal{X}} f(x) \\ &\quad \text{s.t.} \inf_{\mathbb{P} \in \mathcal{P}} \mathbb{P}(\omega \in S(x)) \geq 1 - \alpha . \end{aligned} \tag{1.7}$$

Similar to problem (1.6), $\mathcal{P}$ is the ambiguity set. Hence, $\inf_{\mathbb{P} \in \mathcal{P}} \mathbb{P}(\omega \in S(x))$ is the infimum of the probability of the event $\omega \in S(x)$. Since we expect $\omega \in S(x)$ to occur with a high probability, this infimum of probability corresponds to the worst-case probability. Therefore, the robust chance constraint $\inf_{\mathbb{P} \in \mathcal{P}} \mathbb{P}(\omega \in S(x)) \geq 1 - \alpha$ dictates that even under the worst-case probability distribution within the ambiguity set $\mathcal{P}$, the probability of $\omega \in S(x)$ must be at least $1 - \alpha$. For the optimal solution x^* of (1.7) and the real probability distribution $\mathbb{P}^*$, we have $\mathbb{P}^*(\omega \in S(x^*)) \geq 1 - \alpha$ provided that $\mathbb{P}^* \in \mathcal{P}$.

Compared to SP problems (1.1) and (1.2), DRO problems (1.6) and (1.7) incorporate the idea of RO, enhancing robustness by considering the worst-case probability distributions within the ambiguity sets. Consequently, the construction of the ambiguity set is crucial in DRO. An appropriate ambiguity set construction should meet the following requirements: First, it can be established based on available information about the uncertainty. Second, the ambiguity set should have a high probability of including the real probability distribution, ensuring feasibility and theoretically guaranteeing the robustness of the solution. Third, the ambiguity set should not be too large as to contain significantly worse probability distributions than the real probability distribution, preventing overly conservative solutions. Finally, the formulation of the ambiguity set should yield a tractable optimization problem so that it can be effectively solved. In the following, we will briefly introduce three popular types of ambiguity sets, which will be explored in more detail in later chapters.

1.2.3.1 Moment-Based Ambiguity Sets

Suppose the sample space of the random variable ω is Ω. The expectation vector[1] (first-order moment) μ_0 and the covariance matrix (second-order moment) Σ_0 of ω can be estimated based on the N sample data $\omega_1, \omega_2, \ldots, \omega_N$, i.e.,

$$\mu_0 := \frac{1}{N}\sum_{n=1}^{N}\omega_n\ , \quad \Sigma_0 := \frac{1}{N}\sum_{n=1}^{N}(\omega_n - \mu_0)(\omega_n - \mu_0)^\top .$$

Construct the ambiguity set $\mathcal{P}$ to be the set consisting of all the probability distributions on Ω that have the same first-order and second-order moments (Scarf 1958) as follows:

$$\mathcal{P} := \left\{ \mathbb{P} \in \mathcal{M}(\Omega) \;\middle|\; \begin{array}{l} \mathbb{E}^{\mathbb{P}}[\omega] = \mu_0 \\ \mathbb{E}^{\mathbb{P}}[(\omega - \mu_0)(\omega - \mu_0)^\top] = \Sigma_0 \end{array} \right\}, \tag{1.8}$$

where $\mathcal{M}(\Omega)$ represents the set of all probability distributions on Ω.

In fact, a large amount of data is needed to accurately estimate the first-order and second-order moments. Therefore, when the sample data are insufficient, it is difficult to ensure that the ambiguity set $\mathcal{P}$ defined in (1.8) contains the real probability distribution. Another type of moment-based ambiguity set allows the first-order and second-order moments to vary within a small range (Delage 2010), i.e.,

$$\mathcal{P} := \left\{ \mathbb{P} \in \mathcal{M}(\Omega) \;\middle|\; \begin{array}{l} (\mathbb{E}^{\mathbb{P}}[\omega] - \mu_0)^\top \Sigma_0^{-1} (\mathbb{E}^{\mathbb{P}}[\omega] - \mu_0) \le \gamma_1 \\ \mathbb{E}^{\mathbb{P}}[(\omega - \mu_0)(\omega - \mu_0)^\top] \preceq \gamma_2 \Sigma_0 \end{array} \right\}, \tag{1.9}$$

where Σ_0 is assumed to be positive definite. The parameters $\gamma_1 \ge 0$ and $\gamma_2 \ge 1$ determine the size of the ambiguity set $\mathcal{P}$. Smaller values of γ_1 and γ_2 result in smaller $\mathcal{P}$. The notation $\preceq$ represents matrix inequality, and $\mathbb{E}^{\mathbb{P}}[(\omega - \mu_0)(\omega - \mu_0)^\top] \preceq \gamma_2 \Sigma_0$ denotes that the matrix $\gamma_2 \Sigma_0 - \mathbb{E}^{\mathbb{P}}[(\omega - \mu_0)(\omega - \mu_0)^\top]$ is positive semidefinite. The first constraint in (1.9) indicates that the expectation under the real probability

[1] Vectors in this book are column vectors by default.

distribution lies within an ellipsoid centered at μ_0. The second constraint defines the range of the second-order central moment of ω by a positive semidefinite cone. Consequently, the ambiguity set in (1.9) accounts for the estimation uncertainties of both first-order and second-order moments.

Nevertheless, moment-based ambiguity sets have their drawbacks. Since they rely solely on moment information, their ability to accurately describe probability distributions is limited. In some cases, distinct probability distributions may share identical first-order and second-order moments.

Example 1.1 (*Probability distributions with the same first- and second-order moments*) The following three probability distributions on the real number field $\mathbb{R}$ have the same first-order moment (mean) 0 and second-order moment (variance) 1, but clearly, they are different probability distributions:

$$\begin{cases} \mathbb{P}_1(-1) = 1/2 \,, \\ \mathbb{P}_1(1) = 1/2 \,, \end{cases} \quad \begin{cases} \mathbb{P}_2(-2) = 1/8 \,, \\ \mathbb{P}_2(0) = 3/4 \,, \\ \mathbb{P}_2(2) = 1/8 \,, \end{cases} \quad \begin{cases} \mathbb{P}_3(-5) = 1/50 \,, \\ \mathbb{P}_3(0) = 24/25 \,, \\ \mathbb{P}_3(5) = 1/50 \,. \end{cases}$$

Therefore, moment-based ambiguity sets could result in excessive conservativeness because they are too broad. To alleviate this issue, some studies incorporate unimodality into the ambiguity sets (Li et al. 2016), which leverages the shape information of the probability distribution, leading to less conservative solutions.

1.2.3.2 f-Divergence Ambiguity Set

In contrast to moment-based ambiguity sets, another approach for constructing ambiguity sets involves measuring the difference between probability distributions in some ways and defining the ambiguity set as the collection of probability distributions whose difference from the empirical probability distribution falls within a given range. Both f-divergence and Wasserstein-distance ambiguity sets are constructed following this principle, as depicted in Fig. 1.2. According to probability theory, as the amount of independent and identically distributed sample data increases, the empirical probability distribution converges to the real probability distribution (Vaart 1998). Thus, the real probability distribution is expected to lie close to the empirical probability distribution.

f-divergence is a tool that can measure the difference between two probability distributions. Suppose $\mathbb{P}$ and $\mathbb{P}'$ are two continuous probability distributions on Ω with probability density functions p and p', respectively. The f-divergence from $\mathbb{P}'$ to $\mathbb{P}$ is defined as:

$$D_f(\mathbb{P}'||\mathbb{P}) := \mathbb{E}^{\mathbb{P}}\left[f\left(\frac{p'(\omega)}{p(\omega)}\right)\right] = \int_{\Omega} f\left(\frac{p'(\omega)}{p(\omega)}\right) p(\omega)\mathrm{d}\omega \,,$$

where $f : \mathbb{R} \to \mathbb{R}$ is a function that satisfies the following conditions and is convex on $\mathbb{R}^+$:

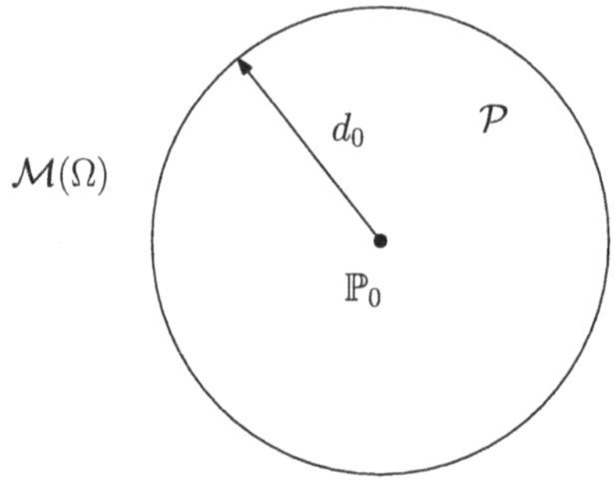

Fig. 1.2 The ambiguity set $\mathcal{P}$ based on f-divergence or Wasserstein distance can be described as a ball centered at the empirical probability distribution $\mathbb{P}_0$ with radius d_0 on the space $\mathcal{M}(\Omega)$ of probability distributions. The difference between probability distributions is measured by f-divergence or Wasserstein distance

$$\begin{cases} f(1) = 0 \,, \\ 0 \cdot f\left(\frac{y}{0}\right) := \begin{cases} y \cdot \lim\limits_{t \to +\infty} \dfrac{f(t)}{t} \,, & \text{if } y > 0 \,, \\ 0 \,, \text{ if } y < 0 \,, \end{cases} \\ f(y) = +\infty \,, \text{ if } y < 0 \,. \end{cases}$$

Kullback-Leibler (KL) divergence is one of the most popular types of f-divergence, named after Solomon Kullback and Richard Leibler (1951). KL divergence is also referred to as relative entropy and has many applications in research involving information theory. For KL divergence, the function f is defined as $f_{\mathrm{KL}}(t) = t \log t - t + 1$ for $t \geq 0$. Consequently, the KL divergence from continuous $\mathbb{P}'$ to continuous $\mathbb{P}$ is as follows:

$$D_{\mathrm{KL}}(\mathbb{P}' || \mathbb{P}) := \int_{\Omega} p'(\omega) \log\left(\frac{p'(\omega)}{p(\omega)}\right) \mathrm{d}\omega \,.$$

KL divergence is always nonnegative. It equals zero if and only if the two probability distributions are identical as measures. A larger KL divergence $D_{\mathrm{KL}}(\mathbb{P}' || \mathbb{P})$ reflects a more significant difference from probability distributions $\mathbb{P}'$ to $\mathbb{P}$. Therefore, KL divergence can distinguish probability distributions provided that the two probability distributions are different and the KL divergence is well-defined. However, KL divergence is not symmetric, meaning that $D_{\mathrm{KL}}(\mathbb{P}' || \mathbb{P})$ and $D_{\mathrm{KL}}(\mathbb{P} || \mathbb{P}')$ are generally different. This characteristic prevents KL divergence from being a distance.

The empirical probability distribution $\mathbb{P}_0$ constructed by the sample data is usually discrete, such as the one in (1.3). In this circumstance, the discrete version of KL divergence should be applied, which is defined as:

$$D_{\mathrm{KL}}(\mathbb{P}'||\mathbb{P}) := \sum_{\omega:\mathbb{P}(\omega)>0} \mathbb{P}'(\omega) \log\left(\frac{\mathbb{P}'(\omega)}{\mathbb{P}(\omega)}\right) ,$$

where the discrete probability distributions $\mathbb{P}$ and $\mathbb{P}'$ should satisfy the condition that whenever $\mathbb{P}'(\omega) > 0$, we have $\mathbb{P}(\omega) > 0$. In other words, the support set of $\mathbb{P}'$ is a subset of the support set of $\mathbb{P}$, expressed as $\mathbb{P}' \in \mathcal{M}(\mathrm{supp}(\mathbb{P}))$, where the notation $\mathrm{supp}(\cdot)$ represents the support set of the probability distribution enclosed within the parentheses.

The ambiguity set based on KL divergence is defined as follows:

$$\mathcal{P} := \{\mathbb{P} \in \mathcal{M}(\mathrm{supp}(\mathbb{P}_0)) \mid D_{\mathrm{KL}}(\mathbb{P}||\mathbb{P}_0) \leq d_0\} ,$$

where $\mathrm{supp}(\mathbb{P}_0) = \{\omega_1, \omega_2, \ldots, \omega_N\}$ coincides with the set of sample data. The parameter $d_0 \geq 0$ controls the size of the ambiguity set $\mathcal{P}$. A larger value of d_0 results in a larger ambiguity set. Ambiguity sets based on other types of f-divergence can be defined similarly.

One of the limitations of the KL divergence ambiguity sets is that this method mainly suits random variables with discrete probability distributions. As empirical probability distributions are usually discrete probability distributions, to make the KL divergence well-defined, the ambiguity sets can contain only discrete probability distributions supported on the support set of the empirical probability distribution. This means the random variables are assumed to take values only within the range of the sample data. However, for general multi-dimensional continuous random variables, such as renewable power generation and load demand, this assumption significantly restricts the range of the random variable, making the model less applicable to practical situations, especially when the sample data are scarce. Therefore, in such application scenarios, the KL divergence ambiguity sets may not include the real probability distributions, hampering the effectiveness guarantee of the optimal solutions with respect to the real probability distribution.

1.2.3.3 Wasserstein-Distance Ambiguity Set

Wasserstein distance can measure the distance between two probability distributions on the same sample space Ω, without requirements of their support sets. The definition of Wasserstein 1-distance is as follows:

$$d_{\mathrm{W}}(\mathbb{P}_1, \mathbb{P}_2) := \inf_{\mathbb{P}_{1,2} \in \mathcal{J}(\mathbb{P}_1, \mathbb{P}_2)} \int_{\Omega \times \Omega} \|\omega - \omega'\| \mathrm{d}\mathbb{P}_{1,2}(\omega, \omega') ,$$

where $\mathbb{P}_1, \mathbb{P}_2 \in \mathcal{M}_1(\Omega)$, i.e., they are probability distributions on Ω that have finite first-order moments. The meaning of this definition will be explained in later chapters. Wasserstein distance is always nonnegative and equals zero if and only if the two probability distributions are identical. Moreover, Wasserstein distance is symmetric and satisfies the triangle inequality, making it a distance function for probability distributions.

Recall that the KL divergence ambiguity set comprises probability distributions near the empirical probability distribution. The idea is similar here. Using Wasserstein distance to measure closeness, the Wasserstein-distance ambiguity set centered at the empirical probability distribution $\mathbb{P}_0$ with radius $d_0 > 0$ is defined as:

$$\mathcal{P} := \{\mathbb{P} \in \mathcal{M}_1(\Omega) \mid d_W(\mathbb{P}, \mathbb{P}_0) \leq d_0\} \ . \tag{1.10}$$

Therefore, the ambiguity set $\mathcal{P}$ comprises all the probability distributions in $\mathcal{M}_1(\Omega)$ whose Wasserstein distance to $\mathbb{P}_0$ is no larger than d_0. Because $\mathcal{M}_1(\Omega)$ does not impose requirements on the continuity or discreteness of the probability distributions, the ambiguity set defined in (1.10) includes general probability distributions, particularly covering both discrete and continuous probability distributions.

Given the limited amount of sample data, there is typically some difference between the empirical probability distribution $\mathbb{P}_0$ and the real probability distribution. However, as long as the parameter d_0 is appropriately chosen, the real probability distribution will reside within the ambiguity set $\mathcal{P}$ with a high probability. DRO considers the worst-case probability distribution in the ambiguity set, accounting for the inexactness of the empirical probability distribution by incorporating some margin. There are theoretical results on how to properly choose the parameter d_0 (Fournier and Guillin 2015), which will be further discussed in later chapters.

1.3 Energy Storage in Power Systems

In this section, we first present the applications of ES in power systems, followed by an introduction to the models used for ES operation.

1.3.1 Usages of Energy Storage in Power Systems

For an extended period, commercial ES applied in power systems was mainly represented by pumped ES. Nowadays, other emerging ES technologies have matured enough for commercial use, such as electrochemical ES, compressed air ES, heat ES, and flywheel ES. Among them, pumped ES, electrochemical ES, and compressed air ES are suitable for large-scale ES applications. Different types of ES exhibit various core features, such as round service efficiency, self-discharge rate, unit cost, lifespan, power and energy densities, response time, environmental impact, and so on. Readers can refer to Mitali et al. (2022) for a survey on ES technologies.

ES can play a crucial role in renewable-integrated power systems in several ways: First, for short-term volatility of renewable generation, such as fluctuations on the scale of seconds or minutes, fast-responding ES systems can help flatten the renewable generation power curve through rapid charging and discharging, thereby facilitating frequency regulation within the system. Second, when it comes to the discrepancy between renewable generation and load demand curves, ES can transfer energy throughout the day, enabling peak shaving and valley filling to maintain system balance. Third, in cases where renewable generation and load demand exhibit seasonal characteristics, some types of ES can assist with seasonal power and energy balance in power systems. Fourth, ES enhances power system flexibility and reduces reserve requirements, enabling more efficient and adaptable energy management. A comprehensive overview of the application areas of ES in power systems can be found in Kebede et al. (2022).

Thanks to the advancements in ES technology and the increasing integration of renewable generation in recent years, ES has become an essential approach for addressing the challenges of renewable-integrated power systems. Its rapid proliferation is expected to grow even faster under the goal of carbon neutrality. According to the data and analysis by the International Energy Agency (IEA) (International Energy Agency 2021), the large-scale integration of wind and PV power has led to a more than 60% increase in worldwide investment in grid-connected ES in 2020, with electrochemical ES capacity reaching 5 GW. If the goal of carbon neutrality is realized by 2050, the worldwide ES capacity must reach 585 GW, of which 120 GW is estimated to be electrochemical ES.

1.3.2 Models of Energy Storage Operation

The presence of ES introduces temporal correlation in the electricity energy, making it essential to consider a time window during the storage operation, which may be a day or a week. In a fundamental operation model, the time window is discretized into time periods, and the electricity power values are assumed constant within each period. We define the set of time periods as $\mathcal{T} := \{1, 2, \dots, T\}$, where T represents the total number of time periods and each period has a duration of Δ_t. The basic model for ES operation can then be described as follows:

$$0 \le p_t^c \le P^c \ , \ 0 \le p_t^d \le P^d \ , \ \forall t \in \mathcal{T} \ , \tag{1.11a}$$

$$p_t^c \text{ and } p_t^d \text{ cannot both be positive} \ , \ \forall t \in \mathcal{T} \ , \tag{1.11b}$$

$$e_t = e_{t-1} + \eta^c p_t^c \Delta_t - \frac{p_t^d \Delta_t}{\eta_d} \ , \ \forall t \in \mathcal{T} \ , \tag{1.11c}$$

$$\underline{\omega} E \le e_t \le \overline{\omega} E \ , \ \forall t \in \mathcal{T} \ , \tag{1.11d}$$

$$e_T = e_0 \ . \tag{1.11e}$$

In constraint (1.11a), p_t^c and p_t^d represent the charging and discharging power at period t, whereas P^c and P^d are the charging and discharging power capacities of the ES. Constraint (1.11a) therefore sets limits on charging and discharging power.

ES cannot charge and discharge at the same time, which is enforced in constraint (1.11b). Constraint (1.11c) defines the dynamics of stored energy, where e_t denotes the stored energy at the end of period t, while η_c and η_d represent the charging and discharging efficiencies. According to this constraint, the stored energy will increase (decrease) when the ES is charging (discharging). Constraint (1.11d) establishes bounds for stored energy, where $\underline{\omega}$ and $\overline{\omega}$ serve as the lower and upper bounds for the state-of-charge (SoC), and E is the energy capacity of the ES. To enable continuous operation, constraint (1.11e) requires that the initial and final stored energy are equal.

In the ES operation model (1.11), the variables p_t^c, p_t^d, and e_t for $t \in \mathcal{T}$ can be adjusted and optimized within the feasible region to achieve objectives such as economic efficiency and curtailment reduction. Constraints (1.11a) and (1.11c)–(1.11e) are already linear, which are easy to manage in mathematical optimization. Therefore, we will focus on constraint (1.11b). With nonnegative p_t^c and p_t^d, constraint (1.11b) can be mathematically expressed as:

$$p_t^c p_t^d = 0 \ , \ \forall t \in \mathcal{T} \ . \tag{1.12}$$

This constraint is referred to as the *charging and discharging complementarity constraint*.

Although the formulation in (1.12) appears simple, this constraint is nonlinear and results in a nonconvex feasible region. However, it can be further linearized using auxiliary binary variables. Specifically, we introduce a binary variable z_t to represent the charging and discharging status during period t: If the ES is charging, then $z_t = 1$; otherwise, $z_t = 0$. Then the constraint group of (1.11a) and (1.11b) is equivalent to:

$$0 \le p_t^c \le z_t P^c \ , \ 0 \le p_t^d \le (1 - z_t) P^d \ , \ \forall t \in \mathcal{T} \ , \tag{1.13a}$$

$$z_t \in \{0, 1\} \ , \ \forall t \in \mathcal{T} \ . \tag{1.13b}$$

In (1.13), when $z_t = 0$, we have $p_t^c = 0$ and $0 \le p_t^d \le P^d$. Conversely, when $z_t = 1$, we have $0 \le p_t^c \le P^c$ and $p_t^d = 0$.

Utilizing this formulation, constraints (1.11c)–(1.11e) and (1.13) form a model of ES operation. In this model, p_t^c, p_t^d, and e_t are the continuous variables, whereas variable z_t is binary. All the constraints in this model are linear.

The model incorporating binary variables is frequently used in ES research, as it is typically tractable when the problem scale remains manageable. However, there are two primary challenges associated with binary variables. Firstly, the complexity of handling binary variables risks exponential explosion. Secondly, binary variables can be particularly difficult to address in embedded optimization problems, such as those arising in two-stage planning–operation problems. Therefore, some studies try to relax the charging and discharging complementarity constraint. Recently, a linear model for ES operation was proposed in Shen et al. (2021), which is outlined below:

$$\frac{p_t^c}{P^c} + \frac{p_t^d}{P^d} \le 1 \ , \ \forall t \in \mathcal{T} \ , \ (1.11a) \ , \ (1.11c)\text{–}(1.11e) \ . \tag{1.14}$$

In model (1.14), the variables are still p_t^c, p_t^d, and e_t, without the involvement of binary variables. $p_t^c > 0$ and $p_t^d > 0$ may occur simultaneously. Nonetheless, an operation strategy can be derived such that the ES will not charge and discharge at the same time: Previously, we assumed the power within a period to be constant. Now, allowing the power to change within a period, we can consider p_t^c and p_t^d as the average charging and discharging power in period t. To achieve the values of p_t^c and p_t^d, charge (discharge) the ES for a time duration of $p_t^c \Delta_t / P^c$ ($p_t^d \Delta_t / P^d$). Since $p_t^c / P^c + p_t^d / P^d \leq 1$, it follows that $p_t^c \Delta_t / P^c + p_t^d \Delta_t / P^d \leq \Delta_t$, ensuring that the length of a period is sufficient to first charge and then discharge the ES. By employing this strategy, the ES will not charge and discharge simultaneously, and any variable realization within the feasible region of model (1.14) can be physically implemented.

1.4 Energy Storage Sizing Under Different Scenarios

ES planning is an important problem for ES applications in power systems. It involves determining which type of ES to use, where to deploy, and the required capacity. By analyzing application requirements and assessing the features of different types of ES, the most suitable ES type can be selected. ES can be installed on the generation side, grid side, and distribution/demand side. Based on geographical conditions, candidate sites can be chosen for ES deployment. Then as long as the capacities at each site are determined, the planning strategy is completed. Our primary focus here is on the ES sizing problem in renewable-integrated power systems, which exhibits the following characteristics.

First, the size of ES has a significant impact on the economic aspects of renewable-integrated power systems. Pumped ES, the most mature type of ES with the largest installed capacity, faces inherent limitations: Its siting is heavily restricted by geographic conditions, and it typically requires a relatively long construction time. In contrast, electrochemical ES equipment is more flexible and can be installed faster, contributing to its increasing application in recent years. Among various electrochemical ES types, lithium–ion batteries lead the way. As technology advances, the price of electrochemical ES has continued to decrease in recent years, but the unit capacity cost remains high. In 2020, the unit cost of a 4-hour lithium–ion battery ES system was approximately 350 US dollars per kWh (Cole et al. 2021). For future tendency, the unit cost of lithium–ion batteries in 2030 was estimated to be roughly two-thirds of that in 2020 (Mongird et al. 2020). Therefore, ES sizing decisions have a critical impact on the overall economy, making it essential to study the optimal sizing of ES in renewable-integrated power systems.

Second, renewable power generation such as wind and PV power has substantial volatility and randomness. The electricity power from renewable resources has not only intrinsic uncertainty and uncontrollability but also prominent nonlinear and nonstationary characteristics. For example, Fig. 1.3 depicts typical 7-day PV and wind power curves in Qinghai province, China. The random fluctuations of both wind and PV power are clearly intense, with distinct peak and valley time zones that

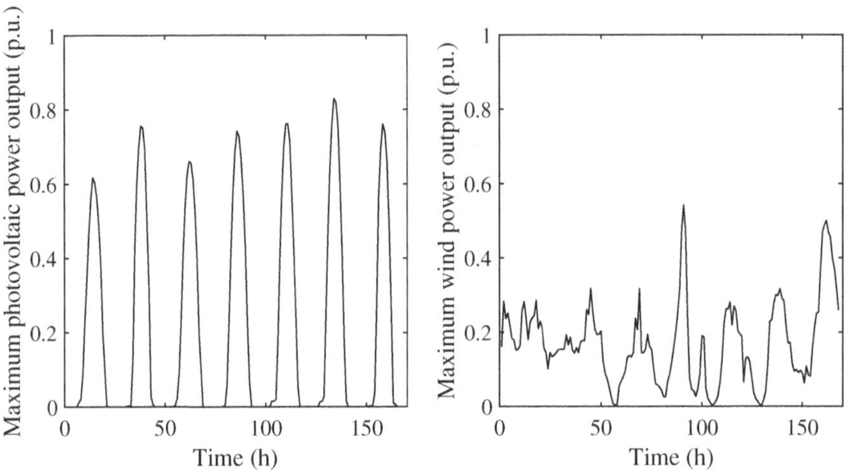

Fig. 1.3 Typical 7-day photovoltaic (PV) and wind power curves, which show the per unit (p.u.) values and the base value is the installed capacity

differ from the general load demand curve. Therefore, when determining the size of ES in renewable-integrated power systems, it is necessary to take into account the uncertainty of renewable power generation. Additionally, during the planning stage, decision-makers often lack access to enough accurate data regarding renewable energy generation and load demand. Hence, there is no precise probability distribution available to describe the uncertainties arising from renewable energy generation, which in turn affects the effectiveness of the ES sizing strategy derived from optimization.

Third, the operation problem of power systems with ES is usually large-scale and involves high-dimensional random variables. The sizing problem in power systems has a planning–operation two-stage form. The inclusion of ES allows for the temporal transfer of electrical energy, making it necessary for the operation problem to consider multiple time periods in order to model the SoC dynamics of the ES. As a result, the second-stage problem, i.e., the operation optimization problem, is large-scale, with dimensions of the involved random variables being proportional to the number of time periods. Therefore, the optimization method for the ES sizing problem should be capable of handling two-stage optimization, where the second-stage problem is large-scale and features high-dimensional random variables.

In conclusion, studying the ES sizing problem in power systems with renewable power generation is essential to fully utilize the functions and advantages of ES, support the safe and economical operation of power systems, and facilitate the transition towards green and low-carbon power systems. Furthermore, addressing crucial questions such as modeling the uncertainties of renewable power generation and load demand in the optimal sizing problem, considering the inexactness of empirical probability distributions, and handling large-scale operation optimization problems alongside high-dimensional random variables, remains of vital importance.

To provide insight into ES sizing, we will review literature related to the topic according to the equipment locations of the ES, which can be divided into three categories: the generation side, the grid side, and the distribution/demand side.

1.4.1 Generation Side

Deploying ES systems at wind farms or PV stations can help reduce the short-term uncertainty of renewable power generation, smooth the power output curve, mitigate the negative impact of renewable power uncertainty on the power grid, and facilitate the integration of renewable energy. Some studies have considered the ES sizing problem in wind farms. Empirical probability distribution was generated based on the historical short-term wind power forecast errors in Bludszuweit and Dominguez-Navarro (2011), a probabilistic method was proposed for ES sizing to decrease the uncertainty of wind power output, and the simulation showed that the ES capacity should change as the forecast accuracy of wind power varies. The spatial-temporal correlation between multiple neighboring wind farms was studied in Wang et al. (2019), and then a rolling operation strategy was proposed and a scenario-based optimal sizing method was developed for ES. A capacity determination and a control strategy were proposed in Wan et al. (2021) for a supercapacitor–battery hybrid ES system to economically smooth the power output of a wind farm, where the wind power scenarios were generated based on probabilistic forecast and multivariate Gauss copula and the ES size was optimized using a two-stage SP model.

For the ES sizing problem in PV stations, the energy and power capacities of the ES inside a large-scale PV power plant were optimized using linear programming (LP) in Saez-de-Ibarra et al. (2016), the model predictive control method was applied to deal with the forecast errors of the PV power and the operation strategy of the PV–storage integrated station in the electricity market was also optimized. The power generation, the peak load support ability, and the line loss reduction were considered comprehensively in Yang et al. (2018), and then an ES sizing method was proposed for the large-scale PV station equipped with lithium–ion batteries or lead–acid batteries. A PV station equipped with a supercapacitor–battery hybrid ES system was considered in Ma et al. (2019), and a dispatch strategy using the charging and discharging of ES and PV power curtailment was proposed to decrease the fluctuations of the power output, based on which the capacity of the ES was optimized. The location of the attached ES of the PV station may impact the reverse power flow, as elaborated in Unahalekhaka and Sripakarach (2020). In addition, according to the simulation results of Unahalekhaka and Sripakarach (2020), equipping the ES right inside the PV station was the recommended choice rather than connecting the ES to a high-voltage bus. Zhang et al. (2022) studied how to utilize the fast response ability of the pumped storage with variable speed and constant frequency to regulate the real-time fluctuations of the PV stations, and a continuous cyclic revision sizing method of the pumped storage for the PV station was proposed considering the operational constraints.

1.4.2 Grid Side

Grid-side ES serves as a flexible resource in the power grid, providing functions such as smoothing the net load demand curve, shaving peaks and filling valleys, serving as a system reserve, and assisting in frequency regulation. Additionally, grid-side ES operators may participate in the electricity market and generate profits by charging during low-electricity-price periods and discharging when the price is high.

Numerous studies have been conducted aiming at determining the capacities of grid-side ES. A bi-level SP model was proposed in Nasrolahpour et al. (2016) with ES sizing and strategic operating in the upper level and market clearing in the lower level, which considered the load and market uncertainties and was solved by Benders decomposition. The optimal sizing and siting problems of grid-side ES were studied in Javadi et al. (2022) considering the presence of the demand response and time-of-use pricing. The problem was formulated as a mixed-integer linear programming (MILP) problem and solved by particle swarm optimization and genetic algorithm. The ES was planned in Xu et al. (2017) to enhance its profitability in the energy and reserve markets, where the bilevel optimization problem was solved via primal decomposition and cutting-plane method. The optimal planning problem of power-to-gas ES was investigated in Sohrabi et al. (2021), where the lower-level market clearing problem was represented by optimality conditions so that the planning results were obtained via solving MILP.

As renewable penetrated power system emerges and develops, some studies paid more attention to better accommodating renewable energy generation by strategically sizing the grid-side ES. In this category, an ES expansion method to decrease the expected wind curtailment was proposed in Saber et al. (2019), where the wind uncertainty was modeled by scenario analysis. The optimal planning of various ES technologies was studied in Wu et al. (2021) by a three-stage framework, where multiple demand scenarios were considered under uncertainties. Renewable power generation was modeled by Gaussian probability distributions in Hemmati et al. (2017), where ES systems were planned and scheduled for congestion management. Wind power historical data were used to derive the marginal distributions and covariance in Xia et al. (2018), based on which a stochastic cost-benefit analysis model was proposed to size the ES system and solved by point estimation and parallel branch and bound algorithm. The above research works used SP to handle optimization problems with uncertainties. In contrast, a distributionally robust planning method for electrochemical ES was developed in Le et al. (2021) based on KL divergence, where the ES lifespan was also considered. The model was solved using SAA and MILP.

The grid-side ES sizing problem can be considered together with the generation expansion in the grid. For example, the coordinated expansion of transmission and ES systems was studied in Nemati et al. (2021) against physical intentional attacks, and a tri-level model was proposed with planning in the first level, attack in the second level, and re-dispatch in the third level. The same planning problem was investigated in Kazemi and Ansari (2022) considering the power system reliability in contingency conditions by N-1 security constraints and reliability indices. By examining the optimal investment decisions of transmission and ES systems, the relationship between the two elements in power systems was analyzed in Bustos et al. (2018) to conclude that they can be substitutes or complements, depending on the power system characteristics. A tri-level method was proposed in Dvorkin et al. (2018) for the co-planning of merchant electrochemical ES and transmission expansion with the lower level considering market clearing. The security-constrained co-planning was investigated in Gan et al. (2019), where the ES helps the renewable generation accommodation and post-contingency security.

The ES and transmission expansion co-optimization methods mentioned above all used either deterministic models or SP models. The following three studies adopted RO to enhance robustness. Polyhedral uncertainty sets were used in Dehghan and Amjady (2016) to model load demand and wind power uncertainties, where an adaptive RO problem was formulated and solved by a decomposition algorithm. A continuous-time hybrid SP and RO method was developed in Nikoobakht and Aghaei (2019) to plan the transmission lines and ES systems. The co-planning problem was also studied in Wang et al. (2019) and the robust formulation was solved by an improved nested column-and-constraint generation (C&CG) algorithm.

The ES sizing in a power grid can also be combined with generation expansion planning. A two-stage generation–storage optimization method considering renewable energy generation was proposed in Shi et al. (2022) based on time sequence simulation. RO based on polyhedral uncertainty sets was adopted in Zhou et al. (2021) to establish a wind–storage expansion method considering unit commitment and economic dispatch in the operation.

There are also research works optimizing generation, transmission, and ES allocation together in renewable-integrated power grids, where the majority used SP technique: An integrated transmission–generation–storage expansion planning model was proposed in Moradi-Sepahvand and Amraee (2021), where the low-carbon policy was considered and a hierarchical clustering method was used to model the load and renewable uncertainties. Focusing on the same problem, system flexibility was considered in Larsen and Sauma (2021), and the case study showed that adding ES systems may not necessarily help reduce carbon emissions. The proposed co-planning method of the wind farm, ES, and transmission network in Zhang et al. (2020) considers optimal transmission switching and unit commitment, where a decentralized decomposition algorithm was developed to solve the proposed model in a parallel way. The heavy-load operation constraints were considered in the joint planning model in Wu and Jiang (2019). The method proposed in Hamidpour et al. (2019) coordinately plans generation, transmission, and ES expansion under the flexibility and reliability criteria of the system, where the demand response is also considered. Different from these methods using scenario-based SP, a combined SP and RO method was established in Hamidpour et al. (2022) to address the co-planning problem with the presence of load, energy price, wind generation, and network equipment availability uncertainties. Few studies focused on the distributionally robust co-planning problem of generation, transmission, and ES expansion in power grids.

1.4.3 Distribution/Demand Side

Along with the development of distributed energy resources and demand response technologies, distribution networks have been upgrading toward a smarter and more proactive future. ES can play an important role on the distribution/demand side from the perspectives of load leveling, renewable energy integration, energy arbitrage, emission reduction, reactive power support, and so on (Saboori et al. 2017). ES systems were planned in Saboori et al. (2015) to enhance the reliability of the radial distribution network. A community ES planning method was proposed in Sardi et al. (2017) based on cost-effective analysis considering various benefits of community ES deployment. The active distribution system expansion problem was studied in Li et al. (2018), where the renewable energy sources, ES systems, and networks were optimized jointly in a three-level optimization model, and the uncertainties were represented by scenarios. A nonparametric chance-constrained SP method was proposed in Akhavan-Hejazi and Mohsenian-Rad (2018) for ES operation and planning in active distribution networks. The soft open points and reactive power capability of distributed generators were considered in the distributed ES system planning method in Bai et al. (2018). The battery ES system was planned in Nazemi et al. (2020) to enhance the resilience of the distribution network against earthquakes. A multi-objective two-stage optimization method was proposed in Zhang et al. (2021) to plan battery ES systems in a distribution electricity market. RO was adopted in Bozorgavari et al. (2020) for distributed battery ES system planning, where the degradation, energy arbitrage, and operation costs were considered. A tri-level robust planning method was proposed in Zhao et al. (2020) for distributed ES systems under generation and load uncertainties. A robust planning method for battery ES systems was proposed in Wang et al. (2021) to reduce the power imbalance in distribution networks with rooftop PV installation. DRO was used in Li et al. (2023) to optimize the renewable energy generators and ES systems in the distribution network with electricity price, renewable, demand, and carbon intensity uncertainties.

In areas far away from the main grid, such as islands and remote residences, stand-alone microgrids may arise, which are not connected to the main grid. Some stand-alone microgrids are fully powered by renewable energy (Lai and McCulloch 2017), and others also have fossil-fuel power generation, such as diesel generators, fuel cells, microturbines, and combined heat and power units. Stand-alone microgrids have to maintain system reliability on their own. ES can support the secure operation of stand-alone microgrids as a flexible resource. Some studies have considered the ES sizing problem in stand-alone microgrids. A method based on historical solar data was proposed in Lai and McCulloch (2017) to coordinate the sizes of the PV panels, ES system, and anaerobic digestion biogas power plant. Discrete Fourier transform was used in Xiao et al. (2014) to size ES and diesel generators in a stand-alone microgrid. A dual-battery bank scheme was proposed in Neto et al. (2018) for an isolated microgrid, where the first bank stores the renewable energy for load and the second one is responsible for voltage regulation. The thermal inertia of the buildings and cold storage were considered in Dakir et al. (2020), where the sizing and operation of the isolated microgrid were studied. An optimal sizing method of a stand-alone microgrid consisting of PV, diesel generators, and batteries was proposed in Wu et al. (2021), where the reliability was evaluated using the sequential Monte Carlo method based on the rolling horizon. The probabilistic unit commitment was considered in the proposed battery sizing method in Khorramdel et al. (2016), where the microgrid was either stand-alone or grid-connected. A method based on SP was proposed in Alharbi and Bhattacharya (2018) for a battery ES system in a stand-alone microgrid, which was solved by MILP problems. These studies are mainly based on traditional SP techniques, either by historical data or parametric probability distribution.

1.5 Overview of the Book

This book is dedicated to introducing DRO theory and demonstrating its usage and effects by the applications in ES sizing. This chapter provides a general introduction, highlighting the advantages of using DRO in uncertain decision-making environments, the usages and operation models of ES in power systems, and existing ES sizing techniques under various scenarios. The book is then divided into two main parts, i.e., theory and application.

Part I: Theory (Chaps. 2–6). This part aims to offer an extensive tutorial on the mathematical foundations of state-of-the-art DRO theory. It contains preliminary (Chap. 2), a basic introduction to the general DRO theory (Chap. 3), and detailed explanations of three specific DRO formulations (Chaps. 4–6). Chapter 2 summarizes some fundamental concepts and results in probability, risk, and optimization theories, which are mathematical preliminaries of the DRO theory and applications. Most results are briefly reviewed without proof, but the readers can find the standard textbooks in the references. The power flow models in power systems are also introduced. In Chap. 3, the core ideas and formulations of distributionally robust uncertainties are presented first. Two-stage DRO problems are then modeled, and some reformulation techniques are explained. The theory in Chap. 3 applies to general DRO problems with any type of ambiguity set. Chapters 4–6 discuss the DRO theory under moment-based, f-divergence, and Wasserstein-distance ambiguity sets, respectively, where these are the three most common types of ambiguity sets in the cutting-edge DRO techniques. Their motivations, characteristics, application scopes, and reformulation methodologies are introduced in detail. Substantial references are also provided.

Part II: Application (Chaps. 7–13). This part applies the DRO theory to various ES sizing problems. Each chapter in this part models a distributionally robust ES sizing problem, develops a solution method using the theory, and conducts case studies to demonstrate the effectiveness of the proposed approach. Chapter 7 studies the renewable–storage sizing in fully renewable stand-alone microgrids. Chapter 8 deals with the transmission–storage sizing for remote large-scale renewable power plants. Chapter 9 studies the storage sizing in power networks to reduce renewable generation curtailment. Chapter 10 covers the renewable–transmission–storage sizing towards low-carbon power systems. Chapter 11 investigates the wind–storage sizing considering the wake effect. Chapter 12 focuses on the PV–storage sizing for electric vehicle (EV) charging stations on highways. Chapter 13 studies the energy hub sizing in multi-carrier energy networks. Applications introduced in this part cover a wide range of ES usages and scenarios, offering effective solutions to ES sizing problems. Various types of distributionally robust uncertainties and ambiguity sets are adopted in these practical problems, bridging the gap between mathematical theories and industrial applications.

References

Akhavan-Hejazi H, Mohsenian-Rad H (2018) Energy storage planning in active distribution grids: a chance-constrained optimization with non-parametric probability functions. IEEE Trans Smart Grid 9(3):1972–1985

Alharbi H, Bhattacharya K (2018) Stochastic optimal planning of battery energy storage systems for isolated microgrids. IEEE Trans Sustain Energy 9(1):211–227

Bai L, Jiang T, Li F et al (2018) Distributed energy storage planning in soft open point based active distribution networks incorporating network reconfiguration and DG reactive power capability. Appl Energy 210:1082–1091

Ben-Tal A, Ghaoui LE, Nemirovski A (2009) Robust optimization. Princeton series in applied mathematics, Princeton University Press, Princeton

Bertsimas D, Sim M (2004) The price of robustness. Oper Res 52(1):35–53

Bludszuweit H, Dominguez-Navarro JA (2011) A probabilistic method for energy storage sizing based on wind power forecast uncertainty. IEEE Trans Power Syst 26(3):1651–1658

Bozorgavari SA, Aghaei J, Pirouzi S et al (2020) Robust planning of distributed battery energy storage systems in flexible smart distribution networks: a comprehensive study. Renew Sustain Energy Rev 123:109739

Bustos C, Sauma E, de la Torre S et al (2018) Energy storage and transmission expansion planning: substitutes or complements? IET Gener Transm Distrib 12(8):1738–1746

Cole W, Frazier AW, Augustine C (2021) Cost projections for utility-scale battery storage: 2021 update. Technical report, National Renewable Energy Laboratory (NREL)

Dakir S, Boukas I, Lemort V et al (2020) Sizing and operation of an isolated microgrid with building thermal dynamics and cold storage. IEEE Trans Ind Appl 56(5):5375–5384

Dehghan S, Amjady N (2016) Robust transmission and energy storage expansion planning in wind farm-integrated power systems considering transmission switching. IEEE Trans Sustain Energy 7(2):765–774

Delage E, Ye Y (2010) Distributionally robust optimization under moment uncertainty with application to data-driven problems. Oper Res 58(3):595–612

Dvorkin Y, Fernández-Blanco R, Wang Y et al (2018) Co-planning of investments in transmission and merchant energy storage. IEEE Trans Power Syst 33(1):245–256

Fournier N, Guillin A (2015) On the rate of convergence in Wasserstein distance of the empirical measure. Probab Theory Relat Fields 162:707–738

Gan W, Ai X, Fang J et al (2019) Security constrained co-planning of transmission expansion and energy storage. Appl Energy 239:383–394

Goh J, Sim M (2010) Distributionally robust optimization and its tractable approximations. Oper Res 58(4-part-1):902–917

Hamidpour H, Aghaei J, Pirouzi S et al (2019) Flexible, reliable, and renewable power system resource expansion planning considering energy storage systems and demand response programs. IET Renew Power Gener 13(11):1862–1872

Hamidpour H, Aghaei J, Pirouzi S et al (2022) Coordinated expansion planning problem considering wind farms, energy storage systems and demand response. Energy 239:122321

Hemmati R, Saboori H, Jirdehi MA (2017) Stochastic planning and scheduling of energy storage systems for congestion management in electric power systems including renewable energy resources. Energy 133:380–387

International Energy Agency (IEA) (2021) Energy storage. Technical report, IEA, Paris, France

Javadi MS, Gough M, Mansouri SA et al (2022) A two-stage joint operation and planning model for sizing and siting of electrical energy storage devices considering demand response programs. Int J Electrical Power & Energy Syst 138:107912

Kazemi M, Ansari MR (2022) An integrated transmission expansion planning and battery storage systems placement - a security and reliability perspective. Int J Electr Power & Energy Syst 134:107329

Kebede AA, Kalogiannis T, Van Mierlo J et al (2022) A comprehensive review of stationary energy storage devices for large scale renewable energy sources grid integration. Renew Sustain Energy Rev 159:112213

Khorramdel H, Aghaei J, Khorramdel B et al (2016) Optimal battery sizing in microgrids using probabilistic unit commitment. IEEE Trans Ind Inf 12(2):834–843

Kullback S, Leibler RA (1951) On information and sufficiency. Ann Math Stat 22(1):79–86

Lai CS, McCulloch MD (2017) Sizing of stand-alone solar PV and storage system with anaerobic digestion biogas power plants. IEEE Trans Ind Electron 64(3):2112–2121

Larsen M, Sauma E (2021) Economic and emission impacts of energy storage systems on power-system long-term expansion planning when considering multi-stage decision processes. J Energy Stor 33:101883

Le J, Liao X, Zhang L et al (2021) Distributionally robust chance constrained planning model for energy storage plants based on Kullback–Leibler divergence. Energy Rep 7:5203–5213

Li R, Wang W, Xia M (2018) Cooperative planning of active distribution system with renewable energy sources and energy storage systems. IEEE Access 6:5916–5926

Li J, Xu Z, Liu H et al (2023) A Wasserstein distributionally robust planning model for renewable sources and energy storage systems under multiple uncertainties. IEEE Trans Sustain Energy 14(3):1346–1356

Li B, Jiang R, Mathieu JL (2016) Distributionally robust risk-constrained optimal power flow using moment and unimodality information. In: 2016 IEEE 55th conference on decision and control (CDC), pp 2425–2430

Ma W, Wang W, Wu X et al (2019) Optimal allocation of hybrid energy storage systems for smoothing photovoltaic power fluctuations considering the active power curtailment of photovoltaic. IEEE Access 7:74787–74799

Mitali J, Dhinakaran S, Mohamad A (2022) Energy storage systems: a review. Energy Stor Saving 1(3):166–216

Mongird K, Viswanathan V, Alam J et al (2020) 2020 grid energy storage technology cost and performance assessment. Technical report, U.S. Department of Energy

Moradi-Sepahvand M, Amraee T (2021) Integrated expansion planning of electric energy generation, transmission, and storage for handling high shares of wind and solar power generation. Appl Energy 298:117137

Nasrolahpour E, Kazempour SJ, Zareipour H et al (2016) Strategic sizing of energy storage facilities in electricity markets. IEEE Trans Sustain Energy 7(4):1462–1472

Nazemi M, Moeini-Aghtaie M, Fotuhi-Firuzabad M et al (2020) Energy storage planning for enhanced resilience of power distribution networks against earthquakes. IEEE Trans Sustain Energy 11(2):795–806

Nemati H, Latify MA, Yousefi GR (2021) Tri-level coordinated transmission and electrical energy storage systems expansion planning under physical intentional attacks. J Energy Stor 42:103095

Neto PBL, Saavedra OR, de Souza Ribeiro LA (2018) A dual-battery storage bank configuration for isolated microgrids based on renewable sources. IEEE Trans Sustain Energy 9(4):1618–1626

Nikoobakht A, Aghaei J (2019) Integrated transmission and storage systems investment planning hosting wind power generation: continuous-time hybrid stochastic/robust optimisation. IET Gener Transm Distrib 13(21):4870–4879
Saber H, Moeini-Aghtaie M, Ehsan M et al (2019) A scenario-based planning framework for energy storage systems with the main goal of mitigating wind curtailment issue. Int J Electr Power & Energy Syst 104:414–422
Saboori H, Hemmati R, Jirdehi MA (2015) Reliability improvement in radial electrical distribution network by optimal planning of energy storage systems. Energy 93:2299–2312
Saboori H, Hemmati R, Ghiasi SMS et al (2017) Energy storage planning in electric power distribution networks - a state-of-the-art review. Renew Sustain Energy Rev 79:1108–1121
Saez-de-Ibarra A, Milo A, Gaztañaga H et al (2016) Co-optimization of storage system sizing and control strategy for intelligent photovoltaic power plants market integration. IEEE Trans Sustain Energy 7(4):1749–1761
Sardi J, Mithulananthan N, Gallagher M et al (2017) Multiple community energy storage planning in distribution networks using a cost-benefit analysis. Appl Energy 190:453–463
Scarf HE (1958) A min-max solution of an inventory problem. Studies in the mathematical theory of inventory and production. Stanford University Press, Stanford, CA
Shapiro A, Dentcheva D, Ruszczynski A (2021) Lectures on stochastic programming: modeling and theory, 3rd edn. Society for industrial and applied mathematics, Philadelphia, PA
Shen Z, Wei W, Wu D et al (2021) Modeling arbitrage of an energy storage unit without binary variables. CSEE J Power Energy Syst 7(1):156–161
Shi Z, Wang W, Huang Y et al (2022) Simultaneous optimization of renewable energy and energy storage capacity with the hierarchical control. CSEE J Power Energy Syst 8(1):95–104
Sohrabi F, Vahid-Pakdel M, Mohammadi-Ivatloo B et al (2021) Strategic planning of power to gas energy storage facilities in electricity market. Sustain Energy Technol Assess 46:101238
Unahalekhaka P, Sripakarach P (2020) Reduction of reverse power flow using the appropriate size and installation position of a BESS for a PV power plant. IEEE Access 8:102897–102906
Vaart AWvd (1998) Asymptotic statistics. Cambridge series in statistical and probabilistic mathematics, Cambridge University Press, Cambridge, UK
Wan C, Qian W, Zhao C et al (2021) Probabilistic forecasting based sizing and control of hybrid energy storage for wind power smoothing. IEEE Trans Sustain Energy 12(4):1841–1852
Wang C, Teng Q, Liu X et al (2019) Optimal sizing of energy storage considering the spatial-temporal correlation of wind power forecast errors. IET Renew Power Gener 13(4):530–538
Wang S, Geng G, Jiang Q (2019) Robust co-planning of energy storage and transmission line with mixed integer recourse. IEEE Trans Power Syst 34(6):4728–4738
Wang B, Zhang C, Dong ZY et al (2021) Improving hosting capacity of unbalanced distribution networks via robust allocation of battery energy storage systems. IEEE Trans Power Syst 36(3):2174–2185
Wu X, Jiang Y (2019) Source-network-storage joint planning considering energy storage systems and wind power integration. IEEE Access 7:137330–137343
Wu Y, Zhang T, Zhong K et al (2021) Optimal planning of energy storage technologies considering thirteen demand scenarios from the perspective of electricity grid: a three-stage framework. Energy Convers Manage 229:113789
Wu X, Zhao W, Wang X et al (2021) An MILP-based planning model of a photovoltaic/diesel/battery stand-alone microgrid considering the reliability. IEEE Trans Smart Grid 12(5):3809–3818
Xia S, Chan K, Luo X et al (2018) Optimal sizing of energy storage system and its cost-benefit analysis for power grid planning with intermittent wind generation. Renew Energy 122:472–486
Xiao J, Bai L, Li F et al (2014) Sizing of energy storage and diesel generators in an isolated microgrid using discrete Fourier transform (DFT). IEEE Trans Sustain Energy 5(3):907–916
Xu B, Wang Y, Dvorkin Y et al (2017) Scalable planning for energy storage in energy and reserve markets. IEEE Trans Power Syst 32(6):4515–4527

Yang Y, Ye Q, Tung LJ et al (2018) Integrated size and energy management design of battery storage to enhance grid integration of large-scale PV power plants. IEEE Trans Ind Electron 65(1):394–402

Zhang C, Cheng H, Liu L et al (2020) Coordination planning of wind farm, energy storage and transmission network with high-penetration renewable energy. Int J Electr Power Energy Syst 120:105944

Zhang C, Qiu J, Yang Y et al (2021) Trading-oriented battery energy storage planning for distribution market. Int J Electr Power Energy Syst 129:106848

Zhang S, Xiang Y, Liu J et al (2022) A regulating capacity determination method for pumped storage hydropower to restrain PV generation fluctuations. CSEE J Power Energy Syst 8(1):304–316

Zhao B, Ren J, Chen J et al (2020) Tri-level robust planning-operation co-optimization of distributed energy storage in distribution networks with high PV penetration. Appl Energy 279:115768

Zhou Y, Zhai Q, Yuan W et al (2021) Capacity expansion planning for wind power and energy storage considering hourly robust transmission constrained unit commitment. Appl Energy 302:117570

Part I
Theory

Chapter 2
Preliminary

Abstract This chapter reviews some preliminary that will be used in later chapters. First, we introduce useful definitions of vectors, sets, and functions. Then, the basics of probability and risk measures are summarized. Subsequently, optimization theories are reviewed, including convex optimization, conic programs and duality, multi-objective optimization, stochastic programming, and robust optimization. Some results are stated without proof and references are given for readers interested in the details. Besides the mathematical preliminary, we also summarize some popular power flow models, which are useful in power system planning and operation.

2.1 Vectors, Sets, and Functions

In this section, we first recall the definitions of norm and distance function, then review different types of sets and the notions of continuity and Lipschitz continuity. Finally, some useful functions are defined.

2.1.1 Norm and Distance Function

Norms can be defined on vector spaces. Metrics or distance functions can be defined on sets. Nonetheless, we mainly focus on M-dimensional *Euclidean space*:

$$\mathbb{R}^M := \{(x_1, \ldots, x_M) \mid x_m \in \mathbb{R}\ ,\ m = 1, \ldots, M\}\ .$$

2.1.1.1 Norm

Definition 2.1 (*Norm*) For a Euclidean space $\mathbb{R}^M$, a *norm* on it is a function $\|\cdot\| : \mathbb{R}^M \to \mathbb{R}$ with the following properties:

R. Xie and W. Wei, *Distributionally Robust Optimization and its Applications in Power System Energy Storage Sizing*,
https://doi.org/10.1007/978-981-97-2566-3_2

1. (Triangle inequality) For any $a, b \in \mathbb{R}^M$, we have $\|a + b\| \leq \|a\| + \|b\|$.
2. (Absolute homogeneity) For any $a \in \mathbb{R}^M$ and $t \in \mathbb{R}$, we have $\|ta\| = |t|\|a\|$.
3. (Point-separating) For any $a \in \mathbb{R}^M$, $\|a\| = 0$ implies $a = 0$, the zero vector.

An important type of norm on $\mathbb{R}^M$ is the p-norm.

Definition 2.2 (*p-norm*) Assume $p \geq 1$ is a constant. The *p-norm* $\|\cdot\|_p$ on $\mathbb{R}^M$ is defined by:

$$\|x\|_p := \left(\sum_{m=1}^{M} |x_m|^p\right)^{\frac{1}{p}}, \quad \forall x = (x_1, \ldots, x_M) \in \mathbb{R}^M .$$

The ∞-*norm* $\|\cdot\|_\infty$ on $\mathbb{R}^M$ is defined as:

$$\|x\|_\infty := \max\{|x_1|, \ldots, |x_m|\}, \quad \forall x = (x_1, \ldots, x_M) \in \mathbb{R}^M .$$

A norm is related to its dual norm.

Definition 2.3 (*Dual norm*) Assume that $\|\cdot\|$ is a norm on $\mathbb{R}^M$. Then its *dual norm* $\|\cdot\|_* : \mathbb{R}^M \to \mathbb{R}$ is defined by:

$$\|x\|_* := \sup_{y \in \mathbb{R}^M : \|y\| \leq 1} x^\top y .$$

Proposition 2.1 *On $\mathbb{R}^M$, the dual norm $\|\cdot\|_*$ of a norm $\|\cdot\|$ is also a norm. Moreover, the dual norm of $\|\cdot\|_*$ is $\|\cdot\|$. In particular, for any $t \in \mathbb{R}$,*

$$\|tx\|_* = \sup_{y : \|y\| \leq |t|} x^\top y , \quad \|tx\| = \sup_{y : \|y\|_* \leq |t|} x^\top y ,$$

and the supremum can be attained.

Three popular p-norms are 1-norm, 2-norm, and ∞-norm.

Example 2.1 (*2-norm*) 2-norm is the most commonly used norm on the Euclidean space $\mathbb{R}^M$ and also called the *Euclidean norm*, which is:

$$\|x\|_2 := \sqrt{\sum_{m=1}^{M} |x_m|^2} , \quad \forall x \in \mathbb{R}^M .$$

2-norm is closely related to the inner product, i.e., $\|x\|_2^2 = x^\top x$, $\forall x \in \mathbb{R}^M$. Moreover, 2-norm is self-dual, which means the dual norm of 2-norm is still 2-norm.

Example 2.2 (*1-norm and* ∞*-norm*) The definition of 1-norm is:

$$\|x\|_1 := \sum_{m=1}^{M} |x_m| \ , \ \forall x \in \mathbb{R}^M \ .$$

∞-norm has been defined in Definition 2.2, i.e., $\|x\|_\infty = \max\{|x_m| \mid m = 1, \ldots, M\}$.

1-norm and ∞-norm are dual norms to each other. They are sometimes adopted in the modeling process for the convenience of deriving tractable reformulations of the optimization problem. The reason is that the constraints bounding 1-norm and ∞-norm from above are equivalent to linear constraints: For any $x \in \mathbb{R}^M$ and $t \in \mathbb{R}$,

$$\|x\|_1 \le t \iff \begin{cases} \exists s \in \mathbb{R}^M \ , \ \text{s.t. } s_1 + \cdots + s_M \le t \ , \\ -s_m \le x_m \le s_m \ , \ m = 1, \ldots, M \ , \end{cases} \tag{2.1a}$$

$$\|x\|_\infty \le t \iff -t \le x_m \le t \ , \ m = 1, \ldots, M \ . \tag{2.1b}$$

Nonetheless, these norms are equivalent in the sense of Proposition 2.2. In fact, all the norms on $\mathbb{R}^M$ are equivalent.

Proposition 2.2 (Equivalence of 1-norm, 2-norm, and ∞-norm on $\mathbb{R}^M$) *For any* $x = (x_1, \ldots, x_M) \in \mathbb{R}^M$, *we have:*

1. $\|x\|_2 \le \|x\|_1 \le \sqrt{M}\|x\|_2$.
2. $\|x\|_\infty \le \|x\|_1 \le M\|x\|_\infty$.
3. $\|x\|_\infty \le \|x\|_2 \le \sqrt{M}\|x\|_\infty$.

2.1.1.2 Distance Function

Definition 2.4 (*Distance function or metric*) Given a nonempty set S, a *distance function* or *metric* on S is a function $d : S \times S \to \mathbb{R}$ with the following properties:

1. (Nonnegativity) For any $a, b \in S$, $d(a, b) \ge 0$.
2. (Point-separating) For any $a, b \in S$, $d(a, b) = 0$ if and only if $a = b$.
3. (Symmetry) For any $a, b \in S$, $d(a, b) = d(b, a)$.
4. (Triangle inequality) For any $a, b, c \in S$, $d(a, c) \le d(a, b) + d(b, c)$.

Comparing Definition 2.4 with Definition 2.1, we can see that there is a similarity between the requirements of distance functions and norms. Proposition 2.3 says that a distance function can be constructed based on a given norm.

Proposition 2.3 *Assume $\|\cdot\|$ is a norm on $\mathbb{R}^M$ and S is a nonempty subset of $\mathbb{R}^M$. Define function $d : S \times S \to \mathbb{R}$ by*

$$d(x, y) = \|x - y\| \ , \ \forall x, y \in S \ .$$

Then d is a distance function on S.

Although many distance functions are constructed based on norms, this is not the only way to define a distance function. A simple example is the so-called discrete metric.

Example 2.3 (*Discrete metric*) Let S be a nonempty set. The *discrete metric* on S is a function $d_{\mathrm{D}} : S \times S \to \mathbb{R}$ defined as follows:

$$\forall a, b \in S \ , \ d_{\mathrm{D}}(a, b) := \begin{cases} 0 \ , & \text{if } a = b \ , \\ 1 \ , & \text{if } a \neq b \ . \end{cases} \tag{2.2}$$

The discrete metric d_{D} defined above is a distance function on S. Moreover, if $S = \mathbb{R}$, the discrete metric d_{D} cannot be constructed by any norm.

Under the discrete metric, the distance between two distinct points is always 1. Although the discrete metric is a distance function according to the definition, it only tells us whether two points are the same but gives no information on how far they are from each other.

Distance functions can also be defined on spaces other than Euclidean spaces. For example, Wasserstein distance will be defined on the space of probability distributions in Definition 6.1 in Sect. 6.1.1.

Definition 2.5 (*Distance between sets*) Let $d : S \times S \to \mathbb{R}$ be a distance function. Assume S_1 and S_2 are nonempty subsets of S. The distance between S_1 and S_2 is denoted by $d(S_1, S_2)$ and defined as follows:

$$d(S_1, S_2) := \inf_{a \in S_1, b \in S_2} d(a, b) \ .$$

Definition 2.5 is consistent with Definition 2.4 because $d(\{a\}, \{b\}) = d(a, b)$ for $a, b \in S$. Furthermore, the distance between a point and a set is $d(a, S_2) = \inf_{b \in S_2} d(a, b)$.

It should be noted that the distance between sets is nonnegative and symmetric, but it does not have the properties of point-separating and triangle inequality. Disjoint sets can have zero distance. For example, the two sets $\{(x, y) \in \mathbb{R}^2 \mid x > 0, y \geq 1/x\}$ and $\{(x, y) \in \mathbb{R}^2 \mid x > 0, y \leq -1/x\}$ are disjoint but their distance is zero. To test the triangle inequality, consider $S_1 = (-\infty, -2]$, $S_2 = [-1, 1]$, and $S_3 = [2, +\infty)$, then $d(S_1, S_3) > d(S_1, S_2) + d(S_2, S_3)$, which shows the triangle inequality does not hold.

2.1.2 *Different Types of Sets*

We list the definitions of some set types in the Euclidean space here, some of which can be generalized to metric spaces (spaces equipped with a distance function).

Definition 2.6 (*Open set*) Suppose that S is a subset of $\mathbb{R}^M$ and d is a distance function on $\mathbb{R}^M$. Then S is called an *open set* in $\mathbb{R}^M$ if for any $a \in S$, there exists $\varepsilon > 0$ such that $b \in S$ whenever $d(a, b) < \varepsilon$.

Definition 2.7 (*Closed set*) Suppose $S \subset \mathbb{R}^M$. Then S is called a *closed set* in $\mathbb{R}^M$ if $\mathbb{R}^M \setminus S$ is an open set.

The union of open sets is open. The intersection of closed sets is closed.

In $\mathbb{R}^M$, if S is closed and a sequence $\{x_1, \ldots, x_n, \ldots\}$ in S converges to x_0, i.e., $x_n \in S, n \in \mathbb{N}_+$ and $d(x_n, x_0) \to 0$ as $n \to \infty$, then $x_0 \in S$.

Definition 2.8 (*Closure*) Suppose $S \subset \mathbb{R}^M$. Then the *closure* of S (denoted by $\mathrm{cl}(S)$) in $\mathbb{R}^M$ is the intersection of all the closed sets containing S in $\mathbb{R}^M$.

In other words, $\mathrm{cl}(S)$ is the smallest closed set containing S. If a sequence $\{x_1, \ldots, x_n, \ldots\}$ in S converges to x_0, then $x_0 \in \mathrm{cl}(S)$. In fact, $\mathrm{cl}(S)$ consists of the points in S and the limit points of S.

Definition 2.9 (*Bounded set*) A set S in the Euclidean space $\mathbb{R}^M$ is called a *bounded set* if there is a constant $N > 0$ such that $\|x\| < N, \forall x \in S$.

Definition 2.10 (*Bounded function*) Assume $f : S \to \mathbb{R}$ is a function. Then f is a *bounded function* if there is a constant $M \geq 0$ such that for any $x \in S$, we have $|f(x)| \leq M$.

Definition 2.11 (*Compact set in Euclidean space*) Suppose $S \subset \mathbb{R}^M$. Then S is called a *compact set* in $\mathbb{R}^M$ if it is closed and bounded.

If $\{x_1, \ldots, x_n, \ldots\}$ is a sequence in a compact set S, then it has a convergent subsequence and the limit point x_0 is in S, i.e., there is a sequence $n_1 < n_2 < \cdots < n_k < \cdots$ in $\mathbb{N}_+$ such that $\|x_{n_k} - x_0\| \to 0$ as $k \to \infty$. This property is called *Bolzano-Weierstrass theorem*.

2.1.3 *Continuity and Lipschitz Continuity*

Definition 2.12 (*Continuous function*) Assume $f : S \to \mathbb{R}$ is a function on a nonempty set $S \subset \mathbb{R}^M$, $x \in S$, and $\|\cdot\|$ is a norm on $\mathbb{R}^M$. Then f is *continuous* at x if for any $\varepsilon > 0$ there is some $\delta > 0$ so that $|f(x) - f(y)| < \varepsilon$ for any $y \in S$ subject to $\|x - y\| < \delta$. If f is continuous at all the points in S, f is called a *continuous function*.

Proposition 2.4 (Norm is continuous) *Let $\|\cdot\| : \mathbb{R}^M \to \mathbb{R}$ be a norm. Then $\|\cdot\|$ is a continuous function on $\mathbb{R}^M$ with respect to the norm $\|\cdot\|$.*

Distance functions are not necessarily to be continuous. For example, the discrete metric d_D defined in (2.2) is not continuous as a function on $\mathbb{R} \times \mathbb{R} = \mathbb{R}^2$.

Definition 2.13 (*Lipschitz continuous*) Let $f : S \to \mathbb{R}$ be a function on a nonempty set $S \subset \mathbb{R}^M$. Let $\|\cdot\|$ be a norm on $\mathbb{R}^M$. Then f is *Lipschitz continuous* if there is a constant $L \in \mathbb{R}$ so that

$$|f(x) - f(y)| \leq L\|x - y\| \ , \ \forall x, y \in S \ .$$

L is called a *Lipschitz constant* of f and f is called L-Lipschitz continuous.

The Lipschitz constant of a Lipschitz continuous function is not unique. In fact, if L is a Lipschitz constant of f, L' is also a Lipschitz constant so long as $L' \geq L$.

Assume $f : S \to \mathbb{R}$ is a Lipschitz continuous function, then 0 is a Lipschitz constant of f if and only if f is a constant function on S, i.e., there is a constant $C \in \mathbb{R}$ such that $f(x) = C, \forall x \in S$.

Proposition 2.5 *Let $f : S \to \mathbb{R}$ be a Lipschitz continuous function. Then f is continuous.*

Lipschitz continuity is stronger than continuity. A continuous function defined on a compact set is not necessarily to be Lipschitz continuous, such as the function $f : [0, 2] \to \mathbb{R}$ defined by $f(x) = \sqrt{1 - (x - 1)^2}$.

Proposition 2.6 *Let $f : S \to \mathbb{R}$ be a Lipschitz continuous function on a nonempty bounded set $S \subset \mathbb{R}^M$. Then f is bounded.*

Definition 2.14 (*Lipschitz modulus*) Assume $f : S \to \mathbb{R}$ is a function on S, where $\varnothing \neq S \subset \mathbb{R}^M$. Then

$$\text{Lip}(f) := \sup_{x,y \in S, x \neq y} \frac{|f(x) - f(y)|}{\|x - y\|} \ ,$$

is called the *Lipschitz modulus* of f.

According to the above definition of Lipschitz modulus, it is the supremum of the slopes and

$$\text{Lip}(f) = \inf\{L \in \mathbb{R} \mid L \text{ is a Lipschitz constant of } f\} \ ,$$

i.e., the smallest Lipschitz constant. If f is not Lipschitz continuous, $\text{Lip}(f) = +\infty$.

2.1.4 Some Useful Functions

Definition 2.15 (*Identity function*) Let S be a nonempty set. The *identity function* of S is the function $i_S : S \to S$ defined by:

$$i_S(a) = a \ , \ \forall a \in S \ .$$

Definition 2.16 (*Positive part and negative part*) The *positive part* function $(\cdot)_+ : \mathbb{R} \to \mathbb{R}$ and the *negative part* function $(\cdot)_- : \mathbb{R} \to \mathbb{R}$ are defined as follows:

$$(x)_+ = \begin{cases} x \ , \text{ if } x \geq 0 \ , \\ 0 \ , \text{ if } x < 0 \ , \end{cases} \quad (x)_- = \begin{cases} 0 \ , & \text{if } x > 0 \ , \\ -x \ , & \text{if } x \leq 0 \ . \end{cases}$$

It can be verified that $x = (x)_+ - (x)_-$ and $|x| = (x)_+ + (x)_-$ for any $x \in \mathbb{R}$.

Definition 2.17 (*Indicator function*) For $S \subset \Omega$, denote the *indicator function* of S on Ω by $1_S : \Omega \to \mathbb{R}$ and its definition is:

$$1_S(\omega) = \begin{cases} 1 \ , \text{ if } \omega \in S \ , \\ 0 \ , \text{ if } \omega \in \Omega \setminus S \ . \end{cases} \tag{2.3}$$

Definition 2.18 (*Characteristic function*) For $S \subset \Omega$, the *characteristic function* of S on Ω is $\chi_S : \Omega \to (-\infty, +\infty]$ defined by:

$$\chi_S(\omega) = \begin{cases} 0 \ , & \text{if } \omega \in S \ , \\ +\infty \ , & \text{if } \omega \in \Omega \setminus S \ . \end{cases}$$

Definition 2.19 (*Diameter of a set*) Let S be a nonempty set with a distance function $d : S \times S \to \mathbb{R}$ on it. Then the *diameter* of the set S is defined as follows:

$$\text{diam}(S) := \sup_{a,b \in S} d(a, b) \ .$$

For a nonempty set S and a norm $\| \cdot \|$ on $\mathbb{R}^M$, S is a bounded set if and only if $\text{diam}(S) < +\infty$.

2.2 Probability

In this section, we summarize some useful definitions and results in basic probability theory, including probability distribution, random variable, expectation, variance, joint probability distribution, and conditional probability distribution.

2.2.1 *Probability Distribution*

We introduce the definition of probability distribution and its basic properties in this subsection.

2.2.1.1 Definition

Definition 2.20 (*σ-algebra*) Let Ω be a nonempty set. A *σ-algebra* $\mathcal{F}$ on Ω is a collection of subsets of Ω with the following properties:

1. $\varnothing \in \mathcal{F}$ and $\Omega \in \mathcal{F}$.
2. ($\mathcal{F}$ is closed under complementation) $S \in \mathcal{F}$ implies $\Omega \setminus S \in \mathcal{F}$.
3. ($\mathcal{F}$ is closed under a countable number of unions) $S_1, S_2, \ldots, S_n, \ldots \in \mathcal{F}$ implies $\bigcup_{n=1}^{+\infty} S_n \in \mathcal{F}$.

Borel σ-algebra consists of all the sets that can be formed from open sets by countable unions, countable intersections, and complements. The set in the Borel σ-algebra is called a *Borel (measurable) set*.

Definition 2.21 (*Probability space and probability distribution*) A *probability space* is a triple $(\Omega, \mathcal{F}, \mathbb{P})$, where Ω is a nonempty set, $\mathcal{F}$ is a σ-algebra on Ω, and the function $\mathbb{P} : \mathcal{F} \to \mathbb{R}$ satisfies:

1. (Nonnegativity) $\mathbb{P}(S) \geq 0, \forall S \in \mathcal{F}$.
2. (Unit measure) $\mathbb{P}(\Omega) = 1$.
3. (Countable additivity) Let $S_1, S_2, \ldots, S_n, \ldots$ be a countable number of disjoint sets in $\mathcal{F}$ so that $S_n \cap S_{n'} = \varnothing$ for all $n \neq n'$. Then $\mathbb{P}(\bigcup_{n=1}^{+\infty} S_n) = \sum_{n=1}^{+\infty} \mathbb{P}(S_n)$.

Here $\mathbb{P}$ is called a *probability distribution* and $(\Omega, \mathcal{F})$ is called a *sample space*.

For $\varnothing \neq \Omega \subset \mathbb{R}^M$, we often let $\mathcal{F}$ be the Borel σ-algebra on Ω. In this circumstance, we sometimes omit $\mathcal{F}$ and call Ω as the sample space.

Definition 2.22 (*Measure*) Let Ω be a nonempty set and let $\mathcal{F}$ be a σ-algebra on Ω. A function $\mu : \mathcal{F} \to (-\infty, +\infty]$ is called a *measure* if it satisfies:

1. (Nonnegativity) $\mu(S) \geq 0, \forall S \in \mathcal{F}$.
2. (Null empty set) $\mu(\varnothing) = 0$.
3. (Countable additivity) Let $S_1, S_2, \ldots, S_n, \ldots$ be a countable number of disjoint sets in $\mathcal{F}$ so that $S_n \cap S_{n'} = \varnothing$ for all $n \neq n'$. Then $\mu(\bigcup_{n=1}^{+\infty} S_n) = \sum_{n=1}^{+\infty} \mu(S_n)$.

Let $(\Omega, \mathcal{F}, \mathbb{P})$ be a probability space, then the probability distribution $\mathbb{P}$ is a measure.

Definition 2.23 (*Random variable*) Let $(\Omega, \mathcal{F}, \mathbb{P})$ be a probability space. A *random variable* X following the probability distribution $\mathbb{P}$ is a measurable function $X : \Omega \to \mathbb{R}^M$ and for any set $S \in \mathcal{F}$, it is defined that

$$\mathbb{P}(X \in S) := \mathbb{P}(\{\omega \in \Omega \mid X(\omega) \in S\}) .$$

X is called an M-dimensional random variable.

In Definition 2.23, $X : \Omega \to \mathbb{R}^M$ is a *measurable function*, which means that for any Borel set S in $\mathbb{R}^M$, the inverse image $\{\omega \in \Omega \mid X(\omega) \in S\}$ is *measurable*, i.e., it is an element of the set $\mathcal{F}$.

If the sample space Ω is already a subset of $\mathbb{R}^M$, the identity function leads to a random variable ω and $\mathbb{P}(\omega \in S) = \mathbb{P}(S)$ for any measurable S.

Definition 2.24 (*Discrete probability distribution*) A probability distribution $\mathbb{P}$ on $\Omega \subset \mathbb{R}^M$ is called a *discrete probability distribution* if there is a countable set $\{\omega_1, \omega_2, \ldots, \omega_n, \ldots\} \subset \Omega$ such that $\sum_n \mathbb{P}(\omega_n) = 1$.

Definition 2.25 (*Continuous probability distribution and probability density function*) A probability distribution $\mathbb{P}$ on Ω is called a *continuous probability distribution* if there is a *probability density function* (PDF) $p : \Omega \to [0, +\infty]$ such that for any measurable $S \subset \Omega$,

$$\mathbb{P}(S) = \int_S p(\omega) \mathrm{d}\omega .$$

Here are some basic properties of probability distributions:

Lemma 2.1 (Monotonicity of probability) *Let $(\Omega, \mathcal{F}, \mathbb{P})$ be a probability space. Assume $S_1, S_2 \in \mathcal{F}$ and $S_1 \subset S_2$, then $\mathbb{P}(S_1) = \mathbb{P}(S_2) - \mathbb{P}(S_2 \setminus S_1)$ and $\mathbb{P}(S_1) \leq \mathbb{P}(S_2)$.*

Lemma 2.2 *Let $(\Omega, \mathcal{F}, \mathbb{P})$ be a probability space. Assume that $A \in \mathcal{F}$ satisfies $\mathbb{P}(A) = 1$. Then for any $S \in \mathcal{F}$, $\mathbb{P}(S) = \mathbb{P}(S \cap A)$.*

Proposition 2.7 (Bonferroni inequality) *Let $(\Omega, \mathcal{F}, \mathbb{P})$ be a probability space. Assume $S_1, S_2, \ldots, S_N \in \mathcal{F}$, then*

$$\mathbb{P}(S_1 \cup S_2 \cup \cdots \cup S_N) \leq \mathbb{P}(S_1) + \mathbb{P}(S_2) + \cdots + \mathbb{P}(S_N) .$$

2.2.1.2 Cumulative Distribution Function

Definition 2.26 (*Cumulative distribution function on* $\mathbb{R}$) Assume $\mathbb{P}$ is a probability distribution defined on the sample space $\mathbb{R}$. Then its *cumulative distribution function* (CDF) $F : \mathbb{R} \to [0, 1]$ is defined by $F(x) = \mathbb{P}((-\infty, x])$, $\forall x \in \mathbb{R}$.

Proposition 2.8 (Properties of CDF) *Suppose* $F : \mathbb{R} \to [0, 1]$ *is the CDF of probability distribution* $\mathbb{P}$ *on* $\mathbb{R}$. *Then*

1. F *is nondecreasing.*
2. F *is right-continuous, i.e.,* $\lim_{x \to x_0+} F(x) = F(x_0), \forall x_0 \in \mathbb{R}$.
3. $\lim_{x \to -\infty} F(x) = 0$ *and* $\lim_{x \to +\infty} F(x) = 1$.

Definition 2.27 (*Generalized inverse distribution function*) Let $F : \mathbb{R} \to [0, 1]$ be the CDF of probability distribution $\mathbb{P}$ on $\mathbb{R}$. Then the *generalized inverse distribution function* is denoted by $F^{-1} : [0, 1] \to [-\infty, +\infty]$ and given by:

$$F^{-1}(x) := \inf\{y \in \mathbb{R} \mid F(y) \geq x\} \ , \ \forall x \in [0, 1] \ .$$

Proposition 2.9 (Properties of generalized inverse distribution function) *Let* $\mathbb{P}$ *be a probability distribution on* $\mathbb{R}$. *Assume* F *is its CDF and* F^{-1} *is the corresponding generalized inverse distribution function. Let* $\mathbb{P}_0$ *be the uniform distribution on* $[0, 1]$. *Then*

1. F^{-1} *is nondecreasing.*
2. *For any* $x \in \mathbb{R}$ *and* $p \in [0, 1]$, *we have*

$$F^{-1}(p) \leq x \iff p \leq F(x) \ ,$$
$$F^{-1}(p) > x \iff p > F(x) \ .$$

3. *For any* $x \in \mathbb{R}$,

$$\mathbb{P}_0(\{p \in [0, 1] \mid F^{-1}(p) \leq x\}) = F(x) = \mathbb{P}((-\infty, x]) \ .$$

The third property indicates that if the random variable X follows the uniform distribution on $[0, 1]$, then $F^{-1}(X)$ follows the probability distribution $\mathbb{P}$. This property can be used to generate random variables following any given probability distribution on $\mathbb{R}$.

Definition 2.28 (*CDF on* $\mathbb{R}^M$) Let $\mathbb{P}$ be a probability distribution on $\mathbb{R}^M$, then its *cumulative distribution function* (CDF) is a function $F : \mathbb{R}^M \to [0, 1]$ satisfying that for any $x = (x_1, \ldots, x_M) \in \mathbb{R}^M$,

$$F(x) = \mathbb{P}(\{y = (y_1, \ldots, y_M) \in \mathbb{R}^M \mid y_m \leq x_m \ , \ m = 1, \ldots, M\}) \ . \qquad (2.4)$$

According to measure theory (Rudin 1987), a probability distribution $\mathbb{P}$ on $\mathbb{R}^M$ can be uniquely determined by its CDF F.

2.2.1.3 Absolute Continuity

Definition 2.29 (*Absolutely continuous*) Assume that $\mathbb{P}_1$ and $\mathbb{P}_2$ are two probability distributions on $(\Omega, \mathcal{F})$. Then we say $\mathbb{P}_2$ is *absolutely continuous* with respect to $\mathbb{P}_1$ and write $\mathbb{P}_2 \ll \mathbb{P}_1$ if for any $S \in \mathcal{F}$,

$$\mathbb{P}_1(S) = 0 \implies \mathbb{P}_2(S) = 0 ,$$

or equivalently,

$$\mathbb{P}_2(S) > 0 \implies \mathbb{P}_1(S) > 0 .$$

When $\mathbb{P}_2 \ll \mathbb{P}_1$, there is a function denoted by $\mathrm{d}\mathbb{P}_2/\mathrm{d}\mathbb{P}_1 : \Omega \to \mathbb{R}$, which represents the density of $\mathbb{P}_2$ with respect to $\mathbb{P}_1$. To be specific,

$$\mathbb{P}_2(S) = \int_S \frac{\mathrm{d}\mathbb{P}_2}{\mathrm{d}\mathbb{P}_1}(\omega)\mathrm{d}\mathbb{P}_1(\omega) , \ \forall S \in \mathcal{F} .$$

The function $\mathrm{d}\mathbb{P}_2/\mathrm{d}\mathbb{P}_1 : \Omega \to \mathbb{R}$ is called *Radon-Nikodym derivative* in measure theory.

For example, continuous probability distributions are exactly those probability distributions that are absolutely continuous with respect to the Lebesgue measure and the PDF is the corresponding Radon-Nikodym derivative. A simpler example is that every probability distribution is absolutely continuous with respect to itself and the Radon-Nikodym derivative equals 1 everywhere.

2.2.2 *Expectation and Variance*

Definition 2.30 (*Expectation*) Suppose that $X : \Omega \to \mathbb{R}$ is a random variable on the probability space $(\Omega, \mathcal{F}, \mathbb{P})$. If $\int_\Omega |X(\omega)|\mathrm{d}\mathbb{P}(\omega)$ is well-defined and has a finite value, then we say *expectation* $\mathbb{E}^{\mathbb{P}}[X(\omega)]$ exists and

$$\mathbb{E}^{\mathbb{P}}[X(\omega)] := \int_\Omega X(\omega)\mathrm{d}\mathbb{P}(\omega) .$$

For the multi-dimensional case, i.e., $X : \Omega \to \mathbb{R}^M$, we can similarly define the expectation $\mathbb{E}^{\mathbb{P}}[X(\omega)]$, which is element-wise and also called the first-order moment of X.

For discrete probability distribution $\mathbb{P}$ on the sample space $\{\omega_1, \omega_2, \ldots, \omega_n, \ldots\}$,

$$\mathbb{E}^{\mathbb{P}}[X(\omega)] = \sum_n X(\omega_n)\mathbb{P}(\omega_n) .$$

For continuous probability distribution $\mathbb{P}$ with PDF p,

$$\mathbb{E}^{\mathbb{P}}[X(\omega)] = \int\limits_{\Omega} X(\omega)p(\omega)\mathrm{d}\omega \ .$$

Definition 2.31 (*Conditional expectation*) Suppose that $X : \Omega \to \mathbb{R}$ is a random variable on the probability space $(\Omega, \mathcal{F}, \mathbb{P})$. Assume $S \in \mathcal{F}$ and $\mathbb{P}(S) > 0$. If $\mathbb{E}^{\mathbb{P}}[X(\omega)]$ exists, then the *conditional expectation* $\mathbb{E}^{\mathbb{P}}[X(\omega)|S]$ is defined by:

$$\mathbb{E}^{\mathbb{P}}[X(\omega)|S] := \frac{\mathbb{E}^{\mathbb{P}}[X(\omega)1_S(\omega)]}{\mathbb{P}(S)} \ ,$$

where $1_S : \Omega \to \mathbb{R}$ is the indicator function of S on Ω.

Lemma 2.3 *Let $X : \Omega \to \mathbb{R}$ be a function and $\mathbb{P}$ be a probability distribution on Ω. Assume $\mathbb{P}(X(\omega) \neq 0) = 0$, then $\mathbb{E}^{\mathbb{P}}[X(\omega)] = 0$.*

Lemma 2.4 (Monotonicity of expectation) *Suppose that $X, Y : \Omega \to \mathbb{R}$ are random variables on the probability space $(\Omega, \mathcal{F}, \mathbb{P})$; $\mathbb{E}^{\mathbb{P}}[X(\omega)]$ and $\mathbb{E}^{\mathbb{P}}[Y(\omega)]$ both exist and $X(\omega) \leq Y(\omega), \forall \omega \in \Omega$. Then $\mathbb{E}^{\mathbb{P}}[X(\omega)] \leq \mathbb{E}^{\mathbb{P}}[Y(\omega)]$.*

Proposition 2.10 *Suppose that $X : \Omega \to \mathbb{R}$ is a random variable on the probability space $(\Omega, \mathcal{F}, \mathbb{P})$ so that $X(\omega) > 0, \forall \omega \in S$, where S is a measurable subset of Ω and $\mathbb{P}(S) > 0$. Then $\int_S X(\omega)\mathrm{d}\mathbb{P}(\omega) > 0$.*

Proof Let $S_n = \{\omega \in S \mid f(\omega) \geq 1/n\}, n \in \mathbb{N}_+$. Then S_n is measurable for any $n \in \mathbb{N}_+$, $S_1 \subset S_2 \subset \ldots \subset S_n \subset \ldots$, and $S = \bigcup_{n \in \mathbb{N}_+} S_n$. By measure theory, $\lim_{n \to +\infty} \mathbb{P}(S_n) = \mathbb{P}(S) > 0$, then there exists $N \in \mathbb{N}_+$ so that $\mathbb{P}(S_N) > 0$. Using Lemma 2.4, we have

$$\int\limits_{S} f(\omega)\mathrm{d}\mathbb{P}(\omega) \geq \int\limits_{S_N} f(\omega)\mathrm{d}\mathbb{P}(\omega) \geq \int\limits_{S_N} \frac{1}{N}\mathrm{d}\mathbb{P}(\omega) = \frac{\mathbb{P}(S_N)}{N} > 0 \ .$$

□

The linearity of expectation follows from the linearity of integrals. In other words,

$$\mathbb{E}^{\mathbb{P}}[tX(\omega) + sY(\omega)] = t\mathbb{E}^{\mathbb{P}}[X(\omega)] + s\mathbb{E}^{\mathbb{P}}[Y(\omega)] \ ,$$

for $t, s \in \mathbb{R}$ if all the expectations above exist.

Example 2.4 (*Represent probability using expectation*) Let $(\Omega, \mathcal{F}, \mathbb{P})$ be a probability space and $S \in \mathcal{F}$. We want to express $\mathbb{P}(S)$ by an expectation. Let 1_S be the indicator function of S, i.e., $1_S(\omega) = 1$, if $\omega \in S$; $1_S(\omega) = 0$, otherwise. Then

$$\mathbb{E}^{\mathbb{P}}[1_S(\omega)] = \int\limits_{\Omega} 1_S(\omega)\mathrm{d}\mathbb{P}(\omega) = \int\limits_{S} \mathrm{d}\mathbb{P}(\omega) = \mathbb{P}(S) \ .$$

Theorem 2.1 (Markov's inequality) *Let $X : \Omega \to \mathbb{R}$ be a nonnegative random variable on the probability space $(\Omega, \mathcal{F}, \mathbb{P})$ and $t > 0$, then*

$$\mathbb{P}(X(\omega) \geq t) \leq \frac{\mathbb{E}^{\mathbb{P}}[X(\omega)]}{t} .$$

Definition 2.32 (*Variance*) Suppose that $X : \Omega \to \mathbb{R}$ is a random variable on the probability space $(\Omega, \mathcal{F}, \mathbb{P})$. Assume its expectation $\mathbb{E}^{\mathbb{P}}[X(\omega)] = \mu \in \mathbb{R}$, then its *variance* is denoted by $\mathrm{Var}(X)$ and defined by:

$$\mathrm{Var}(X) := \mathbb{E}^{\mathbb{P}}[(X(\omega) - \mu)^2] .$$

Theorem 2.2 (Chebyshev's inequality) *Let $X : \Omega \to \mathbb{R}$ be a random variable on the probability space $(\Omega, \mathcal{F}, \mathbb{P})$ such that $\mathbb{E}^{\mathbb{P}}[X(\omega)] = \mu$ and $Var(X) = \sigma^2 > 0$. Then for any $t > 0$,*

$$\mathbb{P}(|X(\omega) - \mu| \geq t\sigma) \leq \frac{1}{t^2} .$$

Definition 2.33 (*Covariance*) Suppose that $X, Y : \Omega \to \mathbb{R}^M$ are random variables on the probability space $(\Omega, \mathcal{F}, \mathbb{P})$. Assume their expectations $\mathbb{E}^{\mathbb{P}}[X(\omega)] = \mu_X \in \mathbb{R}^M$ and $\mathbb{E}^{\mathbb{P}}[Y(\omega)] = \mu_Y \in \mathbb{R}^M$, then the *covariance* of X and Y is denoted by $\mathrm{Cov}(X, Y)$ and defined by:

$$\mathrm{Cov}(X, Y) := \mathbb{E}^{\mathbb{P}}[(X(\omega) - \mu_X)(Y(\omega) - \mu_Y)^\top] ,$$

where the expectations are element-wise.

According to the above definition,

$$\mathrm{Cov}(X, X) = \mathbb{E}^{\mathbb{P}}[(X(\omega) - \mu_X)(X(\omega) - \mu_X)^\top] ,$$

which is called the *covariance matrix* or the second-order moment of X.

2.2.3 *Joint Probability Distribution and Conditional Probability Distribution*

For sample space $(\Omega, \mathcal{F})$, we use $\mathcal{M}(\Omega)$ to denote the set of all probability distributions on Ω.

Definition 2.34 (*Joint probability distribution*) Let $\mathbb{P}_1 \in \mathcal{M}(\Omega_1)$ and $\mathbb{P}_2 \in \mathcal{M}(\Omega_2)$. $\Omega_1 \times \Omega_2$ denotes the product set of Ω_1 and Ω_2. If $\Pi \in \mathcal{M}(\Omega_1 \times \Omega_2)$ and for any measurable $S_1 \subset \Omega_1$ and $S_2 \subset \Omega_2$, we have

$$\Pi(S_1, \Omega_2) = \mathbb{P}_1(S_1) , \quad \Pi(\Omega_1, S_2) = \mathbb{P}_2(S_2) , \tag{2.5}$$

then Π is a *joint probability distribution* with *marginal probability distributions* $\mathbb{P}_1$ and $\mathbb{P}_2$, which is denoted by $\Pi \in \mathcal{J}(\mathbb{P}_1, \mathbb{P}_2)$.

Example 2.5 (*Product distribution*) Assume $\mathbb{P}_1 \in \mathcal{M}(\Omega_1)$ and $\mathbb{P}_2 \in \mathcal{M}(\Omega_2)$. Define a measure $\mathbb{P}_1 \times \mathbb{P}_2$ on $\Omega_1 \times \Omega_2$ such that

$$(\mathbb{P}_1 \times \mathbb{P}_2)(S_1 \times S_2) = \mathbb{P}_1(S_1) \times \mathbb{P}_2(S_2) \ , \ \ \forall S_1 \in \mathcal{F}_1 \ , \ \ S_2 \in \mathcal{F}_2 \ .$$

Then it can be verified that $\mathbb{P}_1 \times \mathbb{P}_2 \in \mathcal{J}(\mathbb{P}_1, \mathbb{P}_2)$. $\mathbb{P}_1 \times \mathbb{P}_2$ is called the *product distribution* of $\mathbb{P}_1$ and $\mathbb{P}_2$.

Theorem 2.3 (Fubini's theorem for probability distributions) *Suppose $\mathbb{P}_1$ and $\mathbb{P}_2$ are probability distributions on Ω_1 and Ω_2, respectively. $f : \Omega_1 \times \Omega_2 \to \mathbb{R}$ is a measurable function with $\int_{\Omega_1 \times \Omega_2} |f(\omega_1, \omega_2)| \mathrm{d}(\mathbb{P}_1 \times \mathbb{P}_2)(\omega_1, \omega_2) < +\infty$, then*

$$\begin{aligned}
&\int\limits_{\Omega_1} \left(\int\limits_{\Omega_2} f(\omega_1, \omega_2) \mathrm{d}\mathbb{P}_2(\omega_2) \right) \mathrm{d}\mathbb{P}_1(\omega_1) \\
= &\int\limits_{\Omega_2} \left(\int\limits_{\Omega_1} f(\omega_1, \omega_2) \mathrm{d}\mathbb{P}_1(\omega_1) \right) \mathrm{d}\mathbb{P}_2(\omega_2) \\
= &\int\limits_{\Omega_1 \times \Omega_2} f(\omega_1, \omega_2) \mathrm{d}(\mathbb{P}_1 \times \mathbb{P}_2)(\omega_1, \omega_2) \ .
\end{aligned}$$

Lemma 2.5 gives another way to characterize joint probability distributions.

Lemma 2.5 *Suppose $\mathbb{P}_1 \in \mathcal{M}(\Omega_1)$, $\mathbb{P}_2 \in \mathcal{M}(\Omega_2)$, and Π is a measure (not necessarily a probability distribution) defined on $\Omega_1 \times \Omega_2$. Then $\Pi \in \mathcal{J}(\mathbb{P}_1, \mathbb{P}_2)$ if and only if for all functions $f_1 : \Omega_1 \to \mathbb{R}$ and $f_2 : \Omega_2 \to \mathbb{R}$ so that $\mathbb{E}^{\mathbb{P}_1}[|f_1(\omega)|]$ and $\mathbb{E}^{\mathbb{P}_2}[|f_2(\omega)|]$ exist (i.e., is well-defined and finite), we have:*

$$\begin{aligned}
&\int\limits_{\Omega_1 \times \Omega_2} (f_1(\omega_1) + f_2(\omega_2)) \mathrm{d}\Pi(\omega_1, \omega_2) \\
= &\int\limits_{\Omega_1} f_1(\omega_1) \mathrm{d}\mathbb{P}_1(\omega_1) + \int\limits_{\Omega_2} f_2(\omega_2) \mathrm{d}\mathbb{P}_2(\omega_2) \ .
\end{aligned} \tag{2.6}$$

Proof 1. When $\Pi \in \mathcal{J}(\mathbb{P}_1, \mathbb{P}_2)$,

$$\int\limits_{\Omega_1 \times \Omega_2} (f_1(\omega_1) + f_2(\omega_2)) \mathrm{d}\Pi(\omega_1, \omega_2) \tag{2.7a}$$

$$= \int\limits_{\Omega_1 \times \Omega_2} f_1(\omega_1) \mathrm{d}\Pi(\omega_1, \omega_2) + \int\limits_{\Omega_1 \times \Omega_2} f_2(\omega_2) \mathrm{d}\Pi(\omega_1, \omega_2) \tag{2.7b}$$

$$= \int_{\Omega_1} f_1(\omega_1)\mathbb{P}_2(\Omega)\mathrm{d}\mathbb{P}_1(\omega_1) + \int_{\Omega_2} f_2(\omega_2)\mathbb{P}_1(\Omega)\mathrm{d}\mathbb{P}_2(\omega_2) \tag{2.7c}$$

$$= \int_{\Omega_1} f_1(\omega_1)\mathrm{d}\mathbb{P}_1(\omega_1) + \int_{\Omega_2} f_2(\omega_2)\mathrm{d}\mathbb{P}_2(\omega_2) \,, \tag{2.7d}$$

where from (2.7b) to (2.7c) we use $\int_{\Omega_2} \mathrm{d}\Pi(\omega_1, \omega_2) = \mathrm{d}\Pi(\omega_1, \Omega_2) = \mathbb{P}_2(\Omega_2)\mathrm{d}\mathbb{P}_1(\omega_1)$ by (2.5).

2. Assume that (2.6) holds for all functions $f_1 : \Omega_1 \to \mathbb{R}$ and $f_2 : \Omega_2 \to \mathbb{R}$ subject to $\mathbb{E}^{\mathbb{P}_1}[|f_1(\omega)|]$ and $\mathbb{E}^{\mathbb{P}_2}[|f_2(\omega)|]$ exist. Let S_2 be an arbitrary measurable subset of Ω_2; let $f_1 = 0$, the zero function; let f_2 be the indicator function of S_2. Then $\mathbb{E}^{\mathbb{P}_1}[|f_1(\omega)|] = 0$, $\mathbb{E}^{\mathbb{P}_2}[|f_2(\omega)|] = \mathbb{P}_2(S_2) \leq 1$, and

$$\int_{\Omega_1} f_1(\omega_1)\mathrm{d}\mathbb{P}_1(\omega_1) + \int_{\Omega_2} f_2(\omega_2)\mathrm{d}\mathbb{P}_2(\omega_2) = \int_{\Omega_2} f_2(\omega_2)\mathrm{d}\mathbb{P}_2(\omega_2) = \mathbb{P}_2(S_2) \,,$$

$$\int_{\Omega_1\times\Omega_2} (f_1(\omega_1) + f_2(\omega_2))\mathrm{d}\Pi(\omega_1, \omega_2)$$
$$= \int_{\Omega_1\times\Omega_2} f_2(\omega_2)\mathrm{d}\Pi(\omega_1, \omega_2) = \Pi(\Omega_1 \times S_2) \,,$$

by the fact that $f_2(\omega_2) = 1$ if and only if $(\omega_1, \omega_2) \in \Omega_1 \times S_2$. Using (2.6), we have $\Pi(\Omega_1, S_2) = \mathbb{P}_2(S_2)$. The result $\Pi(S_1, \Omega_2) = \mathbb{P}_1(S_1)$ follows similarly. Since S_1 and S_2 are arbitrary, $\Pi \in \mathcal{J}(\mathbb{P}_1, \mathbb{P}_2)$ according to the definition of joint probability distribution. □

Remark 2.1 In fact, the functions $f_1 : \Omega_1 \to \mathbb{R}$ and $f_2 : \Omega_2 \to \mathbb{R}$ in Lemma 2.5 can be restricted to be continuous and bounded functions. In this case, part 2 of the proof needs modification. The indicator function is bounded but not continuous. Nonetheless, it can be approximated by a sequence of bounded and continuous functions. The convergence of the integrals is guaranteed by the dominated convergence theorem from measure theory. In conclusion, $\Pi \in \mathcal{J}(\mathbb{P}_1, \mathbb{P}_2)$ if and only if (2.6) holds for all bounded and continuous functions $f_1 : \Omega_1 \to \mathbb{R}$ and $f_2 : \Omega_2 \to \mathbb{R}$. This conclusion will be used in Sect. 6.1.3.

Definition 2.35 (*Conditional probability distribution*) Consider two probability spaces $(\Omega_1, \mathcal{F}_1, \mathbb{P}_1)$ and $(\Omega_2, \mathcal{F}_2, \mathbb{P}_2)$. Assume $\Pi \in \mathcal{J}(\mathbb{P}_1, \mathbb{P}_2)$. Then the *conditional probability distributions* $\mathbb{P}_{1|2}$ and $\mathbb{P}_{2|1}$ satisfy that for any $S_1 \in \mathcal{F}_1$ and $S_2 \in \mathcal{F}_2$,

$$\Pi(S_1, S_2) = \mathbb{P}_{1|2}(S_1|S_2)\mathbb{P}_2(S_2) = \int_{S_2} \mathbb{P}_{1|2}(S_1|\omega_2)\mathrm{d}\mathbb{P}_2(\omega_2) , \tag{2.8a}$$

$$\Pi(S_1, S_2) = \mathbb{P}_{2|1}(S_2|S_1)\mathbb{P}_1(S_1) = \int_{S_1} \mathbb{P}_{2|1}(S_2|\omega_1)\mathrm{d}\mathbb{P}_1(\omega_1) . \tag{2.8b}$$

It can be proven that the conditional probability distribution in Definition 2.35 is well-defined. Moreover, it has a precise meaning. For example, $\mathbb{P}_{1|2}(S_1|S_2)$ is the conditional probability of the event $\omega_1 \in S_1$ provided the condition that $\omega_2 \in S_2$.

The relationship in (2.8) can be expressed using differentials as follows:

$$\mathrm{d}\Pi(\omega_1, \omega_2) = \mathrm{d}\mathbb{P}_{1|2}(\omega_1|\omega_2)\mathrm{d}\mathbb{P}_2(\omega_2) = \mathrm{d}\mathbb{P}_{2|1}(\omega_2|\omega_1)\mathrm{d}\mathbb{P}_1(\omega_1) ,$$

which can be denoted by $\Pi = \mathbb{P}_{1|2}\mathbb{P}_2 = \mathbb{P}_{2|1}\mathbb{P}_1$. Note that these notations are different from those of product distributions. However, they are related in the following sense: $\mathbb{P}_2 \times \mathbb{P}_1 = \mathbb{P}_{2|1}\mathbb{P}_1$ provided that $\mathbb{P}_{2|1}(\cdot|\omega_1) = \mathbb{P}_2(\cdot)$ for any $\omega_1 \in \Omega_1$. In this case, the conditional probability distribution $\mathbb{P}_{2|1}$ is independent of ω_1, which means that the underlying probability distributions of ω_2 and ω_1 are independent and then the joint probability distribution of $\mathbb{P}_1$ and $\mathbb{P}_2$ is the product distribution.

Remark 2.2 In Definition 2.35, $\mathbb{P}_{1|2}(\cdot|S_2)$ and $\mathbb{P}_{2|1}(\cdot|S_1)$ are probability distributions if $S_1 \in \mathcal{F}_1$, $S_2 \in \mathcal{F}_2$, $\mathbb{P}_1(S_1) > 0$, and $\mathbb{P}_2(S_2) > 0$.

Theorem 2.4 (Bayes' theorem) *In Definition 2.35, for any $S_1 \in \mathcal{F}_1$ and $S_2 \in \mathcal{F}_2$ such that $\mathbb{P}_2(S_2) > 0$,*

$$\mathbb{P}_{1|2}(S_1|S_2) = \frac{\mathbb{P}_{2|1}(S_2|S_1)\mathbb{P}_1(S_1)}{\mathbb{P}_2(S_2)} .$$

Theorem 2.5 (Total probability formula) *In Definition 2.35, for any $S_1 \in \mathcal{F}_1$,*

$$\mathbb{P}_1(S_1) = \int_{\Omega_2} \mathbb{P}_{1|2}(S_1|\omega_2)\mathrm{d}\mathbb{P}_2(\omega_2) .$$

Moreover, if $\Omega_2 = S_2^{(1)} \cup \cdots \cup S_2^{(n)} \cup \cdots$ and $S_2^{(1)}, \ldots, S_2^{(n)}, \ldots$ are disjoint with each other, then

$$\mathbb{P}_1(S_1) = \sum_n \mathbb{P}_{1|2}\left(S_1|S_2^{(n)}\right)\mathbb{P}_2\left(S_2^{(n)}\right) .$$

Lemma 2.6 *Let $(\Omega_1, \mathcal{F}_1, \mathbb{P}_1)$ be a probability space and $(\Omega_2, \mathcal{F}_2)$ be a sample space. Assume that any $\omega_1 \in \Omega_1$ is associated with a probability distribution $\mathbb{P}_{\omega_1}$ on $(\Omega_2, \mathcal{F}_2)$. Assume all the involved functions are integrable. Define Π on the measurable sets of $\Omega_1 \times \Omega_2$ so that*

$$\Pi(S_1, S_2) = \int_{S_1} \left(\int_{S_2} \mathrm{d}\mathbb{P}_{\omega_1}(\omega_2) \right) \mathrm{d}\mathbb{P}_1(\omega_1) \ , \ \ \forall S_1 \in \mathcal{F}_1 \ , \ \ S_2 \in \mathcal{F}_2 \ .$$

Then Π is a joint probability distribution with marginal probability distributions $\mathbb{P}_1$ and $\mathbb{P}_2 : \mathcal{F}_2 \to \mathbb{R}$ defined by $\mathbb{P}_2(S_2) = \Pi(\Omega_1, S_2), \forall S_2 \in \mathcal{F}_2$. In addition, the conditional probability distribution satisfies $\mathrm{d}\mathbb{P}_{2|1}(\omega_2|\omega_1) = \mathrm{d}\mathbb{P}_{\omega_1}(\omega_2)$.

Proposition 2.11 ("Composition" of joint probability distributions) *Let $\mathbb{P}_1$, $\mathbb{P}_2$, and $\mathbb{P}_3$ be probability distributions on Ω_1, Ω_2, and Ω_3, respectively. Let $\Pi_1 \in \mathcal{J}(\mathbb{P}_1, \mathbb{P}_2)$ and $\Pi_2 \in \mathcal{J}(\mathbb{P}_2, \mathbb{P}_3)$. The conditional probability distribution $\mathbb{P}_{1|2}$ is defined by Definition 2.35 using Π_1 and $\mathbb{P}_2$. Then the function $\Pi_3 : \mathcal{F}(\Omega_1 \times \Omega_3) \to \mathbb{R}$ defined as follows is a joint probability distribution of $\mathbb{P}_1$ and $\mathbb{P}_3$:*

$$\Pi_3(S) := \int_S \int_{\Omega_2} \mathbb{P}_{1|2}(\mathrm{d}\omega_1|\omega_2)\mathrm{d}\Pi_2(\omega_2, \omega_3) \ , \ \ \forall S \in \mathcal{F}(\Omega_1 \times \Omega_3) \ , \tag{2.9}$$

where $\mathcal{F}(\Omega_1 \times \Omega_3)$ is the set of measurable subsets of $\Omega_1 \times \Omega_3$.

In the integral in (2.9), $\omega_2 \in \Omega_2$ is integrated first, and then $(\omega_1, \omega_3) \in S$ is integrated. Joint probability distributions can be regarded as mass transport plans and the "composition" of joint probability distributions corresponds to the composition of mass transport plans. Please refer to Sects. 6.1.2 and 6.1.4 for details.

Proof 1. Prove that $\Pi_3 \in \mathcal{M}(\Omega_1 \times \Omega_3)$.

(a) Nonnegativity of Π_3 follows from the nonnegativity of $\mathbb{P}_{1|2}$ and Π_2.

(b) Prove that Π_3 is a unit measure: By Fubini's theorem and Remark 2.2, we have:

$$\begin{aligned}
\Pi_3(\Omega_1 \times \Omega_3) &= \int_{\Omega_1 \times \Omega_3} \int_{\Omega_2} \mathbb{P}_{1|2}(\mathrm{d}\omega_1|\omega_2)\mathrm{d}\Pi_2(\omega_2, \omega_3) \\
&= \int_{\Omega_2 \times \Omega_3} \int_{\Omega_1} \mathbb{P}_{1|2}(\mathrm{d}\omega_1|\omega_2)\mathrm{d}\Pi_2(\omega_2, \omega_3) \\
&= \int_{\Omega_2 \times \Omega_3} \mathrm{d}\Pi_2(\omega_2, \omega_3) = 1 \ .
\end{aligned}$$

(c) Prove that Π_3 satisfies countable additivity: Assume $S_1, S_2, \ldots, S_n, \ldots$ are measurable subsets of $\Omega_1 \times \Omega_3$ and they are disjoint with each other. Then by the property of integral,

$$\Pi_3\left(\bigcup_{n=1}^{+\infty} S_n\right) = \int\limits_{\bigcup_{n=1}^{+\infty} S_n} \int\limits_{\Omega_2} \mathbb{P}_{1|2}(\mathrm{d}\omega_1|\omega_2)\mathrm{d}\Pi_2(\omega_2, \omega_3)$$
$$= \sum_{n=1}^{+\infty} \int\limits_{S_n} \int\limits_{\Omega_2} \mathbb{P}_{1|2}(\mathrm{d}\omega_1|\omega_2)\mathrm{d}\Pi_2(\omega_2, \omega_3) = \sum_{n=1}^{+\infty} \Pi_3(S_n) .$$

Therefore, $\Pi_3 \in \mathcal{M}(\Omega_1 \times \Omega_3)$.

2. Prove that $\Pi_3 \in \mathcal{J}(\mathbb{P}_1, \mathbb{P}_3)$: Assume $S_1 \in \mathcal{F}_1$. Using Fubini's theorem and the definitions of joint probability distribution and conditional probability distribution, we have:

$$\Pi_3(S_1, \Omega_3) = \int\limits_{S_1 \times \Omega_3} \int\limits_{\Omega_2} \mathbb{P}_{1|2}(\mathrm{d}\omega_1|\omega_2)\mathrm{d}\Pi_2(\omega_2, \omega_3)$$
$$= \int\limits_{\Omega_2} \int\limits_{S_1 \times \Omega_3} \mathbb{P}_{1|2}(\mathrm{d}\omega_1|\omega_2)\mathrm{d}\Pi_2(\omega_2, \omega_3)$$
$$= \int\limits_{\Omega_2} \mathbb{P}_{1|2}(S_1|\omega_2)\mathrm{d}\Pi_2(\omega_2, \Omega_3)$$
$$= \int\limits_{\Omega_2} \mathbb{P}_{1|2}(S_1|\omega_2)\mathrm{d}\mathbb{P}_2(\omega_2) = \mathbb{P}_1(S_1) .$$

Similarly, we can prove that $\Pi_3(\Omega_1, S_3) = \mathbb{P}_3(S_3)$ for any $S_3 \in \mathcal{F}_3$. Therefore, $\Pi_3 \in \mathcal{J}(\mathbb{P}_1, \mathbb{P}_3)$.

□

2.3 Risk Measure

This section introduces some popular risk measures. In this section, we assume that ω follows the probability distribution $\mathbb{P}$ on the sample space $(\Omega, \mathcal{F})$; the real-valued measurable function $g(\omega)$ represents the loss, which is a random variable. Let $\mathcal{L} := \{g : \Omega \to \mathbb{R} \mid g \text{ is measurable}\}$ be the set consisting of such functions.

Definition 2.36 (*Risk measure*) A *risk measure* under probability distribution $\mathbb{P}$ is a mapping $r^{\mathbb{P}} : \mathcal{L} \to (-\infty, +\infty]$.

Definition 2.37 (*Coherent risk measure* Artzner et al. 1999) The risk measure $r^{\mathbb{P}} : \mathcal{L} \to (-\infty, +\infty]$ is called a *coherent risk measure* if it satisfies the following properties:

1. (Normalized) $r^{\mathbb{P}}(0) = 0$, where 0 in the parentheses represents the zero function.
2. (Monotone) If $g_1, g_2 \in \mathcal{L}$ and $g_1(\omega) \leq g_2(\omega)$, $\forall \omega \in \Omega$, then $r^{\mathbb{P}}(g_1) \leq r^{\mathbb{P}}(g_2)$.

3. (Translation invariant) Assume that $g_1, g_2 \in \mathcal{L}$, $a \in \mathbb{R}$, and $g_2(\omega) = g_1(\omega) + a, \forall \omega \in \Omega$. Then $r^{\mathbb{P}}(g_2) = r^{\mathbb{P}}(g_1) + a$.
4. (Subadditive) If $g_1, g_2 \in \mathcal{L}$, then $r^{\mathbb{P}}(g_1 + g_2) \le r^{\mathbb{P}}(g_1) + r^{\mathbb{P}}(g_2)$.
5. (Positive homogeneous) If $g \in \mathcal{L}$ and $a \ge 0$, then $r^{\mathbb{P}}(ag) = a \cdot r^{\mathbb{P}}(g)$.

Definition 2.38 (*Convex risk measure* Föllmer and Schied 2002) The risk measure $r^{\mathbb{P}} : \mathcal{L} \to (-\infty, +\infty]$ is called a *convex risk measure* if it is monotone, translation invariant, and satisfies convexity: For any $g_1, g_2 \in \mathcal{L}$ and $\lambda \in [0, 1]$,

$$r^{\mathbb{P}}(\lambda g_1 + (1-\lambda) g_2) \le \lambda r^{\mathbb{P}}(g_1) + (1-\lambda) r^{\mathbb{P}}(g_2) \ . \tag{2.10}$$

The combination of subadditivity and positive homogeneity implies the convexity in (2.10). A coherent risk measure must be a convex risk measure, but the inverse direction is not true.

In the following, we introduce the definitions and properties of value-at-risk (VaR), conditional value-at-risk (CVaR), and Shortfall risk.

2.3.1 Value-at-Risk

Definition 2.39 (*VaR*) The *value-at-risk* (VaR) of the loss $g(\omega)$ with parameter $\beta \in (0, 1)$ is defined as follows:

$$\beta\text{-VaR}^{\mathbb{P}}(g(\omega)) := \inf \{ r \in \mathbb{R} \mid \mathbb{P}[g(\omega) \le r] \ge \beta \} \ .$$

Thus, β-VaR$^{\mathbb{P}}(g(\omega))$ is the minimum threshold r such that the probability of the random event $g(\omega) \le r$ is no smaller than β. Larger β-VaR$^{\mathbb{P}}(g(\omega))$ means the loss threshold to guarantee the probability β is larger, so the risk is more severe.

According to Definitions 2.27 and 2.39, if $\Omega = \mathbb{R}$, F is the CDF of $\mathbb{P}$, and F^{-1} is the generalized inverse distribution function, then β-VaR$^{\mathbb{P}}(\omega) = F^{-1}(\beta)$.

Lemma 2.7 (Equivalence of VaR constraint and chance constraint) *For any probability space* $(\Omega, \mathcal{F}, \mathbb{P})$, $t \in \mathbb{R}$, *and* $\beta \in (0, 1)$,

$$\beta\text{-}VaR^{\mathbb{P}}(g(\omega)) \le t \iff \mathbb{P}[g(\omega) \le t] \ge \beta \ .$$

Proof If $\mathbb{P}[g(\omega) \le t] \ge \beta$, then $t \ge \inf \{ r \in \mathbb{R} \mid \mathbb{P}[g(\omega) \le r] \ge \beta \} = \beta$-VaR$^{\mathbb{P}}$ $(g(\omega))$. Therefore, we only need to prove the opposite direction. Assume β-VaR$^{\mathbb{P}}$ $(g(\omega)) \le t$, then

$$\forall \varepsilon > 0 \ , \ \exists r < t + \varepsilon \ , \ \text{ s.t. } \mathbb{P}[g(\omega) \le r] \ge \beta \ ,$$

and

$$\mathbb{P}[g(\omega) \le t] = \lim_{r \to t+} \mathbb{P}[g(\omega) \le r] \ge \beta ,$$

which completes the proof. □

By the same approach as that in the above proof, we can show that the infimum in the definition of VaR can be attained. Therefore, β-VaR$^{\mathbb{P}}(g(\omega))$ is the minimum r such that $\mathbb{P}[g(\omega) \le r] \ge \beta$.

Proposition 2.12 (Properties of VaR) *For any probability space* $(\Omega, \mathcal{F}, \mathbb{P})$ *and* $\beta \in (0, 1)$, β-*VaR*$^{\mathbb{P}}$ *is a normalized, monotone, translation invariant, and positive homogeneous risk measure.*

Example 2.6 (*VaR is not a convex risk measure*) VaR is neither convex nor subadditive in general, and thus it is not coherent. For example, let $\Omega = [0, 1]$ and $\mathbb{P}$ be the uniform distribution on $[0, 1]$. Assume $g_1(\omega) = \omega$, $g_2(\omega) = -\omega$, $\forall \omega \in [0, 1]$. Then β-VaR$^{\mathbb{P}}(g_1) = \beta$, β-VaR$^{\mathbb{P}}(g_2) = \beta - 1$, and β-VaR$^{\mathbb{P}}(g_1/2 + g_2/2) = \beta$-VaR$^{\mathbb{P}}(0) = 0$. Therefore, β-VaR$^{\mathbb{P}}(g_1)/2 + \beta$-VaR$^{\mathbb{P}}(g_2)/2 = \beta - 1/2 < \beta$-VaR$^{\mathbb{P}}$ $(g_1/2 + g_2/2)$ when $\beta \in (0, 1/2)$.

2.3.2 Conditional Value-at-Risk

Intuitively speaking, CVaR of parameter β is the mean of the $(1 - \beta)$-tail of the loss distribution (Rockafellar and Uryasev 2002).

Definition 2.40 (*CVaR*) The *conditional value-at-risk* (CVaR) with parameter $\beta \in (0, 1)$ is defined by:

$$\beta\text{-CVaR}^{\mathbb{P}}(g(\omega)) := \frac{\mathbb{E}^{\mathbb{P}}[g(\omega) 1_{g(\omega) > \text{VaR}}(\omega)] + \text{VaR} \cdot (\mathbb{P}(g(\omega) \le \text{VaR}) - \beta)}{1 - \beta} , \tag{2.11}$$

where VaR represents β-VaR$^{\mathbb{P}}(g(\omega))$.

Remark 2.3 When $\mathbb{P}(g(\omega) = \text{VaR}) = 0$ (for instance, the probability of any point is 0 under a continuous probability distribution),

$$\beta\text{-CVaR}^{\mathbb{P}}(g(\omega)) = \frac{\mathbb{E}^{\mathbb{P}}[g(\omega) 1_{g(\omega) \ge \text{VaR}}(\omega)]}{1 - \beta} = \mathbb{E}^{\mathbb{P}}[g(\omega) | g(\omega) \ge \text{VaR}] ,$$

which is the conditional expectation of the loss when the loss is at least VaR.

In the formula (2.11), the case $g >$ VaR and a part of the case $g =$ VaR are counted, so that the total probability of the considered cases is:

$$\mathbb{P}(g(\omega) > \beta\text{-VaR}^{\mathbb{P}}(g)) + \big(\mathbb{P}(g(\omega) \le \beta\text{-VaR}^{\mathbb{P}}(g)) - \beta\big) = 1 - \beta \ .$$

Therefore, (2.11) calculates the mean of the $(1-\beta)$-tail of g. It shows that larger CVaR values indicate larger risks. Another direct result from the definition is that CVaR must be no smaller than the corresponding VaR, as Lemma 2.8 states.

Lemma 2.8 (VaR is no larger than CVaR) *For any probability distribution $\mathbb{P}$ on $(\Omega, \mathcal{F})$, measurable $g : \Omega \to \mathbb{R}$, and $\beta \in (0, 1)$,*

$$\beta\text{-}CVaR^{\mathbb{P}}(g(\omega)) \ge \beta\text{-}VaR^{\mathbb{P}}(g(\omega)) \ .$$

Theorem 2.6 (Calculation of CVaR, Theorem 1 in Rockafellar and Uryasev 2000) *For any probability space $(\Omega, \mathcal{F}, \mathbb{P})$, measurable $g : \Omega \to \mathbb{R}$, and $\beta \in (0, 1)$,*

$$\beta\text{-}CVaR^{\mathbb{P}}(g(\omega)) = \min_{r \in \mathbb{R}} \left\{ r + \frac{1}{1-\beta} \mathbb{E}^{\mathbb{P}}[\max\{g(\omega) - r, 0\}] \right\} \ . \tag{2.12}$$

Moreover, the minimum can be attained at $r = \beta\text{-}VaR^{\mathbb{P}}(g(\omega))$.

Proof For any $r \in \mathbb{R}$,

$$\begin{aligned} &\mathbb{E}^{\mathbb{P}}[g(\omega) 1_{g(\omega)>r}(\omega)] + r(\mathbb{P}(g(\omega) \le r) - \beta) \\ =&\mathbb{E}^{\mathbb{P}}[\max\{g(\omega), r\}] - r\mathbb{P}(g(\omega) \le r) + r(\mathbb{P}(g(\omega) \le r) - \beta) \\ =&r + \mathbb{E}^{\mathbb{P}}[\max\{g(\omega) - r, 0\}] - r\beta \\ =&r(1-\beta) + \mathbb{E}^{\mathbb{P}}[\max\{g(\omega) - r, 0\}] \ . \end{aligned}$$

Then using the definition of CVaR in Definition 2.40, we only need to prove that the minimum of (2.12) is attained at $r = \beta\text{-VaR}^{\mathbb{P}}(g(\omega))$. When $\mathbb{E}^{\mathbb{P}}[(g(\omega))_+] = +\infty$, both sides of (2.12) equal $+\infty$. In the following, we consider the case $\mathbb{E}^{\mathbb{P}}[(g(\omega))_+] < +\infty$. Define function $G : \mathbb{R} \to \mathbb{R}$ by $G(r) := \mathbb{E}^{\mathbb{P}}[\max\{g(\omega) - r, 0\}]$, $\forall r \in \mathbb{R}$. Then $G(r)$ is continuous and convex. Moreover,

$$\begin{aligned} \lim_{\varepsilon \to 0+} \frac{G(r+\varepsilon) - G(r)}{\varepsilon} &= -\mathbb{P}(g(\omega) > r) \ , \\ \lim_{\varepsilon \to 0-} \frac{G(r+\varepsilon) - G(r)}{\varepsilon} &= -\mathbb{P}(g(\omega) \ge r) \ , \end{aligned}$$

both of which tend to -1 as $r \to -\infty$ and tend to 0 as $r \to +\infty$. Therefore, the function $H(r) := r + G(r)/(1-\beta)$ is continuous and convex. Moreover, it first decreases and then increases on $\mathbb{R}$. Therefore, it attains the minimum at a point r such that

$$\lim_{\varepsilon\to 0+}\frac{H(r+\varepsilon)-H(r)}{\varepsilon}\geq 0\,,\ \lim_{\varepsilon\to 0-}\frac{H(r+\varepsilon)-H(r)}{\varepsilon}\leq 0\,,$$

which is equivalent to:

$$\mathbb{P}(g(\omega)\leq r)\geq\beta\,,\ \mathbb{P}(g(\omega)<r)\leq\beta\,.$$

Because $r=\beta\text{-VaR}^{\mathbb{P}}(g(\omega))$ satisfies the above requirements, the minimum of (2.12) is attained at $r=\beta\text{-VaR}^{\mathbb{P}}(g(\omega))$. □

Proposition 2.13 (CVaR is coherent) *For any probability space* $(\Omega,\mathcal{F},\mathbb{P})$ *and* $\beta\in(0,1)$, *β-CVaR$^{\mathbb{P}}$ is a coherent risk measure and thus a convex risk measure.*

Proof It is obvious by the definition that β-CVaR$^{\mathbb{P}}$ is normalized, monotone, translation invariant, and positive homogeneous. The subadditivity follows from the combination of positive homogeneity and convexity. The convexity can be proven using the formulation of CVaR in Theorem 2.6, because the right side of (2.12) is convex in g. □

Example 2.7 (*CVaR is not convex in the probability distribution*) CVaR is convex in the loss function but not convex in the underlying probability distribution. Here is an example: Assume $\Omega=\mathbb{R}$ and $g(\omega)=\omega$, $\forall\omega\in\mathbb{R}$. Let $\mathbb{P}_0$ be the uniform distribution on $[-2,1]$. Let $\mathbb{P}_1=\delta_{-2}/2+\mathbb{P}_0/2$, where δ_{-2} is the Dirac distribution that has probability 1 at $\omega=-2$. Let $\mathbb{P}_2=\delta_0$, the Dirac distribution at 0. Suppose $\beta=2/3$ and $\mathbb{P}_3=4\mathbb{P}_1/5+\mathbb{P}_2/5$. Then via direct calculation, we have $\beta\text{-CVaR}^{\mathbb{P}_1}(g(\omega))=\beta\text{-CVaR}^{\mathbb{P}_2}(g(\omega))=0$, whereas $\beta\text{-CVaR}^{\mathbb{P}}(g(\omega))>0$.

2.3.3 Shortfall Risk

Definition 2.41 (*Shortfall risk*) Let constant ϱ represent the largest acceptable loss and let function $l:\mathbb{R}\to\mathbb{R}$ be a nonconstant nondecreasing convex function to measure the loss degree. The *Shortfall risk* with parameters ϱ and l is defined as:

$$\text{SR}^{\mathbb{P}}_{l,\varrho}(g(\omega)):=\inf\left\{r\in\mathbb{R}\mid\mathbb{E}^{\mathbb{P}}[l(g(\omega)-r)]\leq\varrho\right\}\,.$$

Thus, $\text{SR}^{\mathbb{P}}_{l,\varrho}(g(\omega))$ is the minimum value to be put into so that the expectation of the loss degree is no larger than ϱ. This value will decrease for smaller risks.

Proposition 2.14 (Properties of Shortfall risk) *In Definition 2.41,* $SR^{\mathbb{P}}_{l,\varrho}$ *is a convex risk measure. Moreover, Shortfall risk is convex in the probability distribution, i.e., for arbitrary probability distributions* $\mathbb{P}_1$ *and* $\mathbb{P}_2$ *on* $(\Omega,\mathcal{F})$, $g\in\mathcal{L}$, *and* $\lambda\in[0,1]$,

$$\text{SR}^{\lambda\mathbb{P}_1+(1-\lambda)\mathbb{P}_2}_{l,\varrho}\leq\lambda\cdot\text{SR}^{\mathbb{P}_1}_{l,\varrho}+(1-\lambda)\cdot\text{SR}^{\mathbb{P}_2}_{l,\varrho}\,.$$

Proof Because $l(t)$ is a nonconstant, nondecreasing, and nonconvex real-valued function on $\mathbb{R}$, it must have a positive slope as t approaches $+\infty$. As a consequence, $\mathrm{SR}^{\mathbb{P}}_{l,\varrho}(g(\omega)) > -\infty$ always holds.

It easily follows from Definition 2.41 that $\mathrm{SR}^{\mathbb{P}}_{l,\varrho}$ is monotone, translation invariant, and convex in the probability distribution. We need to prove its convexity in the loss function to show it is a convex risk measure. To this end, assume that $g_1, g_2 : \Omega \to \mathbb{R}$ are measurable and $\lambda \in [0, 1]$, then by the convexity of l,

$$\begin{aligned}
&\mathbb{E}^{\mathbb{P}}\left[l\left(\lambda g_1(\omega) + (1-\lambda)g_2(\omega) - (\lambda r_1 + (1-\lambda)r_2)\right)\right] \\
&= \mathbb{E}^{\mathbb{P}}\left[l\left(\lambda\left(g_1(\omega) - r_1\right) + (1-\lambda)\left(g_2(\omega) - r_2\right)\right)\right] \\
&\leq \lambda \mathbb{E}^{\mathbb{P}}\left[l\left(g_1(\omega) - r_1\right)\right] + (1-\lambda)\mathbb{E}^{\mathbb{P}}\left[l\left(g_2(\omega) - r_2\right)\right] \ , \ \forall r_1, r_2 \in \mathbb{R} \ ,
\end{aligned}$$

which implies:

$$\mathrm{SR}^{\mathbb{P}}_{l,\varrho}(\lambda g_1 + (1-\lambda)g_2) \leq \lambda \cdot \mathrm{SR}^{\mathbb{P}}_{l,\varrho}(g_1) + (1-\lambda) \cdot \mathrm{SR}^{\mathbb{P}}_{l,\varrho}(g_2) \ .$$

Therefore, $\mathrm{SR}^{\mathbb{P}}_{l,\varrho}$ is a convex risk measure. □

2.4 Convex Optimization

We review some important definitions and conclusions in the convex optimization theory to help readers recall this useful tool. Interpretations rather than proofs are given for a major part of the results. Readers can refer to textbooks in convex optimization theory such as Boyd and Vandenberghe (2004) for more details.

2.4.1 Convex Sets

Definition 2.42 (*Convex set*) Assume S is a subset of $\mathbb{R}^M$. S is called a *convex set* if for any $x, y \in S$ and $\lambda \in [0, 1]$, we have:

$$\lambda x + (1-\lambda)y \in S \ .$$

The form $\sum_{i=1}^n \lambda_i x_i$ for some $\lambda_i \geq 0, \forall i, \sum_{i=1}^n \lambda_i = 1$ is called a *convex combination* of $x_i, \forall i$. The convex combination of x and y positions on the line segment connecting the points x and y in $\mathbb{R}^M$. Therefore, a nonempty set in $\mathbb{R}^M$ is convex if and only if any convex combination of two points in it still lies in the set. An illustration of convex sets is in Fig. 2.1.

Remark 2.4 The notion of the convex set in Definition 2.42 can be extended to subsets of vector spaces. For example, the space $\mathcal{M}$ of signed measures on a nonempty

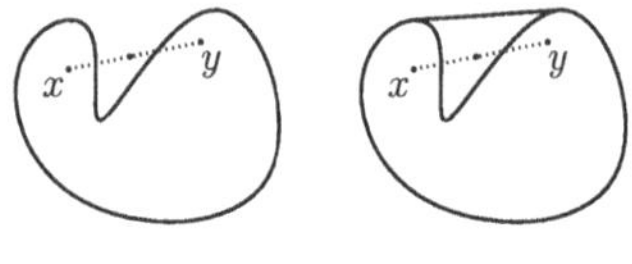

Fig. 2.1 The set on the left side is not convex, since the convex combination $\lambda x + (1 - \lambda)y$ of x and y can be outside of the set. Its convex hull is on the right side, where the convex combinations of the points in the convex hull are still in it

set Ω equipped with a σ-algebra $\mathcal{F}$ is a vector space in the following sense: A measure $\mu \in \mathcal{M}$ can be multiplied by a scalar $c \in \mathbb{R}$ to get a new measure $c\mu \in \mathcal{M}$ defined by $(c\mu)(S) = c \cdot \mu(S), \forall S \in \mathcal{F}$. The addition of two measures $\mu_1, \mu_2 \in \mathcal{M}$ is defined by $(\mu_1 + \mu_2)(S) = \mu_1(S) + \mu_2(S), \forall S \in \mathcal{F}$. Consider the subset $\mathcal{P}$ of $\mathcal{M}$ consisting of all the probability distributions. It can be easily verified that for any $\mathbb{P}_1, \mathbb{P}_2 \in \mathcal{P}$ and $\lambda \in [0, 1]$, the convex combination $\lambda\mathbb{P}_1 + (1 - \lambda)\mathbb{P}_2$ is still a probability distribution in $\mathcal{P}$. Therefore, $\mathcal{P}$ is a convex set in the vector space $\mathcal{M}$.

Definition 2.43 (*Affine function*) A function $f : \mathbb{R}^{M_1} \to \mathbb{R}^{M_2}$ is an *affine function* if there is a matrix $A \in \mathbb{R}^{M_2 \times M_1}$ and a vector $b \in \mathbb{R}^{M_2}$ such that

$$f(x) = Ax + b \ , \ \forall x \in \mathbb{R}^{M_1} \ .$$

Example 2.8 (*Affine functions*) Here are some special cases of affine functions.

1. A *scaling function* $f : \mathbb{R}^M \to \mathbb{R}^M$ is given by $f(x) = kx, \forall x \in \mathbb{R}^M$ for some constant $k \in \mathbb{R}$.
2. A *translation function* $f : \mathbb{R}^M \to \mathbb{R}^M$ is defined by $f(x) = x + b, \forall x \in \mathbb{R}^M$ for some vector $b \in \mathbb{R}$.

Proposition 2.15 (Operations preserving the convexity of sets)

1. *(Intersection preserves convexity) The intersection of arbitrarily many convex sets is convex.*
2. *(Affine image and inverse image preserve convexity) The image of a convex set under an affine function is convex. The inverse image of a convex set under an affine function is convex.*

Proposition 2.15 can be easily proven using the definition of convex sets. The first conclusion follows from the fact that a convex combination of two points in the intersection of these convex sets is contained in each convex set among them and hence is still in the intersection. The reason for the second conclusion is that the line segment between two points is mapped into the line segment between the two image points under an affine function, so the inclusion relation of convex combinations is preserved.

Example 2.9 (*Convex sets*) The following are some commonly used convex sets.

1. A *singleton* is a set with only one point. It is clear that its convex combination is still the point, so singletons are convex sets.

2. A *halfspace* S is the solution set of a linear inequality, i.e., $S = \{x \in \mathbb{R}^M \mid a^\top x \leq b\}$ with some vector $0 \neq a \in \mathbb{R}^M$ and scalar $b \in \mathbb{R}$, which is the inverse image of $(-\infty, b]$ under the affine function $f(x) = a^\top x$. It is easy to see that any convex combination of points in $(-\infty, b]$ is still in $(-\infty, b]$. Thus, $(-\infty, b]$ is convex and then S is convex.
3. A *hyperplane* S in $\mathbb{R}^M$ is the solution set of a linear equation, i.e., $S = \{x \in \mathbb{R}^M \mid a^\top x = b\} = \{x \in \mathbb{R}^M \mid a^\top x \leq b\} \cap \{x \in \mathbb{R}^M \mid -a^\top x \leq -b\}$ for $0 \neq a \in \mathbb{R}^M$ and $b \in \mathbb{R}$, which is the intersection of two halfspaces and hence is convex.
4. A *polyhedron* S is the solution set of some linear inequalities, i.e., $S = \{x \in \mathbb{R}^M \mid Ax \leq b\}$. Clearly, S is the intersection of some halfspaces, which indicates that it is convex.
5. The (closed) *ball* in $\mathbb{R}^M$ centered at $x_0 \in \mathbb{R}^M$ with radius $r \geq 0$ under norm $\|\cdot\|$ is defined by $B(x_0, r) := \{x \in \mathbb{R}^M \mid \|x - x_0\| \leq r\}$.
6. A symmetric real $M \times M$ matrix A is called *positive semidefinite* and denoted by $A \in \mathbb{S}_+^M$ or $A \succeq 0$ if $x^\top Ax \geq 0, \forall x \in \mathbb{R}^M$. A is called *positive definite* and denoted by $A \in \mathbb{S}_{++}^M$ or $A \succ 0$ if $x^\top Ax > 0, \forall 0 \neq x \in \mathbb{R}^M$. According to linear algebra theory, a positive definite matrix is always invertible. An *ellipsoid* is a set in the form of $\{x \in \mathbb{R}^M \mid (x - x_0)^\top A^{-1}(x - x_0) \leq 1\}$, where x_0 is the center point and $A \in \mathbb{S}_{++}^M$. Ellipsoids are convex.

Definition 2.44 (*Convex hull*) Let S be a set in $\mathbb{R}^M$. The *convex hull* of S is denoted by $\mathrm{conv}(S)$, which is the set of all convex combinations of points in S, i.e.,

$$\mathrm{conv}(S) := \left\{ \sum_{i=1}^{n} \lambda_i x_i \mid n \in \mathbb{N}_+, x_i \in S, \lambda_i \geq 0, \forall i, \sum_{i=1}^{n} \lambda_i = 1 \right\}.$$

The convex hull of several points is illustrated in Fig. 2.2, which is a polyhedron.

The idea of a convex hull is to find the smallest convex set to contain the original set. On the one hand, any convex set containing S should include all the convex combinations of points in S, therefore it should be at least as large as $\mathrm{conv}(S)$. On the other hand, $\mathrm{conv}(S)$ itself is convex: Suppose $y, y' \in \mathrm{conv}(S)$, then they can be represented as convex combinations $y = \sum_{i=1}^{n} \lambda_i x_i$ and $y' = \sum_{i=1}^{n'} \lambda'_i x'_i$. For any $\lambda \in [0, 1]$, $\lambda y + (1 - \lambda) y' = \sum_{i=1}^{n} \lambda \lambda_i x_i + \sum_{i=1}^{n'} (1 - \lambda) \lambda'_i x'_i$ is a convex combination of $x_1, \ldots, x_n, x'_1, \ldots, x'_{n'}$. Thus, $\lambda y + (1 - \lambda) y' \in \mathrm{conv}(S)$ and therefore $\mathrm{conv}(S)$ is a convex set. To sum up, we have the following conclusion to characterize the convex hull:

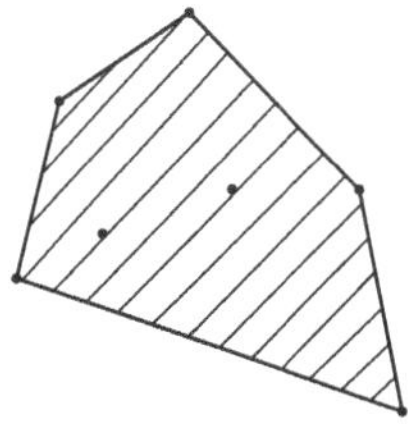

Fig. 2.2 The convex hull of a finite number of points is a polyhedron (*shaded area*)

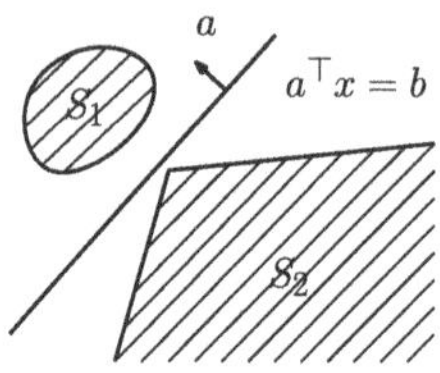

Fig. 2.3 S_1 is compact and convex. S_2 is closed and convex. They can be strictly separated by the hyperplane $\{x \in \mathbb{R}^2 \mid a^\top x = b\}$

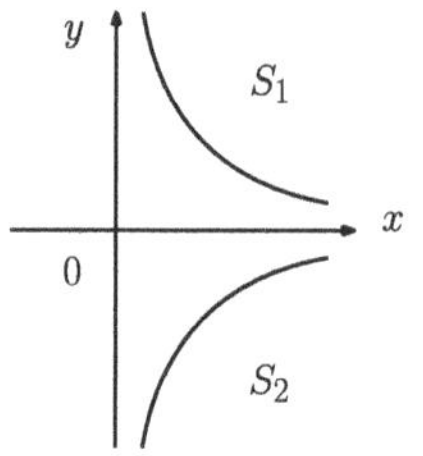

Fig. 2.4 $S_1 := \{(x, y) \in \mathbb{R}^2 \mid x > 0, y \geq 1/x\}$ and $S_2 := \{(x, y) \in \mathbb{R}^2 \mid x > 0, y \leq -1/x\}$. They are disjoint closed and convex sets. They can be separated by the hyperplane $y = 0$ (note that a hyperplane in $\mathbb{R}^2$ is a line), but cannot be strictly separated by any hyperplane

Proposition 2.16 (Characterizing convex hull) *The convex hull conv(S) is the smallest convex set that contains S.*

Proposition 2.17 (Interior and closure of a convex set) *Suppose S is a convex set in $\mathbb{R}^M$, then its interior int(S) and closure cl(S) are also convex sets.*

Theorem 2.7 says that two convex sets can be separated by a hyperplane, which means the two sets are on different sides of the hyperplane. Figures 2.3 and 2.4 give examples of hyperplanes separating two nonempty convex sets.

Theorem 2.7 (Hyperplane separation theorem) *Let S_1 and S_2 be two nonempty convex sets in $\mathbb{R}^M$ with $S_1 \cap S_2 = \varnothing$. Then there is a vector $a \in \mathbb{R}^M$ and a number $b \in \mathbb{R}$ such that $a \neq 0$ and $a^\top x_1 \geq b \geq a^\top x_2, \forall x_1 \in S_1, x_2 \in S_2$. The hyperplane $\{x \in \mathbb{R}^M \mid a^\top x = b\}$ is called a* separating hyperplane *for S_1 and S_2.*

Furthermore, if S_1 is compact and S_2 is closed, then they can be strictly separated, i.e., there is $0 \neq a \in \mathbb{R}^M$ and $b \in \mathbb{R}$ such that $a^\top x_1 > b > a^\top x_2, \forall x_1 \in S_1, x_2 \in S_2$.

The notion of separating hyperplanes is related to supporting hyperplanes.

Definition 2.45 (*Supporting hyperplane*) Suppose $S \subset \mathbb{R}^M$, $x_0 \in \partial S$, $a \in \mathbb{R}^M$, and $a \neq 0$. If $a^\top x \leq a^\top x_0, \forall x \in S$, then call the hyperplane $\{x \in \mathbb{R}^M \mid a^\top x = a^\top x_0\}$ a *supporting hyperplane* of S at x_0.

Note that in the definition of supporting hyperplane, the set S is not required to be convex. However, it is easy to see that there does not always exist a supporting

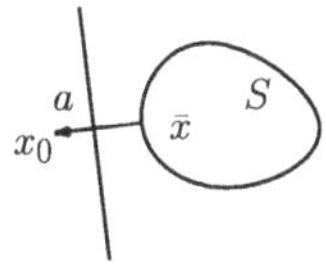

Fig. 2.5 S is a closed convex set in $\mathbb{R}^M$ and x_0 is a point outside of S. Then we can find a point $\bar{x} \in S$ such that $\|x_0 - \bar{x}\|_2$ reaches the minimum distance from x_0 to S. Some hyperplanes perpendicular to the direction $a := x_0 - \bar{x}$ can strictly separate x_0 and S

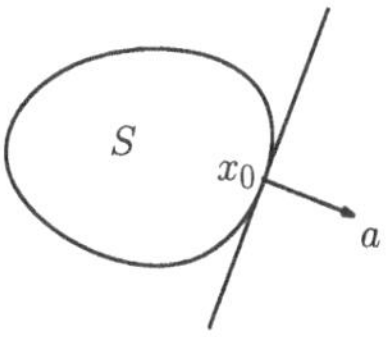

Fig. 2.6 S is a convex set and $x_0 \in \partial S$. Then we can find a direction a and a supporting hyperplane that is perpendicular to a and goes through x_0 so that S is on one side of the supporting hyperplane

hyperplane at a point $x \in \partial S$ if S is not convex. We will see soon that when S is convex, at any $x \in \partial S$, there is a supporting hyperplane.

A singleton is a set containing only one point, so it is nonempty, convex, and compact. Applying the hyperplane separation theorem to the case where one of the convex sets is a singleton, we have the following result, which is illustrated in Figs. 2.5 and 2.6.

Corollary 2.1 (Separating a point and a convex set)

1. *Let S be a closed convex set in $\mathbb{R}^M$ and $x_0 \notin S$. Then there is a vector $a \in \mathbb{R}^M$ such that $a \neq 0$ and $a^\top x_0 > \sup_{x \in S} a^\top x$.*
2. *Let S be a convex set in $\mathbb{R}^M$ and $x_0 \notin int(S)$. Then there is a vector $a \in \mathbb{R}^M$ such that $a \neq 0$ and $a^\top x_0 \geq a^\top x, \forall x \in S$.*

In the proof of the second conclusion, the hyperplane separation theorem is applied to $\{x_0\}$ and $\text{int}(S)$. Moreover, if $x_0 \in \partial S$, $\{x \in \mathbb{R}^M \mid a^\top x = a^\top x_0\}$ is a supporting hyperplane of S at x_0. Therefore, every boundary point of a convex set has a supporting hyperplane across it.

2.4.2 Convex Functions

Definition 2.46 (*Convex function*) Let $f : S \to \mathbb{R}$ be a function defined on the convex set S. We say that f is a *convex function* if for any $x_1, x_2 \in S$ and $\lambda \in [0, 1]$,

$$f(\lambda x_1 + (1 - \lambda)x_2) \leq \lambda f(x_1) + (1 - \lambda) f(x_2) . \tag{2.13}$$

f is called *strictly convex* if for any $\lambda \in (0, 1)$ and $x_1, x_2 \in S$ with $x_1 \neq x_2$,

$$f(\lambda x_1 + (1-\lambda)x_2) < \lambda f(x_1) + (1-\lambda) f(x_2) \,. \tag{2.14}$$

In addition, f is called *concave* if $-f$ is a convex function.

The notion of convex functions can be applied to *extended-value functions*, which are functions on the Euclidean space that take values in $[-\infty, +\infty]$. Let $f : S \to \mathbb{R}$ be a function defined on $S \subset \mathbb{R}^M$. Its extended-value extension function $\tilde{f} : \mathbb{R}^M \to [-\infty, +\infty]$ is defined by $\tilde{f}(x) = f(x), \forall x \in S$ and $\tilde{f}(x) = +\infty, \forall x \in \mathbb{R}^M \setminus S$. Then $\tilde{f}$ is convex if and only if $\tilde{f}(\lambda x + (1-\lambda)x') \leq \lambda \tilde{f}(x) + (1-\lambda)\tilde{f}(x'), \forall x, x' \in \mathbb{R}^M, \lambda \in [0, 1]$, which is equivalent to that S is a convex set and f is a convex function. It is worth noting that the summation $(+\infty) + (-\infty)$ is undefined.

Definition 2.47 (*Proper function*) An extended-value function $f : \mathbb{R}^M \to [-\infty, +\infty]$ is called *proper* if $f(x) > -\infty, \forall x \in \mathbb{R}^M$ and there exists $x_0 \in \mathbb{R}^M$ such that $f(x_0) \in \mathbb{R}$.

Definition 2.48 (*Graph and epigraph*) Let $f : S \to \mathbb{R}$ be a function. Then the *graph* of f is the set $\text{graph}(f) := \{(x, f(S)) \mid x \in S\}$. The *epigraph* of f is the set above the graph, i.e., $\text{epi}(f) := \{(x, y) \mid x \in S, y \geq f(x)\}$. Both graph$(f)$ and epi(f) are subsets of $S \times \mathbb{R}$.

As illustrated in Fig. 2.7, if f is convex, then the convex combination of two points on graph(f) will be above graph(f), which means epi(f) is convex. The converse is also true.

Proposition 2.18 (Convex function and convex epigraph) *Suppose $f : S \to \mathbb{R}$ is a function on the convex set S, then f is a convex function if and only if its epigraph epi(f) is a convex set.*

Since the convexity of a set can be checked by whether the line segment between two points in the set is still in the set, by Proposition 2.18, we can see that for a function $f : S \to \mathbb{R}$ defined on a convex set $S \subset \mathbb{R}^M$, f is a convex function if and only if its restriction to any line is convex, i.e., for any $x_0 \in S$ and $v \in \mathbb{R}^M$, the function g defined by $g(t) = f(x_0 + tv)$ on $\{t \in \mathbb{R} \mid x_0 + tv \in S\}$ is convex.

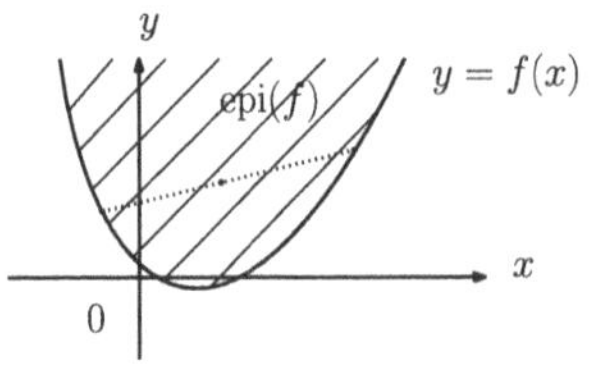

Fig. 2.7 The epigraph epi(f) of $f : \mathbb{R} \to \mathbb{R}$ is the subset of $\mathbb{R}^2$ above graph(f), i.e., the *shaded area* above the curve $y = f(x)$. When f is a convex function, epi(f) is a convex set

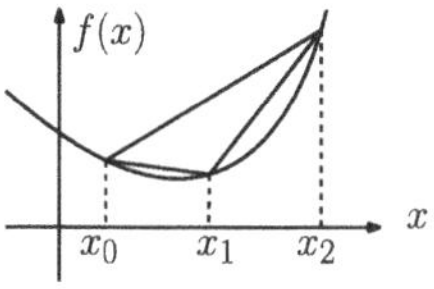

Fig. 2.8 The function $f : \mathbb{R} \to \mathbb{R}$ is a convex function. Consider the points $x_0 < x_1 < x_2$. The slopes of the lines passing through $(x_0, f(x_0))$, $(x_1, f(x_1))$, and $(x_2, f(x_2))$ satisfy (2.15)

The convexity of an extended-value function cannot be defined directly using (2.13) because the value of $(+\infty) + (-\infty)$ is undefined. Nonetheless, an extended-value function $f : \mathbb{R}^M \to [-\infty, +\infty]$ is called *convex* if its epigraph $\text{epi}(f) := \{(x, y) \mid x \in \mathbb{R}^M, y \in \mathbb{R}, y \geq f(x)\}$ is a convex set in $\mathbb{R}^{M+1}$.

Definition 2.49 (*Closed function*) Let $f : \mathbb{R}^M \to [-\infty, +\infty]$ be a extended-value function. It is called a *closed function* if $\text{epi}(f)$ is a closed set in $\mathbb{R}^{M+1}$.

Proposition 2.19 (Slopes of a convex function) *Let $f : S \to \mathbb{R}$ be a convex function. Then for any $x_0, x_1, x_2 \in S$ with $x_2 - x_0 = k(x_1 - x_0) \neq 0$ and $k > 1$, we have:*

$$\frac{f(x_1) - f(x_0)}{\|x_1 - x_0\|} \leq \frac{f(x_2) - f(x_0)}{\|x_2 - x_0\|} \leq \frac{f(x_2) - f(x_1)}{\|x_2 - x_1\|} .$$

The 1-dimensional case is shown in Fig. 2.8, where the conclusion can be stated as follows: For any $x_0, x_1, x_2 \in S \subset \mathbb{R}$ with $x_0 < x_1 < x_2$, we have:

$$\frac{f(x_1) - f(x_0)}{x_1 - x_0} \leq \frac{f(x_2) - f(x_0)}{x_2 - x_0} \leq \frac{f(x_2) - f(x_1)}{x_2 - x_1} . \tag{2.15}$$

If we let x_0 and x_2 approach x_1 from the left and right sides, respectively, we have the following result about the left and right derivatives:

Corollary 2.2 (Derivative of a convex function) *Let $f : S \to \mathbb{R}$ be a convex function defined on $S \subset \mathbb{R}$. Then for any $x_0 \in int(S)$, the following two limits are both finite and satisfy $a \leq b$:*

$$a := \lim_{\Delta x \to 0-} \frac{f(x_0 + \Delta x) - f(x_0)}{\Delta x} , \quad b := \lim_{\Delta x \to 0+} \frac{f(x_0 + \Delta x) - f(x_0)}{\Delta x} .$$

Moreover, for any $t \in [a, b]$, we have:

$$f(x) \geq f(x_0) + t(x - x_0) , \quad \forall x \in S .$$

Such $t \in [a, b]$ is called a subderivative *of f at x_0. The interval $[a, b]$ is called the* subdifferential *of f at x_0 and denoted by $\partial f(x_0)$.*

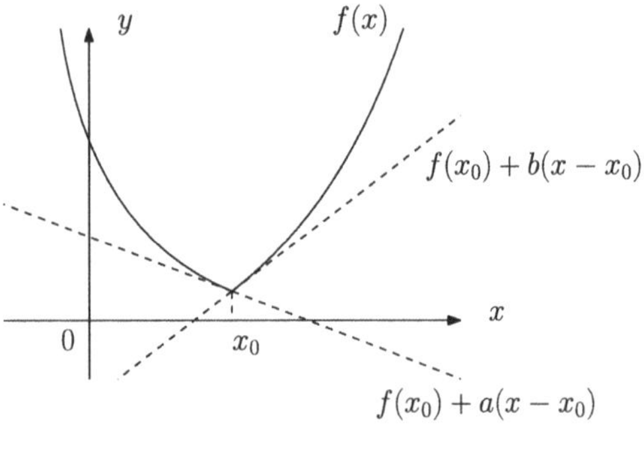

Fig. 2.9 The subdifferential of the convex function $f : S \to \mathbb{R}$ at x_0 is $[a, b]$. For any $t \in [a, b]$ and $x \in S$, $f(x) \geq f(x_0) + t(x - x_0)$

For a convex function on $S \subset \mathbb{R}$, Fig. 2.9 shows the subderivatives at a point in the interior. Subderivative and subdifferential can also be defined for multi-variable convex functions, as Definition 2.50 indicates.

Definition 2.50 (*Subderivative and subdifferential*) Let $f : S \to \mathbb{R}$ be a convex function defined on $S \subset \mathbb{R}^M$. Consider a point $x_0 \in S$. A vector $v \in \mathbb{R}^M$ is called a *subderivative* of f at x_0 if

$$f(x) \geq f(x_0) + v^\top (x - x_0)\ , \ \forall x \in S\ . \tag{2.16}$$

The set consisting of all the subderivative of f at x_0 is called the *subdifferential* of f at x_0 and denoted by $\partial f(x_0)$.

Remark 2.5 If f is furthermore strictly convex in Definition 2.50, we have:

$$f(x) > f(x_0) + v^\top (x - x_0)\ , \ \forall x \neq x_0\ , \ x \in S\ , \ v \in \partial f(x_0)\ .$$

Recall from the hyperplane separation theorem in Theorem 2.7 that for any point on the boundary of a convex set, there exists a supporting hyperplane of the set across the point. By applying this result to epi(f) of a convex function f, we can find an element in the subdifferential:

Proposition 2.20 (Subdifferential of a convex function at an interior point) *Assume $f : S \to \mathbb{R}$ is a convex function with $S \subset \mathbb{R}^M$. When $x_0 \in int(S)$, the subdifferential $\partial f(x_0)$ is a nonempty convex set.*

The convexity of $\partial f(x_0)$ can be directly verified using the definition in (2.16).

When the function f has some differential properties, its convexity can be checked using the derivatives. To this end, we first review the notions of gradient, Hessian matrix, and Jacobian matrix.

Definition 2.51 (*Gradient, Hessian matrix, Jacobian matrix*) For a real-valued C^1 (continuously differentiable, which means that all the first-order partial derivatives exist and are continuous as functions) function $f : S \to \mathbb{R}$ on $S \subset \mathbb{R}^M$ and a point $x \in \text{int}(S)$, the *gradient* of f at x is:

$$\nabla f(x) := \begin{pmatrix} \frac{\partial f}{\partial x_1}(x) \\ \frac{\partial f}{\partial x_2}(x) \\ \vdots \\ \frac{\partial f}{\partial x_M}(x) \end{pmatrix} .$$

If f is furthermore a C^2 (twice continuously differentiable, which means that all the second-order partial derivatives exist and are continuous), then the *Hessian matrix* of f at x is a matrix with the element

$$\nabla^2 f(x)_{ij} = \frac{\partial^2 f}{\partial x_i \partial x_j}(x) , \ \forall i, j = 1, 2, \ldots, M .$$

For a vector-valued function $f := (f_1, f_2, \ldots, f_m)$ with $f_1, f_2, \ldots, f_m : S \to \mathbb{R}$ being C^1 functions on $S \subset \mathbb{R}^M$ and a point $x \in \text{int}(S)$, the *Jacobian matrix* of f at x is:

$$\nabla f(x) := (\nabla f_1(x) \ \ \nabla f_2(x) \ \ \ldots \ \ \nabla f_m(x)) .$$

Notice that when $f : S \to \mathbb{R}$ is C^1, we have $\partial f(x) = \{\nabla f(x)\}, \forall x \in \text{int}(S)$, which can be easily seen by restricting f to a line across x. Then the following result is a natural consequence.

Theorem 2.8 (First-order condition for convex functions) *Assume S is an open subset of $\mathbb{R}^M$ and $f : S \to \mathbb{R}$ is a C^1 function. Then f is convex if and only if*

$$f(y) \geq f(x) + \nabla f(x)^\top (y - x) , \ \forall x, y \in S . \tag{2.17}$$

In fact, the function $f(x) + \nabla f(x)^\top (y - x)$ of y is the first-order Taylor approximation of f at x. When f is C^2, the second-order Taylor approximation is $f(x) + \nabla f(x)^\top (y - x) + (y - x)^\top (\nabla^2 f(x))(y - x)/2$. Then we have the following second-order condition for convex functions, which can be interpreted as that C^2 convex functions have upward curvature everywhere.

Theorem 2.9 (Second-order condition for convex functions) *Assume S is open in $\mathbb{R}^M$ and $f : S \to \mathbb{R}$ is C^2. Then f is convex if and only if $\nabla^2 f(x)$ is positive semidefinite, i.e., $\nabla^2 f(x) \succeq 0$ for any $x \in S$.*

Example 2.10 (*Convex functions*) Here are some examples of convex functions, whose convexity can be easily checked by the original definition, the first-order condition, or the second-order condition.

1. An affine function $f : \mathbb{R}^M \to \mathbb{R}$ is convex and concave.
2. A norm function $\|\cdot\| : \mathbb{R}^M \to \mathbb{R}$ is convex.
3. For any $a \in \mathbb{R}$, the exponential function $f : \mathbb{R} \to \mathbb{R}$ defined by $f(x) = \exp(ax)$ is convex.

4. When $a \geq 1$ or $a < 0$, the power function $f(x) = x^a$ on $(0, +\infty)$ is convex. When $0 \leq a \leq 1$, $f(x) = x^a$ is concave on $(0, +\infty)$.
5. $f(x) = \log(x)$ is concave on $(0, +\infty)$. $g(x) = x\log(x)$ is convex on $(0, +\infty)$.
6. If $A \succeq 0$, then the quadratic function $f(x) = x^\top Ax/2 + b^\top x + c$ is convex on the Euclidean space.

The convexity of functions is preserved under some operations:

Proposition 2.21 (Operations preserving the convexity of functions)

1. *(Positive weighted sum) Assume f_1 and f_2 are convex functions on the convex set S and $\alpha_1, \alpha_2 > 0$, then $f := \alpha_1 f_1 + \alpha_2 f_2$ is convex on S.*
2. *(Affine composition) Assume $f : S \to \mathbb{R}$ is convex on $S \subset \mathbb{R}^M$, then the function $g(x) = f(Ax + b)$ is convex on $\{x \mid Ax + b \in S\}$.*
3. *(Pointwise maximum) Assume f_i is convex on S for any $i \in \mathcal{I}$, then $f(x) = \max_{i\in\mathcal{I}} f_i(x)$ is convex on S.*
4. *Assume $f : S_1 \times S_2 \to \mathbb{R}$ is convex, then the function $g : S_1 \to \mathbb{R}$ defined by $g(x) := \inf_{y\in S_2} f(x, y)$ is also convex.*

The positive weighted sum case can be easily checked using the definition. Recall from Proposition 2.18 that f is convex if and only if epi(f) is convex. Then the affine composition and pointwise maximum cases follow from Proposition 2.15. The pointwise maximum case is also illustrated in Fig. 2.10.

Proposition 2.22 (Level sets of a convex function) *Let $f : S \to \mathbb{R}$ be a function. For any $\alpha \in \mathbb{R}$, the set $\{x \in S \mid f(x) \leq \alpha\}$ is called a* level set. *If f is a convex function, then for any $\alpha \in \mathbb{R}$, the level set is either empty or convex.*

However, the converse is not true, which means a function whose nonempty level sets are all convex may not be a convex function. Figure 2.11 gives a counterexample.

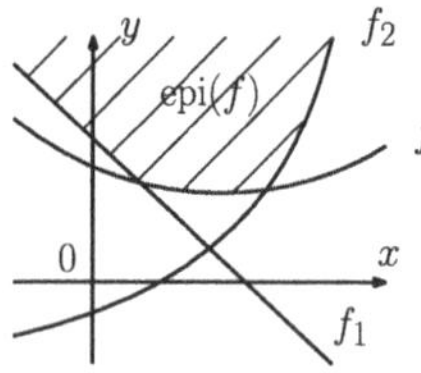

Fig. 2.10 The function f is the pointwise maximum of the convex functions f_1, f_2, and f_3 on $\mathbb{R}$. The epigraph of f is epi$(f_1) \cap$ epi$(f_2) \cap$ epi(f_3). Therefore, f is also convex

Fig. 2.11 The nonempty level sets of function f are all convex sets, but f is not a convex function

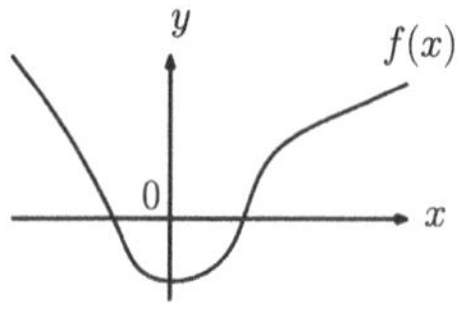

Jensen's inequality can be regarded as a generalization of the inequality (2.13) in Definition 2.46: $f(\lambda x_1 + (1 - \lambda x_2)) \leq \lambda f(x_1) + (1 - \lambda) f(x_2), \forall x_1, x_2 \in S, \lambda \in [0, 1]$, where S is a convex set. To begin with, we can switch to the convex combinations of multiple points: $f(\sum_i \lambda_i x_i) \leq \sum_i \lambda_i f(x_i), \forall x_i \in S, \lambda_i \in [0, 1], \sum_i \lambda_i = 1$. In addition, the coefficients λ_i can be interpreted as the probability associated with the point x_i, because $\lambda_i \in [0, 1], \forall i$ and $\sum_i \lambda_i = 1$. For example, in (2.13), the probability distribution is defined by $\mathbb{P}(x_1) = \lambda$ and $\mathbb{P}(x_2) = 1 - \lambda$. To this end, we have Jensen's inequality as follows:

Theorem 2.10 (Jensen's inequality) *Assume that* $(\Omega, \mathcal{F}, \mathbb{P})$ *is a probability space with* $\Omega \subset \mathbb{R}^M$. *Suppose that* $X : \Omega \to S$ *is a random variable following the probability distribution* $\mathbb{P}$ *and* $\mathbb{E}^{\mathbb{P}}[X]$ *exists. Let* $f : S \to \mathbb{R}$ *be a convex function. Then the* Jensen's inequality *holds:*

$$f\left(\mathbb{E}^{\mathbb{P}}[X]\right) \leq \mathbb{E}^{\mathbb{P}}[f(X)] .$$

Moreover, if f *is strictly convex, we have:*

$$f\left(\mathbb{E}^{\mathbb{P}}[X]\right) = \mathbb{E}^{\mathbb{P}}[f(X)] \iff \mathbb{P}\left(X = \mathbb{E}^{\mathbb{P}}[X]\right) = 1 . \tag{2.18}$$

2.4.3 Convex Optimization Problems

A *mathematical programming* problem can be written in the following form:

$$\begin{aligned} p^* := \min_{x \in \mathcal{X}} \ & f_0(x) \\ \text{s.t.} \ & f_i(x) \leq 0 , \ \forall i \in \mathcal{I}_1 , \\ & h_i(x) = 0 , \ \forall i \in \mathcal{I}_2 , \end{aligned} \tag{2.19}$$

where $f_0, f_i, \forall i \in \mathcal{I}_1$, and $h_i, \forall i \in \mathcal{I}_2$ are real-valued functions on $\mathcal{X}$. The "min" notation means that problem (2.19) tries to minimize the *objective function* $f_0(x)$. The constraints include the inequality constraints $f_i(x) \leq 0, \forall i \in \mathcal{I}_1$ and the equality constraints $h_i(x) = 0, \forall i \in \mathcal{I}_2$. p^* is the optimal value that will be defined below. There are some basic concepts for the mathematical programming problem (2.19):

1. The *feasible region* is $\{x \in \mathcal{X} \mid f_i(x) \leq 0, \forall i \in \mathcal{I}_1, h_i(x) = 0, \forall i \in \mathcal{I}_2\}$, which is the intersection of $\mathcal{X}$ and the region where the constraints hold.
2. The *optimal value* p^* is defined as $p^* := \inf\{f_0(x) \mid x \in \mathcal{X}, f_i(x) \leq 0, \forall i \in \mathcal{I}_1, h_i(x) = 0, \forall i \in \mathcal{I}_2\}$, which is the infimum of $f_0(x)$ when x varies in the feasible region. Therefore, $p^* \in [-\infty, +\infty]$.
3. When the feasible region is nonempty, the problem is called a *feasible problem*. Any point in the feasible region is called a *feasible solution*. When the problem is feasible, $p^* \in [-\infty, +\infty)$. If furthermore $p^* \in \mathbb{R}$, the problem is called *bounded below*. Otherwise, it is called *unbounded below*.

4. When the feasible region is empty, the problem is called *infeasible* and $p^* = +\infty$.
5. A feasible x is called a (globally) *optimal solution* if $f_0(x) = p^*$. Optimal solutions may not exist because the problem may be infeasible and even if it is feasible, the optimal value may not be attained.
6. A feasible x is called a *locally optimal solution* if there is a neighborhood $\mathcal{N}$ of x so that x is optimal for the problem $\min_{x \in \mathcal{X} \cap \mathcal{N}} f_0(x)$, s.t. $f_i(x) \leq 0, \forall i \in \mathcal{I}_1, h_i(x) = 0, \forall i \in \mathcal{I}_2$. The neighborhood $\mathcal{N}$ can be a ball centered at x. A globally optimal solution must be a locally optimal solution, but the converse is not true.

A convex optimization problem minimizes a convex function over a convex feasible region. Therefore, if problem (2.19) is a convex optimization problem, $\mathcal{X}$ should be convex and functions f_0 and $f_i, \forall i \in \mathcal{I}_1$ should be convex. Since $h_i(x) = 0$ is equivalent to $h_i(x) \leq 0$ and $-h_i(x) \leq 0, \forall i \in \mathcal{I}_2$, $h_i(x)$ should be both convex and concave. Thus, they are affine. In conclusion, the definition of convex optimization problems is as follows:

Definition 2.52 (*Convex optimization*) A mathematical programming problem is called a *convex optimization* problem if it can be written as:

$$\begin{aligned} \min_{x \in \mathcal{X}} \ & f_0(x) \\ \text{s.t.} \ & f_i(x) \leq 0 \ , \ \ \forall i \in \mathcal{I} \ , \\ & Ax = b \ , \end{aligned} \tag{2.20}$$

where $\mathcal{X} \subset \mathbb{R}^M$ is a convex set; f_0 and $f_i, \forall i \in \mathcal{I}$ are convex functions on $\mathcal{X}$.

Convexity is a desirable property in optimization because it makes locally optimal solutions globally optimal. Consider problem (2.20). Denote the feasible region by $\mathcal{D}$. If $x_0 \in \mathcal{D}$ is not a globally optimal solution, then there exists $x_1 \in \mathcal{D}$ such that $f_0(x_1) < f_0(x_0)$. Consider the line segment L with endpoints x_0 and x_1. By the convexity of $\mathcal{D}$, we have $L \subset \mathcal{D}$. By the convexity of f_0, we have $f_0(x) < f_0(x_0), \forall x \in L \setminus \{x_0\}$. Since any neighborhood of x_0 intersects L, x_0 cannot be a locally optimal solution. To sum up, we have the following result:

Theorem 2.11 *In convex optimization problems, locally optimal solutions are globally optimal.*

Therefore, we can focus on finding locally optimal solutions when solving convex optimization problems. Derivatives are useful tools to investigate local behaviors, which play an essential role in the solving process when the problem has differential properties.

Example 2.11 (*Unconstrained convex optimization*) Assume the optimization problem is $\min_{x \in \mathcal{X}} f_0(x)$, where $\mathcal{X}$ is an open convex set in $\mathbb{R}^M$ and $f_0 : \mathcal{X} \to \mathbb{R}$ is a convex C^1 function. Suppose $x_0 \in \mathcal{X}$ is optimal. Then $f(x) \geq f(x_0), \forall x \in \mathcal{X}$ and all the directional derivatives at x_0 are nonnegative, which leads to $\nabla f(x_0) = 0$.

Conversely, suppose $\nabla f(x_0) = 0$, then $f(x) \geq f(x_0)$ according to the first-order condition in Theorem 2.8. Therefore, $x_0 \in \mathcal{X}$ is optimal if and only if $\nabla f(x_0) = 0$. However, it is worth noticing that such an optimization problem is not necessary to have an optimal solution. The problem $\min_{x>0} 1/x$ is an instance that does not have an optimal solution.

In order to find the locally optimal solutions to constrained optimization problems, we turn to Lagrange duality and Karush-Kuhn-Tucker (KKT) conditions in the following subsections.

2.4.4 Lagrange Duality

Consider the general mathematical programming problem:

$$\begin{aligned} p^* := \min_{x \in \mathcal{X}} \ & f_0(x) \\ \text{s.t. } & f_i(x) \leq 0 \ , \ \ \forall i \in \mathcal{I}_1 \ , \\ & h_i(x) = 0 \ , \ \ \forall i \in \mathcal{I}_2 \ , \end{aligned} \tag{2.21}$$

where $\varnothing \neq \mathcal{X} \subset \mathbb{R}^M$, $|\mathcal{I}_1| = I_1$ (i.e., the number of elements in $\mathcal{I}_1$ is I_1), and $|\mathcal{I}_2| = I_2$. The *Lagrangian function* $L : \mathcal{X} \times \mathbb{R}^{I_1} \times \mathbb{R}^{I_2} \to \mathbb{R}$ is defined by:

$$L(x, \lambda, \nu) := f_0(x) + \sum_{i \in \mathcal{I}_1} \lambda_i f_i(x) + \sum_{i \in \mathcal{I}_2} \nu_i h_i(x) \ , \tag{2.22}$$

where $\lambda_i, i \in \mathcal{I}_1$ and $\nu_i, i \in \mathcal{I}_2$ are *Lagrange multipliers*. The *Lagrange dual function* $g : \mathbb{R}^{I_1} \times \mathbb{R}^{I_2} \to [-\infty, +\infty)$ is defined by:

$$g(\lambda, \nu) := \inf_{x \in \mathcal{X}} L(x, \lambda, \nu) \ . \tag{2.23}$$

The idea of the Lagrange dual function is to construct a lower bound for the optimal value p^* using a weighted sum of the objective and constraint functions. When $\lambda \geq 0$ and x_0 is feasible, we have $f_i(x_0) \leq 0, \forall i \in \mathcal{I}_1$ and $h_i(x_0) = 0, \forall i \in \mathcal{I}_2$. Therefore, $L(x_0, \lambda, \nu) \leq f_0(x_0)$. Then $g(\lambda, \nu) = \inf_{x \in \mathcal{X}} L(x, \lambda, \nu) \leq \inf_{x \in \mathcal{D}} L(x, \lambda, \nu) \leq \inf_{x \in \mathcal{D}} f_0(x) = p^*$, which means that the Lagrange dual function indeed gives a lower bound for p^*.

Another desired property of the Lagrange dual function g is that it is concave, although we do not assume convexity for $\mathcal{X}$, $f_0(x)$, $f_i(x)$, and $h_i(x)$: Note that $-g(\lambda, \nu) = -\inf_{x \in \mathcal{X}} L(x, \lambda, \nu) = \sup_{x \in \mathcal{X}}(-L(x, \lambda, \nu))$ can be regarded as the pointwise supremum of some affine functions, so $-g(\lambda, \nu)$ is convex. To sum up, we have the following results:

Proposition 2.23 (Properties of Lagrange dual function)

1. *The Lagrange dual function is concave.*
2. *If* $\lambda \geq 0$, *then* $g(\lambda, \nu) \leq p^*$.

Since $g(\lambda, \nu) \leq p^*, \forall \lambda \geq 0, \nu \in \mathbb{R}^{l_2}$, we can take the supremum to get the best lower bound, which leads to the so-called *Lagrange dual problem*:

$$\begin{aligned} d^* := \max_{\lambda, \nu} \; & g(\lambda, \nu) \\ \text{s.t. } & \lambda \geq 0 , \end{aligned} \tag{2.24}$$

where the optimal value is denoted by d^*. Note that maximizing a concave function is equivalent to minimizing a convex function, so problem (2.24) is convex. Based on the above analysis, we have:

Theorem 2.12 (Properties of Lagrange dual problem)

1. *Lagrange dual problems are convex optimization problems.*
2. *(*Weak duality*) The optimal value* d^* *of the Lagrange dual problem is no larger than the optimal value* p^* *of the* primal problem *(i.e., the original mathematical programming problem).*

Because Lagrange dual problems are convex, they are easier to handle compared with general mathematical programming problems. The weak duality $d^* \leq p^*$ always holds no matter whether the primal problem is convex or not. $d^* = p^*$ is called *strong duality* but it is not necessary to hold even for convex optimization problems.

Another way to understand the duality is by the saddle point of the Lagrangian function. The definition of saddle point is as follows, which is also illustrated in Fig. 2.12:

Definition 2.53 (*Saddle point*) Suppose $F : A \times B \to \mathbb{R}$ is a function and $(a^*, b^*) \in A \times B$. Then (a^*, b^*) is called a *saddle point* of F on $A \times B$ if

$$F(a^*, b) \leq F(a^*, b^*) \leq F(a, b^*) , \ \forall (a, b) \in A \times B .$$

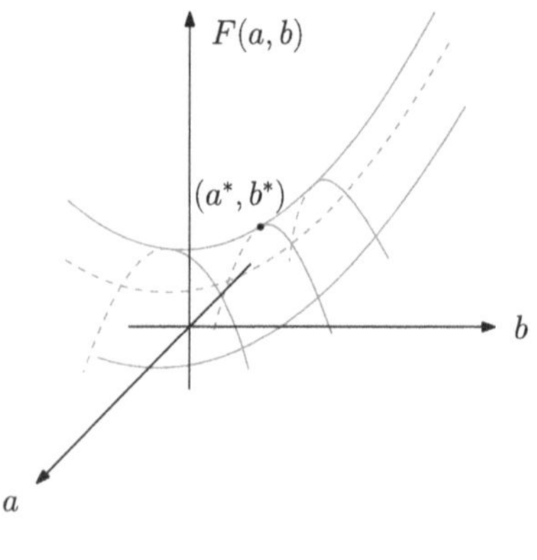

Fig. 2.12 (a^*, b^*) is a saddle point of the function $F : \mathbb{R} \times \mathbb{R} \to \mathbb{R}$ because $F(a^*, b) \leq F(a^*, b^*) \leq F(a, b^*), \forall a, b \in \mathbb{R}$. The shape of the surface is like a saddle. The saddle point is the lowest in one direction and the highest in another direction

Theorem 2.13 *Assume* (a^*, b^*) *is a saddle point of the function* $F : A \times B \to \mathbb{R}$. *Then*

$$\inf_{a \in A} \sup_{b \in B} F(a, b) = \sup_{b \in B} \inf_{a \in A} F(a, b) = F(a^*, b^*) . \tag{2.25}$$

Clearly, $\inf_{a \in A} F(a, b) \leq F(a, b), \forall a \in A, b \in B$, so $\sup_{b \in B} \inf_{a \in A} F(a, b) \leq \sup_{b \in B} F(a, b)$ and then $\sup_{b \in B} \inf_{a \in A} F(a, b) \leq \inf_{a \in A} \sup_{b \in B} F(a, b)$. By the definition of saddle points, we have $\inf_{a \in A} \sup_{b \in B} F(a, b) \leq \sup_{b \in B} F(a^*, b) = F(a^*, b^*) = \inf_{a \in A} F(a, b^*) \leq \sup_{b \in B} \inf_{a \in A} F(a, b)$. Thus, (2.25) is satisfied.

Now let us return to the Lagrangian function. By the definition of Lagrange dual function and dual problem,

$$\sup_{\lambda \geq 0, \nu} \inf_{x \in \mathcal{X}} L(x, \lambda, \nu) = \sup_{\lambda \geq 0, \nu} g(\lambda, \nu) = d^* .$$

For $x \in \mathcal{X}$,

$$\sup_{\lambda \geq 0, \nu} L(x, \lambda, \nu) = \begin{cases} f_0(x) , \text{ if } f_i(x) \leq 0 , \ \forall i \in \mathcal{I}_1 , \ h_i(x) = 0 , \ \forall i \in \mathcal{I}_2 , \\ +\infty , \ \text{ otherwise} , \end{cases}$$

so

$$\inf_{x \in \mathcal{X}} \sup_{\lambda \geq 0, \nu} L(x, \lambda, \nu) = \inf_{x \in \mathcal{X}, f_i(x) \leq 0, \forall i \in \mathcal{I}_1, h_i(x) = 0, \forall i \in \mathcal{I}_2} f_0(x) = p^* .$$

To sum up, $p^* = \inf_{x \in \mathcal{X}} \sup_{\lambda \geq 0, \nu} L(x, \lambda, \nu)$ is the primal problem and $d^* = \sup_{\lambda \geq 0, \nu} \inf_{x \in \mathcal{X}} L(x, \lambda, \nu)$ is the dual problem. We know that $d^* \leq p^*$. The strong duality $d^* = p^*$ is equivalent to that $\inf_{x \in \mathcal{X}}$ and $\sup_{\lambda \geq 0, \nu}$ can be interchanged for the Lagrangian function $L(x, \lambda, \nu)$. Therefore, if the Lagrangian function has a saddle point, then $d^* = p^*$ by Theorem 2.13, and the saddle point is primal- and dual-optimal.

Proposition 2.24 *For the general mathematical programming problem* (2.21), *if* $(x^*, (\lambda^*, \nu^*))$ *is a saddle point of the Lagrangian function* $L : \mathcal{X} \times (\mathbb{R}_+)^{I_1} \times \mathbb{R}^{I_2} \to \mathbb{R}$ *defined in* (2.22), *i.e.,*

$$L(x, \lambda^*, \nu^*) \geq L(x^*, \lambda^*, \nu^*) \geq L(x^*, \lambda, \nu) , \ \forall x \in \mathcal{X} , \ \lambda \geq 0 , \ \nu \in \mathbb{R}^{I_2} ,$$

then x^* *is optimal for the primal problem* (2.21), (λ^*, ν^*) *is optimal for the Lagrange dual problem* (2.24), *and* $p^* = d^* = L(x^*, \lambda^*, \nu^*)$.

The saddle point of the Lagrangian function is a sufficient condition for a general mathematical programming problem to have strong duality. In contrast, Slater's condition is a famous sufficient condition for a convex optimization problem to possess strong duality. Consider the following convex optimization problem (i.e., $\mathcal{X} = \mathbb{R}^M$ in (2.20)):

$$\begin{aligned} \min_{x \in \mathbb{R}^M} \ & f_0(x) \\ \text{s.t.} \ & f_i(x) \le 0 \ , \ \forall i \in \mathcal{I} \ , \\ & Ax = b \ . \end{aligned} \tag{2.26}$$

Definition 2.54 (*Slater's condition*) If problem (2.26) is strictly feasible, i.e., there exists $x_0 \in \mathbb{R}^M$ such that $f_i(x_0) < 0, \forall i \in \mathcal{I}$ and $Ax_0 = b$, then we say that *Slater's condition* holds for this problem.

Theorem 2.14 (Strong duality) *If problem* (2.26) *satisfies Slater's condition, then* $d^* = p^*$ *holds. If furthermore* $p^* > -\infty$, *then* d^* *is attained.*

Slater's condition can be generalized. In order to state the generalized version, we first introduce the definitions of the affine hull and relative interior. The definition of the affine hull is similar to that of the convex hull in Definition 2.44.

Definition 2.55 (*Affine hull*) Let S be a nonempty set in $\mathbb{R}^M$, then the *affine hull* aff(S) of S is the set consisting of all the *affine combinations* of points in S:

$$\text{aff}(S) := \left\{ \sum_{i=1}^{n} \lambda_i x_i \mid n \in \mathbb{N}_+, x_i \in S, \lambda_i \in \mathbb{R}, \forall i, \sum_{i=1}^{n} \lambda_i = 1 \right\} .$$

Note that the only difference between affine combination and convex combination is that $\lambda_i \in \mathbb{R}, \forall i$. The affine combinations of two points are on the line through the two points. In addition, similar to the convex hull, the affine hull is the smallest affine set (all the affine combinations of the points in the set are still in the set) that contains the original set.

Definition 2.56 (*Relative interior*) Let S be a nonempty set in $\mathbb{R}^M$. Then its *relative interior* relint(S) is the interior of S in its affine hull aff(S). In other words,

$$\text{relint}(S) := \{x \in S \mid \exists \varepsilon > 0, \text{s.t. } B(x, \varepsilon) \cap \text{aff}(S) \subset S\} \ ,$$

where $B(x, \varepsilon) = \{y \in \mathbb{R}^M \mid \|y - x\| \le \varepsilon\}$.

Slater's condition can be generalized to be that the relative interior of the feasible region is nonempty, i.e., relint$(\mathcal{D}) \neq \varnothing$, where $\mathcal{D} = \{x \in \mathbb{R}^M \mid f_i(x) \le b, \forall i \in \mathcal{I}, Ax = b\}$. When problem (2.26) is strictly feasible, relint$(\mathcal{D}) \neq \varnothing$, but the converse does not hold. Providing that relint$(\mathcal{D}) \neq \varnothing$, $d^* = p^*$ holds in problem (2.26).

The strong duality of convex optimization problems can be used to check the existence of saddle points or the interchange of "min" and "max" operations. A commonly used property is as follows:

Theorem 2.15 (Classical minimax theorem, Proposition 5.5.4 in Bertsekas 2009) *Suppose that* X *and* Z *are nonempty convex sets in* $\mathbb{R}^N$ *and* $\mathbb{R}^M$, *respectively. The function* $\phi : X \times Z \to \mathbb{R}$ *satisfies:*

1. *For any $z \in Z$, the function $\phi(\cdot, z) : X \to \mathbb{R}$ is convex and closed.*
2. *For any $x \in X$, the function $-\phi(x, \cdot) : Z \to \mathbb{R}$ is convex and closed.*

Define function $t : \mathbb{R}^N \to (-\infty, +\infty]$ by:

$$t(x) := \begin{cases} \sup_{z \in Z} \phi(x, z) \ , & \text{if } x \in X \ , \\ +\infty \ , & \text{if } x \notin X \ . \end{cases}$$

Assume that t is proper and for any $\gamma \in \mathbb{R}$, the level set $\{x \mid t(x) \le \gamma\}$ is compact. Then

$$\sup_{z \in Z} \inf_{x \in X} \ \phi(x, z) = \inf_{x \in X} \sup_{z \in Z} \ \phi(x, z) \ .$$

2.4.5 Karush-Kuhn-Tucker Conditions

We want to characterize the condition for optimal solutions. In this regard, we investigate the case where strong duality $d^* = p^*$ holds and both d^* and p^* are attained, i.e.,

$$\max_{\lambda \ge 0, \nu} \ g(\lambda, \nu) = g(\lambda^*, \nu^*) = f_0(x^*) = \begin{cases} \min\limits_{x \in \mathcal{X}} \ f_0(x) \\ \quad \text{s.t. } f_i(x) \le 0 \ , \ \ \forall i \in \mathcal{I}_1 \ , \\ \qquad\quad h_i(x) = 0 \ , \ \ \forall i \in \mathcal{I}_2 \ , \end{cases}$$

where $\lambda^* \ge 0$, $f_i(x^*) \le 0, \forall i \in \mathcal{I}_1$, and $h_i(x^*) = 0, \forall i \in \mathcal{I}_2$. Then

$$\begin{aligned} f_0(x^*) &= g(\lambda^*, \nu^*) = \inf_{x \in \mathcal{X}} \ L(x, \lambda^*, \nu^*) \\ &\le L(x^*, \lambda^*, \nu^*) = f_0(x^*) + \sum_{i \in \mathcal{I}_1} \lambda_i^* f_i(x^*) + \sum_{i \in \mathcal{I}_2} \nu_i^* h_i(x^*) \le f_0(x^*) \ . \end{aligned} \tag{2.27}$$

Therefore, the above two inequalities are actually equations, so $L(x^*, \lambda^*, \nu^*) = \inf_{x \in \mathcal{X}} L(x, \lambda^*, \nu^*)$ and $\sum_{i \in \mathcal{I}_1} \lambda_i^* f_i(x^*) = 0$. Because $\lambda_i^* \ge 0$ and $f_i(x^*) \le 0$ for any $i \in \mathcal{I}_1$, we have $\lambda_i^* f_i(x^*) = 0, \forall i \in \mathcal{I}_1$, which is called *complementary slackness*. Then $\lambda_i^* > 0$ implies $f_i(x^*) = 0$, and $f_i(x^*) < 0$ implies $\lambda_i^* = 0$.

Assume that $\mathcal{X} = \mathbb{R}^M$, the entire Euclidean space. In addition, f_0, $f_i, \forall i \in \mathcal{I}_1$, and $h_i, \forall i \in \mathcal{I}_2$ are C^1 functions. Then $L(x, \lambda^*, \nu^*)$ is a C^1 function of x on $\mathbb{R}^M$. Because x^* minimizes $L(x, \lambda^*, \nu^*)$, we have $\nabla_x L(x, \lambda^*, \nu^*) = 0$, i.e., $\nabla f_0(x^*) + \sum_{i \in \mathcal{I}_1} \lambda_i^* \nabla f_i(x^*) + \sum_{i \in \mathcal{I}_2} \nu_i^* \nabla h_i(x^*) = 0$. Based on the above discussion, KKT conditions are defined as follows:

Definition 2.57 (*Karush-Kuhn-Tucker (KKT) conditions*) The *Karush-Kuhn-Tucker (KKT) conditions* of the problem

$$\min_{x \in \mathbb{R}^M} f_0(x)$$
$$\text{s.t. } f_i(x) \leq 0\ ,\ \forall i \in \mathcal{I}_1\ ,\ h_i(x) = 0\ ,\ \forall i \in \mathcal{I}_2\ ,$$

with C^1 functions f_0, $f_i(x)$, $\forall i \in \mathcal{I}_1$, and $h_i(x)$, $\forall i \in \mathcal{I}_2$ are the following constraints of primal variable x and dual variables λ, ν:

1. (Primal feasibility) $f_i(x) \leq 0, \forall i \in \mathcal{I}_1, h_i(x) = 0, \forall i \in \mathcal{I}_2$.
2. (Dual feasibility) $\lambda \geq 0$.
3. (Complementary slackness) $\lambda_i f_i(x) = 0, \forall i \in \mathcal{I}_1$.
4. (Gradient of the Lagrangian function with respect to x is 0)

$$\nabla f_0(x) + \sum_{i \in \mathcal{I}_1} \lambda_i \nabla f_i(x) + \sum_{i \in \mathcal{I}_2} \nu_i \nabla h_i(x) = 0\ .$$

We have analyzed that the KKT conditions are necessary under the assumption of strong duality and attained primal- and dual-optimal solutions.

Theorem 2.16 *Under the assumptions in Definition 2.57, if the strong duality holds and x, λ, ν are optimal, then x, λ, ν satisfy the KKT conditions.*

The KKT conditions of the convex optimization problem (2.26) with convex and C^1 functions f_0 and $f_i, i \in \mathcal{I}$ is:

$$\begin{cases} f_i(x) \leq 0\ ,\ \forall i \in \mathcal{I}\ ,\ Ax = b\ , \\ \lambda \geq 0\ , \\ \lambda_i f_i(x) = 0\ ,\ \forall i \in \mathcal{I}\ , \\ \nabla f_0(x) + \sum_{i \in \mathcal{I}} \lambda_i \nabla f_i(x) + A^\top \nu = 0\ . \end{cases} \tag{2.28}$$

Theorem 2.17 says that the KKT conditions are sufficient for the optimality in a convex optimization problem.

Theorem 2.17 *Assume f_0 and $f_i, i \in \mathcal{I}$ are C^1; x, λ, and ν satisfy the KKT conditions* (2.28). *Then they are optimal for the convex optimization problem* (2.26) *and the Lagrange dual problem. In addition, the strong duality holds.*

Proof Since problem (2.26) is convex, the Lagrangian function is convex with respect to x, i.e., $L(ax + (1-a)x', \lambda, \nu) \leq aL(x, \lambda, \nu) + (1-a)L(x', \lambda, \nu), \forall a \in [0, 1], x, x' \in \mathbb{R}^M, \lambda \geq 0$. By the optimality condition of the unconstrained convex optimization problem in Example 2.11, x minimizes $L(x, \lambda, \nu)$ for fixed $\lambda \geq 0$ and ν if and only if $\nabla L_x(x, \lambda, \nu) = 0$. Therefore, when x, λ, and ν satisfy the KKT conditions (2.28), using a deduction like (2.27), we have $d^* = p^*$ and x, λ, and ν are optimal. Hence, the conclusions of Theorem 2.17 hold. □

The conclusion of Theorem 2.17 is not surprising because, for convex optimization problems, KKT points mean exactly saddle points, as pointed out in Proposition 2.25. Recall from Proposition 2.24 that saddle points are sufficient for strong duality.

Proposition 2.25 *Consider the convex optimization problem* (2.26). *Then* x^*, λ^*, *and* ν^* *satisfy the KKT conditions* (2.28) *if and only if* $(x^*, (\lambda^*, \nu^*))$ *is a saddle point of the Lagrangian function* $L : \mathbb{R}^M \times (\mathbb{R}_+)^{l_1} \times \mathbb{R}^{l_2} \to \mathbb{R}$ *with* $L(x, \lambda, \nu) = f_0(x) + \sum_{i \in \mathcal{I}} \lambda_i f_i(x) + (Ax - b)^\top \nu$.

Proof When x^*, λ^*, and ν^* satisfy the KKT conditions, x^* minimizes the convex function $L(x, \lambda^*, \nu^*)$ of x because $\nabla f_0(x^*) + \sum_{i \in \mathcal{I}} \lambda_i^* \nabla f_i(x^*) + A^\top \nu^* = 0$. Thus, $L(x, \lambda^*, \nu^*) \geq L(x^*, \lambda^*, \nu^*) = f_0(x^*) \geq L(x^*, \lambda, \nu), \forall x, \lambda \geq 0, \nu$, which means $(x^*, (\lambda^*, \nu^*))$ is a saddle point of L. Conversely, assume $(x^*, (\lambda^*, \nu^*))$ is a saddle point. $Ax^* - b$ must be 0, otherwise $L(x^*, \lambda, \nu) = f_0(x^*) + \sum_{i \in \mathcal{I}} \lambda_i f_i(x^*) + (Ax^* - b)^\top \nu$ can be arbitrarily large by choosing ν. Similarly, we have $f_i(x^*) \leq 0, \forall i \in \mathcal{I}$. Thus, x^* is feasible. By $L(x^*, \lambda^*, \nu^*) \geq L(x^*, \lambda, \nu), \forall \lambda \geq 0, \nu$, we have $\lambda^* \geq 0$, and $\lambda_i^* f_i(x^*) = 0, \forall i \in \mathcal{I}$. By $L(x, \lambda^*, \nu^*) \geq L(x^*, \lambda^*, \nu^*)$ and L is C^1 with respect to x, we have $\nabla f_0(x^*) + \sum_{i \in \mathcal{I}} \lambda_i^* \nabla f_i(x^*) + A^\top \nu^* = 0$. Thus, the KKT conditions hold. □

Roughly speaking, Theorem 2.18 says that when Slater's condition holds, the KKT conditions constitute a necessary and sufficient optimal condition.

Theorem 2.18 *Assume* f_0 *and* $f_i, i \in \mathcal{I}$ *are* C^1 *functions and the convex optimization problem* (2.26) *satisfies Slater's condition. Then* x *is optimal if and only if there exist* λ *and* ν *so that* x, λ, *and* ν *satisfy the KKT conditions* (2.28).

By Theorem 2.17, we only need to show that when x is optimal, there exist λ and ν to satisfy the KKT conditions. Since Slater's condition holds, $d^* = p^* > -\infty$ and d^* is attained by some λ, ν. Then according to the discussions before the introduction of KKT conditions in Definition 2.57, x, λ, and ν satisfy the KKT conditions.

In conclusion, KKT conditions are closely related to the optimal solutions of mathematical programming problems with C^1 objective and constraint functions, and they are especially effective for convex optimization problems. KKT conditions can be regarded as a generalization of the condition $\nabla f_0(x) = 0$ for unconstrained problems.

2.4.6 Conjugate Function and Fenchel Duality

We have introduced the Lagrange dual problem, which can find a lower bound for the optimal value using a convex optimization problem no matter whether the original problem is convex or not. Conjugate function, also known as Legendre–Fenchel transformation, is another helpful tool, which transforms a general function into a convex function. We review some important definitions and results about the conjugate function and Fenchel duality in this subsection, whose details can be found in Bertsekas (2009).

Definition 2.58 (*Conjugate function*) Let $f : S \to \mathbb{R}$ be a function defined on a nonempty set $S \subset \mathbb{R}^M$. Then the *conjugate function* of f is the extended-value function $f^* : (-\infty, +\infty]$ defined by:

$$f^*(y) := \sup_{x \in S} \left\{ y^\top x - f(x) \right\} \ , \ \forall y \in \mathbb{R}^M \ . \tag{2.29}$$

Remark 2.6 When $\tilde{f} : \mathbb{R}^M \to [-\infty, +\infty]$ is an extended-value function, its conjugate function $\tilde{f}^* : \mathbb{R}^M \to [-\infty, +\infty]$ is also defined by (2.29). Note that if $\tilde{f}$ is extended from f, then for any $y \in \mathbb{R}^M$,

$$\begin{aligned}(\tilde{f})^*(y) = \sup_{x \in \mathbb{R}^M} \{y^\top x - \tilde{f}(x)\} &= \max \left\{ \sup_{x \in S}\{y^\top x - \tilde{f}(x)\}, \ \sup_{x \in \mathbb{R}^M \setminus S} \{y^\top x - \tilde{f}(x)\} \right\} \\ &= \max \left\{ \sup_{x \in S}\{y^\top x - f(x)\}, -\infty \right\} = \sup_{x \in S}\{y^\top x - f(x)\} = f^*(y) \ ,\end{aligned}$$

which means $f^* = (\tilde{f})^*$ as functions, so we do not distinguish them hereinafter.

The meaning of the conjugate function value is illustrated in Fig. 2.13. An immediate result of the definition is Fenchel's inequality:

Theorem 2.19 (Fenchel's inequality) *Let $f : \mathbb{R}^M \to [-\infty, +\infty]$ be an extended-value function and $f^* : \mathbb{R}^M \to [-\infty, +\infty]$ be its conjugate function. Then for any $x, y \in \mathbb{R}^M$ such that $f(x) + f^*(y)$ is well-defined (i.e., not $(+\infty) + (-\infty)$), we have:*

$$x^\top y \le f(x) + f^*(y) \ .$$

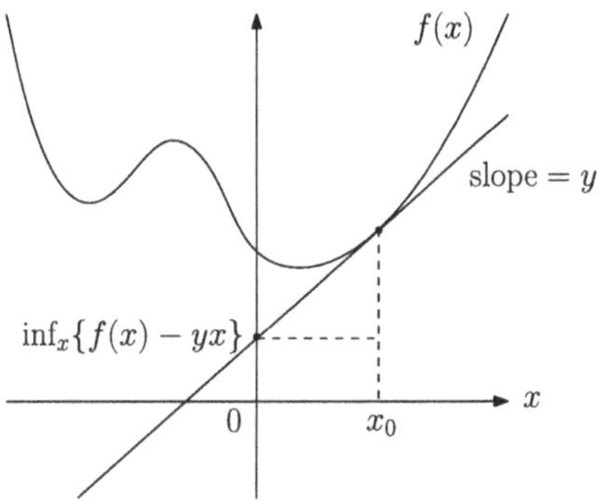

Fig. 2.13 f is a function on $\mathbb{R}$. Suppose a line with the slope $y \in \mathbb{R}$ is a supporting hyperplane of epi(f) at $(x_0, f(x_0))$. Then the intersection of the line with the line $x = 0$ has value $\inf_{x \in \mathbb{R}}\{f(x) - yx\} = -f^*(y)$

Recall from Sect. 2.4.2 that for a point $x_0 \in S \subset \mathbb{R}^M$, a vector $y \in \mathbb{R}^M$ is a subderivative of f at x_0 if $f(x) \geq f(x_0) + y^\top(x - x_0), \forall x \in S$. Combining it with the definition of conjugate function, we have:

$$\begin{aligned} y \in \partial f(x_0) &\iff f(x) \geq f(x_0) + y^\top(x - x_0) \ , \ \forall x \in S \\ &\iff y^\top x_0 - f(x_0) \geq y^\top x - f(x) \ , \ \forall x \in S \\ &\iff f^*(y) = y^\top x_0 - f(x_0) \ , \end{aligned}$$

which bridges the subderivative and the conjugate function.

For arbitrary $y, y' \in \mathbb{R}^M$ and $\lambda \in [0, 1]$,

$$\begin{aligned} f^*(\lambda y + (1-\lambda)y') &= \sup_{x \in S} \left\{(\lambda y + (1-\lambda)y')^\top x - f(x)\right\} \\ &= \sup_{x \in S} \left\{\lambda(y^\top x - f(x)) + (1-\lambda)(y'^\top x - f(x))\right\} \\ &\leq \lambda \sup_{x \in S} \left\{y^\top x - f(x)\right\} + (1-\lambda) \sup_{x \in S} \left\{y'^\top x - f(x)\right\} \\ &= \lambda f^*(y) + (1-\lambda) f^*(y') \ , \end{aligned}$$

so f^* is a convex function.

f^* is also closed. In other words, suppose that $\{y_n \mid n \in \mathbb{N}_+\} \subset \mathbb{R}^M$ and $\{z_n \mid n \in \mathbb{N}_+\} \subset \mathbb{R}$ satisfy $\lim_{n\to\infty} y_n = y_0 \in \mathbb{R}^M$, $\lim_{n\to\infty} z_n = z_0 \in \mathbb{R}$, and $z_n \geq f^*(y_n) = \sup_{x \in S}\{y_n^\top x - f(x)\}$, $\forall n \in \mathbb{N}_+$, then $z_0 \geq f^*(y_0) = \sup_{x \in S}\{y_0^\top x - f(x)\}$. The reason is that for any fixed $x_0 \in S$,

$$z_0 = \lim_{n\to\infty} z_n \geq \lim_{n\to\infty} \sup_{x \in S}\{y_n^\top x - f(x)\} \geq \lim_{n\to\infty} \{y_n^\top x_0 - f(x_0)\} = y_0^\top x_0 - f(x_0) \ ,$$

and then the result follows. To sum up, we have the following conclusion:

Proposition 2.26 *A conjugate function is convex and closed.*

Example 2.12 (*Represent Lagrange dual function using a conjugate function*) Consider a linear constrained mathematical programming problem as follows:

$$\begin{aligned} &\min_{x \in \mathcal{X}} f_0(x) \\ &\text{s.t. } A_1 x \leq b_1 \ , \ A_2 x = b_2 \ . \end{aligned}$$

The Lagrangian function is:

$$L(x, \lambda, \nu) = f_0(x) + \lambda^\top(A_1 x - b_1) + \nu^\top(A_2 x - b_2) \ .$$

The Lagrange dual function is:

$$\begin{aligned} g(\lambda, \nu) &= \inf_{x\in\mathcal{X}} L(x, \lambda, \nu) \\ &= \inf_{x\in\mathcal{X}}\{f_0(x) + \lambda^\top A_1 x + \nu^\top A_2 x\} - b_1^\top \lambda - b_2^\top \nu \\ &= -\sup_{x\in\mathcal{X}}\{(-A_1^\top \lambda - A_2^\top \nu)^\top x - f_0(x)\} - b_1^\top \lambda - b_2^\top \nu \\ &= -f_0^*(-A_1^\top \lambda - A_2^\top \nu) - b_1^\top \lambda - b_2^\top \nu \,. \end{aligned}$$

Therefore, if the conjugate function f_0^* is known, the Lagrange dual function of the linear constrained problem can be easily obtained.

Another application of conjugate functions is the support function.

Definition 2.59 (*Support function*) Let S be a set in $\mathbb{R}^M$. Then the *support function* of S is defined as the conjugate function of the characteristic function of S in $\mathbb{R}^M$.

In other words, the support function $h_S : \mathbb{R}^M \to \mathbb{R}$ of S is as follows:

$$h_S(y) = \chi_S^*(y) = \sup_{x\in\mathbb{R}^M}\{y^\top x - \chi_S(x)\} = \sup_{x\in S}\ y^\top x\ , \ \ \forall y \in \mathbb{R}^M\ .$$

For a fixed $y \in \mathbb{R}^M$, we have $y^\top x \le h_S(y), \forall x \in S$. If S is bounded, then $y^\top x_0 = h_S(y)$ for some $x_0 \in \partial S$ and the hyperplane $\{x \in \mathbb{R}^M \mid y^\top x = h_S(y)\}$ is a supporting hyperplane of S at x_0.

Definition 2.60 (*Biconjugate function*) The conjugate function f^* of an extended-value function $f : \mathbb{R}^M \to [-\infty, +\infty]$ also has its own conjugate function, i.e., $(f^*)^*$, which is denoted by f^{**} and called the *biconjugate function* of f.

According to the definition,

$$\begin{aligned} f^{**}(x) &= \sup_{y\in\mathbb{R}^M}\{x^\top y - f^*(y)\} \\ &= \sup_{y\in\mathbb{R}^M}\left\{x^\top y - \sup_{x'\in\mathbb{R}^M}\{y^\top x' - f(x')\}\right\} \\ &\le \sup_{y\in\mathbb{R}^M}\left\{x^\top y - (y^\top x - f(x))\right\} = f(x)\ ,\ \ \forall x \in \mathbb{R}^M\ . \end{aligned}$$

Therefore, we have the following result:

Proposition 2.27 *Let $f : \mathbb{R}^M \to [-\infty, +\infty]$ be an extended-value function, then its biconjugate function $f^{**} : \mathbb{R}^M \to [-\infty, +\infty]$ is a convex and closed function and $f^{**}(x) \le f(x), \forall x \in \mathbb{R}^M$.*

In fact, $f^{**} = f$ holds under some conditions. Since f^{**} is convex and closed, the convexity and closeness of f are necessary conditions for $f^{**} = f$. The necessary and sufficient condition of $f^{**} = f$ is the content of the Fenchel-Moreau theorem, whose details can be found in Borwein and Lewis (2005). Before that, we introduce the definition of lower semi-continuous functions.

Definition 2.61 (*Lower semi-continuous function*) An extended-value function $f : \mathbb{R}^M \to [-\infty, +\infty]$ is called *lower semi-continuous* if one of the following equivalent conditions holds:

1. $\liminf_{x \to x_0} f(x) \geq f(x_0), \forall x_0 \in \mathbb{R}^M$.
2. The epigraph $\text{epi}(f) = \{(x, y) \mid x \in \mathbb{R}^M, y \geq f(x)\}$ is closed as a set in $\mathbb{R}^M \times \mathbb{R}$.

A function f is upper semi-continuous if and only if $-f$ is lower semi-continuous.

Theorem 2.20 (Fenchel-Moreau theorem) *Let $f : \mathbb{R}^M \to [-\infty, +\infty]$ be an extended-value function, then $f^{**} = f$ if and only if one of the following holds:*

1. *f is proper, lower semi-continuous, and convex.*
2. *$f(x) = +\infty, \forall x \in \mathbb{R}^M$.*
3. *$f(x) = -\infty, \forall x \in \mathbb{R}^M$.*

The characteristic function χ_S is proper if S is nonempty; χ_S is lower semi-continuous if S is closed; χ_S is convex if S is convex. Therefore, we have:

Corollary 2.3 *Let S be a nonempty closed convex set in $\mathbb{R}^M$. Then its characteristic function is the conjugate function of its support function, i.e., $\chi_S = h_S^*$.*

Example 2.13 (*Conjugate and biconjugate functions*) These are some examples of closed, proper, and convex functions and their conjugate and biconjugate functions. The 1-dimensional cases are drawn in Fig. 2.14.

1. Assume $f(x) = a^\top x + b, \forall x \in \mathbb{R}^M$. Then

$$f^*(y) = \sup_{x \in \mathbb{R}^M} \{y^\top x - (a^\top x + b)\} = \begin{cases} -b \ , & \text{if } y = a \ , \\ +\infty \ , & \text{if } y \neq a \ , \end{cases}$$
$$f^{**}(x) = a^\top x + b \ , \ \forall x \in \mathbb{R}^M \ .$$

2. Assume $f(x) = \|x\|, \forall x \in \mathbb{R}^M$ and $\|\cdot\|_*$ is the dual norm of $\|\cdot\|$. Then

$$f^*(y) = \sup_{x \in \mathbb{R}^M} \{y^\top x - \|x\|\} = \begin{cases} 0 \ , & \text{if } \|y\|_* \leq 1 \ , \\ +\infty \ , & \text{if } \|y\|_* > 1 \ , \end{cases}$$
$$f^{**}(x) = \sup_{y : \|y\|_* \leq 1} x^\top y = \|x\| \ , \ \forall x \in \mathbb{R}^M \ .$$

In fact, f^* is the characteristic function of the set $\{y \in \mathbb{R}^M \mid \|y\|_* \leq 1\}$, which is the closed unit ball in $\mathbb{R}^M$ under the norm $\|\cdot\|_*$. Therefore, $f^{**} = f = \|\cdot\|$ is the support function.

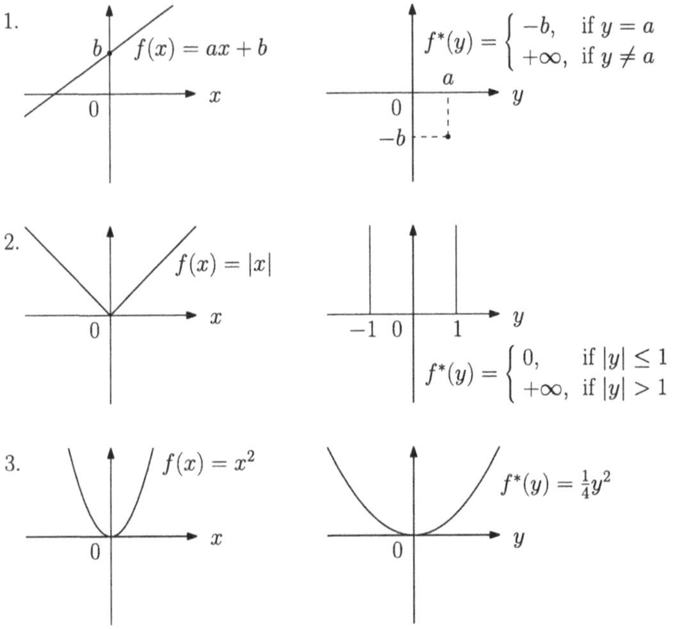

Fig. 2.14 1-dimensional examples of closed, proper, and convex functions and their conjugate functions. The original functions are the same as the biconjugate functions

3. Assume $f(x) = x^\top x, \forall x \in \mathbb{R}^M$. Then using the optimality condition of unconstrained problems with smooth objective functions in Example 2.11, we have:

$$f^*(y) = \sup_{x \in \mathbb{R}^M} \{y^\top x - x^\top x\} = \frac{1}{4} y^\top y \ , \ \forall y \in \mathbb{R}^M \ ,$$

$$f^{**}(x) = \sup_{y \in \mathbb{R}^M} \left\{ x^\top y - \frac{1}{4} y^\top y \right\} = x^\top x \ , \ \forall x \in \mathbb{R}^M \ .$$

Fenchel duality is an important optimization framework that utilizes conjugate functions. Consider the following convex optimization problem:

$$p^* = \min_{x \in \mathbb{R}^M} f_1(x) + f_2(Ax)$$
$$\text{s.t. } x \in S_1 \ , \ Ax \in S_2 \ , \tag{2.30}$$

where A is a $N \times M$ matrix; $f_1 : S_1 \to \mathbb{R}$ and $f_2 : S_2 \to \mathbb{R}$ are closed convex functions on convex sets $S_1 \subset \mathbb{R}^M$ and $S_2 \subset \mathbb{R}^N$, respectively; the feasible region $\mathcal{D} = \{x \in \mathbb{R}^M \mid x \in S_1, Ax \in S_2\} \neq \varnothing$. Let $x_1 = x$ and $x_2 = Ax$, then $x_2 = Ax_1$ and problem (2.30) can be equivalently transformed into:

$$p^* = \min_{(x_1, x_2) \in S_1 \times S_2} f_1(x_1) + f_2(x_2)$$
$$\text{s.t. } x_2 = Ax_1 \ ,$$

which is an equality-constrained convex optimization problem. The Lagrangian function is:

$$L(x_1, x_2, \nu) = f_1(x_1) + f_2(x_2) + \nu^\top (x_2 - Ax_1) \ .$$

The Lagrange dual function is:

$$\begin{aligned}
g(\nu) &= \inf_{(x_1, x_2) \in S_1 \times S_2} L(x_1, x_2, \nu) \\
&= \inf_{(x_1, x_2) \in S_1 \times S_2} \{f_1(x_1) + f_2(x_2) + \nu^\top (x_2 - Ax_1)\} \\
&= \inf_{x_1 \in S_1} \{f_1(x_1) - \nu^\top Ax_1\} + \inf_{x_2 \in S_2} \{f_2(x_2) + \nu^\top x_2\} \\
&= - \sup_{x_1 \in S_1} \{\nu^\top Ax_1 - f_1(x_1)\} - \sup_{x_2 \in S_2} \{-\nu^\top x_2 - f_2(x_2)\} \\
&= -f_1^*(A^\top \nu) - f_2^*(-\nu) \ .
\end{aligned}$$

The Lagrange dual problem is:

$$d^* = \max_{\nu \in \mathbb{R}^N} \left\{-f_1^*(A^\top \nu) - f_2^*(-\nu)\right\} \ ,$$

which is equivalent to the following problem after changing the sign:

$$-d^* = \min_{\nu \in \mathbb{R}^N} \left\{f_1^*(A^\top \nu) + f_2^*(-\nu)\right\} \ . \tag{2.31}$$

Note that problem (2.31) has a similar form to the primal problem (2.30).

Recall the properties of the Lagrange dual problem and Slater's condition from Theorems 2.12 and 2.14 in Sect. 2.4.4, and also the deduction process of the KKT conditions in Sect. 2.4.5, then we have the following results:

Theorem 2.21 (Fenchel duality) *For the primal problem* (2.30) *and the dual problem* (2.31)*, we have:*

1. *(Weak duality)* $d^* \le p^*$.
2. *(Strong duality under Slater's condition) If the primal problem* (2.30) *satisfies Slater's condition, i.e., there exists* $x_0 \in \mathbb{R}^M$ *such that* $x_0 \in int(S_1 \cap \{x \in \mathbb{R}^M \mid Ax \in S_2\})$*, then* $d^* = p^*$*. If furthermore* $p^* > -\infty$*, then* d^* *is attained at some* ν^*.
3. *(Optimality condition)* $d^* = p^*$*,* x^* *is primal-optimal, and* ν^* *is dual-optimal if and only if* x^* *minimizes* $f_1(x) - \nu^{*\top} Ax$ *and* Ax^* *minimizes* $f_2(y) + \nu^{*\top} y$.

2.5 Conic Programs and Duality

This section is about conic programs and duality. The notions of cones, dual cones, and generalized inequalities are first introduced. Then the duality of general conic programs is presented, followed by discussions on three commonly used conic programs: linear programming (LP), second-order cone programming (SOCP), and semidefinite programming (SDP) problems. Finally, the problem of moments is briefly introduced. The contents of this section mainly come from Boyd and Vandenberghe (2004), Ben-Tal and Nemirovski (2001), Alizadeh and Goldfarb (2003), Shapiro (2001).

2.5.1 *Cones and Dual Cones*

A cone is a nonempty set in an Euclidean space such that for any nonzero point in the set, the ray starting from 0 through the point is also in the set.

Definition 2.62 (*Cone*) Let K be a nonempty set in $\mathbb{R}^M$. Then K is called a *cone* if for any $x \in K$ and $\lambda \geq 0$, we have $\lambda x \in K$.

By definition, $0 \in K$ for any cone K in $\mathbb{R}^M$. We have introduced convex combinations and affine combinations. The notion of conic combination is related to and different from these two combinations. A *conic combination* of N points $x_1, x_2, \ldots, x_N$ is $\lambda_1 x_1 + \lambda_2 x_2 + \cdots + \lambda_N x_N$ with $\lambda_1, \lambda_2, \ldots, \lambda_N \geq 0$. A convex cone is a nonempty set in a Euclidean space that is closed under conic combinations:

Definition 2.63 (*Convex cone*) Let K be a nonempty set in $\mathbb{R}^M$. Then K is called a *convex cone* if any conic combination of the points in K is still in K.

Proposition 2.28 *A nonempty set in $\mathbb{R}^M$ is a convex cone if and only if it is a cone and also a convex set.*

The above proposition explains the definition of convex cone. Note that any convex combination is a conic combination, so a convex cone must be a convex set. Conversely, $\sum_{n=1}^N \lambda_n x_n = (\sum_{n=1}^N \lambda_n x_n / \sum_{n=1}^N \lambda_n)(\sum_{n=1}^N \lambda_n)$ is a positive multiplication of the convex combination $\sum_{n=1}^N \lambda_n x_n / \sum_{n=1}^N \lambda_n$ of $x_1, \ldots, x_N$ when $\lambda_n \geq 0, \forall n$ and $\sum_{n=1}^N \lambda_n > 0$. Therefore, a cone being a convex set must be a convex cone.

Definition 2.64 (*Proper cone*) A convex cone K in $\mathbb{R}^M$ is called a *proper cone* if:

1. K is closed as a set in $\mathbb{R}^M$.
2. K is *solid*, which means $\text{int}(K) \neq \varnothing$.
3. K is *pointed*, which means that K does not contain any line, i.e., if $0 \neq x \in K$, then $-x \notin K$.

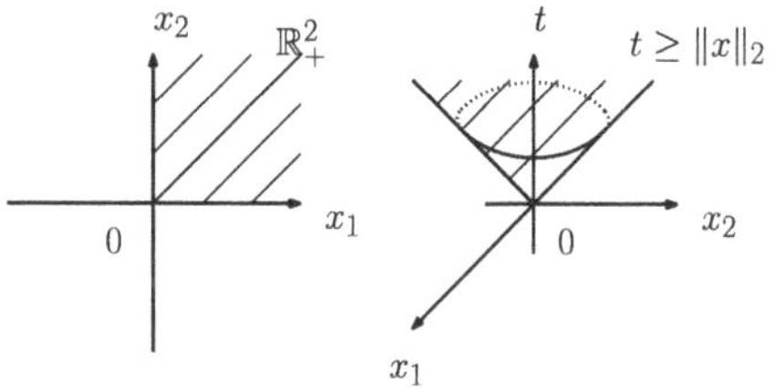

Fig. 2.15 The nonnegative orthant $\mathbb{R}^2_+$ is on the left side and the second-order cone $\mathbb{L}^3 = \{(t, x) \in \mathbb{R}^3 \mid t \geq \|x\|_2\}$ is on the right side. They are both proper cones

For a proper cone K in $\mathbb{R}^M$, $0 \in K$ but $0 \notin \text{int}(K)$ because K is pointed. In addition, $\text{int}(K) \cup \{0\}$ is a convex cone, which can be checked using the definition.

Example 2.14 (*Cones*) Here are some examples of cones.

1. The Euclidean space $\mathbb{R}^M$ is a cone in $\mathbb{R}^M$. It is a closed and solid convex cone but not pointed.
2. The nonempty polyhedron $\{x \in \mathbb{R}^M \mid Ax \geq 0\}$ is a closed convex cone but may not be solid or pointed.
3. The *nonnegative orthant* $\mathbb{R}^M_+ := \{x \in \mathbb{R}^M \mid x \geq 0\}$ is a proper cone in $\mathbb{R}^M$.
4. The set $\mathbb{L}^{M+1} := \{(t, x) \in \mathbb{R} \times \mathbb{R}^M \mid t \geq \|x\|_2\}$ is a proper cone in $\mathbb{R}^{M+1}$, which is called the *second-order cone* or *Lorentz cone*.
5. The set $\mathbb{S}^M_+$ consisting of all the $M \times M$ real-valued symmetric positive semidefinite matrices is a proper cone in the vector space of all the $M \times M$ real-valued symmetric matrices. This vector space is $M(M+1)/2$-dimensional. $\mathbb{S}^M_+$ is a proper cone in $\mathbb{R}^{M(M+1)/2}$, which is called the *positive semidefinite cone*.

The nonnegative orthant $\mathbb{R}^2_+$ and the second-order cone in $\mathbb{R}^3$ are depicted in Fig. 2.15.

Every cone is associated with a dual cone:

Definition 2.65 (*Dual cone*) Let K be a cone in $\mathbb{R}^M$, then the *dual cone* of K is denoted by K^* and defined as follows:

$$K^* := \{y \in \mathbb{R}^M \mid x^\top y \geq 0, \forall x \in K\}.$$

A 2-dimensional dual cone is illustrated in Fig. 2.16. If $K^* = K$, the cone K is called a *self-dual cone*. The nonnegative orthant $\mathbb{R}^M_+$, the second-order cone $\mathbb{L}^M$, and the positive semidefinite cone $\mathbb{S}^M_+$ are self-dual cones, which can be checked using the definitions.

If $y_1, y_2, \ldots, y_N \in K^*$ and $\lambda_1, \lambda_2, \ldots, \lambda_N \geq 0$, then $x^\top(\sum_{n=1}^N \lambda_n y_n) \geq 0$, so K^* is closed under conic combination and thus it is a convex cone. By the definition, it is also clear that K^* is closed. Hence, we have the following result:

Proposition 2.29 *Let K be a cone in $\mathbb{R}^M$, then its dual cone K^* is a closed convex cone in $\mathbb{R}^M$.*

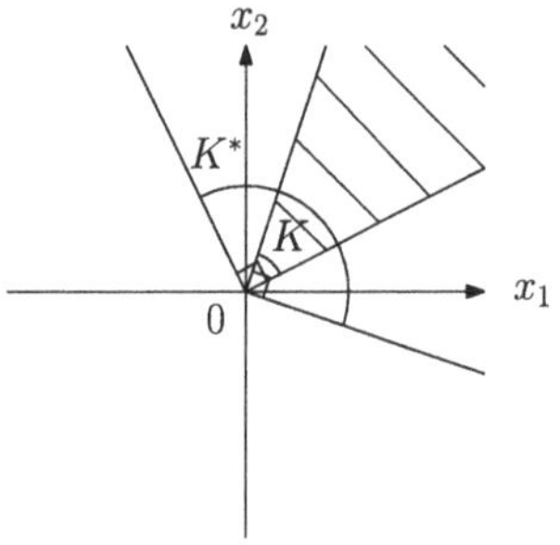

Fig. 2.16 Assume that K is a proper cone in $\mathbb{R}^2$. Its dual cone K^* consists of all the points whose inner product with the points in K is nonnegative. K^* is also a proper cone in $\mathbb{R}^2$

The dual cone K^* of K also has its own dual cone K^{**}. By definition,

$$K^{**} = \{x \in \mathbb{R}^M \mid y^\top x \geq 0, \forall y \in K^*\} \supset K ,$$

because $x^\top y \geq 0, \forall x \in K, y \in K^*$. Similar to the case of conjugate functions, $K^{**} = K$ holds when K satisfies some conditions. By Proposition 2.29, we know that K^{**} must be closed and convex. Conversely, when K is closed and convex, assume $x_0 \notin K$. Then by the hyperplane separation theorem in Corollary 2.1, there is a vector $a \in \mathbb{R}^M$ so that $a^\top x_0 > \sup_{x \in K} a^\top x$. Since K is a cone, we have $a^\top x \leq 0, \forall x \in K$ and $a^\top x_0 > 0$. Therefore, $-a \in K^*$ and $x_0 \notin K^{**}$. Hence, $K^{**} \subset K$, and thus $K^{**} = K$. To sum up, we have the following conclusions:

Proposition 2.30 *Let K be a cone in $\mathbb{R}^M$, then $K \subset K^{**}$. Moreover, $K^{**} = K$ if and only if K is closed and convex. In particular, if K is a proper cone, then $K^{**} = K$.*

There is another important property of the dual cones of proper cones:

Proposition 2.31 *Dual cones of proper cones are proper.*

Proof Assume K is a proper cone in $\mathbb{R}^M$. By Proposition 2.29, we only need to show that K^* is solid and pointed.

Assume $0 \neq y \in K^*$. Because K is solid, there exists $x_0 \in \text{int}(K)$, which implies that there is a small number $\varepsilon > 0$ such that $x_0 - \varepsilon y \in K$. Therefore, $(x_0 - \varepsilon y)^\top y \geq 0$, and thus $x_0^\top y > 0$. Hence, $-y \notin K^*$, which shows that K^* is pointed.

Suppose K^* is not solid, then because it is convex, its affine hull $\text{aff}(K^*)$ has a dimension smaller than M. Therefore, K^{**} includes the orthogonal complement of $\text{aff}(K^*)$ and hence $K = K^{**}$ is not pointed, which is a contradiction. Therefore, K^* is solid and this completes the proof. □

We then introduce two useful facts about the products of cones:

Proposition 2.32 *Let $K_1, K_2, \ldots, K_N$ be cones in $\mathbb{R}^{M_1}, \mathbb{R}^{M_2}, \ldots, \mathbb{R}^{M_N}$, respectively. Then $K_1 \times K_2 \times \cdots \times K_N$ is a cone in $\mathbb{R}^{M_1+M_2+\cdots+M_N}$ and its dual cone $(K_1 \times K_2 \times \cdots \times K_N)^* = K_1^* \times K_2^* \times \cdots \times K_N^*$.*

Proof When $(x_1, x_2, \ldots, x_N) \in K_1 \times K_2 \times \cdots \times K_N$, $x_i \in K_i, \forall i$, so for any $\lambda \geq 0$, $\lambda x_i \in K_i, \forall i$, and then $\lambda(x_1, x_2, \ldots, x_N) \in K_1 \times K_2 \times \cdots \times K_N$. Therefore, $K_1 \times K_2 \times \cdots \times K_N$ is a cone.

Assume $y = (y_1, y_2, \ldots, y_N) \in (K_1 \times K_2 \times \cdots \times K_N)^*$, where $y_i \in \mathbb{R}^{M_i}$. Consider $x = (x_1, 0, \ldots, 0) \in K_1 \times K_2 \times \cdots \times K_N$. Since $y_1^\top x_1 = y^\top x \geq 0$ and $x_1 \in K_1$ is arbitrary, we have $y_1 \in K_1^*$. Similarly, $y_i \in K_i^*, \forall i$ and hence $(K_1 \times K_2 \times \cdots \times K_N)^* \subset K_1^* \times K_2^* \times \cdots \times K_N^*$.

Conversely, for $y = (y_1, y_2, \ldots, y_N) \in K_1^* \times K_2^* \times \cdots \times K_N^*$, $y^\top x = y_1^\top x_1 + y_2^\top x_2 + \cdots + y_N^\top x_N \geq 0$ for $x \in K_1 \times K_2 \times \cdots \times K_N$. This completes the proof. □

Proposition 2.33 *Assume $K_1, K_2, \ldots, K_N$ are proper cones in $\mathbb{R}^{M_1}, \mathbb{R}^{M_2}, \ldots, \mathbb{R}^{M_N}$, respectively. Then $K_1 \times K_2 \times \cdots \times K_N$ is a proper cone in $\mathbb{R}^{M_1+M_2+\cdots+M_N}$.*

Proof When $K_1, K_2, \ldots, K_N$ are proper cones, $K_1 \times K_2 \times \cdots \times K_N$ is clearly a closed convex cone. $\text{int}(K_1) \times \text{int}(K_2) \times \cdots \times \text{int}(K_N) \subset \text{int}(K_1 \times K_2 \times \cdots \times K_N)$, so $K_1 \times K_2 \times \cdots \times K_N$ is solid. To see that it is also pointed, note that if $x = (x_1, x_2, \ldots, x_N)$ and $-x$ are both in $K_1 \times K_2 \times \cdots \times K_N$, then $x_i = 0, \forall i$ and thus $x = 0$. □

By these facts about product cones, we can merge multiple constraints $x_i \in K_i, \forall i$ into one constraint $x \in K$ and the latter one is still a constraint requiring a vector to be in a cone.

2.5.2 *Generalized Inequalities and Conic Programs*

For any $x, y \in \mathbb{R}$, either $x \leq y$ or $y \leq x$ holds, which means two numbers in $\mathbb{R}$ are always comparable. In a Euclidean space $\mathbb{R}^M$ with $M > 1$, we also want to compare points. In this case, we have to define what the meaning of being "larger" is. We may not obtain an order between arbitrary two points, but a partial ordering.

Definition 2.66 (*Partial ordering*) A (non-strict) *partial ordering* $\leq$ on a set S is a relation that satisfies the following requirements:

1. (Reflexivity) $a \leq a, \forall a \in S$.
2. (Antisymmetry) If $a \leq b$ and $b \leq a$, then a is the same element as b in S.
3. (Transitivity) If $a \leq b$ and $b \leq c$, then $a \leq c$.

A *strict partial ordering* $<$ on S is a relation that satisfies:

1. (Irreflexivity) $a \nless a, \forall a \in S$.
2. (Asymmetry) If $a < b$, then $b \nless a$.
3. (Transitivity) If $a < b$ and $b < c$, then $a < c$.

The definition of partial ordering does not require that there is an order between arbitrary two elements in S. In other words, $a \npreceq b$ and $b \npreceq a$ can happen simultaneously and so do $a \nprec b$ and $b \nprec a$. Partial orderings in Euclidean spaces can be defined using proper cones.

Definition 2.67 (*Generalized inequality*) Let K be a proper cone in $\mathbb{R}^M$. Then the *generalized inequalities* associated to K are a non-strict partial ordering $\preceq_K$ and a strict partial ordering $\prec_K$ on $\mathbb{R}^M$ defined by:

$$x \preceq_K y \iff y - x \in K ,$$
$$x \prec_K y \iff y - x \in \operatorname{int}(K) ,$$

where $x, y \in \mathbb{R}^M$ are arbitrary.

To see that $\preceq_K$ is indeed a non-strict partial ordering when K is a proper cone, note that reflexivity follows from $0 \in K$, antisymmetry is because K is pointed, and transitivity is due to that K is a convex cone which is closed under conic combinations. $\prec_K$ is a strict partial ordering because $0 \notin \operatorname{int}(K)$ and that $\operatorname{int}(K) \cup \{0\}$ is a convex cone.

Proposition 2.34 (Properties of generalized inequalities) *Let K be a proper cone in $\mathbb{R}^M$, then the following properties hold for generalized inequalities $\preceq_K$ and $\prec_K$, where $x, y, z, u \in \mathbb{R}^M$ and $\lambda \in \mathbb{R}$ are arbitrary.*

1. *If $x \prec_K y$, then $x \preceq_K y$.*
2. *(Additivity of $\preceq_K$) If $x \preceq_K y$ and $z \preceq_K u$, then $x + z \preceq_K y + u$.*
3. *(Additivity of $\prec_K$) If $x \prec_K y$ and $z \preceq_K u$, then $x + z \prec_K y + u$.*
4. *(Homogeneity of $\preceq_K$) If $x \preceq_K y$ and $\lambda \geq 0$, then $\lambda x \preceq_K \lambda y$.*
5. *(Homogeneity of $\prec_K$) If $x \prec_K y$ and $\lambda > 0$, then $\lambda x \prec_K \lambda y$.*
6. *If sequences $\{x_i \mid i \in \mathbb{N}_+\}$ and $\{y_i \mid i \in \mathbb{N}_+\}$ in $\mathbb{R}^M$ satisfy that $\lim_{i\to\infty} x_i = x_0$, $\lim_{i\to\infty} y_i = y_0$, and $x_i \preceq_K y_i, \forall i \in \mathbb{N}_+$, then $x_0 \preceq_K y_0$.*
7. *If $x \prec_K y$, then there exists $\varepsilon > 0$ such that $x + z \prec_K y$ provided that $z \in \mathbb{R}^M$ and $\|z\| < \varepsilon$.*

These properties are natural generalizations from those of the usual inequalities on $\mathbb{R}$. These properties are closely related to the requirements of a proper cone. For example, additivity and homogeneity are because K is a convex cone; the last property comes from the notion of the interior. Therefore, we need proper cones to define general inequalities on $\mathbb{R}^M$.

Example 2.15 (*Generalized inequalities*) Recall the proper cones in Example 2.14. We have the following examples of generalized inequalities:

1. $\mathbb{R}_+ = [0, +\infty)$ is a proper cone in $\mathbb{R}$. The generalized inequality $\preceq_{\mathbb{R}_+}$ is exactly the usual ordering $\leq$ on $\mathbb{R}$.

2. The nonnegative orthant $\mathbb{R}_+^M$ is a proper cone in $\mathbb{R}^M$. The generalized inequality $\preceq_{\mathbb{R}_+^M}$ is the componentwise inequality on $\mathbb{R}^M$. In other words, $x \preceq_{\mathbb{R}_+^M} y$ if and only if $x_i \le y_i, i = 1, 2, \ldots, M$, where $x = (x_1, x_2, \ldots, x_M)$, $y = (y_1, y_2, \ldots, y_M) \in \mathbb{R}^M$. Similarly, $x \prec_{\mathbb{R}_+^M}$ if and only if $x_i < y_i, i = 1, 2, \ldots, M$. We write $x \le y$ and $x < y$ for componentwise inequality on $\mathbb{R}^M$.
3. The second-order cone $\mathbb{L}^{M+1} = \{(t, x) \in \mathbb{R} \times \mathbb{R}^M \mid t \ge \|x\|_2\}$ is a proper cone in $\mathbb{R}^{M+1}$. Thus, $(t, x) \preceq_{\mathbb{L}^{M+1}} (s, y)$ if and only if $s - t \ge \|y - x\|_2$, and $(t, x) \prec_{\mathbb{L}^{M+1}} (s, y)$ if and only if $s - t > \|y - x\|_2$.
4. The positive semidefinite cone $\mathbb{S}_+^M$ is a proper cone in the space of $M \times M$ real-valued symmetric matrices. For such matrices X and Y, $X \preceq_{\mathbb{S}_+^M} Y$ if and only if $Y - X$ is positive semidefinite, and $X \prec_{\mathbb{S}_+^M} Y$ if and only if $Y - X$ is positive definite. For conciseness, we denote them by $X \preceq Y$ and $X \prec Y$, respectively.

A *conic program* with a general inequality has the following form:

$$
\begin{aligned}
p^* = \min_{x \in \mathbb{R}^M} \ & c^\top x \\
\text{s.t.} \ & Ax = b \ , \ x \succeq_K 0 \ ,
\end{aligned} \tag{2.33}
$$

where $c \in \mathbb{R}^M$, $A \in \mathbb{R}^{N \times M}$, and $b \in \mathbb{R}^N$ are coefficients. We assume that the rows of A are linearly independent and K is a proper cone in $\mathbb{R}^M$. Since K is convex, the feasible region $\mathcal{D} = \{x \in K \mid Ax = b\} = K \cap \{x \in \mathbb{R}^M \mid Ax = b\}$ is convex. Thus, conic program (2.33) is convex.

Proposition 2.35 *A conic program is a convex optimization problem.*

2.5.3 Duality of Conic Programs

We have introduced Lagrange duality in Sect. 2.4.4 focusing on the constraints constructed by usual inequalities. We are about to extend it to the case of generalized inequalities.

Recall that the idea of Lagrange duality is to find a lower bound for the objective function by combining the functions in the objective and constraints. Consider conic program (2.33). The Lagrangian function $L : \mathbb{R}^M \times \mathbb{R}^M \times \mathbb{R}^N \to \mathbb{R}$ is:

$$
L(x, \lambda, \nu) = c^\top x - \lambda^\top x - \nu^\top (Ax - b) \ .
$$

When x is feasible, i.e., $x \in K$ and $Ax = b$, we hope that $L(x, \lambda, \nu) \le c^\top x$ holds, which is equivalent to $\lambda^\top x \ge 0$. Recall the definition of dual cones in Definition 2.65. So long as $\lambda \in K^*$ and $x \in K$, we have $\lambda^\top x \ge 0$. Therefore, instead of letting $\lambda \ge 0$, we require $\lambda \in K^*$, i.e., $\lambda \succeq_{K^*} 0$.

From now on, we follow the procedures of Lagrange duality in Sect. 2.4.4. The Lagrange dual function $g : \mathbb{R}^M \times \mathbb{R}^N \to [-\infty, +\infty)$ is:

$$g(\lambda, \nu) = \inf_{x \in \mathbb{R}^M} L(x, \lambda, \nu) ,$$

which is concave and $g(\lambda, \nu) \leq p^*, \forall \lambda \in K^*, \nu \in \mathbb{R}^N$. In fact, we can calculate $g(\lambda, \nu)$ as follows:

$$\begin{aligned} g(\lambda, \nu) &= \inf_{x \in \mathbb{R}^M} \{c^\top x - \lambda^\top x - \nu^\top (Ax - b)\} \\ &= b^\top \nu + \inf_{x \in \mathbb{R}^M} \{(c - \lambda - A^\top \nu)^\top x\} \\ &= \begin{cases} b^\top \nu , \text{ if } A^\top \nu + \lambda = c , \\ -\infty , \text{ if } A^\top \nu + \lambda \neq c . \end{cases} \end{aligned}$$

Then the dual problem $\max_{\lambda \in K^*, \nu \in \mathbb{R}^N} g(\lambda, \nu)$ is equivalent to:

$$\begin{aligned} d^* = \max_{\lambda, \nu} \ & b^\top \nu \\ \text{s.t. } & A^\top \nu + \lambda = c , \ \lambda \succeq_{K^*} 0 . \end{aligned} \tag{2.34}$$

Suppose that x^* and (λ^*, ν^*) are primal- and dual-feasible, respectively. In other words, $x^* \in K$, $Ax^* = b$, $\lambda^* \in K^*$, and $A^\top \nu^* + \lambda^* = c$, then

$$\begin{aligned} b^\top \nu^* &= b^\top \nu^* + (c - A^\top \nu^* - \lambda^*)^\top x^* \\ &= c^\top x^* - (Ax^* - b)^\top \nu^* - (\lambda^*)^\top x^* \leq c^\top x^* . \end{aligned}$$

Therefore, $b^\top \nu^* \leq c^\top x^*$ and the equality holds if and only if $(\lambda^*)^\top x^* = 0$.

Theorem 2.22 (Duality of conic programs) *For conic program* (2.33) *and its dual problem* (2.34), *we have:*

1. *The weak duality holds, i.e.,* $d^* \leq p^*$.
2. *If* x^* *and* (λ^*, ν^*) *are primal- and dual-feasible and the complementary slackness holds, i.e,* $(\lambda^*)^\top x^* = 0$, *then we have the strong duality* $d^* = p^*$ *and* (x^*, λ^*, ν^*) *is optimal. Conversely, if* (x^*, λ^*, ν^*) *is optimal and* $d^* = p^*$, *then* $(\lambda^*)^\top x^* = 0$.
3. *The KKT conditions are:*

$$\begin{cases} Ax = b , \ x \succeq_K 0 , \\ A^\top \nu + \lambda = c , \ \lambda \succeq_{K^*} 0 , \\ \lambda^\top x = 0 . \end{cases}$$

   If x^* *and* (λ^*, ν^*) *satisfy the KKT conditions, then* $d^* = p^*$ *and* (x^*, λ^*, ν^*) *is optimal.*
4. *The* primal–dual optimality conditions *are:*

$$\begin{cases} Ax = b , \ x \succeq_K 0 , \\ A^\top \nu + \lambda = c , \ \lambda \succeq_{K^*} 0 , \\ c^\top x = b^\top \nu . \end{cases}$$

If x^* and (λ^*, ν^*) satisfy the primal–dual optimality conditions, then $d^* = p^*$ and (x^*, λ^*, ν^*) is optimal.

We can also deduce the dual of problem (2.34). First, rewrite it into a minimizing problem:

$$\begin{aligned}\tilde{p}^* = \min_{\nu} \quad & -b^\top \nu \\ \text{s.t.}\ & c - A^\top \nu \succeq_{K^*} 0 \,.\end{aligned}$$

The Lagrangian function and the Lagrange dual function are as follows:

$$L(\nu, x) = -b^\top \nu - x^\top (c - A^\top \nu) \,.$$

$$g(x) = \inf_{\nu} L(\nu, x) = \begin{cases} -c^\top x \,, & \text{if } Ax = b \,, \\ -\infty \,, & \text{if } Ax \neq b \,. \end{cases}$$

where $x \in K^{**} = K$. Therefore, the Lagrange dual problem of problem (2.34) is:

$$\begin{aligned}\tilde{d}^* = \max_{x} \quad & -c^\top x \\ \text{s.t.}\ & Ax = b \,,\ x \succeq_K 0 \,.\end{aligned}$$

This problem is equivalent to the primal problem (2.33). Hence, for conic programs, the dual of the dual problem is the primal problem and we have $-\tilde{d}^* = p^*$. Combine it with the weak duality result, then $d^* = -\tilde{p}^* \leq -\tilde{d}^* = p^*$.

We are interested in the condition for strong duality $d^* = p^*$. Recall Slater's condition from Definition 2.54. It is natural to conjecture that $d^* = p^*$ when the problems are strictly feasible. This is true.

Theorem 2.23 (Strong duality of conic programs) *For conic program* (2.33) *and its dual program* (2.34)*, we have:*

1. *If problem* (2.33) *is strictly feasible, i.e., there exists $x_0 \in int(K)$ such that $Ax_0 = b$, then $d^* = p^*$. If furthermore $p^* > -\infty$, then d^* is attained.*
2. *If problem* (2.34) *is strictly feasible, i.e., there exists $\nu_0 \in \mathbb{R}^N$ such that $c - A^\top \nu_0 \in int(K^*)$, then $d^* = p^*$. If furthermore $d^* < +\infty$, then p^* is attained.*

The above conclusions basically are consequences of the separation properties of disjoint convex sets. We sketch the proof of the second statement. Assume that there exists $\nu_0 \in \mathbb{R}^N$ such that $c - A^\top \nu_0 \in \text{int}(K^*)$. An immediate observation is $d^* > -\infty$. If $d^* = +\infty$, then by weak duality we have $p^* = +\infty = d^*$. Therefore, we can assume d^* is finite.

First, consider the case $b = 0$. Since the dual problem (2.34) is feasible, $d^* = 0$ and we only need to prove that $p^* \leq 0$. Because $x = 0$ is feasible for the primal problem (2.33), we have $p^* \leq c^\top 0 = 0$.

Second, consider the case $c = 0$. Because $-A^\top 0 = 0 \in K^*$, we have $d^* \geq b^\top 0 = 0$. If $d^* > 0$, then there is ν^* such that $b^\top \nu^* > 0$ and $-A^\top \nu^* \in K^*$. Multiplying ν^* by large numbers, we conclude $d^* = +\infty$. Therefore, we only need to prove that when $d^* = 0$, we have $p^* = 0$ and it is attained. Note that by the primal problem (2.33), $p^* = 0$, if there is $x_0 \in K$ such that $Ax_0 = b$; otherwise, $p^* = +\infty$. Assume $p^* = +\infty$, then the closed convex sets $\{Ax - b \mid x \in K\}$ and $\{0\}$ are disjoint. By the hyperplane separation theorem, there exists $a \neq 0$ such that $a^\top(Ax - b) < 0, \forall x \in K$. Let $x = 0$, then we have $a^\top b > 0$. Let $\|x\|_2 \to +\infty$, then $a^\top Ax \leq 0, \forall x \in K$. Therefore, $-A^\top a \in K^*$ and $b^\top a > 0$, which indicates $d^* = +\infty$. Now we can conclude that $p^* = 0$ is attained if $d^* = 0$.

Finally, consider the case where $b \neq 0$ and $c \neq 0$. To make use of the hyperplane separation theorem, define a set $S := \{c - A^\top \nu \mid b^\top \nu \geq d^*\}$. Note that here the only requirement for ν is $b^\top \nu \geq d^*$ and ν is not necessary to be feasible for the dual problem (2.34). It is clear that $S \neq \varnothing$. In fact, $S \cap \text{int}(K^*) = \varnothing$. To see this, assume $c - A^\top \nu_0 \in \text{int}(K^*)$ for some $b^\top \nu_0 \geq d^*$. Then there is a small $\varepsilon > 0$ such that $c - A^\top(\nu_0 + \varepsilon b) \in K^*$ and $b^\top(\nu_0 + \varepsilon b) > b^\top \nu_0 \geq d^*$, which contradicts the optimality of d^*. Therefore, $S \cap \text{int}(K^*) = \varnothing$ and by the hyperplane separation theorem, there exists $a \in \mathbb{R}^M$ such that $a \neq 0$ and $a^\top y \geq a^\top(c - A^\top \nu)$ provided that $y \in K^*$ and $b^\top \nu \geq d^*$. Therefore, $a^\top y \geq 0, \forall y \in K^*$, i.e., $a \in (K^*)^* = K$. In addition,

$$a^\top(c - A^\top \nu) \leq 0\ ,\ \ \forall \nu\ ,\ \ \text{s.t. } b^\top \nu \geq d^*\ . \tag{2.35}$$

Because $c - A^\top \nu_0 \in \text{int}(K^*)$ and $0 \neq a \in K$, we have $a^\top(c - A^\top \nu_0) > 0$. Hence, $(Aa)^\top \nu_0 < a^\top c \leq (Aa)^\top \nu$ for any ν with $b^\top \nu \geq d^*$. Therefore, $Aa \neq 0$. Then by (2.35), there is some $\mu > 0$ such that $Aa = \mu b$. Let $x = a/\mu$, then $x \in K$ and $Ax = Aa/\mu = b$, which means x is feasible for the primal problem (2.33). The objective value $c^\top x = c^\top a/\mu \leq (Aa)^\top \nu/\mu = b^\top \nu$ whenever $b^\top \nu \geq d^*$. Since we can find ν so that $b^\top \nu = d^*$, we have $p^* \leq c^\top x \leq d^*$. Then by weak duality, $p^* = d^*$ is attained at x. This completes the proof.

When the strong duality holds and d^* is attained at ν^*, $p^* = d^* = b^\top \nu^*$. According to this equation, the optimal dual variable can reflect how the optimal value will change when the parameter b in the constraints changes. In this regard, the optimal dual variable shows the sensitivity of the optimal value with respect to constraint perturbations and it is called *shadow price* sometimes.

In the subsequent subsections, we will discuss three kinds of commonly used conic programs, i.e., LP, SOCP, and SDP. After that, the conic programs in Euclidean space will be generalized, based on which the infinite-dimensional conic programs will be introduced.

2.5.4 Linear Programs

2.5.4.1 Linear Programs and Duality

An LP problem with decision variable $x \in \mathbb{R}^M$ is a mathematical program with affine objective and constraint functions, i.e.,

$$\begin{aligned} \min_{x \in \mathbb{R}^M} \ & c^\top x + d \\ \text{s.t. } & Ax + b \le 0 \ , \ \ Gx + h = 0 \ , \end{aligned} \tag{2.36}$$

where $A \in \mathbb{R}^{N_1 \times M}, b \in \mathbb{R}^{N_1}, c \in \mathbb{R}^M, d \in \mathbb{R}, G \in \mathbb{R}^{N_2 \times M}, h \in \mathbb{R}^{N_2}$ are coefficients. Therefore, an LP problem optimizes an affine objective function over a polyhedral feasible region.

LP problems have a standard form as follows:

$$\begin{aligned} p^* := \max_{x \in \mathbb{R}^M} \ & c^\top x \\ \text{s.t. } & Ax = b \ , \ \ x \ge 0 \ , \end{aligned} \tag{2.37}$$

where $A \in \mathbb{R}^{N \times M}$, $b \in \mathbb{R}^N$, and $c \in \mathbb{R}^M$ are coefficients.

Example 2.16 (*LP standard form*) LP (2.36) can be equivalently transformed into the standard form in (2.37) by changing decision variables: Let $x = y - z$ with $y \ge 0$ and $z \ge 0$. Let $s = -Ax - b \ge 0$. Note that minimizing $c^\top x$ is equivalent to maximizing $-c^\top x$ and the constant d does not have an impact on the optimal solution. Then (2.36) is equivalent to the following program:

$$\begin{aligned} \max_{(y,z,s) \in \mathbb{R}^{2M+N_1}} \ & \begin{pmatrix} -c \\ c \\ 0 \end{pmatrix}^\top \begin{pmatrix} y \\ z \\ s \end{pmatrix} \\ \text{s.t. } & \begin{pmatrix} A & -A & I_{N_1} \\ G & -G & 0 \end{pmatrix} \begin{pmatrix} y \\ z \\ s \end{pmatrix} = \begin{pmatrix} -b \\ -h \end{pmatrix} , \ \ \begin{pmatrix} y \\ z \\ s \end{pmatrix} \ge 0 \ , \end{aligned}$$

where I_{N_1} is the $N_1 \times N_1$-dimensional identity matrix.

LP problems belong to conic programs. To see this, consider the nonnegative orthant $\mathbb{R}_+^M$. The generalized inequality corresponding to this proper cone is exactly the componentwise inequality, i.e.,

$$x = (x_1, x_2, \dots, x_M) \succeq_{\mathbb{R}_+^M} 0 \iff x_i \ge 0 \ , \ i = 1, 2, \dots, M \iff x \ge 0 \ .$$

Therefore, LP (2.37) is a conic program.

The Lagrange dual problem of LP standard form (2.37) is:

$$d^* := \min_{\nu \in \mathbb{R}^N} b^\top \nu$$
$$\text{s.t. } A^\top \nu \geq c \,. \tag{2.38}$$

As a corollary of the duality results of conic programs in Theorems 2.22 and 2.23, the weak duality, complementary slackness condition, KKT conditions, primal–dual optimality conditions, and strong duality of LP are summarized in Theorem 2.24. Note that the strong duality results here are stronger than that of a general conic program: First, the feasible region of a feasible LP always has a nonempty relative interior, so Slater's condition naturally holds in this case. Second, a feasible and bounded LP always has an optimal solution.

Theorem 2.24 (LP duality) *Consider LP problem* (2.37) *and its dual program* (2.38), *then*

1. *(Weak duality)* $p^* \leq d^*$.
2. *(Complementary slackness condition) For a pair of primal- and dual-feasible solutions* (x^*, ν^*), *if the complementary slackness condition, i.e.,* $(A^\top \nu^* - c)^\top x^* = 0$, *holds, then* $p^* = d^*$ *and* (x^*, ν^*) *is optimal. Conversely, when* (x^*, ν^*) *is optimal, we have* $(A^\top \nu^* - c)^\top x^* = 0$.
3. *(KKT conditions and primal–dual optimality conditions) The KKT conditions consist of the primal and dual feasibility constraints and the complementary slackness condition as follows:*

$$\begin{cases} Ax = b \,, \; x \geq 0 \,, \\ A^\top \nu \geq c \,, \\ (A^\top \nu - c)^\top x = 0 \,. \end{cases}$$

The primal–dual optimality conditions are similar except that the complementary slackness condition is replaced by the strong duality as follows:

$$\begin{cases} Ax = b \,, \; x \geq 0 \,, \\ A^\top \nu \geq c \,, \\ c^\top x = b^\top \nu \,. \end{cases}$$

(x^*, ν^*) *is optimal* $\iff$ *the KKT conditions are satisfied* $\iff$ *the primal–dual optimality conditions are satisfied. If this is the case, then* $p^* = d^*$.
4. *(Strong duality) The primal and dual LP problems fall into one of the following four cases:*
(a) Both LP problems are feasible and bounded. In this case, $p^* = d^*$ *is finite and there exist primal- and dual-optimal solutions* (x^*, ν^*).
(b) The primal program is unbounded and the dual program is infeasible. In this case, $p^* = d^* = +\infty$.

(c) The primal program is infeasible and the dual program is unbounded. In this case, $p^* = d^* = -\infty$.
(d) Both LP problems are infeasible. In this case, $-\infty = p^* < d^* = +\infty$.

2.5.4.2 Examples of Linear Programming Application

In the following, we introduce some applications of LP problems.

Example 2.17 (*Minimize 1-norm and minimize* ∞*-norm*) Recall the properties of 1-norm and ∞-norm in (2.1). Then for $\varnothing \neq \mathcal{X} \subset \mathbb{R}^M$,

$$\min_{x \in \mathcal{X}} \|x\|_1 = \begin{cases} \min\limits_{x \in \mathcal{X}, t} \ t \\ \text{s.t. } t \geq \|x\|_1 \end{cases} = \begin{cases} \min\limits_{x \in \mathcal{X}, t, s} \ t \\ \text{s.t. } t \geq s_1 + \cdots + s_M \ , \\ \quad - s_m \leq x_m \leq s_m \ , \ m = 1, \ldots, M \ , \end{cases}$$

and

$$\min_{x \in \mathcal{X}} \|x\|_\infty = \begin{cases} \min\limits_{x \in \mathcal{X}, t} \ t \\ \text{s.t. } t \geq \|x\|_\infty \end{cases} = \begin{cases} \min\limits_{x \in \mathcal{X}, t} \ t \\ \text{s.t. } -t \leq x_m \leq t \ , \ m = 1, \ldots, M \ . \end{cases}$$

When $\mathcal{X}$ is a polyhedron, the above problems are linear.

Example 2.18 (*Chebyshev center of a polyhedron*) A Euclidean ball is a ball defined by the 2-norm $\| \cdot \|_2$. A *Chebyshev center* of a nonempty polyhedron is the center of the largest Euclidean ball in the polyhedron. Suppose the polyhedron is:

$$P := \{x \in \mathbb{R}^M \mid a_i^\top x \leq b_i, i = 1, \ldots, N\} ,$$

where $a_i, b_i, i = 1, \ldots, N$ are coefficients. Assume the Euclidean ball is:

$$B(c, r) := \{c + y \mid \|y\|_2 \leq r\} ,$$

where c is the center and r is the radius.

Mathematically, the problem of finding the Chebyshev center can be written as follows:

$$\begin{aligned} &\max_{c \in \mathbb{R}^M, r \geq 0} \ r \\ &\quad \text{s.t. } a_i^\top x \leq b_i \ , \ i = 1, \ldots, N \ , \ \forall x \in B(c, r) \ . \end{aligned} \tag{2.39}$$

Because

$$\begin{aligned} & a_i^\top x \le b_i \ , \ \forall x \in B(c, r) \\ \iff & a_i^\top c + a_i^\top y \le b_i \ , \ \forall y \ , \ \text{s.t. } \|y\|_2 \le r \\ \iff & b_i - a_i^\top c \ge \sup_{y:\|y\|_2 \le r} a_i^\top y = r\|a_i\|_2 \ , \end{aligned}$$

problem (2.39) is equivalent to:

$$\begin{aligned} \max_{c\in\mathbb{R}^M, r\ge 0} \quad & r \\ \text{s.t. } & b_i - a_i^\top c \ge r\|a_i\|_2 \ , \ i = 1, \ldots, N \ , \end{aligned}$$

which is an LP problem.

2.5.4.3 Vertices and Extreme Rays of Polyhedron

We introduce extreme points and extreme rays of polyhedron, which are useful in studying LP problems.

Definition 2.68 (*Vertex of a polyhedron*) Let P be a polyhedron and $x \in P$. x is called a *vertex* or *extreme point* of P if $x = \lambda x_1 + (1-\lambda)x_2$ for $\lambda \in (0, 1)$ and $x_1, x_2 \in P$ implies that $x_1 = x_2 = x$.

Definition 2.69 (*Ray of a polyhedron*) Let P be a nonempty polyhedron in $\mathbb{R}^M$ and $0 \ne r \in \mathbb{R}^M$. Then r is called a *ray* of P if for any $x \in P$ and $\lambda \ge 0$, we have $x + \lambda r \in P$.

Definition 2.70 (*Extreme ray of a polyhedron*) Let P be a polyhedron and r be a ray of P. If $r = \lambda r_1 + (1-\lambda)r_2$ for $\lambda \in (0, 1)$ and rays r_1 and r_2 of P implies that $r_1 = \mu r_2$ for some $\mu > 0$, then r is called an *extreme ray* of P.

The points in a polyhedron can be represented by the vertices and extreme rays.

Theorem 2.25 *Let P be a polyhedron with vertex set $V(P)$ and extreme ray set $R(P)$, then*

$$P = \left\{ x = \sum_{x\in V(P)} \lambda_x x + \sum_{r\in R(P)} \mu_r r \ \middle| \ \begin{array}{l} \lambda_x \ge 0 \ , \ \forall x \in V(P) \ , \ \sum_{x\in V(P)} \lambda_x = 1, \\ \mu_r \ge 0 \ , \ \forall r \in R(P) \end{array} \right\} .$$

A compact polyhedron does not have rays, so it is the set of all convex combinations of the vertices, which is exactly the convex hull of the vertex set. Using the representation of a polyhedron based on vertices and extreme rays, we have the following results about LP problems:

Proposition 2.36 *Suppose $V(P)$ and $R(P)$ are the vertex set and extreme ray set of the nonempty polyhedron P, respectively. Consider the LP problem $\min_{x\in P} c^\top x$. Then the optimal value is finite if and only if there exists a vertex $x^* \in V(P)$ that is optimal. The LP problem is unbounded if and only if there exists an extreme ray $r^* \in R(P)$ such that $c^\top r^* < 0$.*

2.5.4.4 Multi-parametric Linear Programming

Assume that function $g : \mathbb{R}^N \to \mathbb{R}$ is defined by:

$$\begin{aligned} g(b) := \max_{x\in\mathbb{R}^M} \ & c^\top x \\ \text{s.t. } & Ax = b\ ,\ x \geq 0\ , \end{aligned}$$

where $b \in S \subset \mathbb{R}^N$ appears as the right-hand-side parameter of the LP constraint, and the optimal value of the LP problem is used as the value of the function at b. The dual LP problem is:

$$\begin{aligned} \min_{\nu\in\mathbb{R}^N} \ & b^\top \nu \\ \text{s.t. } & A^\top \nu \geq c\ . \end{aligned}$$

We assume that the dual feasible set $P := \{\nu \in \mathbb{R}^N \mid A^\top \nu \geq c\}$ is nonempty and the dual problem has a finite optimal value for any $b \in S$, then the strong duality holds. Thus,

$$g(b) = \min_{\nu : A^\top \nu \geq c} b^\top \nu\ ,\ \forall b \in S\ ,$$

is the pointwise infimum of a set of affine functions $b^\top \nu$, where ν satisfies $A^\top \nu \geq c$. Hence, g is a concave function.

By Proposition 2.36, the optimal value can be achieved at the vertices $\nu_1, \ldots, \nu_K$ of the polyhedral P. Therefore, we have $b^\top \nu = \sum_{k=1}^K \lambda_k b^\top \nu_k \geq \min_k b^\top \nu_k, \forall b \in S$. Consequently,

$$g(b) = \min\{b^\top \nu_k \mid k = 1, \ldots, K\}\ ,\ \forall b \in S\ .$$

Therefore, the function g is the pointwise minimum of a finite number of affine functions. If we can find all the vertices of P, then g will be known. Unfortunately, calculating the vertex set of a polyhedron can be very difficult and suffers from the curse of dimensionality.

The region of b subject to $g(b) = b^\top \nu_k$ is

$$S_k := \{b \in S \mid b^\top \nu_k \le b^\top \nu_l, l = 1, \ldots, K, l \ne k\} .$$

Therefore, as long as S is polyhedral, S_k is either empty or polyhedral for any k, which means that g is piecewise affine. In each piece, there exists k such that $g(b) = b^\top \nu_k$, where the value of ν_k can reflect the sensitivity of the LP optimal value with respect to the right-hand-side parameter b in the constraint. The set S_k is a *critical region*.

2.5.5 Second-Order Cone Programs

Recall that the definition of the second-order cone $\mathbb{L}^M$ in $\mathbb{R}^M$ $(M \ge 2)$ is:

$$\mathbb{L}^M := \{(t, x) \in \mathbb{R} \times \mathbb{R}^{M-1} \mid t \ge \|x\|_2\} . \tag{2.40}$$

We first state and prove the basic property of the second-order cone.

Proposition 2.37 *The second-order cone $\mathbb{L}^M$ is a self-dual proper cone in $\mathbb{R}^M$.*

Proof To show that $\mathbb{L}^M$ is a convex cone, assume $(t_1, x_1), (t_2, x_2), \ldots, (t_N, x_N) \in \mathbb{L}^M$ and $\lambda_1, \lambda_2, \ldots, \lambda_N \ge 0$, then $t_i \ge \|x_i\|_2, \forall i$ and $\sum_{i=1}^N \lambda_i t_i \ge \sum_{i=1}^N \lambda_i \|x_i\|_2 \ge \|\sum_{i=1}^N \lambda_i x_i\|_2$, which means $\sum_{i=1}^N (t_i, x_i) \in \mathbb{L}^M$ and hence $\mathbb{L}^M$ is a convex cone. From a geometrical point of view, we know that $\mathbb{L}^M$ is closed and $\mathrm{int}(\mathbb{L}^M) = \{(t, x) \in \mathbb{R} \times \mathbb{R}^{M-1} \mid t > \|x\|_2\} \ne \varnothing$. To see that $\mathbb{L}^M$ is pointed, note that $(t, x) \in \mathbb{L}^M \implies t \ge 0$ and $t = 0 \implies x = 0$. Therefore, $(t, x), (-t, -x) \in \mathbb{L}^M$ implies $(t, x) = 0$. In conclusion, $\mathbb{L}^M$ is a proper cone.

Consider the dual cone $(\mathbb{L}^M)^* = \{(s, y) \in \mathbb{R} \times \mathbb{R}^{M-1} \mid st + y^\top x \ge 0, \forall t \ge \|x\|_2\}$. $\mathbb{L}^M \subset (\mathbb{L}^M)^*$ is clear: $st \ge \|y\|_2 \|x\|_2 \ge -y^\top x$, so long as $(s, y), (t, x) \in \mathbb{L}^M$. To prove the opposite direction, assume $(s, y) \in (\mathbb{L}^M)^*$. If $\|y\|_2 > 0$, let $(t, x) = (\|y\|_2, -y)$, then $(t, x) \in \mathbb{L}^M$ and $0 \le st + y^\top x = s\|y\|_2 - \|y\|_2^2$, which implies $s \ge \|y\|_2$ and $(s, y) \in \mathbb{L}^M$. If $\|y\|_2 = 0$, let $(t, x) = (|s|, 0) \in \mathbb{L}^M$, then $0 \le st + y^\top x = s|s|$, which implies $s \ge 0$ and $(s, y) = (s, 0) \in \mathbb{L}^M$. In conclusion, $\mathbb{L}^M = (\mathbb{L}^M)^*$ is a self-dual proper cone. □

SOCP problems are conic programs with the product cone of second-order cones:

$$\begin{aligned} p^* = \min_{x \in \mathbb{R}^M} \ & c^\top x \\ \text{s.t. } & Ax = b\ ,\ x \succeq_K 0\ , \end{aligned} \tag{2.41}$$

where $A \in \mathbb{R}^{N \times M}$, $b \in \mathbb{R}^N$, and $c \in \mathbb{R}^M$ are coefficients. The cone

$$K := \mathbb{L}^{M_1} \times \mathbb{L}^{M_2} \times \cdots \times \mathbb{L}^{M_I} ,$$

is the product of some second-order cones, and $M = M_1 + M_2 + \cdots + M_I$.

Note that second-order cones are proper and self-dual. According to Propositions 2.32 and 2.33, K is a proper cone and $K^* = (\mathbb{L}^{M_1})^* \times (\mathbb{L}^{M_2})^* \times \cdots \times (\mathbb{L}^{M_I})^* = \mathbb{L}^{M_1} \times \mathbb{L}^{M_2} \times \cdots \times \mathbb{L}^{M_I} = K$. Therefore, the SOCP problem (2.41) is indeed a conic program and its dual program is:

$$\begin{aligned} d^* = \max_{\nu \in \mathbb{R}^N} \ & b^\top \nu \\ \text{s.t. } & c - A^\top \nu \succeq_K 0 . \end{aligned} \tag{2.42}$$

Partition A and c as follows:

$$A = (\, A_1 \, \ldots \, A_I \,) \ , \ c = \begin{pmatrix} c_1 \\ \vdots \\ c_I \end{pmatrix} ,$$

with $A_i \in \mathbb{R}^{N \times M_i}$, $c_i \in \mathbb{R}^{M_i}$, $i = 1, 2, \ldots, I$. Then $c - A^\top \nu \succeq_K 0$ is equivalent to:

$$c_i - A_i^\top \nu \in \mathbb{L}^{M_i} \ , \ i = 1, 2, \ldots, I \ .$$

Further partition A_i and c_i into:

$$A_i = (\, p_i \ D_i \,) \ , \ c_i = \begin{pmatrix} q_i \\ d_i \end{pmatrix} ,$$

with $p_i \in \mathbb{R}^N$, $D_i \in \mathbb{R}^{N \times (M_i - 1)}$, $q_i \in \mathbb{R}$, and $d_i \in \mathbb{R}^{M_i - 1}$. Then by the definition of second-order cones, $c_i - A_i^\top \nu \in \mathbb{L}^{M_i}$ if and only if $q_i - p_i^\top \nu \geq \|d_i - D_i^\top \nu\|_2$, $i = 1, 2, \ldots, I$. Therefore, the SOCP problem (2.42) is equivalent to:

$$\begin{aligned} \max_{\nu \in \mathbb{R}^N} \ & b^\top \nu \\ \text{s.t. } & q_i - p_i^\top \nu \geq \|d_i - D_i^\top \nu\|_2 \ , \ i = 1, 2, \ldots, I \ . \end{aligned}$$

Similarly, the primal program (2.41) can also be partitioned and written as follows:

$$\begin{aligned} \min_{t_i \in \mathbb{R},\, y_i \in \mathbb{R}^{M_i - 1}, \forall i} \ & \sum_{i=1}^{I} (q_i t_i + d_i^\top y_i) \\ \text{s.t. } & \sum_{i=1}^{I} (p_i t_i + D_i y_i) = b \ , \ t_i \geq \|y_i\|_2 \ , \ i = 1, 2, \ldots, I \ , \end{aligned}$$

where

$$x_i = \begin{pmatrix} t_i \\ y_i \end{pmatrix} , \quad i = 1, 2, \ldots, I , \quad x = \begin{pmatrix} x_1 \\ \vdots \\ x_I \end{pmatrix} .$$

Before introducing the complementary slackness conditions of SOCP problems, we first investigate the meaning of $x, \lambda \in \mathbb{L}^M$ and $x^\top \lambda = 0$.

Proposition 2.38 *Suppose* $x = (t, y), \lambda = (s, \mu) \in \mathbb{L}^M$ *with* $t, s \in \mathbb{R}$ *and* $y, \mu \in \mathbb{R}^{M-1}$, *then*

$$\begin{aligned} x^\top \lambda = 0 &\iff t = \|y\|_2 , \ t\mu + sy = 0 \\ &\iff s = \|\mu\|_2 , \ t\mu + sy = 0 . \end{aligned}$$

Proof Assume $x^\top \lambda = 0$. Because $x, \lambda \in \mathbb{L}^M$ and $(\mathbb{L}^M)^* = \mathbb{L}^M$, we must have $x, \lambda \in \mathbb{L}^M \setminus \text{int}(\mathbb{L}^M)$, which means $t = \|y\|_2$ and $s = \|\mu\|_2$. Therefore, $0 = x^\top \lambda = \|y\|_2 \|\mu\|_2 + y^\top \mu$. By Cauchy-Schwarz inequality, y and μ must be linearly dependent. In other words, there exist $a, b \in \mathbb{R}$ such that $ay + b\mu = 0$ with $a^2 + b^2 \neq 0$. Without loss of generality, assume $a \neq 0$, then $y = -b\mu/a$ with $-b/a \leq 0$. Therefore, $\|y\|_2 = b\|\mu\|_2/a$ and $t\mu + sy = \|y\|_2 \mu + \|\mu\|_2 y = b\|\mu\|_2 \mu/a - b\|\mu\|_2 \mu/a = 0$. Conversely, assume $t = \|y\|_2$, $s = \|\mu\|_2$, and $t\mu + sy = 0$. When $s = 0$, we have $\mu = 0$ and $x^\top \lambda = 0$. When $s \neq 0$, $x^\top \lambda = ts + y^\top \mu = ts - t\mu^\top \mu / s = t(s^2 - \mu^\top \mu) = 0$. Therefore, $x^\top \lambda = 0$ if and only if $s = \|\mu\|_2$ and $t\mu + sy = 0$. □

As a special case of the conic program, SOCP also has duality properties, which directly follow from Theorems 2.22 and 2.23.

Theorem 2.26 (SOCP duality) *The SOCP problem* (2.41) *and its dual program* (2.42) *satisfies:*

1. *(Weak duality)* $d^* \leq p^*$.
2. *(Complementary slackness condition, KKT conditions, and primal–dual optimality conditions) The complementary slackness condition is* $(c - A^\top \nu)^\top x = 0$. *In addition,*

$$\begin{aligned} & (c - A^\top \nu)^\top x = 0 \\ \iff & (c_i - A_i^\top \nu)^\top x_i = 0 , \ i = 1, 2, \ldots, I \\ \iff & \begin{cases} q_i - p_i^\top \nu = \|d_i - D_i^\top \nu\|_2 , \ i = 1, 2, \ldots, I \\ (q_i - p_i^\top \nu) y_i + t_i (d_i - D_i^\top \nu) = 0 , \ i = 1, 2, \ldots, I \end{cases} \\ \iff & \begin{cases} t_i = \|y_i\|_2 , \ i = 1, 2, \ldots, I , \\ (q_i - p_i^\top \nu) y_i + t_i (d_i - D_i^\top \nu) = 0 , \ i = 1, 2, \ldots, I . \end{cases} \end{aligned}$$

The KKT conditions are:

$$\begin{cases} Ax = b \ , \ x \succeq_K 0 \ , \\ c - A^\top \nu \succeq_K 0 \ , \\ (c - A^\top \nu)^\top x = 0 \ . \end{cases}$$

The primal–dual optimality conditions are:

$$\begin{cases} Ax = b \ , \ x \succeq_K 0 \ , \\ c - A^\top \nu \succeq_K 0 \ , \\ c^\top x = b^\top \nu \ . \end{cases}$$

If (x^*, ν^*) *satisfies the KKT conditions or the primal–dual optimality conditions, then they are optimal solutions, and* $d^* = p^*$.

3. *(Strong duality)*
   (a) If primal program (2.41) *is strictly feasible, i.e.,*

$$\exists \tilde{t}_i \in \mathbb{R} \ , \ \tilde{y}_i \in \mathbb{R}^{M_i - 1} \ , \ i = 1, 2, \ldots, I \ ,$$

$$\text{s.t.} \ \sum_{i=1}^{I} (p_i \tilde{t}_i + D_i \tilde{y}_i) = b \ , \ \tilde{t}_i > \|\tilde{y}_i\|_2 \ , \ i = 1, 2, \ldots, I \ ,$$

   then $d^* = p^*$. *If furthermore* $p^* > -\infty$, *then* d^* *is attained.*
   (b) If dual program (2.42) *is strictly feasible, i.e.,*

$$\exists \tilde{\nu} \in \mathbb{R}^N \ , \ \text{s.t.} \ q_i - p_i^\top \tilde{\nu} > \|d_i - D_i^\top \tilde{\nu}\|_2 \ , \ i = 1, 2, \ldots, I \ ,$$

   then $d^* = p^*$. *If furthermore* $d^* < +\infty$, *then* p^* *is attained.*

SOCP can model various mathematical programming problems. To begin with, LP can be regarded as a special case of SOCP: The second-order cone $\mathbb{L}^1$ is exactly the nonnegative orthant $\mathbb{R}_+$ in $\mathbb{R}$. In addition, the second-order cone $\mathbb{L}^2 = \{(t, x) \in \mathbb{R} \times \mathbb{R} \mid t \geq |x|\}$ in $\mathbb{R}^2$ can be obtained by rotating the nonnegative orthant $\mathbb{R}_+^2$ around 0, as illustrated in Fig. 2.17. This relation can be expressed algebraically by changing variables, i.e., $t \geq |x|$ is equivalent to $y \geq 0$, $s \geq 0$, where $y = (t + x)/\sqrt{2}$ and $s = (t - x)/\sqrt{2}$. Therefore, LP belongs to SOCP. Conversely, if in the SOCP problem, the cones are all $\mathbb{L}^1$ and $\mathbb{L}^2$, then this program can be represented by an LP problem. For higher-dimensional second-order cones, the generalized inequalities $x \succeq_{\mathbb{L}^M} 0 \, (M \geq 3)$ can be approximated by a set of linear inequalities, and the approximation can be arbitrarily accurate by increasing the number of linear inequalities (Ben-Tal and Nemirovski 2001).

An important category of mathematical programming problems that can be modeled by SOCP is the convex quadratically constrained quadratic programming problems, which have the following form:

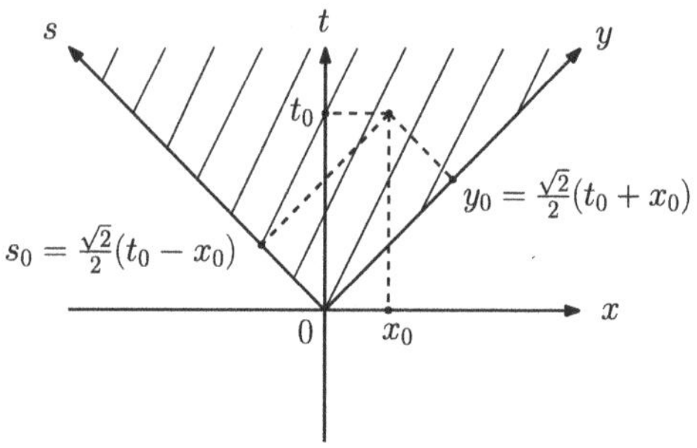

Fig. 2.17 We can obtain the second-order cone $\mathbb{L}^2$ by rotating the nonnegative orthant $\mathbb{R}^2_+$ around 0 (*shaded area*)

$$\begin{aligned}\min_{x \in \mathbb{R}^M} \; & q_0(x) := x^\top Q_0 x + a_0^\top x + \beta_0 \\ \text{s.t.} \; & q_i(x) := x^\top Q_i x + a_i^\top x + \beta_i \le 0 \, , \; i = 1, 2, \ldots, N \, ,\end{aligned}$$

where $Q_i \succeq 0, i = 0, 1, \ldots, N$ are positive semidefinite matrices. If $Q_i = 0$, $i = 0, 1, \ldots, N$, the problem degenerates to an LP problem. For the general case, first, observe that the objective function can be assumed to be linear according to the following equivalent transformation:

$$\begin{aligned}\min_{(x,u) \in \mathbb{R}^M \times \mathbb{R}} \; & \begin{pmatrix} 0 \\ 1 \end{pmatrix}^\top \begin{pmatrix} x \\ u \end{pmatrix} \\ \text{s.t.} \; & \begin{pmatrix} x \\ u \end{pmatrix}^\top \begin{pmatrix} Q_0 & 0 \\ 0 & 0 \end{pmatrix} \begin{pmatrix} x \\ u \end{pmatrix} + \begin{pmatrix} a_0 \\ -1 \end{pmatrix}^\top \begin{pmatrix} x \\ u \end{pmatrix} + \beta_0 \le 0 \, , \\ & \begin{pmatrix} x \\ u \end{pmatrix}^\top \begin{pmatrix} Q_i & 0 \\ 0 & 0 \end{pmatrix} \begin{pmatrix} x \\ u \end{pmatrix} + \begin{pmatrix} a_i \\ 0 \end{pmatrix}^\top \begin{pmatrix} x \\ u \end{pmatrix} + \beta_i \le 0 \, , \; i = 1, 2, \ldots, N \, ,\end{aligned}$$

where the square matrices are still positive semidefinite. Second, note that a positive semidefinite matrix $Q_i \in \mathbb{R}^{M \times M}$ with rank $N \le M$ can be decomposed as $Q_i = B_i^\top B_i$ for some $B_i \in \mathbb{R}^{N \times M}$ (Lay et al. 2020). Utilizing

$$a_i^\top x + \beta_i = \left(\frac{a_i^\top x + \beta_i + 1}{2} \right)^2 - \left(\frac{1 - a_i^\top x - \beta_i}{2} \right)^2 ,$$

the convex quadratic constraint

$$\begin{aligned} & q_i(x) = x^\top B_i^\top B_i x + a_i^\top x + \beta_i \le 0 \\ \iff & \left(\frac{1 - a_i^\top x - \beta_i}{2} \right)^2 \ge (B_i x)^\top (B_i x) + \left(\frac{a_i^\top x + \beta_i + 1}{2} \right)^2 \\ \iff & \begin{pmatrix} \frac{1 - a_i^\top x - \beta_i}{2} \\ B_i x \\ \frac{a_i^\top x + \beta_i + 1}{2} \end{pmatrix} \in \mathbb{L}^{N+2} \, . \end{aligned}$$

In conclusion, convex quadratically constrained quadratic programs can be written as SOCP to solve.

More detailed reviews of the SOCP modeling techniques can be found in Ben-Tal and Nemirovski (2001), Alizadeh and Goldfarb (2003).

2.5.6 Semidefinite Programs

Recall that the positive semidefinite cone is defined as:

$$\mathbb{S}_+^M := \{A \in \mathbb{S}^M \mid x^\top A x \geq 0, \forall x \in \mathbb{R}^M\} . \tag{2.43}$$

$\mathbb{S}_+^M$ is a cone in the space of $M \times M$ real symmetric matrices, i.e., $\mathbb{S}^M := \{A \in \mathbb{R}^{M\times M} \mid A^\top = A\}$. This space is "equivalent" to $\mathbb{R}^{M(M+1)/2}$ because a symmetric matrix can be uniquely determined by the elements on and above the diagonal and vice versa.

The inner product in $\mathbb{S}^M$ is defined by:

$$\langle A, B\rangle := \operatorname{tr}(A^\top B) , \ \forall A, B \in \mathbb{S}^M , \tag{2.44}$$

where $\operatorname{tr}(\cdot)$ is the *trace* operator which sums up the diagonal elements of the matrix in the parentheses. Denote the element on the ith row and jth column of matrix A by A_{ij}, then

$$\langle A, B\rangle = \operatorname{tr}(A^\top B) = \sum_{j=1}^{M}(A^\top B)_{jj} = \sum_{j=1}^{M}\sum_{i=1}^{M}(A^\top)_{ji}B_{ij} = \sum_{i=1}^{M}\sum_{j=1}^{M}A_{ij}B_{ij} ,$$

which is the sum of the element-wise product of the two matrices. Thus, $\langle A, B\rangle = \langle B, A\rangle$ and $\langle A, A\rangle \geq 0$ for any $A, B \in \mathbb{S}^M$. Moreover, the $\langle\cdot, \cdot\rangle$ relation defined in (2.44) is clearly linear in the first element, so it is indeed an inner product (over the field $\mathbb{R}$) in $\mathbb{S}^M$.

The inner product can be used to define a norm over the space $\mathbb{S}^M$:

$$\|A\|_F := \sqrt{\langle A, A\rangle} = \sqrt{\sum_{i=1}^{M}\sum_{j=1}^{M}A_{ij}^2} ,$$

which is called the *Frobenius norm*. Then the distance between matrices in $\mathbb{S}^M$ can be measured and the limit of matrices can be defined. In other words, for a sequence of matrices $A^{(1)}, A^{(2)}, \ldots, A^{(n)}, \ldots \in \mathbb{S}^M$, it converges to $A^{(0)} \in \mathbb{S}^M$ if $\lim_{n\to\infty} \|A^{(n)} - A^{(0)}\|_F = 0$. In fact, since the number of elements in an $M \times M$ matrix is finite, this convergence is equivalent to $\lim_{n\to\infty}(A^{(n)})_{ij} = (A^{(0)})_{ij}, \forall i, j = 1, 2, \ldots, M$, i.e., the element-wise convergence.

Now we focus on the properties of the positive semidefinite cone $\mathbb{S}_+^M$.

Proposition 2.39 (The interior of the positive semidefinite cone) *The interior of* $\mathbb{S}_+^M$ *in* $\mathbb{S}^M$ *is:*

$$int(\mathbb{S}_+^M) = \{A \in \mathbb{S}^M \mid x^\top Ax > 0\ ,\ \forall x \in \mathbb{R}^M \setminus \{0\}\} =: \mathbb{S}_{++}^M\ ,$$

which is the set of positive definite matrices in $\mathbb{S}^M$.

Proposition 2.40 $\mathbb{S}_+^M$ *is a proper cone in* $\mathbb{S}^M$.

Proof 1. $\mathbb{S}_+^M$ is a convex cone: Assume $A^{(1)}, \ldots, A^{(n)} \in \mathbb{S}_+^M$ and $\lambda_1, \ldots, \lambda_N \in \mathbb{R}_+$. Then $(\sum_{i=1}^N \lambda_i A^{(i)})^\top = \sum_{i=1}^N \lambda_i (A^{(i)})^\top = \sum_{i=1}^M \lambda_i A^{(i)}$ is symmetric and $x^\top(\sum_{i=1}^N \lambda_i A^{(i)})x = \sum_{i=1}^N \lambda_i x^\top A^{(i)} x \geq 0$ for any $x \in \mathbb{R}^M$.
2. $\mathbb{S}_+^M$ is closed: Assume $A^{(1)}, \ldots, A^{(n)}, \ldots \in \mathbb{S}_+^M$, $A^{(0)} \in \mathbb{S}^M$, and $\lim_{n\to\infty} A^{(n)} = A^{(0)}$. Then for any $x \in \mathbb{R}^M$, $x^\top A^{(0)} x = \lim_{n\to\infty} x^\top A^{(n)} x \geq 0$. Therefore, $A^{(0)} \in \mathbb{S}_+^M$.
3. $\mathbb{S}_+^M$ is solid: It is clear that the identity matrix $I_M \in \text{int}(\mathbb{S}_+^M)$.
4. $\mathbb{S}_+^M$ is pointed: Assume $A, -A \in \mathbb{S}_+^M$. Then for any $x \in \mathbb{R}^M$, $x^\top Ax \geq 0$ and $-x^\top Ax \geq 0$, which imply $x^\top Ax = 0$. Let x be the M-dimensional vector with the ith element being 1 and 0 elsewhere, then $0 = x^\top Ax = A_{ii}, i = 1, 2, \ldots, M$. For $i \neq j$, let x be the vector with $x_i = 1$, $x_j = -1$, and $x_k = 0$ for $k \neq i, j$. Then $0 = x^\top Ax = A_{ii} - A_{ij} - A_{ji} + A_{jj} = -2A_{ij}$, i.e., $A_{ij} = 0, \forall i \neq j$. Therefore, $A = 0$.

In conclusion, $\mathbb{S}_+^M$ is a proper cone in $\mathbb{S}^M$. □

We need an important and famous result in linear algebra to further investigate the property of $\mathbb{S}_+^M$:

Theorem 2.27 (Cholesky decomposition of positive semidefinite matrix) *Let* $A \in \mathbb{S}_+^M$, *then there exists a lower triangular matrix* L *such that* $A = LL^\top$.

Proposition 2.41 *If* $A, B \in \mathbb{S}_+^M$, *then* $\langle A, B\rangle \geq 0$.

Proof Assume the Cholesky decomposition of A is $A = LL^\top$ with a lower triangular matrix L. Then

$$\begin{aligned}\langle A, B\rangle &= \text{tr}(A^\top B) = \text{tr}(LL^\top B)\\ &= \sum_{i=1}^M \sum_{j=1}^M (LL^\top)_{ij} B_{ij} = \sum_{i=1}^M \sum_{j=1}^M B_{ij} \sum_{k=1}^M L_{ik} L_{jk}\\ &= \sum_{k=1}^M \sum_{i=1}^M \sum_{j=1}^M B_{ij} L_{ik} L_{jk} = \sum_{k=1}^M (L_{\cdot k})^\top B L_{\cdot k} \geq 0\ ,\end{aligned}$$

where $L_{\cdot k}$ denotes the kth column of L. Therefore, $\langle A, B\rangle \geq 0$. □

Proposition 2.42 *$\mathbb{S}_+^M$ is a self-dual cone in $\mathbb{S}^M$.*

Proof By Proposition 2.41, it is clear that $\mathbb{S}_+^M \subset (\mathbb{S}_+^M)^*$. To show the oppositive direction, assume $A \in \mathbb{S}^M \setminus \mathbb{S}_+^M$, then there exists $x \in \mathbb{R}^M$ such that $x^\top Ax < 0$. Since $x^\top Ax = \sum_{i=1}^M \sum_{j=1}^M A_{ij}x_ix_j$, we define a matrix $B := xx^\top$, then $B_{ij} = x_ix_j, i, j = 1, 2, \ldots, M, B^\top = B$, and $y^\top By = y^\top xx^\top y = (x^\top y)^2 \geq 0, \forall y \in \mathbb{R}^M$, which imply $B \in \mathbb{S}_+^M$. Therefore, $\operatorname{tr}(A^\top B) = \sum_{i=1}^M \sum_{j=1}^M A_{ij}B_{ij} = \sum_{i=1}^M \sum_{j=1}^M A_{ij} x_ix_j = x^\top Ax < 0$. By Proposition 2.41, we know $A \notin (\mathbb{S}_+^M)^*$. In conclusion, $(\mathbb{S}_+^M)^* = \mathbb{S}_+^M$. □

Based on the proper and self-dual cone $\mathbb{S}_+^M$, the SDP problems have the following formulation:

$$\begin{aligned}
\min_{X^{(i)} \in \mathbb{S}^{M_i}, i=1,2,\ldots,M} \quad & \sum_{i=1}^M \langle C^{(i)}, X^{(i)} \rangle \\
\text{s.t.} \quad & \sum_{i=1}^M \langle A^{(ji)}, X^{(i)} \rangle = b_j \ , \ j = 1, 2, \ldots, N \ , \\
& X^{(i)} \succeq 0 \ , \ i = 1, 2, \ldots, M \ ,
\end{aligned} \tag{2.45}$$

where coefficients $C^{(i)}$ and $A^{(ji)}$ are symmetric matrices with $i = 1, 2, \ldots, M$ and $j = 1, 2, \ldots, N$. The cone is $K = \mathbb{S}_+^{M_1} \times \mathbb{S}_+^{M_2} \times \cdots \times \mathbb{S}_+^{M_M}$ if we regard program (2.45) as a conic program like that in (2.33). The dual program is:

$$\begin{aligned}
\max_{\nu \in \mathbb{R}^N} \quad & b^\top \nu \\
\text{s.t.} \quad & C^{(i)} - \sum_{j=1}^N A^{(ji)} \nu_j \succeq 0 \ , \ i = 1, 2, \ldots, M \ .
\end{aligned} \tag{2.46}$$

In fact, $X^{(i)} \in \mathbb{S}_+^{M_i}, i = 1, 2, \ldots, M$ is equivalent to $\operatorname{diag}(X^{(1)}, X^{(2)}, \ldots, X^{(M)}) \in \mathbb{S}_+^{\sum_{i=1}^M M_i}$, where

$$\operatorname{diag}(X^{(1)}, X^{(2)}, \ldots, X^{(M)}) := \begin{pmatrix} X^{(1)} & 0 & \cdots & 0 \\ 0 & X^{(2)} & \cdots & 0 \\ \vdots & \vdots & \ddots & \vdots \\ 0 & 0 & \cdots & X^{(M)} \end{pmatrix} .$$

Note that this variable transformation is affine. Therefore, the formulation of the SDP problem in (2.41) can be equivalently transformed into the single-variable formulation as follows:

$$p^* = \min_{X \in \mathbb{S}^M} \langle C, X \rangle$$
$$\text{s.t. } \langle A^{(j)}, X \rangle = b_j \; , \; j = 1, 2, \ldots, N \; , \; X \succeq 0 \, . \tag{2.47}$$

Its dual program is:

$$d^* = \max_{\nu \in \mathbb{R}^N} b^\top \nu$$
$$\text{s.t. } C - \sum_{j=1}^{N} A^{(j)} \nu_j \succeq 0 \, . \tag{2.48}$$

The constraint $C - \sum_{j=1}^{N} A^{(j)} \nu_j \succeq 0$ is a linear matrix inequality, which means that it requires a linear combination of symmetric matrices to be positive semidefinite, where the symmetric matrices are parameters and the linear combination coefficients can be variables.

Proposition 2.43 *Assume $A, B \in \mathbb{S}_+^M$. Then*

$$\langle A, B \rangle = 0 \iff AB = 0 \iff BA = 0 \iff \frac{AB + BA}{2} = 0 \, .$$

Proof Assume the Cholesky decomposition forms of A and B are $A = LL^\top$ and $B = \tilde{L}\tilde{L}^\top$, respectively. Then

$$\begin{aligned}
\langle A, B \rangle &= \operatorname{tr}(A^\top B) = \operatorname{tr}(LL^\top \tilde{L}\tilde{L}^\top) \\
&= \sum_{i=1}^{M} \sum_{j=1}^{M} \left(\sum_{k=1}^{M} L_{ik} L_{jk} \right) \left(\sum_{l=1}^{M} \tilde{L}_{il} \tilde{L}_{jl} \right) \\
&= \sum_{i=1}^{M} \sum_{j=1}^{M} \sum_{k=1}^{M} \sum_{l=1}^{M} L_{ik} L_{jk} \tilde{L}_{il} \tilde{L}_{jl} \\
&= \sum_{k=1}^{M} \sum_{l=1}^{M} \left(\sum_{i=1}^{M} L_{ik} \tilde{L}_{il} \right) \left(\sum_{j=1}^{M} L_{jk} \tilde{L}_{jl} \right) \\
&= \sum_{k=1}^{M} \sum_{l=1}^{M} (L_{\cdot k}^\top \tilde{L}_{\cdot l})^2 \geq 0 \, ,
\end{aligned}$$

where $L_{\cdot k}$ is the kth column of L and $\tilde{L}_{\cdot l}$ is the lth column of $\tilde{L}$. Therefore, $\langle A, B \rangle = 0$ if and only if $L^\top \tilde{L} = 0$. On the one hand, when $L^\top \tilde{L} = 0$, we have $AB = LL^\top \tilde{L}\tilde{L}^\top = 0$, $BA = \tilde{L}\tilde{L}^\top LL^\top = 0$, and $(AB + BA)/2 = 0$. On the other hand, note that $\langle A, B \rangle = \operatorname{tr}(A^\top B) = \operatorname{tr}(AB)$, $\langle A, B \rangle = \operatorname{tr}(A^\top B) = \operatorname{tr}(B^\top A) = \operatorname{tr}(BA)$, and then $\langle A, B \rangle = \operatorname{tr}(AB + BA)/2$. Therefore, any one of $AB = 0$, $BA = 0$, and $(AB + BA)/2 = 0$ implies $\langle A, B \rangle = 0$. Hence, the conclusion of Proposition 2.43 holds. □

Theorem 2.28 (SDP duality) *Consider the SDP problem (2.47) and its dual program (2.48), then*

1. *(Weak duality)* $d^* \le p^*$.
2. *(Complementary slackness condition, KKT conditions, and primal–dual optimality conditions) The complementary slackness condition is* $\langle X, C - \sum_{j=1}^{N} A^{(j)} v_j \rangle = 0$. *In addition,*

$$
\begin{aligned}
&\left\langle X, C - \sum_{j=1}^{N} A^{(j)} v_j \right\rangle = 0 \\
\iff & X \left(C - \sum_{j=1}^{N} A^{(j)} v_j \right) = 0 \iff \left(C - \sum_{j=1}^{N} A^{(j)} v_j \right) X = 0 \\
\iff & \frac{X\left(C - \sum_{j=1}^{N} A^{(j)} v_j\right) + \left(C - \sum_{j=1}^{N} A^{(j)} v_j\right) X}{2} = 0 \,.
\end{aligned}
$$

The KKT conditions are:

$$
\begin{cases}
\langle A^{(j)}, X \rangle = b_j \,, \ j = 1, 2, \ldots, N \,, \ X \succeq 0 \,, \\
C - \sum_{j=1}^{N} A^{(j)} v_j \succeq 0 \,, \\
\langle X, C - \sum_{j=1}^{N} A^{(j)} v_j \rangle = 0 \,.
\end{cases}
$$

The primal–dual optimality conditions are:

$$
\begin{cases}
\langle A^{(j)}, X \rangle = b_j \,, \ j = 1, 2, \ldots, N \,, \ X \succeq 0 \,, \\
C - \sum_{j=1}^{N} A^{(j)} v_j \succeq 0 \,, \\
\langle C, X \rangle = b^\top v \,.
\end{cases}
$$

If the primal variable $X \in S^M$ *and the dual variable* $v \in \mathbb{R}^N$ *satisfy the KKT conditions or the primal–dual optimality conditions, then* $d^* = p^*$ *and* X *and* v *are primal- and dual-optimal solutions.*
3. *(Strong duality)*
   (a) Assume that the primal program (2.45) *is strictly feasible, i.e.,*

$$
\exists \tilde{X} \in S_{++}^M \,, \ \text{s.t.}\ \langle A^{(j)}, \tilde{X} \rangle = b_j \,, \ j = 1, 2, \ldots, N \,.
$$

   Then $d^* = p^*$. *If furthermore* $p^* > -\infty$, *then* d^* *can be attained.*
   (b) Assume that the dual program (2.46) *is strictly feasible, i.e.,*

$$
\exists \tilde{v} \in \mathbb{R}^N \,, \ \text{s.t.}\ C - \sum_{j=1}^{N} A^{(j)} \tilde{v}_j \in S_{++}^M \,.
$$

Then $d^* = p^*$. *If furthermore* $d^* < +\infty$, *then* p^* *can be attained.*

Second-order cones can be represented by positive semidefinite cones: For dimension $M = 1$, $\mathbb{L}^1 = \mathbb{S}^1_+ = [0, +\infty)$. For $M > 1$, $\mathbb{L}^M = \{(t, x) \in \mathbb{R} \times \mathbb{R}^{M-1} \mid t \geq \|x\|_2\}$, and we have the following result:

Proposition 2.44 *Assume* $t \in \mathbb{R}$ *and* $x \in \mathbb{R}^{M-1}$ *with* $M > 1$. *Then*

$$\begin{pmatrix} t \\ x \end{pmatrix} \in \mathbb{L}^M \iff \begin{pmatrix} t & x^\top \\ x & tI_{M-1} \end{pmatrix} \succeq 0\,, \tag{2.49}$$

where I_{M-1} *is the* $(M-1) \times (M-1)$*-dimensional identity matrix.*

Proof To see this, let $(s, y) \in \mathbb{R} \times \mathbb{R}^{M-1}$, then

$$\begin{pmatrix} s \\ y \end{pmatrix}^\top \begin{pmatrix} t & x^\top \\ x & tI_{M-1} \end{pmatrix} \begin{pmatrix} s \\ y \end{pmatrix} = s^2 t + 2sx^\top y + ty^\top y\,.$$

Assume that the left side of (2.49) holds, then

$$\begin{aligned} & s^2 t + 2sx^\top y + ty^\top y \\ \geq\ & s^2 t - 2|s|\|x\|_2\|y\|_2 + t\|y\|_2^2 \\ \geq\ & s^2 t - 2|s|t\|y\|_2 + t\|y\|_2^2 = t(s - \|y\|_2)^2 \geq 0\,, \end{aligned}$$

so the right side holds. Conversely, assume the right side of (2.49) holds, then $s^2 t + 2sx^\top y + ty^\top y \geq 0$ for any $(s, y) \in \mathbb{R} \times \mathbb{R}^{M-1}$. Let $s = 1$ and $y = 0$, then we have $t \geq 0$. Let $y = -x$ and $s = \|x\|_2$, then we have $t \geq \|x\|_2$. □

Therefore, SOCP can be expressed by SDP.

2.5.7 *Problem of Moments*

We are interested in the following problem of moments:

$$\begin{aligned} \min_{\mathbb{P} \in \mathcal{P}} \ & \int_\Omega \phi(\omega) \mathrm{d}\mathbb{P}(\omega) \\ \text{s.t.} \ & \int_\Omega \psi_n(\omega) \mathrm{d}\mathbb{P}(\omega) = b_n\ ,\ n = 1, 2, \ldots, N'\ , \\ & \int_\Omega \psi_n(\omega) \mathrm{d}\mathbb{P}(\omega) \geq b_n\ ,\ n = N' + 1, N' + 2, \ldots, N\ , \end{aligned} \tag{2.50}$$

where $\mathcal{P}$ is a set of probability distributions on $(\Omega, \mathcal{F})$; $\phi, \psi_1, \psi_2, \ldots, \psi_N : \Omega \to \mathbb{R}$ are functions that are integrable with respect to any $\mathbb{P} \in \mathcal{P}$; $b_n, n = 1, 2, \ldots, N$ are coefficients. The decision variable of problem (2.50) is the probability distribution $\mathbb{P}$.

If Ω is a finite set $\{\omega_1, \omega_2, \ldots, \omega_M\}$ and $\mathcal{P} = \mathcal{M}(\Omega)$ is the set of all the probability distributions on Ω, then problem (2.50) comes down to an ordinary LP problem:

$$
\begin{aligned}
\min_{p} \ & \sum_{m=1}^{M} p_m \phi(\omega_m) \\
\text{s.t.} \ & \sum_{m=1}^{M} p_m \psi_n(\omega_m) = b_n : \lambda_n \ , \ n = 1, 2, \ldots, N' \ , \\
& \sum_{m=1}^{M} p_m \psi_n(\omega_m) \geq b_n : \lambda_n \ , \ n = N'+1, N'+2, \ldots, N \ , \\
& \sum_{m=1}^{M} p_m = 1 : \lambda_0 \ , \\
& p_m \geq 0 \ , \ m = 1, 2, \ldots, M \ ,
\end{aligned}
\tag{2.51}
$$

where p_m represents $\mathbb{P}(\omega_m)$ for $m = 1, 2, \ldots, M$; $\mathbb{P}$ constructed in this way is a probability distribution if and only if $p_m \geq 0, \forall m$ and $\sum_{m=1}^{M} p_m = 1$. The dual problem of (2.51) is as follows and the duality properties of LP hold for these two problems:

$$
\begin{aligned}
\max_{\lambda} \ & \lambda_0 + \sum_{n=1}^{N} b_n \lambda_n \\
\text{s.t.} \ & \lambda_0 + \sum_{n=1}^{N} \lambda_n \psi_n(\omega_m) \leq \phi(\omega_m) \ , \ m = 1, 2, \ldots, M \ , \\
& \lambda_n \geq 0 \ , \ n = N'+1, N'+2, \ldots, N \ .
\end{aligned}
\tag{2.52}
$$

For the general case, Ω may be an infinite set, then problem (2.50) can have an infinite-dimensional decision variable, whereas the number of constraints is finite. An interesting question is what the duality is in this circumstance. By observing the finite sample space case in (2.52), it is natural to conjecture that the dual problem is somewhat of the following form:

$$
\begin{aligned}
\max_{\lambda} \ & \lambda_0 + \sum_{n=1}^{N} b_n \lambda_n \\
\text{s.t.} \ & \lambda_0 + \sum_{n=1}^{N} \lambda_n \psi_n(\omega) \leq \phi(\omega) \ , \ \forall \omega \in \Omega \ , \\
& \lambda_n \geq 0 \ , \ n = N'+1, N'+2, \ldots, N \ ,
\end{aligned}
\tag{2.53}
$$

where the decision variable (λ, ν) is finite-dimensional and the number of constraints can be infinite. In order to establish this duality, we need to work in infinite-dimensional spaces. Therefore, we first extend the conic duality property in the following, whose details can be found in Shapiro (2001).

Suppose X and Y are vector spaces over $\mathbb{R}$ and they are paired with vector spaces X' and Y', which means that there are bilinear forms $\langle \cdot, \cdot \rangle : X' \times X \to \mathbb{R}$ and $\langle \cdot, \cdot \rangle : Y' \times Y \to \mathbb{R}$. C and K are convex cones in X and Y, respectively. Consider the conic programs of the form

$$\begin{aligned} p^* = \min_{x \in C} & \ \langle c, x \rangle \\ \text{s.t.} & \ Ax - b \in K \ , \end{aligned} \tag{2.54}$$

where the linear map $A : X \to Y$, $b \in Y$, and $c \in X'$ are given parameters. We further assume that for any $y^* \in Y'$, there exists a unique $x^* \in X'$ such that $\langle y^*, Ax \rangle = \langle x^*, x \rangle, \forall x \in X$. In this case, $A : X \to Y$ has an adjoint map $A^* : Y' \to X'$ determined by $\langle y^*, Ax \rangle = \langle A^* y^*, x \rangle, \forall x \in X, y^* \in Y'$.

The Lagrangian function of problem (2.54) is:

$$L(x, y^*) = \langle c, x \rangle - \langle y^*, Ax - b \rangle \ .$$

Note that when $y^* \in K^* = \{y^* \in Y' \mid \langle y^*, y \rangle \geq 0, \forall y \in K\}$ and $Ax - b \in K$, we have $\langle y^*, Ax - b \rangle \geq 0$ and $L(x, y^*) \leq \langle c, x \rangle$. Then the dual program of (2.54) should be $\max_{y^* \in K^*} \inf_{x \in C} L(x, y^*)$. By $\langle y^*, Ax \rangle = \langle A^* y^*, x \rangle$, we have $L(x, y^*) = \langle c - A^* y^*, x \rangle + \langle y^*, b \rangle$. Since $\inf_{x \in C} \langle c - A^* y^*, x \rangle$ equals 0 if $c - A^* y^* \in C^*$ and becomes $-\infty$ otherwise, the dual program is equivalent to:

$$\begin{aligned} d^* = \max_{y^* \in K^*} & \ \langle y^*, b \rangle \\ \text{s.t.} & \ c - A^* y^* \in C^* \ . \end{aligned} \tag{2.55}$$

According to the derivation process above, the weak duality $d^* \leq p^*$ holds. The strong duality is satisfied under some conditions as Proposition 2.45 states.

Proposition 2.45 (Proposition 2.9 Shapiro 2001) *Suppose that X and Y are Banach spaces, $C \subset X$ and $K \subset Y$ are closed convex cones, and $\langle c, \cdot \rangle$ and $A : X \to Y$ are continuous. In addition, the condition*

$$b \in int(A(C) - K) \ , \tag{2.56}$$

holds. Then $p^* = d^*$ in problem (2.54) *and problem* (2.55), *and the optimal solution set of the dual program* (2.55) *is nonempty and bounded.*

In Proposition 2.45, Banach space means complete normed vector space. For example, X is a Banach space if and only if there is a norm $\| \cdot \|$ on the vector space X and for any Cauchy sequence $\{x_n \mid n \in \mathbb{N}_+\}$ in X, there is a limit $x_0 \in X$. In other words, if for any $\varepsilon > 0$, there is $N \in \mathbb{N}$ such that $\|x_m - x_n\| < \varepsilon, \forall m, n > N$, then there is $x_0 \in X$ such that $\lim_{n\to\infty} \|x_n - x_0\| = 0$.

The set $A(C) - K = \{Ax - y \mid x \in C, y \in K\} \subset Y$. The conic program (2.54) is feasible if and only if there is $x \in C$ such that $Ax - b \in K$, which is equivalent to $b \in A(C) - K$. Thus, condition (2.56) is a sufficient condition for the feasibility of problem (2.54). In addition, condition (2.56) is also related to Slater's condition. Note that Slater's condition is $\exists x_0 \in C$ such that $Ax_0 - b \in \text{int}(K)$, which implies $b \in Ax_0 - \text{int}(K) \subset \text{int}(A(C) - K)$. Therefore, condition (2.56) is a relaxation of Slater's condition. Consequently, Slater's condition can also guarantee the strong duality for the problems Proposition 2.45 considers. When $\text{int}(K) \neq \varnothing$, the two conditions are equivalent.

The formulation can handle the existence of equality constraints. For example, the constraints $A_1x - b_1 = 0$, $A_2x + b_2 \in K_2$ can be equivalently written as $Ax - b \in K$ with $K = \{0\} \times K_2$ and $Ax = (A_1x, A_2x)$, where $K_2 \subset Y_2$ is a closed convex cone and $A_1 : X \to Y_1$, $A_2 : X \to Y_2$ are linear mappings. If $\text{int}(C) \neq \varnothing$ and $\text{int}(K_2) \neq \varnothing$, then condition (2.56) is equivalent to:

$$\begin{cases} A_1(X) = Y_1 \ , \\ \exists x_0 \in \text{int}(C) \ , \ \text{s.t. } A_1x_0 - b_1 = 0 \ , \ A_2x_0 - b_2 \in \text{int}(K_2) \ . \end{cases}$$

This condition can be generalized to be that the relative interior of the feasible region is nonempty.

Note that

$$p^* = \inf_{x\in C, Ax-b\in K} \langle c, x\rangle = \inf_{x\in C} \sup_{y^*\in K^*} L(x, y^*) \ ,$$

$$d^* = \sup_{y^*\in K^*, c-A^*y^*\in C^*} \langle y^*, b\rangle = \sup_{y^*\in K^*} \inf_{x\in C} L(x, y^*) \ ,$$

and $L(x, y^*) = \langle c, x\rangle - \langle y^*, Ax - b\rangle = \langle c - A^*y^*, x\rangle + \langle y^*, b\rangle$. Under the assumption of Proposition 2.45, $p^* = d^*$, which can be comprehended as the operation $\inf_{x\in C}$ and $\sup_{y^*\in K^*}$ can be interchanged for $L(x, y^*)$. Recall that $\sup_{y^*} \inf_x(\cdot) \leq \inf_x \sup_{y^*}(\cdot)$ always holds but the opposite direction does not. Proposition 2.45 provides a sufficient condition to maintain equality after interchanging the supremum and infimum operations. In this case, the pair of optimal primal and dual solutions constitute the saddle point of the Lagrangian function $L(x, y^*)$.

Now that we have the extended version of conic strong duality in Proposition 2.45, we return to the problem of moments. On the sample space $(\Omega, \mathcal{F})$, $\phi, \psi_0, \psi_1, \ldots, \psi_N : \Omega \to \mathbb{R}$ are measurable functions, where $\psi_0(\omega) = 1, \forall\omega \in \Omega$. Let $\mathcal{M}(\Omega)$ be the set of probability distributions on $(\Omega, \mathcal{F})$ such that $\phi, \psi_0, \psi_1, \ldots, \psi_N$ are integrable with respect to any $\mathbb{P} \in \mathcal{M}(\Omega)$. Let X be the vector space of signed measures generated by the probability distributions in $\mathcal{M}(\Omega)$. Let X' be the vector space generated by functions $\phi, \psi_0, \psi_1, \ldots, \psi_N$, which means that the elements of X' are the linear combinations of $\phi, \psi_0, \psi_1, \ldots, \psi_N$. In addition, X and X' are paired by

$$\langle f, \mu \rangle := \int_{\Omega} f(\omega) \mathrm{d}\mu(\omega) \ , \ \ \forall \mu \in X \ , \ \ f \in X' \ .$$

Let C be the convex cone generated by $\mathcal{M}(\Omega)$ in X, then C contains all the non-negative measures in X, and thus C is closed. Suppose $b \in \mathbb{R}^{N+1}$, $A : X \to \mathbb{R}^{N+1}$ is given by $A(\mu) = (\langle \psi_0, \mu \rangle, \langle \psi_1, \mu \rangle, \ldots, \langle \psi_N, \mu \rangle), \forall \mu \in X$, and K is a convex cone in $\mathbb{R}^{N+1}$. When $x \in \mathbb{R}^{N+1}$ and $\mu \in X$, $x^\top (A\mu) = \sum_{n=0}^{N} x_n \cdot \langle \psi_n, \mu \rangle = \langle \sum_{n=0}^{N} x_n \psi_n, \mu \rangle$. Therefore, $A^* x = \sum_{n=0}^{N} x_n \psi_n$. Moreover, $\phi - A^* x \in C^* = \{f \in X' \mid \langle f, \mu \rangle \geq 0, \forall \mu \in C\} = \{f \in X' \mid \langle f, \mu \rangle \geq 0, \forall \mu \in \mathcal{M}(\Omega)\}$ is equivalent to $\langle \phi - A^* x, \mu \rangle \geq 0, \forall \mu \in \mathcal{M}(\Omega)$. Then the conic program (2.54) and the dual program (2.55) come down to the following programs, respectively:

$$p^* = \min_{\mu \in C} \langle \phi, \mu \rangle \ , \ \text{ s.t. } A\mu - b \in K \ , \tag{2.57}$$

$$d^* = \max_{x \in K^*} b^\top x \ , \ \text{ s.t. } \left\langle \phi - \sum_{n=0}^{N} x_n \psi_n, \mu \right\rangle \geq 0 \ , \ \forall \mu \in \mathcal{M}(\Omega) \ . \tag{2.58}$$

As mentioned before, in problem (2.58), the decision variable is finite-dimensional while the number of constraints can be infinite. Therefore, problem (2.58) is a linear semi-infinite program.

To apply Proposition 2.45, we add some conditions.

Corollary 2.4 (Proposition 3.4 Shapiro 2001) *Suppose that Ω is a metric space and $\mathcal{F}$ is the Borel σ-algebra on it; ϕ is upper semi-continuous and $\psi_0, \psi_1, \ldots, \psi_N$ are continuous and bounded; $K \subset \mathbb{R}^{N+1}$ is a closed convex cone; condition* (2.56) *holds, i.e., $b \in \text{int}(A(C) - K)$ in $\mathbb{R}^{N+1}$. Then $p^* = d^*$ in problem* (2.57) *and problem* (2.58)*, and the optimal solution set of the dual program* (2.58) *is nonempty and bounded.*

Corollary 2.4 can be applied to the problems of moments in (2.50) and (2.53) with $K = \{0\}^{N'+1} \times \mathbb{R}_+^{N-N'}$ and $K^* = \mathbb{R}^{N'+1} \times \mathbb{R}_+^{N-N'}$, which indicates that under the assumption of Corollary 2.4, the optimal values of the two problems are equal. In this case, condition (2.56) $b \in \text{int}(A(C) - K)$ is equivalent to:

$$\exists \varepsilon > 0, \text{ s.t. } \forall \tilde{b} \in B(b, \varepsilon), \exists \tilde{\mu} \in C, \text{ s.t. } \begin{cases} \displaystyle\int_{\Omega} \psi_n(\omega) \mathrm{d}\tilde{\mu}(\omega) = \tilde{b}_n \ , \ \ 0 \leq n \leq N' \ , \\ \displaystyle\int_{\Omega} \psi_n(\omega) \mathrm{d}\tilde{\mu}(\omega) \geq \tilde{b}_n \ , \ \ N' + 1 \leq n \leq N \ , \end{cases}$$

where $B(b, \varepsilon)$ is the open ball centered at b with radius ε. The equal optimal values are:

$$\inf_{\mu \in C} \sup_{\lambda:\lambda_n \geq 0, N'+1 \leq n \leq N} \left\{ \int_{\Omega} \phi(\omega) \mathrm{d}\mu(\omega) - \sum_{n=0}^{N} \lambda_n \left(\int_{\Omega} \psi_n(\omega) \mathrm{d}\mu(\omega) - b_n \right) \right\}$$

$$= \sup_{\lambda:\lambda_n \geq 0, N'+1 \leq n \leq N} \inf_{\mu \in C} \left\{ \sum_{n=0}^{N} b_n \lambda_n + \int_{\Omega} \left(\phi(\omega) - \sum_{n=0}^{N} \lambda_n \psi_n(\omega) \right) \mathrm{d}\mu(\omega) \right\} ,$$

which can be regarded as interchanging the operations $\inf_{\mu}$ and $\sup_{\lambda}$.

Proposition 2.45 can also be applied to the case of optimizing multiple measures, and the corresponding results can be obtained, which are similar to those in Corollary 2.4.

There is another important result about when $\inf_x$ and $\sup_{\mathbb{P}}$ can be interchanged, where the function is not restricted to the Lagrangian function. Specifically, the following two problems are investigated:

$$\inf_{x \in S} \sup_{\mathbb{P} \in \mathcal{P}} \mathbb{E}^{\mathbb{P}}[\phi(x, \omega)] , \tag{2.59}$$

$$\sup_{\mathbb{P} \in \mathcal{P}} \inf_{x \in S} \mathbb{E}^{\mathbb{P}}[\phi(x, \omega)] . \tag{2.60}$$

Adopting the language of saddle points, we want to know whether the function $h : S \times \mathcal{P} \to \mathbb{R}$ defined by $h(x, \mathbb{P}) = \mathbb{E}^{\mathbb{P}}[\phi(x, \omega)]$ has a saddle point.

Theorem 2.29 (Stochastic saddle point theorem, Theorem 2.1 Shapiro and Kleywegt 2002) *Suppose Ω is a metric space and $\mathcal{F}$ is the Borel σ-algebra on it. $\mathcal{P}$ is a nonempty convex set of probability distributions on $(\Omega, \mathcal{F})$ such that*

1. *For any $\varepsilon > 0$, there is a set $\Omega_\varepsilon \subset \Omega$ so that $\mathbb{P}(\Omega_\varepsilon) > 1 - \varepsilon, \forall \mathbb{P} \in \mathcal{P}$;*
2. *If $\bar{\mathbb{P}}$ and $\{\bar{\mathbb{P}}_1, \bar{\mathbb{P}}_2, \ldots, \bar{\mathbb{P}}_n, \ldots\} \subset \mathcal{P}$ satisfy that $\lim_{n\to\infty} \mathbb{E}^{\bar{\mathbb{P}}_n}[f(\omega)] = \mathbb{E}^{\bar{\mathbb{P}}}[f(\omega)]$ for any integrable function f, then $\bar{\mathbb{P}} \in \mathcal{P}$.*

S is a closed and convex set in $\mathbb{R}^N$. $\phi : \mathbb{R}^N \times \Omega \to \mathbb{R}$ is a function. There is a convex neighborhood V of S such that

1. *For any $x \in V$ and $\mathbb{P} \in \mathcal{P}$, $\phi(x, \cdot)$ is measurable and integrable;*
2. *For any $\omega \in \Omega$, $\phi(\cdot, \omega)$ is convex on V;*
3. *For any $x \in V$, $\sup_{\mathbb{P} \in \mathcal{P}} \mathbb{E}^{\mathbb{P}}[\phi(x, \omega)] < +\infty$.*

x^* is an optimal solution of problem (2.59). $\phi(x, \cdot)$ is bounded and upper semi-continuous on Ω in a neighborhood of x^*. $\phi(x^*, \cdot)$ is bounded and continuous on Ω.

Then there exists $\mathbb{P}^* \in \mathcal{P}$ such that

1. *$(x^*, \mathbb{P}^*)$ is a saddle point of $h(x, \mathbb{P}) = \mathbb{E}^{\mathbb{P}}[\phi(x, \omega)]$;*
2. *$\mathbb{P}^*$ is an optimal solution of problem (2.60);*
3. *Problem (2.59) and (2.60) have the same optimal value.*

Corollary 2.5 *Suppose that $\Omega \subset \mathbb{R}^M$ is compact and $\mathcal{F}$ is its Borel σ-algebra. $\mathcal{P}$ is a nonempty convex set of probability distributions on $(\Omega, \mathcal{F})$. $S \subset \mathbb{R}^N$ is closed* *and convex.* $\phi : \mathbb{R}^N \times \Omega \to \mathbb{R}$ *is continuous and* $\phi(\cdot, \omega)$ *is convex for any* $\omega \in \Omega$. *Problem* (2.59) *has an optimal solution* x^*. *Then problems* (2.59) *and* (2.60) *have the same optimal value.*

2.6 Multi-objective Optimization

This section introduces the basic definitions and solution methods of multi-objective optimization.

2.6.1 *Dominance, Pareto Optimal Solution, and Pareto Frontier*

Sometimes there are multiple goals we want to consider in one optimization task, such as economy, security, and emissions. Such a problem can be modeled as multi-objective optimization.

Suppose that the decision variable x has feasible region $\mathcal{X}$ and there are $N > 1$ objective functions to minimize, i.e., $f_n : \mathcal{X} \to \mathbb{R}$, $n = 1, \ldots, N$. Note that maximizing a function is equivalent to minimizing its negative value, so the general multi-objective optimization problem is written as follows:

$$\min_{x \in \mathcal{X}} \{f_1(x), \ldots, f_N(x)\} . \tag{2.61}$$

Before discussing how to solve problem (2.61), we should first clearly define the "optimal solution" for a multi-objective optimization problem. In the single-objective case, if we minimize a function, then a smaller objective function is better provided that it is feasible. In other words, we can compare two feasible solutions to decide which one is better. For the multi-objective optimization, we also depend on the comparison to obtain the sense of "optimal". Since there are N objectives, the partial ordering $\leq$ on $\mathbb{R}^N$ is adopted to compare the feasible solutions, which leads to the concept of dominance as follows:

Definition 2.71 (*Dominance*) Suppose $x_1, x_2 \in \mathcal{X}$ in problem (2.61). Then x_1 is called to *dominate* x_2 if $f_n(x_1) \leq f_n(x_2)$, $n = 1, \ldots, N$ and there is n_0 such that $f_{n_0}(x_1) < f_{n_0}(x_2)$.

Intuitively, the "optimal" solution is feasible and no other feasible solution is better than it, whose formal definition is as follows:

Definition 2.72 (*Pareto optimal solution and Pareto frontier*) Suppose $x^* \in \mathcal{X}$ in problem (2.61). x^* is called a *Pareto optimal solution* if x^* is not dominated by any $x \in \mathcal{X}$. *Pareto frontier* is the set $\{(f_1(x), \ldots, f_N(x)) \mid x \text{ is Pareto optimal}\}$.

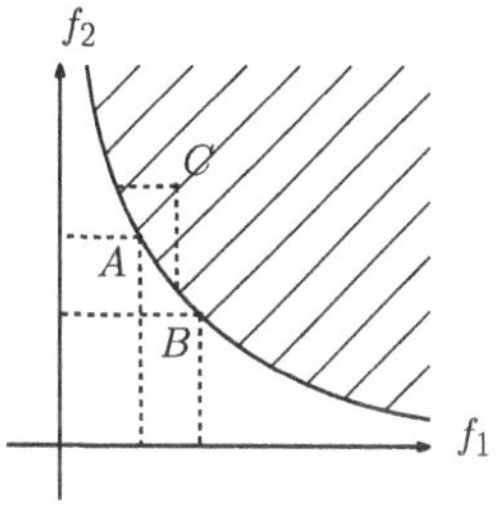

Fig. 2.18 There are two objectives. The *shaded area* is $\{(f_1(x), f_2(x)) \mid x \in \mathcal{X}\}$. A and B are on the Pareto frontier but C is not, which can be verified by Lemma 2.9. A dominates C. B does not dominate C

Here is a more direct criterion to check whether a point is on the Pareto frontier, which follows from the above definitions.

Lemma 2.9 *Suppose $x^* \in \mathcal{X}$ and $\mathcal{Y} = \{(f_1(x), \ldots, f_N(x)) \mid x \in \mathcal{X}\}$. Then $y^* := (f_1(x^*), \ldots, f_N(x^*))$ is on the Pareto frontier of problem* (2.61) *if and only if $\mathcal{Y} \cap (\{y^*\} - \mathbb{R}_+^N) = \{y^*\}$, where $\{y^*\} - \mathbb{R}_+^N = \{y^* - y \mid y \in \mathbb{R}_+^N\} = \{y \in \mathbb{R}^N \mid y \leq y^*\}$.*

Dominance, Pareto optimal solution, and Pareto frontier are illustrated in Fig. 2.18.

2.6.2 Solution Methods

The solution process of multi-objective optimization problem (2.61) tries to find either a Pareto optimal solution or the Pareto frontier. We introduce two commonly used methods, namely, the weighted sum method and the ϵ-constraint method. There are also other popular methods such as lexicographic optimization and evolutionary methods. Readers can refer to Emmerich and Deutz (2018) for more details.

2.6.2.1 Weighted Sum Method

The weighted sum method solves the following single-objective optimization problem:

$$\min_{x \in \mathcal{X}} \sum_{n=1}^{N} w_n f_n(x) , \tag{2.62}$$

where $w_n \geq 0, n = 1, \ldots, N$ are weight parameters with $\sum_{n=1}^{N} w_n = 1$.

Proposition 2.46 *Given $w_n > 0, n = 1, \ldots, N$, suppose that x^* is the optimal solution of problem* (2.62)*. Then x^* is a Pareto optimal solution of problem* (2.61).

Proposition 2.46 can be verified using Lemma 2.9. Changing the weights $w_n \geq 0$, $n = 1, \ldots, N$, we can adjust the preferences between the objectives. If we keep

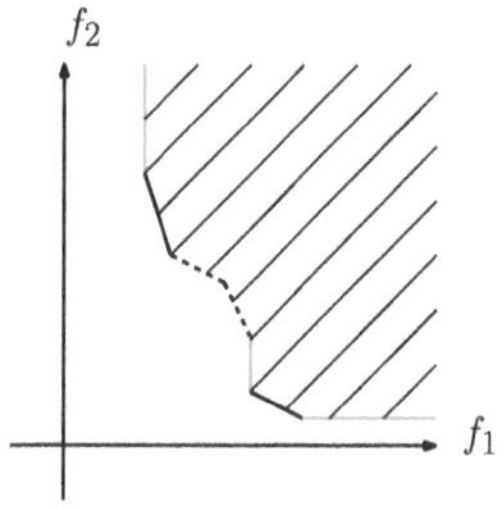

Fig. 2.19 The shaded area is $\{(f_1(x), f_2(x)) \mid x \in \mathcal{X}\}$. The *black thick solid line* is the part of the Pareto frontier that can be found by the weighted sum method, while the *black thick dashed line* denotes the part of the Pareto frontier that cannot

$\sum_{n=1}^{N} w_n = 1$, then the larger w_n, the more important the objective of minimizing $f_n(x)$. However, it should be noted that a drawback of the weighted sum method is that some points on the Pareto frontier cannot be found in this way. An example is given in Fig. 2.19.

2.6.2.2 ϵ-Constraint Method

In the ϵ-constraint method, $N - 1$ objectives are moved into constraints and a single objective is preserved. Without loss of generality, suppose that $f_N(x)$ is the preserved objective function. Given $\epsilon_1, \ldots, \epsilon_{N-1} \in \mathbb{R}$, the ϵ-constraint method solves the following single-objective problem:

$$\begin{aligned} &\min_{x \in \mathcal{X}} \; f_N(x) \\ &\text{s.t.} \; f_n(x) \leq \epsilon_n \; , \; n = 1, \ldots, N-1 \, . \end{aligned} \tag{2.63}$$

By varying the value of ϵ, different points on the Pareto frontier can be obtained. Moreover, different from the weighted sum method, the optimal solution x^* of problem (2.63) may not be Pareto optimal (there may be $x' \in \mathcal{X}$ that dominates x^* with $f_{n_0}(x') < f_{n_0}(x^*) \leq \epsilon_{n_0}$ for some n_0). Nonetheless, if x^* is a Pareto optimal solution, then it must be the optimal solution of problem (2.63) under a suitable value of ϵ. For example, $\epsilon_n = f_n(x^*), n = 1, \ldots, N-1$.

2.7 Stochastic Programming

This section is devoted to the models and solution methods of data-driven stochastic programming (SP). We first introduce the models of expectation and chance constraint in SP, then turn to two-stage SP, which has enormous applications in real problems.

2.7.1 Expectation and Chance Constraint

A SP problem minimizing the expectation has the following form:

$$\min_{x\in\mathcal{X}} \left\{f(x) + \mathbb{E}^{\mathbb{P}}[g(x,\omega)]\right\} , \tag{2.64}$$

where x is the decision variable, $\mathcal{X}$ is the feasible region, $\mathbb{P}$ is the probability distribution that the random variable ω follows, and $g(x, \omega)$ is a function of x and ω that represents loss. Because ω is random, the value of $g(x, \omega)$ is also random even if x is fixed. Therefore, we minimize its expectation in problem (2.64). The expectation can also appear in the constraints, such as:

$$\begin{aligned} \min_{x\in\mathcal{X}} \ & f(x) \\ \text{s.t. } & \mathbb{E}^{\mathbb{P}}[g_k(x,\omega)] \le \gamma_k \ , \ k = 1, \ldots, K \ , \end{aligned}$$

where $g_k(x, \omega)$ are functions containing uncertainty and γ_k, $k = 1, \ldots, K$ are threshold parameters.

Another popular type of SP contains chance constraint, which is:

$$\begin{aligned} \min_{x\in\mathcal{X}} \ & f(x) \\ \text{s.t. } & \mathbb{P}(g(x,\omega) \le 0) \ge \beta \ , \end{aligned} \tag{2.65}$$

where $g(x, \omega) \le 0$ is a desired random event and we want to ensure that the probability of this event is at least β.

In data-driven SP, the probability distribution $\mathbb{P}$ is estimated by some sample data. Suppose that $\omega_1, \ldots, \omega_N$ are N data sampled independently from the random variable ω. The probability distribution of ω can be estimated by the following discrete empirical probability distribution:

$$\mathbb{P} = \frac{1}{N}\sum_{n=1}^{N} \delta_{\omega_n} , \tag{2.66}$$

where δ_{ω_n} denotes the Dirac distribution at ω_n, i.e., the probability distribution with unit mass at ω_n. Therefore, every sample is attached with probability $1/N$. Under this empirical probability distribution, the SP problems (2.64) and (2.65) can be transformed, i.e.,

$$\min_{x\in\mathcal{X}} \left\{f(x) + \mathbb{E}^{\mathbb{P}}[g(x,\omega)]\right\} = \min_{x\in\mathcal{X}} \left\{ f(x) + \frac{1}{N}\sum_{n=1}^{N} g(x,\omega_n) \right\} , \tag{2.67}$$

and

$$\begin{cases} \min_{x \in \mathcal{X}} f(x) \\ \text{s.t. } \mathbb{P}(g(x,\omega) \le 0) \ge \beta \end{cases} = \begin{cases} \min_{x \in \mathcal{X}} f(x) \\ \text{s.t. } \frac{1}{N} \sum_{n=1}^{N} \tilde{g}(x,\omega_n) \ge \beta \end{cases},$$

where $\tilde{g}(x,\omega_n) = 1$ if $g(x,\omega_n) \le 0$ and $\tilde{g}(x,\omega_n) = 0$ if $g(x,\omega_n) > 0$. The function $\tilde{g}$ can be implemented using binary variables. Assume that the constant $M > 0$ satisfies $g(x,\omega) \le M, \forall x \in \mathcal{X}, \omega \in \Omega$. Then for fixed x and ω, $g(x,\omega) \le 0$ if and only if $z = 0$ is feasible for $g(x,\omega) \le Mz, z \in \{0,1\}$. If $g(x,\omega) > 0$, then $g(x,\omega) \le Mz$ and $z \in \{0,1\}$ force $z = 1$. Therefore, the chance-constrained SP (2.65) can be equivalently transformed into mixed-integer programming as follows:

$$\begin{cases} \min_{x \in \mathcal{X}} f(x) \\ \text{s.t. } \mathbb{P}(g(x,\omega) \le 0) \ge \beta \end{cases} = \begin{cases} \min_{x \in \mathcal{X}, z} f(x) \\ \text{s.t. } \frac{1}{N} \sum_{n=1}^{N} (1 - z_n) \ge \beta , \\ \qquad g(x,\omega_n) \le M z_n , \ z_n \in \{0,1\} , \ n = 1, \ldots, N . \end{cases}$$

2.7.2 *Two-Stage Stochastic Programming*

In two-stage SP, the function $g(x,\omega)$ is the optimal value of a second-stage optimization problem. The following is a two-stage linear SP problem:

$$\min_{x \in \mathcal{X}} \left\{ C^\top x + \mathbb{E}^{\mathbb{P}}[g(x,\omega)] \right\} , \tag{2.68}$$

where $g(x,\omega)$ is defined by:

$$\begin{aligned} g(x,\omega) := \min_{y} \ & D(\omega)^\top y \\ \text{s.t. } & A(\omega) y \ge B(\omega) x + H(\omega) . \end{aligned} \tag{2.69}$$

In (2.68), x is the first-stage decision variable, which is also called the "here-and-now" decision, because it is made before the realization of the random variable ω. $\mathcal{X}$ is a polyhedron. $C^\top x$ is the first-stage objective. In the second-stage problem (2.69), y is the second-stage decision variable and also called the "wait-and-see" decision, because it is decided to minimize the second-stage objective after the value of ω is revealed. In problem (2.69), $A(\omega)$, $B(\omega)$, $D(\omega)$, and $H(\omega)$ are the coefficients depending on ω. Two-stage SP can model a large number of practical problems with uncertainties. Such problems make the here-and-now decisions before the uncertainty realization and after the uncertainty is realized and observed, the wait-and-see decisions are made.

Under the sample-based empirical probability distribution in (2.66), using (2.67), the two-stage linear SP problem defined in (2.68) and (2.69) can be equivalently transformed into:

$$
\begin{aligned}
& \min_{x \in \mathcal{X}, y} \; C^\top x + \frac{1}{N} \sum_{n=1}^{N} D(\omega_n)^\top y_n \\
& \text{s.t. } A(\omega_n) y_n \geq B(\omega_n) x + H(\omega_n) \;, \; n = 1, \ldots, N \;,
\end{aligned}
\tag{2.70}
$$

where the decision variables include x and N copies of y, i.e., $y_1, \ldots, y_N$. y_n is the decision variable of the second-stage problem under uncertainty realization ω_n. Because $A(\omega_n)$, $B(\omega_n)$, $D(\omega_n)$, and $H(\omega_n)$ are constant, problem (2.70) is an LP problem, which can be solved by commercial solvers for LP.

Two-stage SP problems with chance constraints can be handled similarly. Suppose that the first-stage problem is (2.65) and $g(x, \omega)$ is defined by (2.69), then the equivalent reformulation is:

$$
\begin{aligned}
& \min_{x \in \mathcal{X}, y, z} \; f(x) \\
& \text{s.t. } \frac{1}{N} \sum_{n=1}^{N} (1 - z_n) \geq \beta \;, \\
& \qquad D(\omega_n)^\top y_n \leq M z_n \;, \; z_n \in \{0, 1\} \;, \; n = 1, \ldots, N \;, \\
& \qquad A(\omega_n) y_n \geq B(\omega_n) x + H(\omega_n) \;, \; n = 1, \ldots, N \;,
\end{aligned}
$$

where M is a large enough positive constant. If $\mathcal{X}$ is a polyhedron and $f(x)$ is linear in x, then the above problem is a mixed-integer linear programming (MILP) problem.

In the equivalent formulation (2.70), the number of variables and the number of constraints grow linearly as the sample number N increases. However, when N is very large, the problem scale may be too large so that (2.70) cannot be efficiently solved. *Benders decomposition* is a way to solve such problems more effectively. From now on, we assume $A(\omega) = A$ and $D(\omega) = D$ are constant coefficients that do not depend on ω. Moreover, assume that problem (2.69) is dual-feasible, i.e., $\Lambda := \{\lambda \mid A^\top \lambda = D, \lambda \geq 0\} \neq \varnothing$. Then the strong duality of LP leads to:

$$
g(x, \omega) = \left\{ \begin{aligned} & \min_{y} \; D^\top y \\ & \text{s.t. } Ay \geq B(\omega) x + H(\omega) : \lambda \end{aligned} \right. = \left\{ \begin{aligned} & \max_{\lambda} \; (B(\omega) x + H(\omega))^\top \lambda \\ & \text{s.t. } A^\top \lambda = D \;, \; \lambda \geq 0 \;. \end{aligned} \right.
$$

Recall the representation of polyhedron by vertices and extreme rays in Sect. 2.5.4. Let $V(\Lambda)$ and $R(\Lambda)$ be the vertex set and the extreme ray set of Λ, respectively. According to Proposition 2.36, for fixed $x \in \mathcal{X}$ and ω_n, $g(x, \omega_n) < +\infty$ if and only if $(B(\omega_n)x + H(\omega_n))^\top r \leq 0, \forall r \in R(\Lambda)$. Moreover, when $g(x, \omega_n) < +\infty$, $g(x, \omega_n) = \max_{\lambda \in V(\Lambda)}(B(\omega_n)x + H(\omega_n))^\top \lambda$. Since problem (2.68) is minimization, we only need to consider x such that $g(x, \omega_n) < +\infty$, $n = 1, \ldots, N$. Therefore, when $\Lambda \neq \varnothing$,

$$\min_{x \in \mathcal{X}} \left\{ C^\top x + \mathbb{E}^{\mathbb{P}} \left[\min_{y: Ay \geq B(\omega)x + H(\omega)} D^\top y \right] \right\} \tag{2.71}$$

$$= \begin{cases} \min\limits_{x \in \mathcal{X}, s} C^\top x + \dfrac{1}{N} \sum\limits_{n=1}^{N} s_n \\ \text{s.t. } (B(\omega_n)x + H(\omega_n))^\top \lambda \leq s_n \ , \ \ \forall \lambda \in V(\Lambda) \ , \ n = 1, \ldots, N \ , \\ \qquad (B(\omega_n)x + H(\omega_n))^\top r \leq 0 \ , \ \ \forall r \in R(\Lambda) \ , \ n = 1, \ldots, N \ , \end{cases} \tag{2.72}$$

where s_n is an auxiliary variable representing $g(x, \omega_n)$. $(B(\omega_n)x + H(\omega_n))^\top \lambda \leq s_n$ is called *Benders optimality cut* and $(B(\omega_n)x + H(\omega_n))^\top r \leq 0$ is called *Benders feasibility cut*. However, it is very difficult to calculate the sets $V(\Lambda)$ and $R(\Lambda)$ and they may have enormous elements, so the formulation in (2.72) cannot be employed directly. Nonetheless, we can select the pivotal vertices and extreme rays to effectively approximate problem (2.72). We use subsets $\mathcal{V} \subset V(\Lambda)$ and $\mathcal{R} \subset R(\Lambda)$ to approximate. Find the most important vertex and extreme ray that are not in the subsets and add them to $\mathcal{V}$ and $\mathcal{R}$, respectively. Then the approximation error will become acceptable after including enough elements in $\mathcal{V}$ and $\mathcal{R}$, and therefore problem (2.71) is solved in an iterative way.

For fixed $x \in \mathcal{X}$ and ω_n, we now introduce the problems to find the most important vertex and extreme ray. In fact, if the problem $\max_{\lambda: A^\top \lambda = D, \lambda \geq 0}(B(\omega_n)x + H(\omega_n))^\top \lambda$ has a finite optimal value, then the commercial LP solver can find a vertex that achieves the optimal value according to Proposition 2.36. This vertex is the most important in the sense that it gives the value of $g(x, \omega_n)$. If the above problem is unbounded, it means that there is an extreme ray r such that $(B(\omega_n)x + H(\omega_n))^\top r > 0$. The extreme ray $r \in R(\Lambda)$ satisfies $A^\top r = 0$ and $r \geq 0$, where the two constraints define a cone. To find the most important extreme ray, we look for the ray that increases $(B(\omega_n)x + H(\omega_n))^\top r$ the most quickly. Therefore, the problem $\max_{r: A^\top r = 0, 0 \leq r \leq 1}(B(\omega_n)x + H(\omega_n))^\top r$ is solved. The constraint $r \leq 1$ is added to restrict r in a bounded region because only the direction of the ray matters. Summing up the above discussions, the algorithm of Benders decomposition for problem (2.71) is in Algorithm 2.1.

Algorithm 2.1 Benders decomposition algorithm for problem (2.71)

Input: Parameters of problem (2.71) and convergence criterion parameter $\varepsilon > 0$.
Output: Solution of problem (2.71).
1: Initiation: Let $\mathcal{V} \leftarrow \varnothing$ and $\mathcal{R} \leftarrow \varnothing$.
2: Master problem: Solve the LP problem:

$$\begin{aligned} \min_{x \in X, s} \ & C^\top x + \frac{1}{N} \sum_{n=1}^{N} s_n \\ \text{s.t.} \ & (B(\omega_n)x + H(\omega_n))^\top \lambda \le s_n \ , \ \forall \lambda \in \mathcal{V} \ , \ n = 1, \ldots, N \ , \\ & (B(\omega_n)x + H(\omega_n))^\top r \le 0 \ , \ \forall r \in \mathcal{R} \ , \ n = 1, \ldots, N \ . \end{aligned}$$

If the problem is infeasible, report infeasibility and terminate. Otherwise, let LB be the optimal value and x^* be an optimal solution.
3: Generating cuts: For $n = 1, \ldots, N$, solve the LP problem:

$$\begin{aligned} \max_{r} \ & (B(\omega_n)x^* + H(\omega_n))^\top r \\ \text{s.t.} \ & A^\top r = 0 \ , \ 0 \le r \le 1 \ , \end{aligned}$$

where 0 and 1 here are all-zero and all-one vectors, respectively. If the optimal value is positive, let r_n^* be an optimal solution, $\mathcal{R} \leftarrow \mathcal{R} \cup \{r_n^*\}$, and $s_n^* \leftarrow +\infty$. Otherwise, solve the LP problem:

$$\begin{aligned} \max_{\lambda} \ & (B(\omega_n)x^* + H(\omega_n))^\top \lambda \\ \text{s.t.} \ & A^\top \lambda = D \ , \ \lambda \ge 0 \ . \end{aligned}$$

Let s_n^* be the optimal value and λ_n^* be an optimal solution. Let $\mathcal{V} \leftarrow \mathcal{V} \cup \{\lambda_n^*\}$.
4: Convergence check: Let $UB \leftarrow C^\top x^* + \sum_{n=1}^{N} s_n^*/N$. If $UB \le LB + \varepsilon$, output solution x^* and terminate. Otherwise, turn to Step 2.

The decision variables of the master problem are x and the N-dimensional vector s. N constraints are added to the master problem in each iteration. In Step 3, the LP problems are separated into samples and therefore they can be solved in parallel. Benders decomposition algorithm is guaranteed to terminate after a finite number of steps and either detect infeasibility or output the optimal solution. Additionally, it also applies to the problem with a more complicated first-stage problem, such as integer x and a more general set X.

2.8 Robust Optimization

This section is devoted to the basic ideas of robust optimization (RO). In the context of RO, decisions are made considering the worst-case scenario. Only the variation range of the uncertainty matters and other probability distribution information is not considered. We first introduce static RO and then discuss two-stage RO.

2.8.1 Static Robust Optimization

The problem

$$\min_{x\in\mathcal{X}}\left\{f(x)+\max_{\omega\in\Omega}g(x,\omega)\right\}, \tag{2.73}$$

belongs to static RO. In problem (2.73), x is the decision variable that we want to optimize. $\mathcal{X}$ is the feasible region of x. $f:\mathcal{X}\to\mathbb{R}$ represents the part of cost depending solely on x. ω is a random variable taking values in Ω. In RO, Ω is called *uncertainty set*. $g:\mathcal{X}\times\Omega\to\mathbb{R}$ is the other part of cost that depends on both x and ω. In contrast to the expectation formulation in SP (2.64), the worst-case cost $g(x,\omega)$ under all possible values of ω is considered in (2.73). By adjusting x, the goal of RO (2.73) is to minimize the worst-case cost.

Uncertainty can also appear in the constraints of static RO. For instance,

$$\begin{aligned}&\min_{x\in\mathcal{X}}\ f(x)\\&\text{s.t. } g(x,\omega)\le 0\ ,\ \forall\omega\in\Omega\ ,\end{aligned} \tag{2.74}$$

is a static RO problem. Note that $g(x,\omega)\le 0,\forall\omega\in\Omega$ can be written as $\max_{\omega\in\Omega}g(x,\omega)\le 0$, which means that the decision variable x satisfies the requirement that the worst-case scenario of $g(x,\omega)$ is no larger than 0. Moreover, problem (2.74) can be regarded as an optimization problem of x such that for each $\omega\in\Omega$, there is a constraint for x. When Ω has infinite elements, problem (2.74) has an infinite number of constraints.

In (2.73) and (2.74), x is decided before the realization of ω and then ω takes the worst-case scenario under fixed x. This is the characteristic of static RO. In the following, we introduce two simple examples of how to solve static RO. Readers can refer to Ben-Tal (2009) for more details about the solution methods.

2.8.1.1 Example: Robust Optimization with a Finite Uncertainty Set

Suppose $\Omega=\{\omega_1,\ldots,\omega_n\}$. Then the RO problem

$$\begin{aligned}&\min_{x\in\mathcal{X}}\left\{f(x)+\max_{\omega\in\Omega}g(x,\omega)\right\}\\&=\min_{x\in\mathcal{X}}\{f(x)+\max\{g(x,\omega_1),\ldots,g(x,\omega_n)\}\}\\&=\begin{cases}\min\limits_{x\in\mathcal{X},s}\ f(x)+s\\ \text{s.t. } s\ge g(x,\omega_n)\ ,\ n=1,\ldots,N\ .\end{cases}\end{aligned} \tag{2.75}$$

Lemma 2.10 (Extend an uncertainty set to its convex hull) *If $g(x, \omega)$ is convex in ω on $conv(\Omega)$ for any $x \in \mathcal{X}$, then*

$$\min_{x\in\mathcal{X}} \left\{ f(x) + \max_{\omega\in\Omega} g(x,\omega) \right\} = \min_{x\in\mathcal{X}} \left\{ f(x) + \max_{\omega\in conv(\Omega)} g(x,\omega) \right\} .$$

Proof The left side is no larger than the right side because $\Omega \subset \mathrm{conv}(\Omega)$. To show the opposite direction, for any $\omega_0 \in \mathrm{conv}(\Omega)$, there are $\omega_i \in \Omega$, $\lambda_i \in [0, 1]$, $\sum_i \lambda_i = 1$ such that $\sum_i \lambda_i \omega_i = \omega_0$, then

$$g(x,\omega_0) = g\left(x, \sum_i \lambda_i \omega_i\right) \le \sum_i \lambda_i g(x,\omega_i) \le \sum_i \lambda_i \max_{\omega\in\Omega} g(x,\omega) = \max_{\omega\in\Omega} g(x,\omega) .$$

The conclusion follows. □

Therefore, for a compact polyhedron uncertainty set P and a function $g(x, \omega)$ convex in ω, if the vertex set $V(P)$ of P is known, then we can replace the uncertainty set with $V(P)$ and use the reformulation in (2.75) to solve the RO problem.

2.8.1.2 Example: Linear Static Robust Optimization

Suppose that $\Omega = \{\omega \mid P\omega \le Q\}$ is nonempty and bounded. $g(x, \omega) = B(x)^\top \omega$, where $B(x)$ is a coefficient vector depending on x. Then by the strong duality of LP, we have:

$$\max_{\omega: P\omega\le Q} B(x)^\top \omega = \min_{\lambda: P^\top\lambda = B(x), \lambda\ge 0} Q^\top \lambda .$$

Therefore, the static RO

$$\begin{aligned} &\min_{x\in\mathcal{X}} \left\{ f(x) + \max_{\omega: P\omega\le Q} B(x)^\top \omega \right\} \\ =&\min_{x\in\mathcal{X}} \left\{ f(x) + \min_{\lambda: P^\top\lambda = B(x), \lambda\ge 0} Q^\top \lambda \right\} \\ =&\left\{ \begin{array}{l} \min_{x\in\mathcal{X}, \lambda} \ f(x) + Q^\top \lambda \\ \text{s.t. } P^\top \lambda = B(x) \ , \ \lambda \ge 0 \ . \end{array} \right. \end{aligned}$$

When $\mathcal{X}$ is polyhedral and $f(x)$ and $B(x)$ are linear in x, the last formulation belongs to LP.

2.8.2 Two-Stage Robust Optimization

Recall from Sect. 2.7.2 that there are here-and-now decisions, uncertainty, and wait-and-see decisions in two-stage SP problems. There is a similar structure in the two-stage RO problems. For example,

$$\min_{x\in\mathcal{X}}\left\{f(x)+\max_{\omega\in\Omega}\min_{y\in\mathcal{Y}(x,\omega)}h(y;x,\omega)\right\},\tag{2.76}$$

is a two-stage RO problem, where x is the here-and-now decision variable with feasible region $\mathcal{X}$. x is decided before observing the realization of the random variable ω. The wait-and-see decision variable is y, whose feasible region $\mathcal{Y}(x,\omega)$ depends on x and ω. y is optimized to minimize $h(y;x,\omega)$, which is a function of y with parameters depending on x and ω. Because of the presence of the wait-and-see decision variable, such a problem is also called adjustable RO.

Another type of two-stage RO problem is as follows:

$$\begin{aligned}&\min_{x\in\mathcal{X}}\ f(x)\\&\ \text{s.t.}\ \min_{y\in\mathcal{Y}(x,\omega)}h(y;x,\omega)\le 0\ ,\ \forall\omega\in\Omega\ .\end{aligned}$$

Two-stage linear RO as follows can be solved by *column-and-constraint generation (C&CG) algorithm* (Zeng and Zhao 2013):

$$\min_{x\in\mathcal{X}}\left\{f(x)+\max_{\omega:P\omega\le Q}\ \min_{y:Ay\ge B\omega+H(x)}D^\top y\right\},\tag{2.77}$$

where A, B, D, P, and Q are constant and $H(x)$ is a coefficient vector depending on x. We assume $\Omega:=\{\omega\mid P\omega\le Q\}$ is nonempty and compact. Moreover, $\Lambda:=\{\lambda\mid A^\top\lambda=D,\lambda\ge 0\}$ is assumed to be nonempty.

Lemma 2.11 *The function* $g(\omega):=\min_{y:Ay\ge B\omega+H}D^\top y$ *is convex.*

Proof Assume that $Ay_1\ge B\omega_1+H$ and $Ay_2\ge B\omega_2+H$. Then for any $\lambda\in[0,1]$, we have:

$$\begin{aligned}A(\lambda y_1+(1-\lambda)y_2)&\ge B(\lambda\omega_1+(1-\lambda)\omega_2)+H\ ,\\D^\top(\lambda y_1+(1-\lambda)y_2)&=\lambda D^\top y_1+(1-\lambda)D^\top y_2\ .\end{aligned}$$

Because y_1 and y_2 can vary within the feasible regions under ω_1 and ω_2, respectively,

$$g(\lambda\omega_1+(1-\lambda)\omega_2)\le\lambda g(\omega_1)+(1-\lambda)g(\omega_2)\ ,$$

which shows that $g(\omega)$ is convex. □

By Lemma 2.11, the value of $\min_{y:Ay \geq B\omega+H(x)} D^\top y$ is convex in ω. Subsequently, we can apply Lemma 2.10 to $\max_{\omega:P\omega \leq Q} \min_{y:Ay \geq B\omega+H(x)} D^\top y$. For a fixed $x \in \mathcal{X}$, the optimal value of $\max_{\omega:P\omega \leq Q} \min_{y:Ay \geq B\omega+H(x)} D^\top y$ can be achieved at a vertex of Ω. Denote the vertex set of Ω by $V(\Omega)$, then

$$
\begin{aligned}
(2.77) &= \min_{x \in \mathcal{X}} \left\{ f(x) + \max_{\omega \in V(\Omega)} \min_{y:Ay \geq B\omega+H(x)} D^\top y \right\} \\
&= \left\{ \begin{array}{l} \min\limits_{x \in \mathcal{X}, \zeta} f(x) + \zeta \\ \text{s.t. } \zeta \geq \min\limits_{y:Ay \geq B\omega+H(x)} D^\top y \ , \ \forall \omega \in V(\Omega) \end{array} \right. \\
&= \left\{ \begin{array}{l} \min\limits_{x \in \mathcal{X}, \zeta, y} f(x) + \zeta \\ \quad \text{s.t. } \zeta \geq D^\top y_\omega \ , \ \forall \omega \in V(\Omega) \ , \\ \qquad Ay_\omega \geq B\omega + H(x) \ , \ \forall \omega \in V(\Omega) \ . \end{array} \right.
\end{aligned} \tag{2.78}
$$

For each $\omega \in V(\Omega)$, there is a variable y_ω. Similar to solving a two-stage SP problem by Benders decomposition, the set $V(\Omega)$ may be difficult to obtain and it may have enormous elements, so we approximate (2.78) by considering a subset $\mathcal{V} \subset V(\Omega)$ consisting of the important vertices of Ω. Here the important vertices are obtained by solving $\max_{\omega:P\omega \leq Q} \min_{y:Ay \geq B\omega+H(x)} D^\top y$.

Fix $x \in \mathcal{X}$. If for any $\omega \in \Omega$, there exists y such that $Ay \geq B\omega + H(x)$, then we can apply LP strong duality and KKT conditions to $\min_{y:Ay \geq B\omega+H(x)} D^\top y$, because we have assumed that $\Lambda \neq \varnothing$, i.e., this LP problem is dual-feasible. Therefore, its KKT condition is:

$$
\left\{ \begin{array}{l} Ay \geq B\omega + H(x) \ , \\ A^\top \lambda = D \ , \ \lambda \geq 0 \ , \\ \lambda^\top (Ay - B\omega - H(x)) = 0 \ . \end{array} \right.
$$

The complementarity condition $\lambda^\top (Ay - B\omega - H(x)) = 0$ is bilinear. To linearize it, we adopt the big-M method: Suppose $M > 0$ is large enough. Then

$$
\begin{aligned}
& Ay \geq B\omega + H(x) \ , \ \lambda \geq 0 \ , \ \lambda^\top (Ay - B\omega - H(x)) = 0 \\
\iff & \left\{ \begin{array}{l} \exists \text{ binary vector } z \ , \ \text{s.t.} \\ 0 \leq Ay - B\omega - H(x) \leq Mz \ , \\ 0 \leq \lambda \leq M(1 - z) \ . \end{array} \right.
\end{aligned}
$$

The reformulation contains binary variable z and linear constraints. Since the KKT conditions are equivalent to optimality for $\min_{y:Ay\geq B\omega+H(x)} D^\top y$ under the adopted assumptions, the optimality testing problem can be reformulated as follows:

$$\max_{\omega:P\omega\leq Q} \min_{y:Ay\geq B\omega+H(x)} D^\top y$$
$$= \left\{ \begin{aligned} &\max_{\omega,y,\lambda,z} D^\top y \\ &\text{s.t. } P\omega \leq Q \ , \ A^\top\lambda = D \ , \ z \text{ is binary} \ , \\ &\qquad 0 \leq Ay - B\omega - H(x) \leq Mz \ , \ 0 \leq \lambda \leq M(1-z) \ . \end{aligned} \right.$$

The above formulation is a MILP problem, which can be solved by commercial solvers.

It is possible that for some $\omega \in \Omega$, the constraint $Ay \geq B\omega + H(x)$ is infeasible. We want to find the most important ω causing this infeasibility. Therefore, consider the following feasibility testing problem:

$$\max_{\omega:P\omega\leq Q} \min_{y,s:Ay+s\geq B\omega+H(x),s\geq 0} 1^\top s \ ,$$

where s is a slack variable to make the "min" problem feasible and 1 is an all-one vector. Since $s \geq 0$, the optimal value of the "min" problem is always nonnegative. Moreover, the optimal value equals 0 if and only if there exists y such that $Ay \geq B\omega + H(x)$ and $s = 0$. This is the reason why this problem is called the feasibility testing problem. Since the "min" problem is feasible and bounded, KKT conditions and the big-M method can be applied to equivalently transform it into a MILP problem, similar to the optimality testing problem.

Therefore, the C&CG algorithm for the two-stage linear RO problem (2.77) is in Algorithm 2.2, where we assume that $\Omega = \{\omega \mid P\omega \leq Q\} \neq \varnothing$, $\Lambda = \{\lambda \mid A^\top\lambda = D, \lambda \geq 0\} \neq \varnothing$, and Ω is compact.

In each iteration, an LP problem and at most two MILP problems are solved, a vertex is added to the set $\mathcal{V}$, and the number of variables and constraints in the master problem increases. Since Ω is a compact polyhedron, $V(\Omega)$ is a finite set, so Algorithm 2.2 stops after a finite number of steps and can either detect infeasibility or find an optimal solution within the error bound.

Algorithm 2.2 C&CG algorithm for problem (2.77)

Input: Parameters of problem (2.77) and convergence criterion parameter $\varepsilon > 0$.
Output: Solution of problem (2.77).
1: Initiation: Let $\mathcal{V} \leftarrow \varnothing$.
2: Master problem: Solve the LP problem:

$$\begin{aligned} \min_{x \in \mathcal{X}, \zeta, y} \quad & f(x) + \zeta \\ \text{s.t.} \quad & \zeta \geq D^\top y_\omega \,, \ \forall \omega \in \mathcal{V} \,, \\ & A y_\omega \geq B\omega + H(x) \,, \ \forall \omega \in \mathcal{V} \,. \end{aligned}$$

If the problem is infeasible, report infeasibility and terminate. Otherwise, let LB be the optimal value and x^* be an optimal solution.
3: Generating cuts: Solve the feasibility testing problem:

$$\begin{aligned} \max_{\omega, y, s, \lambda, z_1, z_2} \quad & 1^\top s \\ \text{s.t.} \quad & P\omega \leq Q \,, \ A^\top \lambda = 0 \,, \ z_1 \text{ and } z_2 \text{ are binary} \,, \\ & 0 \leq Ay + s - B\omega - H(x^*) \leq M z_1 \,, \ 0 \leq \lambda \leq M(1 - z_1) \,, \\ & 0 \leq s \leq M z_2 \,, \ 0 \leq 1 - \lambda \leq M(1 - z_2) \,. \end{aligned}$$

If the optimal value is positive, let ω^* be an optimal solution, $\mathcal{V} \leftarrow \mathcal{V} \cup \{\omega^*\}$, and $\zeta^* \leftarrow +\infty$. Otherwise, solve the optimality testing problem:

$$\begin{aligned} \max_{\omega, y, \lambda, z} \quad & D^\top y \\ \text{s.t.} \quad & P\omega \leq Q \,, \ A^\top \lambda = D \,, \ z \text{ is binary} \,, \\ & 0 \leq Ay - B\omega - H(x^*) \leq M z \,, \ 0 \leq \lambda \leq M(1 - z) \,. \end{aligned}$$

Let ζ^* be the optimal value and ω^* be an optimal solution. Let $\mathcal{V} \leftarrow \mathcal{V} \cup \{\omega^*\}$.
4: Convergence check: Let $UB \leftarrow f(x^*) + \zeta^*$. If $UB \leq LB + \varepsilon$, output solution x^* and terminate. Otherwise, turn to Step 2.

2.9 Power Flow Model of Power Grid

In a power network, the power flowing through the transmission lines is governed by physical rules, i.e., Ohm's and Kirchhoff's laws, which can be considered by power flow models. In this section, we briefly introduce some frequently used power flow models in power system planning and operation optimization, including alternating current (AC) power flow, direct current (DC) power flow, and branch flow models. For a more detailed tutorial, readers can refer to Molzahn and Hiskens (2019).

Throughout this section, we consider a balanced power network and use a single-phase equivalent network representation. We now introduce some notations. Let $j = \sqrt{-1}$, the imaginary unit. Denote the set of buses by S_B. At bus $a \in S_B$, the complex voltage phasor is $\dot{V}_a = V_a \exp(j\theta_a) \in \mathbb{C}$, where $V_a \in \mathbb{R}_+$ and $\theta_a \in (-\pi, \pi]$. In addition, let $v_a = V_a^2$. The phasor angle value depends on the reference. Thus, assume $0 \in S_B$ is the reference bus and $\theta_0 = 0$. The complex power injection at bus

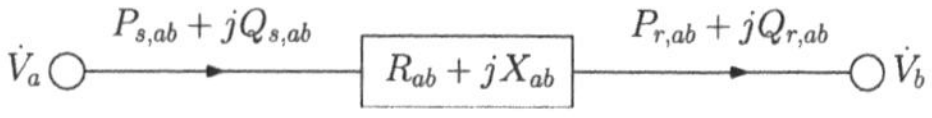

Fig. 2.20 The model of line $(a, b) \in S_L$

$a \in S_B$ is denoted by $\dot{S}_a = P_a + jQ_a \in \mathbb{C}$ with active power injection $P_a \in \mathbb{R}$ and reactive power injection $Q_a \in \mathbb{R}$.

Denote the set of transmission lines by S_L, which is a subset of $S_B \times S_B$. In addition, $(a, b) \in S_L$ represents the line from a to b. Note that in this definition, a line has a reference direction, although the power flowing in it may be negative. For simplicity, we omit the modeling of branch shunt admittance and transformer here, which can be considered without altering the essential power flow model (Molzahn and Hiskens 2019). Assume the series impedance of line $(a, b) \in S_L$ is $R_{ab} + jX_{ab} \in \mathbb{C}$ with resistance $R_{ab} \in \mathbb{R}$ and reactance $X_{ab} \in \mathbb{R}$, and thus the admittance is $G_{ab} + jB_{ab} = 1/(R_{ij} + jX_{ij}) \in \mathbb{C}$ with conductance $G_{ab} \in \mathbb{R}$ and susceptance $B_{ab} \in \mathbb{R}$. The complex current from a to b is denoted by $\dot{I}_{ab} \in \mathbb{C}$. In addition, let $l_{ab} = |\dot{I}_{ab}|^2$. The power flow sending from a to b is $P_{s,ab} + jQ_{s,ab} \in \mathbb{C}$ and the power flow receiving at b from a is $P_{r,ab} + jQ_{r,ab} \in \mathbb{C}$, where $P_{s,ab}, Q_{s,ab}, P_{r,ab}, Q_{r,ab} \in \mathbb{R}$. The model of a line is illustrated in Fig. 2.20.

We use $(\cdot)^*$ to represent the conjugate of a complex number in this section. For example, $(x + jy)^* = x - jy$ for $x, y \in \mathbb{R}$ and $\dot{V}_a^* = V_a \exp(-j\theta_a)$.

2.9.1 AC Power Flow

According to Ohm's law, the current $\dot{I}_{ab}$ in the line $(a, b) \in S_L$ satisfies:

$$\dot{I}_{ab} = \frac{\dot{V}_a - \dot{V}_b}{R_{ab} + jX_{ab}} = (G_{ab} + jB_{ab})(V_a \exp(j\theta_a) - V_b \exp(j\theta_b)) \,. \tag{2.79}$$

Moreover, the complex power can be computed by:

$$P_{s,ab} + jQ_{s,ab} = \dot{V}_a \dot{I}_{ab}^* \,, \tag{2.80a}$$

$$P_{r,ab} + jQ_{r,ab} = \dot{V}_b \dot{I}_{ab}^* \,. \tag{2.80b}$$

Combining (2.79) and (2.80), we then have $\forall (a, b) \in S_L$,

$$P_{s,ab} = G_{ab} V_a^2 - G_{ab} V_a V_b \cos(\theta_a - \theta_b) - B_{ab} V_a V_b \sin(\theta_a - \theta_b) \,, \tag{2.81a}$$

$$Q_{s,ab} = -B_{ab} V_a^2 + B_{ab} V_a V_b \cos(\theta_a - \theta_b) - G_{ab} V_a V_b \sin(\theta_a - \theta_b) \,, \tag{2.81b}$$

$$P_{r,ab} = -G_{ab} V_b^2 + G_{ab} V_a V_b \cos(\theta_a - \theta_b) - B_{ab} V_a V_b \sin(\theta_a - \theta_b) \,, \tag{2.81c}$$

$$Q_{r,ab} = B_{ab} V_b^2 - B_{ab} V_a V_b \cos(\theta_a - \theta_b) - G_{ab} V_a V_b \sin(\theta_a - \theta_b) \,. \tag{2.81d}$$

Therefore, the power loss $P_{l,ab} + jQ_{l,ab}$ in line $(a, b) \in S_L$ is given by:

$$P_{l,ab} = P_{s,ab} - P_{r,ab} = G_{ab}V_a^2 + G_{ab}V_b^2 - 2G_{ab}V_aV_b\cos(\theta_a - \theta_b)\ , \tag{2.82a}$$

$$Q_{l,ab} = Q_{s,ab} - Q_{r,ab} = -B_{ab}V_a^2 - B_{ab}V_b^2 + 2B_{ab}V_aV_b\cos(\theta_a - \theta_b)\ . \tag{2.82b}$$

According to Kirchhoff's law, we have power balance equations at each bus as follows:

$$\forall a \in S_B\ ,\ P_a + \sum_{b:(b,a)\in S_L} P_{r,ba} = \sum_{b:(a,b)\in S_L} P_{s,ab}\ , \tag{2.83a}$$

$$Q_a + \sum_{b:(b,a)\in S_L} Q_{r,ba} = \sum_{b:(a,b)\in S_L} Q_{s,ab}\ . \tag{2.83b}$$

The AC power flow model consists of equations in (2.81) and (2.83). The variables are $V_a, \theta_a, P_{s,ab}, Q_{s,ab}, P_{r,ab}, Q_{r,ab}, P_a$ and Q_a for any $a \in S_B$ and $(a, b) \in S_L$, and they are all real. Equations in (2.83) are linear, but those in (2.81) are nonlinear, which generally makes the AC power flow-based optimization problems nonlinear and nonconvex and thus prevents them from being effectively solved.

2.9.2 DC Power Flow

DC power flow is a linearized approximate power flow model and its applications in power system operation and electricity markets have a long history. DC power flow is derived from the AC power flow model in (2.81) and (2.83) by using the following approximations (Stott et al. 2009):

$$\cos(\theta_a - \theta_b) \approx 1\ ,\ \sin(\theta_a - \theta_b) \approx \theta_a - \theta_b\ ,\ \forall (a, b) \in S_L\ , \tag{2.84a}$$

$$V_a \approx 1\ ,\ -B_{ab} \approx 1/X_{ab}\ ,\ \forall a \in S_B\ ,\ (a, b) \in S_L\ , \tag{2.84b}$$

where $\cos(\theta_a - \theta_b) \approx 1$ and $\sin(\theta_a - \theta_b) \approx \theta_a - \theta_b$ hold true when $\theta_a \approx \theta_b$; $V_a \approx 1$ per unit (p.u.) holds when the voltage is close to the base value at all buses; $-B_{ab} \approx 1/X_{ab}$ if $R_{ab} \ll X_{ab}$, i.e., R_{ab} is very small compared to X_{ab}. The assumptions in (2.84) are true for many transmission networks.

Based on the assumptions in (2.84), the DC power flow model only considers the relationship between active power and voltage angle and neglects line loss, which is as follows:

$$P_{ab} = \frac{\theta_a - \theta_b}{X_{ab}}\ ,\ \forall (a, b) \in S_L\ , \tag{2.85a}$$

$$P_a + \sum_{b:(b,a)\in S_L} P_{ba} = \sum_{b:(a,b)\in S_L} P_{ab}\ ,\ \forall a \in S_B\ . \tag{2.85b}$$

In (2.85a), the power flow in line (a, b) is denoted by P_{ab}, where the loss is neglected, so $P_{ab} = P_{s,ab} = P_{r,ab}$. The power balance at buses is stipulated in (2.85b). In the DC power flow model (2.85), the variables are θ_a, P_{ab}, and P_a for any $a \in S_B$ and $(a, b) \in S_L$. Moreover, the equations are linear, which is tractable for power system optimization. The name "DC power flow" comes from the similarity of the formulation in (2.85a) to a DC circuit (the power and voltage angle in (2.85a) correspond to the current and voltage in a DC circuit), although the model applies to AC power networks.

DC power flow model has an equivalent variant called the Power Transfer Distribution Factor (PTDF) model. There is a group of linear equations in (2.85), based on which the line flow $P_{ab}, \forall(a, b) \in S_L$ can be represented as linear combinations of power injection $P_a, \forall a \in S_B$:

$$P_{ab} = \sum_{c \in S_B} F_{ab,c} P_c \ , \ \forall(a, b) \in S_L \ , \tag{2.86a}$$

$$\sum_{c \in S_B} P_c = 0 \ . \tag{2.86b}$$

In (2.86a), $F_{ab,c}$ is the PTDF from bus c to line (a, b). The PTDF parameters can be calculated using $X_{ab}, \forall(a, b) \in S_L$ and the network topology (Molzahn and Hiskens 2019). The power balance of the entire network is in (2.86b), where the network loss is neglected.

2.9.3 *Branch Flow*

We first derive the general branch flow model and then focus on the radial network case. After that, linearized formulation and second-order cone relaxation of the branch flow model are introduced.

2.9.3.1 General Branch Flow Model

Branch flow model (Baran and Wu 1989a, b) is derived from the AC power flow model in (2.81) and (2.83) by changing variables. Recall the notations $v_a = V_a^2$ and $l_{ab} = |\dot{I}_{ab}|^2$. Then using Ohm's law, we have:

$$\begin{aligned} P_{r,ab} + jQ_{r,ab} &= \dot{V}_b \dot{I}_{ab}^* \\ &= (\dot{V}_a - (R_{ab} + jX_{ab})\dot{I}_{ab})\dot{I}_{ab}^* \\ &= \dot{V}_a \dot{I}_{ab}^* - (R_{ab} + jX_{ab})\dot{I}_{ab}\dot{I}_{ab}^* \\ &= P_{s,ab} + jQ_{s,ab} - (R_{ab} + jX_{ab}) l_{ab} \ , \ \forall(a, b) \in S_L \ . \end{aligned} \tag{2.87}$$

$$\begin{aligned} v_b &= \dot{V}_b \dot{V}_b^* \\ &= (\dot{V}_a - (R_{ab} + jX_{ab})\dot{I}_{ab})(\dot{V}_a^* - (R_{ab} - jX_{ab})\dot{I}_{ab}^*) \\ &= \dot{V}_a \dot{V}_a^* - (R_{ab} - jX_{ab})\dot{V}_a \dot{I}_{ab}^* - (R_{ab} + jX_{ab})\dot{I}_{ab}\dot{V}_a^* + (R_{ab}^2 + X_{ab}^2)\dot{I}_{ab}\dot{I}_{ab}^* \\ &= v_a - 2R_{ab}P_{s,ab} - 2X_{ab}Q_{s,ab} + (R_{ab}^2 + X_{ab}^2)l_{ab} \ , \ \forall (a,b) \in S_L \ . \end{aligned} \tag{2.88}$$

$$l_{ab}v_a = \dot{I}_{ab}\dot{I}_{ab}^* \dot{V}_a \dot{V}_a^* = (\dot{V}_a \dot{I}_{ab}^*)(\dot{V}_a \dot{I}_{ab}^*)^* = P_{s,ab}^2 + Q_{s,ab}^2 \ . \tag{2.89}$$

Therefore,

$$\forall (a,b) \in S_L \ , \ P_{r,ab} = P_{s,ab} - R_{ab}l_{ab} \ , \tag{2.90a}$$

$$Q_{r,ab} = Q_{s,ab} - X_{ab}l_{ab} \ , \tag{2.90b}$$

$$v_b = v_a - 2R_{ab}P_{s,ab} - 2X_{ab}Q_{s,ab} + (R_{ab}^2 + X_{ab}^2)l_{ab} \ , \tag{2.90c}$$

$$l_{ab}v_a = P_{s,ab}^2 + Q_{s,ab}^2 \ . \tag{2.90d}$$

Branch flow model consists of (2.90) and the power balance equations in (2.83), where the variables are v_a, l_{ab}, $P_{s,ab}$, $Q_{s,ab}$, $P_{r,ab}$, and $Q_{r,ab}$ for any $a \in S_B$ and $(a,b) \in S_L$. Moreover, except (2.90d), all equations in branch flow model are linear. The branch flow model is a relaxation of the AC power flow model because in some cases, the voltage angle θ cannot be recovered.

2.9.3.2 Branch Flow Model in a Radial Network

The branch flow model is also called the DistFlow model (Baran and Wu 1989a, b), which is usually used in distribution networks. Distribution networks have a radial topology during operation in most cases. In other words, the undirected graph generated by S_L does not have a circle, as illustrated in Fig. 2.21.

In a radial network, the reference directions of the lines can be chosen properly so that for each bus, there is at most one line sending power to the bus. For such S_L, the power balance equation in (2.83) becomes:

$$P_a + P_{r,ca} = \sum_{b:(a,b)\in S_L} P_{s,ab} \ , \ \text{if } \exists c \in S_B \ , \ \text{s.t. } (c,a) \in S_L \ , \tag{2.91a}$$

$$P_a = \sum_{b:(a,b)\in S_L} P_{s,ab} \ , \ \text{otherwise} \ . \tag{2.91b}$$

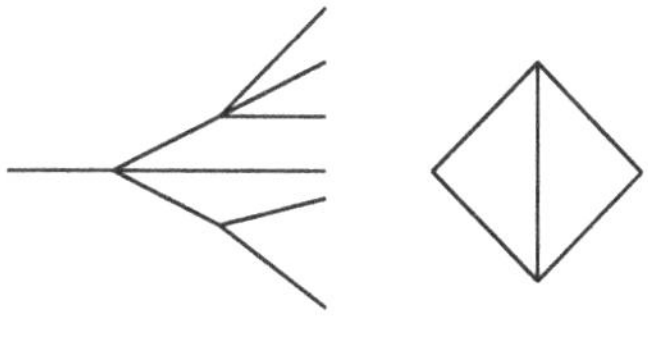

Fig. 2.21 The network on the left is radial while the network on the right is not

Therefore, the branch flow model in a radial network can be written as follows:

$$\forall (a,b) \in S_L \ , \ P_b + P_{s,ab} - R_{ab} l_{ab} = \sum_{c:(b,c)\in S_L} P_{s,bc} \ , \tag{2.92a}$$

$$Q_b + Q_{s,ab} - X_{ab} l_{ab} = \sum_{c:(b,c)\in S_L} Q_{s,bc} \ , \tag{2.92b}$$

$$v_b = v_a - 2R_{ab} P_{s,ab} - 2X_{ab} Q_{s,ab} + (R_{ab}^2 + X_{ab}^2) l_{ab} \ , \tag{2.92c}$$

$$l_{ab} v_a = P_{s,ab}^2 + Q_{s,ab}^2 \ . \tag{2.92d}$$

In (2.92), the variables are v_a, l_{ab}, $P_{s,ab}$, and $Q_{s,ab}$ for any $a \in S_B$ and $(a,b) \in S_L$. In radial networks, the branch flow model is equivalent to the AC power flow model because the voltage angle θ can be recovered.

2.9.3.3 Linearized DistFlow

In branch flow model (2.92) for a radial network, the nonlinearity appears in (2.92d). To make it tractable, a way is to linearize the model. In this regard, neglect all the terms related to l_{ab} and then the linearized DistFlow model (Baran and Wu 1989c) is as follows:

$$\forall (a,b) \in S_L \ , \ P_b + P_{s,ab} = \sum_{c:(b,c)\in S_L} P_{s,bc} \ , \tag{2.93a}$$

$$Q_b + Q_{s,ab} = \sum_{c:(b,c)\in S_L} Q_{s,bc} \ , \tag{2.93b}$$

$$v_b = v_a - 2R_{ab} P_{s,ab} - 2X_{ab} Q_{s,ab} \ . \tag{2.93c}$$

Therefore, linearized DistFlow is a linear approximation of the branch flow model, where the losses $R_{ab} l_{ab}$ and $X_{ab} l_{ab}$ are assumed to be much smaller than the power flows $P_{s,ab}$ and $Q_{s,ab}$.

2.9.4 Second-Order Cone Relaxation of Branch Flow Model

Second-order cone relaxation of branch flow model (Farivar and Low 2013a, b) relaxes the nonlinear equation (2.92d) into:

$$l_{ab} v_a \geq P_{s,ab}^2 + Q_{s,ab}^2 \ , \ \forall (a,b) \in S_L \ , \tag{2.94}$$

which is equivalent to the following second-order cone constraint:

$$\left\| \begin{matrix} 2P_{s,ab} \\ 2Q_{s,ab} \\ l_{ab} - v_a \end{matrix} \right\|_2 \le l_{ab} + v_a \ , \ \forall (a, b) \in S_L \ . \tag{2.95}$$

Therefore, the second-order cone relaxation of branch flow model in a radial network consists of linear equations (2.92a)–(2.92c) and second-order cone constraint (2.95). Since second-order cone constraints are convex, this formulation is tractable in optimization problems. Proposition 3 in Huang et al. (2017) gives a sufficient but mild condition for the second-order cone relaxation to be exact, which means that at the optimal solution point, equality holds in (2.95). The second-order cone constraint (2.95) can be further linearized using the approach in Ben-Tal and Nemirovski (2001). In other words, (2.95) can be approximated by a set of linear constraints.

References

Alizadeh F, Goldfarb D (2003) Second-order cone programming. Math Program 95:3–51

Artzner P, Delbaen F, Eber JM et al (1999) Coherent measures of risk. Math Finance 9(3):203–228

Baran M, Wu F (1989) Optimal capacitor placement on radial distribution systems. IEEE Trans Power Deliv 4(1):725–734

Baran M, Wu F (1989) Optimal sizing of capacitors placed on a radial distribution system. IEEE Trans Power Deliv 4(1):735–743

Baran M, Wu F (1989) Network reconfiguration in distribution systems for loss reduction and load balancing. IEEE Trans Power Deliv 4(2):1401–1407

Ben-Tal A, Nemirovski A (2001) On polyhedral approximations of the second-order cone. Math Oper Res 26(2):193–205

Ben-Tal A, Ghaoui LE, Nemirovski A (2009) Robust optimization. Princeton Series in Applied Mathematics. Princeton University Press, Princeton

Ben-Tal A, Nemirovski A (2001) Lectures on modern convex optimization: analysis, algorithms, and engineering applications. MOS-SIAM Series on Optimization, Society for Industrial and Applied Mathematics, Philadelphia, PA

Bertsekas DP (2009) Convex optimization theory. Athena Scientific, Belmont, Massachusetts

Borwein J, Lewis A (2005) Convex analysis and nonlinear optimization: theory and examples. Springer, New York, NY

Boyd S, Vandenberghe L (2004) Convex optimization. Cambridge University Press, Cambridge

Emmerich MTM, Deutz AH (2018) A tutorial on multiobjective optimization: fundamentals and evolutionary methods. Nat Comput 17:585–609

Farivar M, Low SH (2013) Branch flow model: relaxations and convexification-Part I. IEEE Trans Power Syst 28(3):2554–2564

Farivar M, Low SH (2013) Branch flow model: relaxations and convexification-Part II. IEEE Trans Power Syst 28(3):2565–2572

Föllmer H, Schied A (2002) Convex measures of risk and trading constraints. Finance Stochast 6:429–447

Huang S, Wu Q, Wang J et al (2017) A sufficient condition on convex relaxation of AC optimal power flow in distribution networks. IEEE Trans on Power Syst 32(2):1359–1368

Lay DC, Lay SR, McDonald JJ (2020) Linear algebra and its applications, 6th edn. Pearson, Boston

Molzahn DK, Hiskens IA (2019) A survey of relaxations and approximations of the power flow equations. Found Trends Electr Energy Syst 4(1–2):1–221
Rockafellar R, Uryasev S (2000) Optimization of conditional value-at-risk. J Risk 2(3):21–41
Rockafellar R, Uryasev S (2002) Conditional value-at-risk for general loss distributions. J Bank Finance 26(7):1443–1471
Rudin W (1987) Real and complex analysis, 3rd edn. Series in Higher Mathematics, McGraw-Hill, New York
Shapiro A (2001) On duality theory of conic linear problems. In: Goberna MÁ, López MA (eds) Semi-infinite programming. Nonconvex optimization and its applications, vol 57. Springer, Boston, MA
Shapiro A, Kleywegt A (2002) Minimax analysis of stochastic problems. Optim Methods Softw 17(3):523–542
Stott B, Jardim J, Alsac O (2009) DC power flow revisited. IEEE Trans Power Syst 24(3):1290–1300
Zeng B, Zhao L (2013) Solving two-stage robust optimization problems using a column-and-constraint generation method. Oper. Res Lett 41(5):457–461

Chapter 3
Basic Distributionally Robust Optimization

Abstract This chapter provides an overview of the general theory of distributionally robust optimization (DRO). It begins with the basic ideas of this technique, followed by an explanation of distributionally robust uncertainties, including worst-case expectations, robust chance constraints, and distributionally robust risk measures. Subsequently, a model of two-stage DRO is presented, which will serve as the main optimization framework applied to various energy storage sizing problems in the application part of this book. Finally, some general reformulation methods between different types of distributionally robust uncertainties are explained. These methods are applicable to any type of ambiguity set.

3.1 Basic Ideas

Consider a random variable ω on the sample space Ω. Throughout this chapter, we assume Ω is a nonempty closed set in an Euclidean space. Nonetheless, if Ω is not closed, it can be replaced by its closure. Denote the real or actual probability distribution of the random variable ω by $\mathbb{P}_R$, but $\mathbb{P}_R$ is unknown. Construct the ambiguity set $\mathcal{P}$ according to known information such as the sample data of the random variable and the risk preference of the decision maker so that the decision maker believes the ambiguity set is very likely to contain the real probability distribution, i.e., $\mathbb{P}_R \in \mathcal{P}$. The ambiguity set $\mathcal{P}$ is a subset of $\mathcal{M}(\Omega)$, the set of all probability distributions on Ω. The decision maker does not know $\mathbb{P}_R$ but has confidence for $\mathbb{P}_R \in \mathcal{P}$, so the optimization will be conducted with respect to the worst-case probability distribution in $\mathcal{P}$. For example, suppose $g(x, \omega)$ is a function of the decision variable x and random variable ω which represents the loss. Using distributionally robust optimization (DRO) to minimize $g(x, \omega)$ means minimizing the expectation of $g(x, \omega)$ under the worst-case probability distribution in the ambiguity set $\mathcal{P}$, which is:

$$\min_{x} \sup_{\mathbb{P} \in \mathcal{P}} \mathbb{E}^{\mathbb{P}}[g(x, \omega)] ,$$

R. Xie and W. Wei, *Distributionally Robust Optimization and its Applications in Power System Energy Storage Sizing*,
https://doi.org/10.1007/978-981-97-2566-3_3

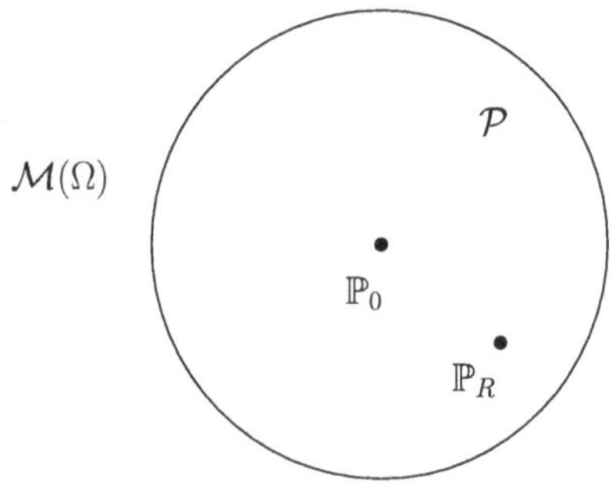

Fig. 3.1 The ambiguity set $\mathcal{P}$ is a set of probability distributions constructed based on the empirical probability distribution $\mathbb{P}_0$. The ambiguity set $\mathcal{P}$ is expected to cover the real probability distribution $\mathbb{P}_R$, although the decision maker does not know the exact information of $\mathbb{P}_R$

where $\mathbb{E}^{\mathbb{P}}[\cdot]$ denotes the operation of taking expectation when the random variable ω follows the probability distribution $\mathbb{P}$. Therefore, $\sup_{\mathbb{P}\in\mathcal{P}} \mathbb{E}^{\mathbb{P}}[g(x,\omega)]$ is the supremum of the expectations of $g(x,\omega)$ when the probability distribution $\mathbb{P}$ varies in $\mathcal{P}$. Minimizing this supremum is exactly minimizing the expectation of $g(x,\omega)$ under the worst-case probability distribution.

On the one hand, DRO takes into account the uncertainty of the probability distribution, so its results can have some robustness even if the sample data are not enough or have bias. On the other hand, it utilizes the information of probability distribution, which leads to less conservative results compared with robust optimization (RO). In data-driven DRO, the empirical probability distribution $\mathbb{P}_0$ is established first based on the sample data, then the ambiguity set $\mathcal{P}$ is constructed around $\mathbb{P}_0$ so that $\mathcal{P}$ contains the real probability distribution $\mathbb{P}_R$ with a high confidence level. The schematic diagram of ambiguity set in data-driven DRO is shown in Fig. 3.1.

If the ambiguity set is the singleton containing the empirical probability distribution $\mathbb{P}_0$, i.e., $\mathcal{P} = \{\mathbb{P}_0\}$, the DRO problem degenerates to be the stochastic programming (SP) problem under the probability distribution $\mathbb{P}_0$ and the ambiguity of probability distribution is no longer considered. Because usually there is more or less some difference between $\mathbb{P}_0$ and the real probability distribution $\mathbb{P}_R$, so $\mathbb{P}_R \in \mathcal{P}$ may not hold if $\mathcal{P} = \{\mathbb{P}_0\}$. When the empirical data are not enough or have some bias, the difference between $\mathbb{P}_0$ and $\mathbb{P}_R$ is nonnegligible, then the SP results cannot guarantee feasibility and optimality.

In the case where the ambiguity set $\mathcal{P}$ consists of all the probability distributions supported on the range of the random variable ω, the DRO problem degenerates to an RO problem, i.e., optimizing with respect to the worst-case scenario of ω, which corresponds to considering the probability distribution with unit mass at ω in DRO. In this circumstance, only the range information of ω is utilized, while the empirical probability distribution and the sample data information are neglected, which takes responsibility for the usual over-conservativeness of RO. From the perspective of DRO, the reason for over-conservativeness is that the chosen ambiguity set $\mathcal{P}$ is too large. A larger $\mathcal{P}$ contains more probability distributions and thus the worst-case probability distribution in $\mathcal{P}$ can be worse. Therefore, the sample data of ω should be fully utilized to construct a suitable ambiguity set, which not only contains the real probability distribution with a high confidence level but also is not too large. In this way, a balance between optimality and robustness can be maintained.

In conclusion, DRO has close relationships with SP and RO. The latter two methods can be regarded as degenerate cases of DRO with special ambiguity sets.

3.2 Distributionally Robust Uncertainties

Uncertainties usually appear in optimization problems as expectations, chance constraints, and risk measures. In the scope of DRO, these forms correspond to worst-case expectations, robust chance constraints, and distributionally robust risk measures, respectively. Among them, robust chance constraints only arise in the constraints of optimization problems, while worst-case expectations and distributionally robust risk measures can both appear in the constraints and the objective functions.

3.2.1 Worst-Case Expectation

Let $g(\omega)$ denote a real-valued measurable function of the random variable ω. When ω follows probability distribution $\mathbb{P}$, the expectation of $g(\omega)$ is $\mathbb{E}^{\mathbb{P}}[g(\omega)] = \int_{\Omega} g(\omega)\mathrm{d}\mathbb{P}(\omega)$ provided that it exists. If the probability distribution $\mathbb{P}$ varies in the ambiguity set $\mathcal{P}$, the corresponding expectation also changes. Consider the expectation under the worst-case probability distribution and call it the worst-case expectation. If $g(\omega)$ is a function reflecting the loss, then a larger loss means a worse case. In this circumstance, the worst-case expectation is:

$$\sup_{\mathbb{P}\in\mathcal{P}} \mathbb{E}^{\mathbb{P}}[g(\omega)] ,$$

which is the supremum of $\mathbb{E}^{\mathbb{P}}[g(\omega)]$ as $\mathbb{P}$ varies in $\mathcal{P}$. If the real probability distribution is contained in $\mathcal{P}$, the loss expectation under the real probability distribution will not exceed the worst-case expectation, so the worst-case expectation gives a conservative estimation for the real loss expectation.

3.2.2 Robust Chance Constraint

Suppose S is a measurable subset of Ω. Denote by $\mathbb{P}(\omega \in S)$ the probability of random event S when ω follows the probability distribution $\mathbb{P}$. Suppose S represents the event where there is no loss or the loss is acceptable. Then the constraint requiring the probability of S to be no smaller than the threshold β is a chance constraint, i.e.,

$\mathbb{P}(\omega \in S) \geq \beta$. In DRO, robust chance constraint means that for any probability distribution in the ambiguity set, the probability of event S is no smaller than β, i.e.,

$$\inf_{\mathbb{P} \in \mathcal{P}} \mathbb{P}(\omega \in S) \geq \beta \,. \tag{3.1}$$

When $\mathcal{P}$ contains the real probability distribution, (3.1) implies that the real probability of S is at least β.

3.2.3 *Distributionally Robust Risk Measure*

Suppose $g(\omega)$ is a real-valued measurable function of ω, which is also a real-valued random variable representing the loss. If the risk measure of $g(\omega)$ with ω following $\mathbb{P}$ is denoted by $r^{\mathbb{P}}(g(\omega))$, the distributionally robust risk measure is the supremum of the risk measure when $\mathbb{P}$ takes values in $\mathcal{P}$, which is:

$$\sup_{\mathbb{P} \in \mathcal{P}} r^{\mathbb{P}}(g(\omega)) \,.$$

Since the expectation $\mathbb{E}^{\mathbb{P}}[g(\omega)]$ can be regarded as a kind of risk measure for the loss, the worst-case expectation $\sup_{\mathbb{P} \in \mathcal{P}} \mathbb{E}^{\mathbb{P}}[g(\omega)]$ is a special case of distributionally robust risk measures.

3.3 Two-Stage Distributionally Robust Optimization

The energy storage (ES) sizing problem in power systems generally has the form of two-stage optimization. The first-stage optimization in the outer layer determines the sizing strategy, where the decision variables are the capacities of the components to be sized, denoted by x. The second-stage optimization in the inner layer is for operation optimization, where the capacity variable x is fixed and the random variable ω representing the renewable power generation and load demand has a specific sample value, hence is also fixed. By adjusting the operation variable y with fixed x and ω, the second-stage optimization minimizes the objective function such as load shedding, renewable power curtailment, and carbon emissions. Thus, the optimal value of the second-stage optimization can be regarded as a function of x and ω, which is uncertain. Through worst-case expectations, robust chance constraints, and distributionally robust risk measures, the uncertainties from the second stage appear in the first-stage optimization, resulting in the two-stage DRO model.

The general mathematical model of the first-stage optimization is as follows:

$$\min_{x\in\mathcal{X}} c^\top x + \sum_{i\in I_1} \sup_{\mathbb{P}\in\mathcal{P}_i} r_i^{\mathbb{P}}[g_i(x,\omega)] \tag{3.2a}$$

$$\text{s.t.} \inf_{\mathbb{P}\in\mathcal{P}_i} \mathbb{P}(\omega \in S_i(x)) \geq \beta_i \ , \ \forall i \in I_2 \ , \tag{3.2b}$$

$$\sup_{\mathbb{P}\in\mathcal{P}_i} r_i^{\mathbb{P}}(g_i(x,\omega)) \leq \gamma_i \ , \ \forall i \in I_3 \ . \tag{3.2c}$$

In (3.2), the set $\mathcal{X}$ is the range of the first-stage decision variable x and $\mathcal{X}$ is bounded in practical problems. c is a constant coefficient vector. The robust chance constraint appears in the constraints, while the distributionally robust risk measures are in both the objective function and the constraints. Note that the worst-case expectation is a special case of the distributionally robust risk measures, so (3.2) can also include the worst-case expectations. There may be various ambiguity sets, loss functions, risk measures, and thresholds of probabilities and risks, so they are distinguished by subscript i. I_1, I_2, and I_3 are index sets, where an empty set means the corresponding component of the constraints or objective function does not come up. The loss function $g_i(x,\omega)$ is defined by the second-stage optimization, so it not only depends on ω but is also influenced by x. Similarly, the random event $S_i(x)$ is a measurable subset of the sample space Ω. $S_i(x)$ is also given by the second stage and depends on x.

The second-stage optimization model defining the loss function $g_i(x,\omega)$ is:

$$g_i(x,\omega) := \min_y D_i^\top y \tag{3.3a}$$

$$\text{s.t.} \ A_i y \geq B_i(x)\cdot\omega + H_i(x) \ , \tag{3.3b}$$

where coefficients D_i and A_i are constant vector or matrix; $B_i(x)$ and $H_i(x)$ depend on and is linear in x, which means their components are linear functions of x. Therefore, $g_i(x,\omega)$ is the optimal value of a linear programming (LP) problem with decision variable y, where x and ω only emerge on the right-hand side of the constraint and the right-hand side is affine when it is regarded as a function of x or a function of ω. The considered loss function should have a finite value for any $x \in \mathcal{X}$ and $\omega \in \Omega$.

For the definition of random event $S_i(x)$, the second-stage mathematical model is:

$$S_i(x) := \{\omega \in \Omega \mid \exists y \ , \ \text{s.t.} \ A_i y \geq B_i(x)\cdot\omega + H_i(x)\} \ . \tag{3.4}$$

In other words, $S_i(x)$ contains all the values of ω that make the inside constraint feasible for some y. Thus, $S_i(x)$ is a set depending on x and $S_i(x) \subset \Omega$.

The considered two-stage DRO model is jointly defined by (3.2)–(3.4). This is a specific type of two-stage DRO problem: The second stage is defined by LP or a set of linear constraints, where the first-stage decision variable and random variable only appear on the right-hand side of the constraints, and the right-hand side is affine with respect to the first-stage decision variable or the random variable. Although the second stage is linear, nonlinearity can be integrated into the first stage by the constraint $x \in \mathcal{X}$. For example, it can be used to enforce a part of x to be integer-valued.

The reason for considering this type of two-stage DRO is: On the one hand, the ES sizing problems can be modeled in this form, so the developed modeling and solution methods can be the foundation for the considered sizing problems. On the other hand, the reformulation methods of the distributionally robust uncertainties in later sections utilize the characteristics of this model.

3.4 Some Reformulation Methods

In this section, we introduce some reformulation methods for distributionally robust uncertainties. The reformulation relationship is summarized in Fig. 3.2.

3.4.1 Reformulating Random Event by Loss Function

The robust chance constraint has been introduced in (3.1). For convenience, it is rewritten here:

$$\inf_{\mathbb{P} \in \mathcal{P}} \mathbb{P}(\omega \in S) \geq \beta \,. \tag{3.5}$$

Suppose a loss function $g : \Omega \to \mathbb{R}$ is defined so that

$$\omega \in S \iff g(\omega) \leq 0 \,, \tag{3.6}$$

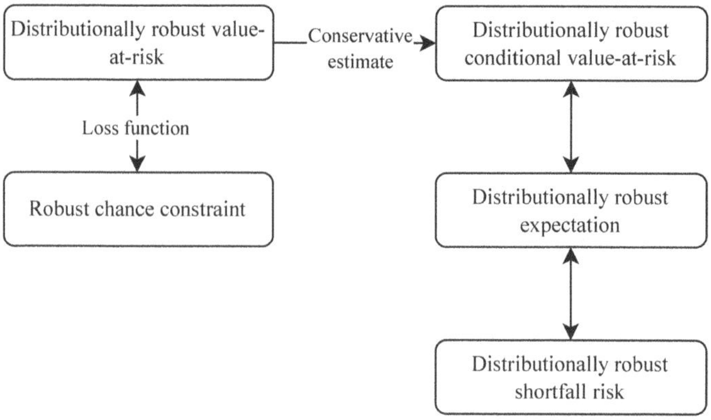

Fig. 3.2 Diagram of the reformulation methods. The two-way arrows represent equivalent reformulation relationships, whereas a one-way arrow indicates that the term at the end can provide a conservative estimate for the term at the beginning

which means that the loss is nonpositive if and only if the desired event $\omega \in S$ happens. Suppose such a loss function $g(\omega)$ has been found with property (3.6), then robust chance constraint (3.5) is equivalent to:

$$\inf_{\mathbb{P} \in \mathcal{P}} \mathbb{P}(g(\omega) \leq 0) \geq \beta . \tag{3.7}$$

Equations (3.5) and (3.7) are two different forms to express robust chance constraints, and they can be interchanged using (3.6). The question left to answer is: *How to implement the equivalent relationship in* (3.6) *by finding suitable random event* S *and loss function* g?

It is obvious that given loss function g, the random event S satisfying (3.6) should be $S = \{\omega \in \Omega \mid g(\omega) \leq 0\}$. Lemma 3.1 states the relationship between the continuity of g and the closeness of S.

Lemma 3.1 *If* (3.6) *holds and the loss function* $g : \Omega \to \mathbb{R}$ *is continuous, then the random event* S *is closed as a set in* Ω.

Proof Suppose $\{x_n \mid n \in \mathbb{N}_+\}$ is a sequence in S that converges to $x_0 \in \Omega$, i.e., $\lim_{n \to +\infty} \|x_n - x_0\| = 0$. Using (3.6), $g(x_n) \leq 0$, $n \in \mathbb{N}_+$. By the continuity of g, $g(x_0) \leq 0$, which exactly means $x_0 \in S$. Since the sequence $\{x_n\}$ is arbitrary, S is closed in Ω. □

Suppose the random event $S \subset \Omega$ is given and now we try to find a loss function $g : \Omega \to \mathbb{R}$ so that (3.6) holds. The choices of g are not unique. For example, when $\omega \in S \iff g(\omega) \leq 0$, it is clear that $\omega \in S \iff (g(\omega))^3 \leq 0$, which means that $\tilde{g}(\omega) := (g(\omega))^3$ is another desired loss function. We provide examples of the desired loss functions for different types of random event S in Sects. 3.4.1.1–3.4.1.3.

In a practical problem, there might be an intuitive way to define the loss function. For instance, if the chance constraint indicates that the probability of zero load shedding is no smaller than some threshold, then we can define the loss function as the minimum load shedding according to the physical meaning because zero load shedding means exactly the minimum load shedding is no larger than 0.

3.4.1.1 General Random Event

An immediate corollary of Lemma 3.1 is that if S is not closed in Ω, there does not exist a continuous loss function g satisfying (3.6). In this regard, we must give up finding a continuous loss function for a general random event S which is not closed. A simple choice of the loss function for (3.6) is the indicator function of $\Omega \backslash S$, which is $g : \Omega \to \mathbb{R}$ subject to:

$$g(\omega) = \begin{cases} 1 \,, & \text{if } \omega \notin S \,, \\ 0 \,, & \text{if } \omega \in S \,. \end{cases}$$

This formulation applies to all kinds of S. However, the indicator function is not continuous, which may burden the robust chance constraint's subsequent reformulation and solution process.

We have assumed that Ω is closed in the Euclidean space. Lemma 3.2 shows that in this circumstance, S is closed in Ω is equivalent to that S is closed in the Euclidean space.

Lemma 3.2 *If $\Omega \subset \mathbb{R}^M$ is closed and $S \subset \Omega$, then S is closed in Ω if and only if S is closed in $\mathbb{R}^M$.*

Proof 1. Assume S is closed in Ω. Suppose the sequence $\{\omega_n \mid n \in \mathbb{N}_+\} \subset S$ and $\omega_0 \in \mathbb{R}^M$ satisfying that $\lim_{n\to+\infty} \omega_n = \omega_0$. Because Ω is closed in $\mathbb{R}^M$ and $S \subset \Omega$, $\{\omega_n \mid n \in \mathbb{N}_+\}$ is also a convergent sequence in Ω, so $\omega_0 \in \Omega$. Then by the assumption that S is closed in Ω, we have $\omega_0 \in S$. Since the sequence is arbitrary, S is closed in $\mathbb{R}^M$.

2. Assume S is closed in $\mathbb{R}^M$. Suppose the sequence $\{\omega_n \mid n \in \mathbb{N}_+\} \subset S$ and $\omega_0 \in \Omega$ satisfying that $\lim_{n\to+\infty} \omega_n = \omega_0$. Then S is closed in $\mathbb{R}^M$ implies that $x_0 \in S$. Because the sequence is arbitrary, we have S is closed in Ω. □

Example 3.1 1. When $\Omega = \mathbb{R}$ and $S = [0, 1)$, S is not closed in Ω, so there is no continuous loss function g satisfying (3.6).

2. If we tentatively allow Ω to be not closed and let $\Omega = [-1, 1)$ and $S = [0, 1)$, then S is closed in Ω. The function $g(\omega) := -\omega$ is continuous and (3.6) holds.

3.4.1.2 Nonempty Closed Random Event

For nonempty closed $S \subset \Omega$, we can always find a continuous loss function $g : \Omega \to \mathbb{R}$ to satisfy (3.6) using the distance function to S defined as follows:

$$g(\omega) := \mathrm{dist}(\omega, S) = \inf_{\omega' \in S} \|\omega - \omega'\| . \tag{3.8}$$

The effectiveness follows from Proposition 3.1, but before that, we prove the infimum in $\inf_{\omega' \in S} \|\omega - \omega'\|$ can be achieved by some $\omega_0 \in S$ provided that S is closed.

Lemma 3.3 *For nonempty and closed $S \subset \Omega$ and $\omega \in \Omega$, there is $\omega_0 \in S$ such that $\inf_{\omega' \in S} \|\omega - \omega'\| = \|\omega - \omega_0\|$.*

Proof Denote $\inf_{\omega' \in S} \|\omega - \omega'\|$ by C. $C < +\infty$ because S is nonempty. Thus, for any $n \in \mathbb{N}_+$, there is $\omega_n \in S$ such that $\|\omega - \omega_n\| < C + 1/n$. Therefore, the sequence $\{\omega_n \mid n \in \mathbb{N}_+\}$ is in the closed ball

$$B(\omega, C + 1) := \{\omega' \in \Omega \mid \|\omega' - \omega\| \le C + 1\} .$$

Because $B(\omega, C + 1)$ is compact, there is a subsequence $\{\omega_{n_k} \mid k \in \mathbb{N}_+\}$ of $\{\omega_n \mid n \in \mathbb{N}_+\}$ and ω_0 subject to $n_1 \le n_2 \le \ldots$ and $\lim_{k\to+\infty} \omega_{n_k} = \omega_0$. Since Ω

is closed, $\omega_0 \in \Omega$. Then $\omega_0 \in S$ because S is closed in Ω. By the continuity of norm, $\|\omega - \omega_0\| = \lim_{k\to+\infty} \|\omega - \omega_{n_k}\| = C$, which completes the proof. □

Proposition 3.1 *For nonempty and closed $S \subset \Omega$ and $g : \Omega \to \mathbb{R}$ defined in (3.8), (3.6) holds.*

Proof It is obvious that when $\omega \in S$, $g(\omega) = \text{dist}(\omega, S) = \|\omega - \omega\| = 0$. For the opposite direction, assume $g(\omega) \leq 0$, then for any $n \in \mathbb{N}_+$, there is $\omega_n \in S$ such that $\|\omega - \omega_n\| < 1/n$. Thus, the sequence $\{\omega_n \mid n \in \mathbb{N}_+\}$ in S converges to ω. Then $\omega \in S$ follows from the hypothesis that S is closed. □

In addition to continuity, the loss function g defined in (3.8) also has convexity if Ω and S are convex sets.

Proposition 3.2 *If Ω is convex and S is a nonempty, closed, and convex subset of Ω, then $g : \Omega \to \mathbb{R}$ defined in (3.8) is a convex function.*

Proof According to the definition of convex functions, we need to prove that for any $\omega_1, \omega_2 \in \Omega$ and $\lambda \in (0, 1)$, we have $g(\lambda\omega_1 + (1-\lambda)\omega_2) \leq \lambda g(\omega_1) + (1-\lambda)g(\omega_2)$.

By Lemma 3.3, there are $\omega_1', \omega_2' \in S$ such that

$$\begin{aligned} g(\omega_1) &= \inf_{\omega' \in S} \|\omega_1 - \omega'\| = \|\omega_1 - \omega_1'\| , \\ g(\omega_2) &= \inf_{\omega' \in S} \|\omega_2 - \omega'\| = \|\omega_2 - \omega_2'\| . \end{aligned}$$

Let

$$\begin{aligned} \omega_3 &:= \lambda\omega_1 + (1-\lambda)\omega_2 , \\ \omega_3' &:= \lambda\omega_1' + (1-\lambda)\omega_2' , \\ \omega_4 &:= \lambda\omega_1 + (1-\lambda)\omega_2' . \end{aligned}$$

The points are illustrated in Fig. 3.3. By direct calculation, we have:

$$\begin{aligned} \omega_3 - \omega_4 &= (1-\lambda)(\omega_2 - \omega_2') , \\ \omega_4 - \omega_3' &= \lambda(\omega_1 - \omega_1') . \end{aligned}$$

Using the triangle inequality and linearity of the norm, we have:

$$\begin{aligned} g(\lambda\omega_1 + (1-\lambda)\omega_2) &= \inf_{\omega' \in S} \|\omega_3 - \omega'\| \\ &\leq \|\omega_3 - \omega_3'\| \\ &\leq \|\omega_3 - \omega_4\| + \|\omega_4 - \omega_3'\| \\ &= (1-\lambda)\|\omega_2 - \omega_2'\| + \lambda\|\omega_1 - \omega_1'\| \\ &= \lambda g(\omega_1) + (1-\lambda)g(\omega_2) , \end{aligned}$$

so g is convex. □

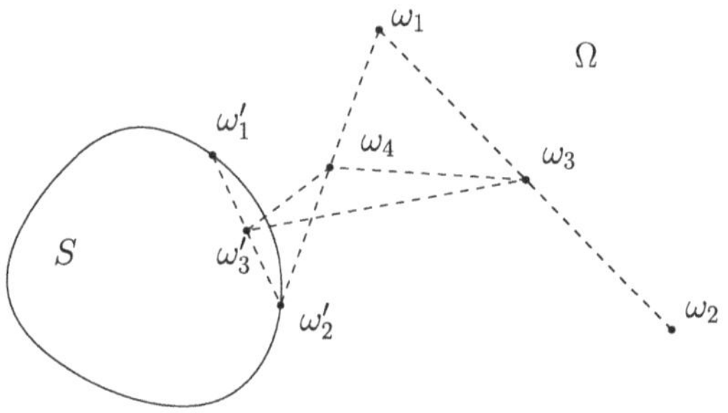

Fig. 3.3 Illustration of the points in the proof of Proposition 3.2

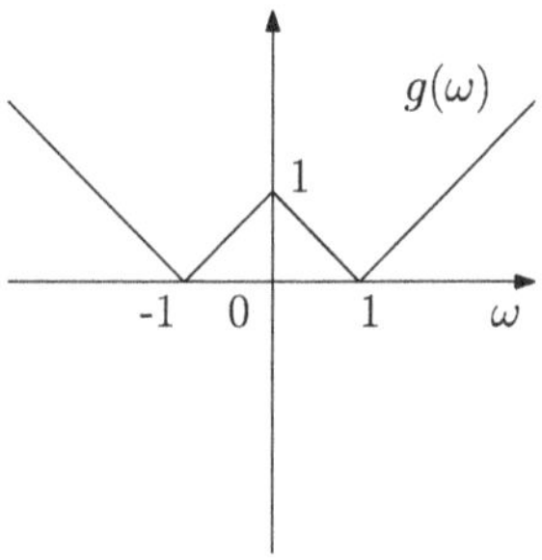

Fig. 3.4 The function defined in (3.9)

Example 3.2 Suppose $\Omega = \mathbb{R}^2$ and $S = B((0,0),1) := \{(x, y) \in \mathbb{R}^2 \mid x^2 + y^2 \leq 1\} \subset \Omega$. S is the closed unit disk centered at the zero point, and therefore it is closed and convex. Using 2-norm, the loss function $g : \mathbb{R}^2 \to \mathbb{R}$ defined in (3.8) is:

$$g(\omega) = g((x, y)) = \begin{cases} 0\,, & \text{if } x^2 + y^2 \leq 1\,, \\ \left\| (x, y) - \frac{(x,y)}{\|(x,y)\|_2} \right\|_2 = \|(x, y)\|_2 - 1\,, & \text{if } x^2 + y^2 > 1\,, \end{cases}$$

which is obviously continuous. Since $g(\omega) = \max\{0, \|\omega\|_2 - 1\}$ is the maximum of two convex functions, g is convex.

Example 3.3 Suppose $\Omega = \mathbb{R}$ and $S = \{-1, 1\} \subset \Omega$. S is a finite set and thus is closed but not convex. The loss function $g : \mathbb{R} \to \mathbb{R}$ defined in (3.8) is:

$$g(\omega) = \begin{cases} |\omega - (-1)| = -\omega - 1\,, & \text{if } \omega \leq -1\,, \\ |\omega - (-1)| = \omega + 1\,, & \text{if } -1 < \omega \leq 0\,, \\ |\omega - 1| = 1 - \omega\,, & \text{if } 0 < \omega \leq 1\,, \\ |\omega - 1| = \omega - 1\,, & \text{if } \omega > 1\,. \end{cases} \tag{3.9}$$

which is continuous but not convex. This function is depicted in Fig. 3.4.

3.4.1.3 Polyhedral Random Event

Suppose S is a nonempty polyhedron described by linear inequalities, i.e., $S := \{\omega \in \Omega \mid A\omega \leq B\} \subset \Omega$, where A and B are coefficients. The loss function can be defined as follows to satisfy (3.6):

$$g(\omega) := \min_{s} \ 1^\top s \tag{3.10a}$$

$$\text{s.t. } s \geq A\omega - B\ ,\ \ s \geq 0\ , \tag{3.10b}$$

where s is the slack variable for the constraint $A\omega \leq B$; 1 and s have the same dimension as B. $\omega \in \Omega$ is a parameter in the LP (3.10), so the optimal value can be regarded as a function of ω. Proposition 3.3 shows the effectiveness of the loss function g in (3.10).

Proposition 3.3 *For any $\omega \in \Omega$, LP (3.10) has a finite nonnegative optimal value. In addition, $\omega \in S = \{\omega \in \Omega \mid A\omega \leq B\}$ if and only if $g(\omega) = 0$.*

Proof Fix $\omega \in \Omega$. We can set s with large enough components so that $s \geq 0$ and $s \geq A\omega - B$, then LP (3.10) is feasible. Since $s \geq 0$, we have $1^\top s \geq 0$. Thus, (3.10) must have a finite nonnegative optimal value.

Assuming $\omega \in S$, we have $A\omega \leq B$, then $s = 0$ is a feasible solution of (3.10), which is also optimal. Thus, $g(\omega) = 0$. Conversely, assume $g(\omega) = 0$. Since $s \geq 0$ and $1^\top s = 0$, the only possible value of s is $s = 0$, which implies $A\omega \leq B$ and thus $\omega \in S$. □

Remark 3.1 From the proof we can see that the loss function g defined in (3.10) is exactly $g(\omega) := 1^\top \max\{A\omega - B, 0\}$, where the maximum notation means component-wise maximum and 0 is the zero vector with the same dimension as B.

Remark 3.2 As (3.10) shows, the loss function g is the optimal value of an LP problem, where ω appears on the right-hand side of the constraint. By the properties of LP, $g(\omega)$ is a piecewise linear and convex function.

Exercise 3.1 Does $g(\omega) := \max\{\max\{A\omega - B, 0\}\}$ satisfy (3.6) for polyhedron $S = \{\omega \in \Omega \mid A\omega \leq B\}$? Can this loss function be written as the optimal value of an LP problem? What are the dimensions of the decision variables in this LP problem?

When Ω is convex, the polyhedron $S = \{\omega \in \Omega \mid A\omega \leq B\}$ is a closed and convex subset, so the distance function in (3.8) is another way to define a convex loss function satisfying (3.6). If the 1-norm $\|\cdot\|_1$ is used, the loss function $g : \Omega \to \mathbb{R}$ is defined as follows:

$$g(\omega) := \inf_{\omega' \in S} \|\omega - \omega'\|_1 \tag{3.11a}$$

$$= \begin{cases} \min_{\omega'} \|\omega - \omega'\|_1 \\ \text{s.t. } A\omega' \le B \end{cases} \tag{3.11b}$$

$$= \begin{cases} \min_{\omega', s} 1^\top s \\ \text{s.t. } A\omega' \le B \ , \ \omega - \omega' \le s \ , \ -(\omega - \omega') \le s \ , \end{cases} \tag{3.11c}$$

where s is an auxiliary variable to represent $|\omega - \omega'|$ with component-wise value. Thus, this loss function is also the optimal value of an LP problem, where the dimension of decision variables (ω', s) is twice the dimension of ω.

Exercise 3.2 Prove the equation from (3.11b) to (3.11c).

The random event $S_i(x)$ defined in (3.4) is:

$$S_i(x) = \{\omega \in \Omega \mid \exists y \ , \ \text{s.t. } A_i y \ge B_i(x) \cdot \omega + H_i(x)\} \ , \tag{3.12}$$

which is a set depending on x. It is the projection of the polyhedron $\{(\omega, y) \mid A_i y \ge B_i(x) \cdot \omega + H_i(x)\}$ to the space of Ω. Note that the projection of a polyhedron is still a polyhedron. Although (3.10) cannot be directly applied, its idea leads to the following loss function:

$$g_i(x, \omega) := \min_{s, y} 1^\top s \tag{3.13a}$$

$$\text{s.t. } A_i y + s \ge B_i(x) \cdot \omega + H_i(x) \ , \ s \ge 0 \ . \tag{3.13b}$$

which satisfies that for any $x \in \mathcal{X}$ and $\omega \in \Omega$, $g_i(x, \omega)$ has a finite value and $\omega \in S_i(x)$ is equivalent to $g_i(x, \omega) = 0$. Note that it can be written in the form of the loss function (3.3) in the two-stage DRO model in Sect. 3.3 as follows:

$$g_i(x, \omega) := \min_{(s, y)} \begin{pmatrix} 1 \\ 0 \end{pmatrix}^\top \begin{pmatrix} s \\ y \end{pmatrix}$$

$$\text{s.t.} \begin{pmatrix} I & A_i \\ I & 0 \end{pmatrix} \begin{pmatrix} s \\ y \end{pmatrix} \ge \begin{pmatrix} B_i(x) \\ 0 \end{pmatrix} \cdot \omega + \begin{pmatrix} H_i(x) \\ 0 \end{pmatrix} ,$$

where I is the identity matrix of suitable size.

Proposition 3.4 *Suppose $S_i(x)$ and $g_i(x, \omega)$ are defined in (3.12) and (3.13), respectively. Then for any $x \in \mathcal{X}$ and $\omega \in \Omega$, $g_i(x, \omega)$ has a finite and nonnegative value. In addition, $\omega \in S_i(x)$ is equivalent to $g_i(x, \omega) = 0$.*

Proof The solution with $y = 0$ and $s = \max\{B_i(x) \cdot \omega + H_i(x), 0\}$ is obviously feasible for (3.13). Because $s \ge 0$ and the goal is to minimize $1^\top s$, we have $g_i(x, \omega)$ is finite and nonnegative.

When $\omega \in S_i(x)$, there is y_0 subject to $A_i y_0 \geq B_i(x) \cdot \omega + H_i(x)$, so the solution with $y = y_0$ and $s = 0$ is feasible for (3.13) and $g_i(x, \omega) \leq 0$. We have proved that $g_i(x, \omega) \geq 0$, so $g_i(x, \omega) = 0$. Conversely, when $g_i(x, \omega) = 0$, suppose $(s, y) = (s_0, y_0)$ is the optimal solution, then $1^\top s_0 = 0$ and $s_0 \geq 0$, which implies $s_0 = 0$. Then $A_i y_0 \geq B_i(x) \cdot \omega + H_i(x)$. According to the definition of $S_i(x)$, we have $\omega \in S_i(x)$. □

3.4.2 *Reformulating Robust Chance Constraint by Distributionally Robust Value-at-Risk Constraint*

Recall the definition of value-at-risk (VaR) in Definition 2.39:

$$\beta\text{-VaR}^{\mathbb{P}}(g(\omega)) := \inf \{r \in \mathbb{R} \mid \mathbb{P}(g(\omega) \leq r) \geq \beta\} . \tag{3.14}$$

The relationship between chance constraint and VaR constraint is elaborated in Lemma 2.7: For any constant $\beta \in [0, 1]$ and $r_0 \in \mathbb{R}$,

$$\mathbb{P}(g(\omega) \leq r_0) \geq \beta \iff \beta\text{-VaR}^{\mathbb{P}}(g(\omega)) \leq r_0 .$$

Exercise 3.3 Is $\beta\text{-VaR}^{\mathbb{P}}(g(\omega)) < r_0$ equivalent to $\mathbb{P}[g(\omega) \leq r_0] > \beta$? Prove it or give a counterexample.

Theorem 3.1 further shows that robust chance constraints and distributionally robust VaR constraints are equivalent.

Theorem 3.1 (Equivalence of robust chance constraint and distributionally robust VaR constraint) *For any ambiguity set $\mathcal{P}$, constant $\beta \in [0, 1]$ and $r_0 \in \mathbb{R}$, we have:*

$$\inf_{\mathbb{P} \in \mathcal{P}} \mathbb{P}(g(\omega) \leq r_0) \geq \beta \iff \sup_{\mathbb{P} \in \mathcal{P}} \beta\text{-}VaR^{\mathbb{P}}(g(\omega)) \leq r_0 . \tag{3.15}$$

Proof Using the properties of infimum and supremum as well as Lemma 2.7,

$$\begin{aligned} & \inf_{\mathbb{P} \in \mathcal{P}} \mathbb{P}(g(\omega) \leq r_0) \geq \beta \\ \iff & \forall \mathbb{P} \in \mathcal{P} , \ \mathbb{P}(g(\omega) \leq r_0) \geq \beta \\ \iff & \forall \mathbb{P} \in \mathcal{P} , \ \beta\text{-VaR}^{\mathbb{P}}(g(\omega)) \leq r_0 \\ \iff & \sup_{\mathbb{P} \in \mathcal{P}} \beta\text{-VaR}^{\mathbb{P}}(g(\omega)) \leq r_0 . \end{aligned}$$

This completes the proof. □

Exercise 3.4 Is $\sup_{\mathbb{P} \in \mathcal{P}} \mathbb{P}(g(\omega) \leq r_0) \geq \beta$ equivalent to $\inf_{\mathbb{P} \in \mathcal{P}} \beta\text{-VaR}^{\mathbb{P}}(g(\omega)) \leq r_0$? Prove it or give a counterexample.

3.4.3 *Conservative Estimate of Value-at-Risk by Conditional Value-at-Risk*

Recall the relation between VaR and conditional value-at-risk (CVaR) from Lemma 2.8:

$$\beta\text{-CVaR}^{\mathbb{P}}(g(\omega)) \geq \beta\text{-VaR}^{\mathbb{P}}(g(\omega)) \; .$$

CVaR gives a conservative estimate for the corresponding VaR if it exists, whereas "conservative" means that CVaR always gives a larger risk value and thus is less optimistic. Considering the supremum with varying probability distributions in the ambiguity set, we have similar results for distributionally robust risks.

Theorem 3.2 (Conservative estimate of distributionally robust VaR by distributionally robust CVaR) *For any ambiguity set $\mathcal{P}$ and loss function $g : \Omega \to \mathbb{R}$,*

$$\sup_{\mathbb{P}\in\mathcal{P}} \beta\text{-}CVaR^{\mathbb{P}}(g(\omega)) \geq \sup_{\mathbb{P}\in\mathcal{P}} \beta\text{-}VaR^{\mathbb{P}}(g(\omega)) \; .$$

3.4.4 *Reformulating Distributionally Robust Conditional Value-at-Risk by Worst-Case Expectation*

Recall from Theorem 2.6 that CVaR equals the optimal value of an optimization problem containing an expectation as follows:

$$\beta\text{-CVaR}^{\mathbb{P}}(g(\omega)) = \min_{r\in\mathbb{R}} \left\{ r + \frac{1}{1-\beta} \mathbb{E}^{\mathbb{P}}[\max\{g(\omega) - r, 0\}] \right\} \; , \tag{3.16}$$

which gives a way to link CVaR and expectation. Based on this property, we are going to derive an equivalent reformulation for distributionally robust CVaR by worst-case expectation.

Theorem 3.3 (Reformulating distributionally robust CVaR by worst-case expectation) *For any ambiguity set $\mathcal{P}$ and loss function $g : \Omega \to \mathbb{R}$,*

$$\sup_{\mathbb{P}\in\mathcal{P}} \beta\text{-}CVaR^{\mathbb{P}}(g(\omega)) = \min_{r\in\mathbb{R}} \left\{ r + \frac{1}{1-\beta} \sup_{\mathbb{P}\in\mathcal{P}} \mathbb{E}^{\mathbb{P}}[g'(r, \omega)] \right\} \; , \tag{3.17}$$

where

$$g'(r, \omega) := \max\{g(\omega) - r, 0\} = \left\{ \begin{array}{l} \min\limits_{s} \; s \\ \text{s.t. } s \geq g(\omega) - r \; , \; s \geq 0 \; . \end{array} \right. \tag{3.18}$$

Proof As per (3.16),

$$\sup_{\mathbb{P}\in\mathcal{P}} \beta\text{-CVaR}^{\mathbb{P}}(g(\omega)) = \sup_{\mathbb{P}\in\mathcal{P}} \min_{r\in\mathbb{R}} \left\{ r + \frac{1}{1-\beta}\mathbb{E}^{\mathbb{P}}[\max\{g(\omega)-r, 0\}] \right\} .$$

By stochastic saddle point theorem, $\sup_{\mathbb{P}\in\mathcal{P}}$ and $\min_{r\in\mathbb{R}}$ can be exchanged, so

$$\begin{aligned}
\sup_{\mathbb{P}\in\mathcal{P}} \beta\text{-CVaR}^{\mathbb{P}}(g(\omega)) &= \min_{r\in\mathbb{R}} \sup_{\mathbb{P}\in\mathcal{P}} \left\{ r + \frac{1}{1-\beta}\mathbb{E}^{\mathbb{P}}[\max\{g(\omega)-r, 0\}] \right\} \\
&= \min_{r\in\mathbb{R}} \left\{ r + \frac{1}{1-\beta} \sup_{\mathbb{P}\in\mathcal{P}} \mathbb{E}^{\mathbb{P}}[\max\{g(\omega)-r, 0\}] \right\} \\
&= \min_{r\in\mathbb{R}} \left\{ r + \frac{1}{1-\beta} \sup_{\mathbb{P}\in\mathcal{P}} \mathbb{E}^{\mathbb{P}}[g'(r,\omega)] \right\} .
\end{aligned}$$

It remains to show the equation in (3.18). Since $s \geq g(\omega) - r$ and $s \geq 0$, the right side is no smaller than the left side. $s = \max\{g(\omega) - r, 0\}$ is feasible for the LP problem on the right side, so $\max\{g(\omega) - r, 0\}$ is no smaller than the optimal value of the LP problem, which completes the proof. □

The distributionally robust CVaR in the objective function can be equivalently transformed into worst-case expectation using (3.17). Note that

$$\min_{x\in X} \min_{r\in\mathbb{R}} \left\{ r + \frac{1}{1-\beta} \sup_{\mathbb{P}\in\mathcal{P}} \mathbb{E}^{\mathbb{P}}[g'(x,r,\omega)] \right\} ,$$

is equivalent to:

$$\min_{x\in X, r\in\mathbb{R}} \left\{ r + \frac{1}{1-\beta} \sup_{\mathbb{P}\in\mathcal{P}} \mathbb{E}^{\mathbb{P}}[g'(x,r,\omega)] \right\} .$$

For a distributionally robust CVaR constraint, Proposition 3.5 shows that it is equivalent to a constraint about a worst-case expectation, which is a direct corollary of Theorem 3.3.

Proposition 3.5 (Equivalence of distributionally robust CVaR constraint and worst-case expectation constraint) *For any ambiguity set $\mathcal{P}$, loss function $g : \Omega \to \mathbb{R}$, and constant $\gamma \in \mathbb{R}$,*

$$\begin{aligned}
&\sup_{\mathbb{P}\in\mathcal{P}} \beta\text{-}CVaR^{\mathbb{P}}(g(\omega)) \leq \gamma \\
\Longleftrightarrow\ &\exists r \in \mathbb{R}\ ,\ \text{s.t. } r + \frac{1}{1-\beta} \sup_{\mathbb{P}\in\mathcal{P}} \mathbb{E}^{\mathbb{P}}[g'(r,\omega)] \leq \gamma ,
\end{aligned}$$

where g' is defined in (3.18).

Let us take a closer look at the function g' defined in (3.18). For the two-stage DRO described in Sect. 3.3, the original loss function g_i is the optimal value of an LP problem, which is in (3.3) and written here again for convenience:

$$g_i(x,\omega) := \min_y \; D_i^\top y \tag{3.19a}$$

$$\text{s.t. } A_i y \geq B_i(x) \cdot \omega + H_i(x) \, , \tag{3.19b}$$

As (3.18) shows, g_i' should be the optimal value of an LP problem where $g_i(x,\omega)$ appears on the right-hand side of the constraint. At first glance, there is an embedding structure of LP problems, but in fact,

$$g_i'(x,r,\omega) = \begin{cases} \min\limits_s \; s \\ \text{s.t. } s \geq g_i(x,\omega) - r \, , \; s \geq 0 \end{cases} \tag{3.20a}$$

$$= \begin{cases} \min\limits_{s,y} \; s \\ \text{s.t. } s \geq D_i^\top y - r \, , \; s \geq 0 \, , \\ \qquad A_i y \geq B_i(x) \cdot \omega + H_i(x) \, , \end{cases} \tag{3.20b}$$

where LP (3.20b) can also be written in the form of (3.19) as follows:

$$\min_{(s,y)} \begin{pmatrix} 1 \\ 0 \end{pmatrix}^\top \begin{pmatrix} s \\ y \end{pmatrix}$$

$$\text{s.t. } \begin{pmatrix} 1 & -D_i^\top \\ 1 & 0 \\ 0 & A_i \end{pmatrix} \begin{pmatrix} s \\ y \end{pmatrix} \geq \begin{pmatrix} 0 \\ 0 \\ B_i(x) \end{pmatrix} \cdot \omega + \begin{pmatrix} -r \\ 0 \\ H_i(x) \end{pmatrix} .$$

Note that (x,r) is regarded as the new first-stage decision variable. Therefore, the new function g' is still the optimal value of a single-layer LP and the structure does not become essentially more complicated after expressing the distributionally robust CVaR by worst-case expectation.

Exercise 3.5 Suppose $g_i(x,\omega)$ defined in (3.19) is finite with given x and ω. Prove that the optimal values of (3.20a) and (3.20b) are both finite and equal to each other.

3.4.5 Reformulating Distributionally Robust Shortfall Risk by Worst-Case Expectation

Using the definition of Shortfall risk in Definition 2.41, the distributionally robust Shortfall risk is:

$$\sup_{\mathbb{P}\in\mathcal{P}} \mathrm{SR}_{l,\lambda}^{\mathbb{P}}(g(\omega)) := \sup_{\mathbb{P}\in\mathcal{P}} \inf \left\{ r \in \mathbb{R} \;\middle|\; \mathbb{E}^{\mathbb{P}}[l(g(\omega)-r)] \leq \lambda \right\} , \tag{3.21}$$

where λ is the maximum acceptable loss and $l : \mathbb{R} \to \mathbb{R}$ is a nonconstant, nondecreasing, and convex function measuring the degree of loss. Theorem 3.4 says that distributionally robust Shortfall risk can be expressed using worst-case expectation (Guo and Xu 2019).

Theorem 3.4 (Reformulating distributionally robust Shortfall risk by worst-case expectation) *Let $\mathcal{P}$ be an ambiguity set, $g(\omega)$ be a real-valued function of ω, $l : \mathbb{R} \to \mathbb{R}$ be a nonconstant, nondecreasing, and convex function, λ be a constant. Then*

$$\sup_{\mathbb{P}\in\mathcal{P}} SR^{\mathbb{P}}_{l,\lambda}(g(\omega)) = \inf\left\{r \in \mathbb{R} \;\middle|\; \sup_{\mathbb{P}\in\mathcal{P}} \mathbb{E}^{\mathbb{P}}[l(g(\omega) - r)] \le \lambda\right\} \tag{3.22a}$$

$$= \left\{\begin{array}{l} \min\limits_{r} r \\ \text{s.t. } \sup\limits_{\mathbb{P}\in\mathcal{P}} \mathbb{E}^{\mathbb{P}}[l(g(\omega) - r)] \le \lambda\,. \end{array}\right. \tag{3.22b}$$

Proof Denote $\sup_{\mathbb{P}\in\mathcal{P}} \mathrm{SR}^{\mathbb{P}}_{l,\lambda}(g(\omega))$ by A. We prove (3.22a) in two steps.

1. Prove that $A \le \inf\left\{r \in \mathbb{R} \;\middle|\; \sup_{\mathbb{P}\in\mathcal{P}} \mathbb{E}^{\mathbb{P}}[l(g(\omega) - r)] \le \lambda\right\}$.

Fix a probability distribution $\mathbb{P}_1 \in \mathcal{P}$. We have:

$$\left\{r \in \mathbb{R} \mid \mathbb{E}^{\mathbb{P}_1}[l(g(\omega) - r)] \le \lambda\right\} \supset \bigcap_{\mathbb{P}\in\mathcal{P}} \left\{r \in \mathbb{R} \mid \mathbb{E}^{\mathbb{P}}[l(g(\omega) - r)] \le \lambda\right\}\,. \tag{3.23}$$

Assume that $r \in \mathbb{R}$ satisfies $\sup_{\mathbb{P}\in\mathcal{P}} \mathbb{E}^{\mathbb{P}}[l(g(\omega) - r)] \le \lambda$, then for any $\mathbb{P} \in \mathcal{P}$, $\mathbb{E}^{\mathbb{P}}[l(g(\omega) - r)] \le \lambda$. Therefore,

$$\bigcap_{\mathbb{P}\in\mathcal{P}} \left\{r \in \mathbb{R} \mid \mathbb{E}^{\mathbb{P}}[l(g(\omega) - r)] \le \lambda\right\} \supset \left\{r \in \mathbb{R} \;\middle|\; \sup_{\mathbb{P}\in\mathcal{P}} \mathbb{E}^{\mathbb{P}}[l(g(\omega) - r)] \le \lambda\right\}\,. \tag{3.24}$$

Then (3.23) and (3.24) imply:

$$\inf\left\{r \in \mathbb{R} \mid \mathbb{E}^{\mathbb{P}_1}[l(g(\omega) - r)] \le \lambda\right\} \le \inf\left\{r \in \mathbb{R} \;\middle|\; \sup_{\mathbb{P}\in\mathcal{P}} \mathbb{E}^{\mathbb{P}}[l(g(\omega) - r)] \le \lambda\right\}\,.$$

Thus, by letting $\mathbb{P}_1$ change in $\mathcal{P}$, we have:

$$\begin{aligned} A &= \sup_{\mathbb{P}\in\mathcal{P}} \inf\left\{r \in \mathbb{R} \mid \mathbb{E}^{\mathbb{P}}[l(g(\omega) - r)] \le \lambda\right\} \\ &\le \inf\left\{r \in \mathbb{R} \;\middle|\; \sup_{\mathbb{P}\in\mathcal{P}} \mathbb{E}^{\mathbb{P}}[l(g(\omega) - r)] \le \lambda\right\}\,. \end{aligned}$$

2. Prove that $A \geq \inf \left\{ r \in \mathbb{R} \;\middle|\; \sup_{\mathbb{P} \in \mathcal{P}} \mathbb{E}^{\mathbb{P}}[l(g(\omega) - r)] \leq \lambda \right\}$.
By (3.21),

$$\forall \mathbb{P} \in \mathcal{P}\,,\; A \geq \inf \left\{ r \in \mathbb{R} \mid \mathbb{E}^{\mathbb{P}}[l(g(\omega) - r)] \leq \lambda \right\}\,, \tag{3.25}$$

which implies that:

$$\forall \mathbb{P} \in \mathcal{P}\,,\; \forall \varepsilon > 0\,,\; \exists \delta < \varepsilon\,,\; \text{s.t. } \mathbb{E}^{\mathbb{P}}[l(g(\omega) - (A + \delta))] \leq \lambda\,. \tag{3.26}$$

Since l is nondecreasing, we have:

$$\forall \mathbb{P} \in \mathcal{P}\,,\; \forall \varepsilon > 0\,,\; \mathbb{E}^{\mathbb{P}}[l(g(\omega) - (A + \varepsilon))] \leq \lambda\,,$$

which implies:

$$\forall \varepsilon > 0\,,\; \sup_{\mathbb{P} \in \mathcal{P}} \mathbb{E}^{\mathbb{P}}[l(g(\omega) - (A + \varepsilon))] \leq \lambda\,.$$

Change the variable using $r = A + \varepsilon$, then

$$\forall r > A\,,\; \sup_{\mathbb{P} \in \mathcal{P}} \mathbb{E}^{\mathbb{P}}[l(g(\omega) - r)] \leq \lambda\,.$$

In other words,

$$\{r \in \mathbb{R} \mid r > A\} \subset \left\{ r \in \mathbb{R} \;\middle|\; \sup_{\mathbb{P} \in \mathcal{P}} \mathbb{E}^{\mathbb{P}}[l(g(\omega) - r)] \leq \lambda \right\}\,.$$

By taking the infimum, we have $A \geq \inf \left\{ r \in \mathbb{R} \;\middle|\; \sup_{\mathbb{P} \in \mathcal{P}} \mathbb{E}^{\mathbb{P}}[l(g(\omega) - r)] \leq \lambda \right\}$. □

Exercise 3.6 1. Prove (3.22b). 2. Prove that (3.25) implies (3.26).

Exercise 3.7 Let $\mathcal{P}$ be an ambiguity set, and $g(\omega)$ be a real-valued function of ω. Prove:

$$\sup_{\mathbb{P} \in \mathcal{P}} \beta\text{-VaR}^{\mathbb{P}}(g(\omega)) = \inf \left\{ r \in \mathbb{R} \;\middle|\; \inf_{\mathbb{P} \in \mathcal{P}} \mathbb{P}(g(\omega) \leq r) \geq \beta \right\}\,.$$

In the following, we assume the function $l : \mathbb{R} \to \mathbb{R}$ is piecewise linear and the loss function is defined by (3.3) in the two-stage DRO model, which is:

$$g_i(x, \omega) := \min_{y} \; D_i^\top y \tag{3.27a}$$

$$\text{s.t. } A_i y \geq B_i(x) \cdot \omega + H_i(x)\,. \tag{3.27b}$$

We will show that under these assumptions, $l(g_i(x, \omega) - r)$ can also be written in the form of the optimal value of an LP problem. Lemma 3.4 gives an expression of the function l.

Lemma 3.4 *Suppose the function $l : \mathbb{R} \to \mathbb{R}$ is nondecreasing, convex, and piecewise linear, then there are constants $C_{1,j} \geq 0$, $C_{0,j} \in \mathbb{R}$ and index set $j \in \mathcal{J}$ such that*

$$l(t) = \max_{j \in \mathcal{J}} \{C_{1,j} \cdot t + C_{0,j}\}, \ \forall t \in \mathbb{R}. \tag{3.28}$$

Conversely, if $l(t) = \sup_{j \in \mathcal{J}} \{C_{1,j} \cdot t + C_{0,j}\}, \forall t \in \mathbb{R}$ for a function $l : \mathbb{R} \to \mathbb{R}$ and constants $C_{1,j} \geq 0$, $C_{0,j} \in \mathbb{R}$, then l is nondecreasing and convex.

Proof 1. Suppose l is nondecreasing, convex, and piecewise linear. l is piecewise linear means that there are $\cdots < t_{-2} < t_{-1} < t_0 < t_1 < t_2 < \cdots < t_j < \cdots$ and $C_{0,j}$, $C_{1,j}$ such that $l(t) = C_{1,j} \cdot t + C_{0,j}, \forall t_j \leq t \leq t_{j+1}$. Let the index set of j be $\mathcal{J}$. Because l is nondecreasing, $C_{1,j} \geq 0, \forall j \in \mathcal{J}$.

For any $t \in \mathbb{R}$, there is $j \in \mathcal{J}$ such that $t_j \leq t \leq t_{j+1}$. If $j' \in \mathcal{J}$ and $j' \neq j$, then $j' < j$ or $j' > j$. Assume $j' < j$. Choose t' and t'' so that $t_{j'} < t' < t'' < t_{j'+1} \leq t_j \leq t$. Since l is convex, we have:

$$\begin{aligned} l(t) &\geq l(t') + \frac{t - t'}{t'' - t'}(l(t'') - l(t')) \\ &= C_{1,j'} \cdot t' + C_{0,j'} + \frac{t - t'}{t'' - t'} \cdot C_{1,j'} \cdot (t'' - t') \\ &= C_{1,j'} \cdot t + C_{0,j'} . \end{aligned}$$

When $j' > j$, a similar result holds. Thus, $l(t) \geq C_{1,j'} \cdot t + C_{0,j'}, \forall j' \neq j$ and $l(t) = C_{1,j} \cdot t + C_{0,j}$, so (3.28) follows.

2. Suppose $l(t) = \sup_{j \in \mathcal{J}} \{C_{1,j} \cdot t + C_{0,j}\}, \forall t \in \mathbb{R}$ and $C_{1,j} \geq 0, \forall j \in \mathcal{J}$.

To show l is nondecreasing, assume $t' > t$. For any $j \in \mathcal{J}$, $C_{1,j} \cdot t' + C_{0,j} \geq C_{1,j} \cdot t + C_{0,j}$ because $C_{1,j} \geq 0$. By taking supremum over $j \in \mathcal{J}$, we have:

$$l(t') = \sup_{j \in \mathcal{J}} \{C_{1,j} \cdot t' + C_{0,j}\} \geq \sup_{j \in \mathcal{J}} \{C_{1,j} \cdot t + C_{0,j}\} = l(t),$$

so l is nondecreasing.

To show l is convex, assume $t < t'$, $0 < \lambda < 1$, and $t'' = \lambda t + (1 - \lambda)t'$. By $l(t'') = \sup_{j \in \mathcal{J}} \{C_{1,j} \cdot t'' + C_{0,j}\}$,

$$\forall \varepsilon > 0, \ \exists j''_\varepsilon \in \mathcal{J}, \ \text{s.t.}\ l(t'') - \varepsilon \leq C_{1,j''_\varepsilon} \cdot t'' + C_{0,j''_\varepsilon} .$$

Note that

$$\begin{aligned}C_{1,j''_\varepsilon} \cdot t'' + C_{0,j''_\varepsilon} &= \lambda(C_{1,j''_\varepsilon} \cdot t + C_{0,j''_\varepsilon}) + (1-\lambda)(C_{1,j''_\varepsilon} \cdot t' + C_{0,j''_\varepsilon}) \\ &\le \lambda \cdot l(t) + (1-\lambda) \cdot l(t') \ .\end{aligned}$$

Let $\varepsilon \to 0$, then $l(t'') \le \lambda \cdot l(t) + (1-\lambda) \cdot l(t')$. Therefore, l is a convex function. □

Remark 3.3 The index set $\mathcal{J}$ is not necessarily finite.

Using the expression of l in (3.28), we have:

$$\begin{aligned}& l(g_i(x,\omega) - r) \\ =\ & \max_{j \in \mathcal{J}} \{C_{1,j} \cdot (g_i(x,\omega) - r) + C_{0,j}\} \\ =\ & \begin{cases} \min\limits_{s}\ s \\ \text{s.t. } C_{1,j} \cdot (g_i(x,\omega) - r) + C_{0,j} \le s\ ,\ \forall j \in \mathcal{J} \end{cases} \\ =\ & \begin{cases} \min\limits_{s,y}\ s \\ \text{s.t. } C_{1,j}(D_i^\top y - r) + C_{0,j} \le s\ ,\ \forall j \in \mathcal{J}\ , \\ \qquad A_i y \ge B_i(x) \cdot \omega + H_i(x)\ , \end{cases}\end{aligned} \tag{3.29}$$

which can be written as follows to be consistent with the form in (3.3) provided that $\mathcal{J}$ is finite.

$$\min_{(s,y)} \begin{pmatrix} 1 \\ 0 \end{pmatrix}^\top \begin{pmatrix} s \\ y \end{pmatrix} \tag{3.30a}$$

$$\text{s.t. } \begin{pmatrix} 1 & -C_1 D_i^\top \\ 0 & A_i \end{pmatrix} \begin{pmatrix} s \\ y \end{pmatrix} \ge \begin{pmatrix} 0 \\ B_i(x) \end{pmatrix} \cdot \omega + \begin{pmatrix} C_0 - rC_1 \\ H_i(x) \end{pmatrix}\ , \tag{3.30b}$$

where C_0 and C_1 are the column vectors of $(C_{0,j};\ j \in \mathcal{J})$ and $(C_{1,j};\ j \in \mathcal{J})$, respectively. The new first-stage decision variable is (x, r).

A general function l can be approximated by a piecewise linear function with finite pieces. In practical problems, the loss function $g(x,\omega)$ has a bounded range, so we only need to approximate l using a piecewise linear function on a bounded set, where the approximation precision can be controlled. Therefore, l being piecewise linear is a mild assumption.

Exercise 3.8 Explain the equivalence of (3.29) and (3.30).

Reference

Guo S, Xu H (2019) Distributionally robust shortfall risk optimization model and its approximation. Math Program 174:473–498

Chapter 4
Moment-Based Distributionally Robust Optimization

Abstract The idea of distributionally robust optimization (DRO) originated in the middle of the 20th century, where the ambiguity set is built with first- and second-order moments constraints. Moment-based DRO not only has the longest history among different DRO techniques but also is one of the most popular and mature ways to model the inexactness of empirical probability distribution for uncertainty-integrated optimization problems. This chapter focuses on the moment-based DRO, typical moment-based ambiguity sets are summarized, the reformulation methods are explained, and finally, the moment-based DRO incorporating unimodality is also introduced.

4.1 Moment-Based Ambiguity Sets

We first present the definitions of two types of moment-based ambiguity sets and then discuss the choices of parameters.

4.1.1 Definitions

Suppose the random variable is denoted by ω. Its sample space $\Omega \subset \mathbb{R}^M$ is a closed set. $\mathcal{F}$ is the Borel σ-algebra on Ω. $\mathcal{M}(\Omega)$ is the set of all probability distributions on $(\Omega, \mathcal{F})$. The first- and second-order moments of ω are estimated using the historical data $\omega_1, \omega_2, \ldots, \omega_N$ by

$$\mu_0 := \frac{1}{N}\sum_{n=1}^{N} \omega_n \ , \ \Sigma_0 := \frac{1}{N}\sum_{n=1}^{N}(\omega_n - \mu_0)(\omega_n - \mu_0)^\top \ . \tag{4.1}$$

We assume Σ_0 is positive definite. In fact, Σ_0 can have zero eigenvalues only when there are linear dependencies between the elements of ω.

R. Xie and W. Wei, *Distributionally Robust Optimization and its Applications in Power System Energy Storage Sizing*,
https://doi.org/10.1007/978-981-97-2566-3_4

The first type of moment-based ambiguity set is constructed by the probability distributions in $\mathcal{M}(\Omega)$ that have first-order moment μ_0 and second-order moment Σ_0 (Scarf 1958).

Definition 4.1 (*Moment-based ambiguity set with exact moments*) The moment-based ambiguity set $\mathcal{P}_{\mathrm{M}}(\Omega, \mu_0, \Sigma_0)$ on sample space Ω with exact first- and second-order moments μ_0 and Σ_0 is defined as follows:

$$\begin{aligned}\mathcal{P}_{\mathrm{M}}(\Omega, \mu_0, \Sigma_0) &:= \left\{\mathbb{P} \in \mathcal{M}(\Omega) \;\middle|\; \begin{array}{l}\mathbb{E}^{\mathbb{P}}[\omega] = \mu_0 \,, \\ \mathbb{E}^{\mathbb{P}}[(\omega - \mu_0)(\omega - \mu_0)^{\top}] = \Sigma_0\end{array}\right\} \\ &= \left\{\mathbb{P} \in \mathcal{M}(\Omega) \;\middle|\; \begin{array}{l}\mathbb{E}^{\mathbb{P}}[\omega] = \mu_0 \,, \\ \mathbb{E}^{\mathbb{P}}[\omega\omega^{\top}] = \Sigma_0 + \mu_0\mu_0^{\top}\end{array}\right\} . \end{aligned} \tag{4.2}$$

The formulation in Definition 4.1 assumes exact knowledge of the first- and second-order moments. However, since they are estimated based on limited historical data, there are estimation errors inevitably. To this end, another moment-based ambiguity set was proposed by Delage and Ye (2010), where the moments are allowed to take values in a small range.

Definition 4.2 (*Moment-based ambiguity set with ambiguous moments*) On sample space Ω, parametrized by size parameters $\gamma_1 \geq 0$ and $\gamma_2 \geq 1$, the moment-based ambiguity set with estimated first- and second-order moments μ_0 and $\Sigma_0 \succ 0$ is defined as follows:

$$\mathcal{P}_{\mathrm{M}}(\Omega, \mu_0, \Sigma_0, \gamma_1, \gamma_2) := \left\{\mathbb{P} \in \mathcal{M}(\Omega) \;\middle|\; \begin{array}{l}(\mathbb{E}^{\mathbb{P}}[\omega] - \mu_0)^{\top}\Sigma_0^{-1}(\mathbb{E}^{\mathbb{P}}[\omega] - \mu_0) \leq \gamma_1 \,, \\ \mathbb{E}^{\mathbb{P}}[(\omega - \mu_0)(\omega - \mu_0)^{\top}] \preceq \gamma_2\Sigma_0\end{array}\right\} . \tag{4.3}$$

In Definition 4.2, the first constraint says the expectation of ω lies in the ellipsoid centered at μ_0, whose size is controlled by the parameter γ_1. The second constraint means $\gamma_2\Sigma_0 - \mathbb{E}^{\mathbb{P}}[(\omega - \mu_0)(\omega - \mu_0)^{\top}]$ is a positive semidefinite matrix. It should be noted that $\mathbb{E}^{\mathbb{P}}[(\omega - \mu_0)(\omega - \mu_0)^{\top}]$ is not the second-order moment of ω because the estimated first-order moment μ_0 may have errors. γ_1 and γ_2 control the size of the ambiguity set together. Larger γ_1 and γ_2 mean a larger ambiguity set. Moreover, the two types of moment-based ambiguity sets satisfy $\mathbb{P}_0 \in \mathcal{P}_{\mathrm{M}}(\Omega, \mu_0, \Sigma_0) \subsetneq \mathcal{P}_{\mathrm{M}}(\Omega, \mu_0, \Sigma_0, 0, 1) \subset \mathcal{P}_{\mathrm{M}}(\Omega, \mu_0, \Sigma_0, \gamma_1, \gamma_2), \forall \gamma_1 \geq 0, \gamma_2 \geq 1$, where $\mathbb{P}_0 := \sum_{n=1}^{N} \delta_{\omega_n}/N$ is the empirical probability distribution with δ_{ω_n} denoting the Dirac distribution at ω_n.

4.1.2 The Choices of Parameters

There are some theoretical results about the convergence rates of the moments, which can facilitate the choices of the parameters γ_1 and γ_2 in the ambiguity set

$\mathcal{P}_{\mathrm{M}}(\Omega, \mu_0, \Sigma_0, \gamma_1, \gamma_2)$ in Definition 4.2. Theorem 4.1 gives the confidence regions of the first- and second-order moments.

Theorem 4.1 (Corollary 3 in Delage and Ye (2010)) *Suppose $\{\omega_i\}_{i=1}^N$ are independently drawn from the probability distribution $\mathbb{P}_*$ on a bounded sample space $\Omega \subset \mathbb{R}^M$. The first- and second-order moments are denoted by μ_* and Σ_*, respectively. The estimated first- and second-order moments are μ_0 and $\Sigma_0 \succ 0$ in (4.1). If for $\delta > 0$, the sample number satisfies*

$$N > \max\left\{(\hat{R}^2+2)^2\left(2+\sqrt{2\log(4/\bar{\delta})}\right)^2, \frac{\left(8+\sqrt{32\log(4/\bar{\delta})}\right)^2}{(\sqrt{\hat{R}+4}-\hat{R})^4}\right\},$$

with

$$\bar{\delta} = 1-\sqrt{1-\delta}\,,\ \hat{R} = \sup_{\omega\in\Omega} \|\Sigma_0^{-1/2}(\omega-\mu_0)\|_2\,,$$

where $\Sigma_0^{-1/2}$ is the unique matrix such that $\Sigma_0^{-1/2}\Sigma_0^{-1/2} = (\Sigma_0^{-1/2})^\top\Sigma_0^{-1/2} = \Sigma_0^{-1}$ for positive definite Σ_0. Then the following constraints are jointly met with a probability no smaller than $1-\delta$:

$$\begin{cases} (\mu_*-\mu_0)\Sigma_*^{-1}(\mu_*-\mu_0) \le \bar{\beta}(\bar{\delta}/2)\,, \\ \dfrac{1}{1+\bar{\alpha}(\bar{\delta}/4)}\Sigma_0 \preceq \Sigma_* \preceq \dfrac{1}{1-\bar{\alpha}(\bar{\delta}/4)-\bar{\beta}(\bar{\delta}/2)}\Sigma_0\,, \end{cases}$$

where

$$\bar{\alpha}(\bar{\delta}/4) = \frac{\bar{R}^2}{\sqrt{N}}\left(\sqrt{1-M/\bar{R}^4}+\sqrt{\log(4/\bar{\delta})}\right),$$

$$\bar{\beta}(\bar{\delta}/2) = \frac{\bar{R}^2}{N}\left(2+\sqrt{2\log(2/\bar{\delta})}\right)^2,$$

$$\bar{R} = \left(1-(\hat{R}^2+2)\frac{2+\sqrt{2\log(4/\bar{\delta})}}{\sqrt{N}}\right)^{-1/2}\hat{R}\,.$$

Based on Theorem 4.1, the parameter choices of $\mathcal{P}_{\mathrm{M}}(\Omega, \mu_0, \Sigma_0, \gamma_1, \gamma_2)$ with a probability guarantee can be derived, and the result is as follows.

Theorem 4.2 (Corollary 4 in Delage and Ye (2010)) *Under the assumptions of Theorem 4.1, set the parameters γ_1 and γ_2 by*

$$\gamma_1 = \frac{\bar{\beta}(\bar{\delta}/2)}{1-\bar{\alpha}(\bar{\delta}/4)-\bar{\beta}(\bar{\delta}/2)}\,,\ \gamma_2 = \frac{1+\bar{\beta}(\bar{\delta}/2)}{1-\bar{\alpha}(\bar{\delta}/4)-\bar{\beta}(\bar{\delta}/2)}\,.$$

Then the probability of $\mathbb{P}_* \in \mathcal{P}_{\mathrm{M}}(\Omega, \mu_0, \Sigma_0, \gamma_1, \gamma_2)$ is no smaller than $1-\delta$.

The probability of $\mathbb{P}_* \in \mathcal{P}_M(\Omega, \mu_0, \Sigma_0, \gamma_1, \gamma_2)$ is comprehended in this way: $\mathbb{P}_*$ is a fixed probability distribution. $\{\omega_i\}_{i=1}^N$ are independently drawn from $\mathbb{P}_*$. The independent sample process can be described by a product distribution. In other words, $(\omega_i; i = 1, 2, \ldots, N)$ is a random variable following the product distribution $\times_{i=1}^N \mathbb{P}_*$. Since the estimated moments μ_0 and Σ_0 are functions of $(\omega_i; i = 1, 2, \ldots, N)$, they are also random variables. Then with given parameters Ω, γ_1, and γ_2, the set $\mathcal{P}_M(\Omega, \mu_0, \Sigma_0, \gamma_1, \gamma_2)$ is a random set of probability distributions. Thus, $\mathbb{P}_* \in \mathcal{P}_M(\Omega, \mu_0, \Sigma_0, \gamma_1, \gamma_2)$ is a random event, and its probability is measured with respect to the product distribution $\times_{i=1}^N \mathbb{P}_*$, so the probability guarantee result in Theorem 4.2 can be denoted by

$$(\times_{i=1}^N \mathbb{P}_*)(\mathbb{P}_* \in \mathcal{P}_M(\Omega, \mu_0, \Sigma_0, \gamma_1, \gamma_2)) \geq 1 - \delta \ .$$

When the above probability guarantee holds for the ambiguity set, we also have a probability guarantee for the worst-case expectation:

$$(\times_{i=1}^N \mathbb{P}_*) \left(\mathbb{E}^{\mathbb{P}_*}[g(\omega)] \leq \sup_{\mathbb{P} \in \mathcal{P}_M(\Omega, \mu_0, \Sigma_0, \gamma_1, \gamma_2)} \mathbb{E}^{\mathbb{P}}[g(\omega)] \right) \geq 1 - \delta \ ,$$

where $g(\omega) : \Omega \to \mathbb{R}$ is a function representing the loss.

Exercise 4.1 Under the assumptions of Theorem 4.2, write down the probability guarantees for robust chance constraints and distributionally robust risk measures.

4.2 Reformulation of Worst-Case Expectation

In this section, we introduce the reformulation methods of worst-case expectation under the two types of moment-based ambiguity sets one by one.

4.2.1 Ambiguity Set of Exact Moments

Suppose $g : \mathcal{X} \times \Omega \to \mathbb{R}$ is a continuous and bounded function of decision variable $x \in \mathcal{X}$ and random variable $\omega \in \Omega$. We first investigate the reformulation of worst-case expectation under the ambiguity set $\mathcal{P}_M(\Omega, \mu_0, \Sigma_0)$ defined in (4.2).

$$\sup_{\mathbb{P}\in\mathcal{P}_{\mathrm{M}}(\Omega,\mu_0,\Sigma_0)} \mathbb{E}^{\mathbb{P}}[g(x,\omega)] \tag{4.4a}$$

$$= \left\{ \begin{array}{l} \sup\limits_{\mathbb{P}\in C} \displaystyle\int_\Omega g(x,\omega)\mathrm{d}\mathbb{P}(\omega) \\ \text{s.t. } \displaystyle\int_\Omega \mathrm{d}\mathbb{P}(\omega) = 1 : \theta \ , \\ \quad\ \displaystyle\int_\Omega \omega \mathrm{d}\mathbb{P}(\omega) = \mu_0 : \phi \ , \\ \quad\ \displaystyle\int_\Omega \omega\omega^\top \mathrm{d}\mathbb{P}(\omega) = \Sigma_0 + \mu_0\mu_0^\top : \Psi \end{array} \right. \tag{4.4b}$$

$$= \left\{ \begin{array}{l} \inf\limits_{\theta\in\mathbb{R},\phi\in\mathbb{R}^M,\Psi\in\mathbb{S}^M} \theta + \mu_0^\top\phi + \langle \Sigma_0 + \mu_0\mu_0^\top, \Psi\rangle \\ \qquad\qquad \text{s.t. } \theta + \omega^\top\phi + \langle\omega\omega^\top, \Psi\rangle \ge g(x,\omega) \ , \ \forall\omega\in\Omega \end{array} \right. \tag{4.4c}$$

$$= \left\{ \begin{array}{l} \inf\limits_{\theta\in\mathbb{R},\phi\in\mathbb{R}^M,\Psi\in\mathbb{S}^M} \left\langle \begin{pmatrix} \Sigma_0 + \mu_0\mu_0^\top & \mu_0 \\ \mu_0^\top & 1 \end{pmatrix}, \begin{pmatrix} \Psi & \phi/2 \\ \phi^\top/2 & \theta \end{pmatrix} \right\rangle \\ \qquad\qquad \text{s.t. } \left\langle \begin{pmatrix} \omega\omega^\top & \omega \\ \omega^\top & 1 \end{pmatrix}, \begin{pmatrix} \Psi & \phi/2 \\ \phi^\top/2 & \theta \end{pmatrix} \right\rangle \ge g(x,\omega) \ , \ \forall\omega\in\Omega \ . \end{array} \right. \tag{4.4d}$$

The equality from (4.4a) to (4.4b) follows from the definition of $\mathcal{P}_{\mathrm{M}}(\Omega, \mu_0, \Sigma_0)$, where C is the conic of all the nonnegative measures on $(\Omega, \mathcal{F})$. (4.4b) is a problem of moments, which has been introduced in Sect. 2.5.7. The empirical probability distribution $\mathbb{P}_0$ is a feasible solution. Since g is continuous and bounded and Σ_0 is positive definite, the strong duality of conic programs in Proposition 2.45 can be applied to problem (4.4b), which leads to (4.4c). In problem (4.4c), the decision variables are θ, ϕ, and Ψ. The objective and constraint functions are linear in these decision variables. Recall the definition of the inner product for symmetric matrices from Sect. 2.5.6, then we have $\langle\Sigma_0, \Psi\rangle = \mathrm{tr}(\Sigma_0^\top\Psi)$ and $\langle\omega\omega^\top, \Psi\rangle = \mathrm{tr}(\omega\omega^\top\Psi) = \omega^\top\Psi\omega$. The number of constraints is infinite when Ω is an infinite set. Problem (4.4c) and problem (4.4d) have the form of a robust optimization (RO) program. However, since Ψ is free, i.e., Ψ is not required to be positive semidefinite, the function $\theta + \omega^\top\phi + \langle\omega\omega^\top, \Psi\rangle - g(x,\omega)$ can be nonconvex in ω even if $g(x,\omega)$ is concave in ω, which may lead to intractable problems.

For a general Ω, the worst-case expectation $\sup_{\mathbb{P}\in\mathcal{P}_{\mathrm{M}}(\Omega,\mu_0,\Sigma_0)} \mathbb{E}^{\mathbb{P}}[g(x,\omega)]$ can be calculated by applying RO solution technique to the equivalent problem (4.4c) or (4.4d). In the following, we discuss three kinds of special sample spaces.

4.2.1.1 Example: Finite Sample Space

When $\Omega = \{\omega_1, \omega_2, \ldots, \omega_N\}$ is a finite set, the worst-case expectation can be further equivalently transformed into:

$$\begin{cases} \inf\limits_{\theta,\phi,\Psi\in\mathbb{S}^M} & \left\langle \begin{pmatrix} \Sigma_0+\mu_0\mu_0^\top & \mu_0 \\ \mu_0^\top & 1 \end{pmatrix}, \begin{pmatrix} \Psi & \phi/2 \\ \phi^\top/2 & \theta \end{pmatrix} \right\rangle \\ \text{s.t.} & \left\langle \begin{pmatrix} \omega_n\omega_n^\top & \omega_n \\ \omega_n^\top & 1 \end{pmatrix}, \begin{pmatrix} \Psi & \phi/2 \\ \phi^\top/2 & \theta \end{pmatrix} \right\rangle \geq g(x,\omega_n)\ ,\ n=1,2,\dots,N\ . \end{cases}$$

If the function $g : \mathcal{X} \times \Omega \to \mathbb{R}$ is the optimal value of a linear programming (LP) problem, i.e., $g(x,\omega) = \min_y \{D^\top y \mid Ay \geq B(x)\cdot\omega + H(x)\}$, where A and D are constant and $B(x)$ and $H(x)$ are linear in x. Then $\sup_{\mathbb{P}\in\mathcal{P}_{\mathrm{M}}(\Omega,\mu_0,\Sigma_0)} \mathbb{E}^{\mathbb{P}}[g(x,\omega)]$ can be calculated by solving the following LP problem:

$$\begin{cases} \inf\limits_{\theta,\phi,\Psi\in\mathbb{S}^M,y} & \left\langle \begin{pmatrix} \Sigma_0+\mu_0\mu_0^\top & \mu_0 \\ \mu_0^\top & 1 \end{pmatrix}, \begin{pmatrix} \Psi & \phi/2 \\ \phi^\top/2 & \theta \end{pmatrix} \right\rangle \\ \text{s.t.} & \left\langle \begin{pmatrix} \omega_n\omega_n^\top & \omega_n \\ \omega_n^\top & 1 \end{pmatrix}, \begin{pmatrix} \Psi & \phi/2 \\ \phi^\top/2 & \theta \end{pmatrix} \right\rangle \geq D^\top y_n\ ,\ n=1,2,\dots,N\ , \\ & Ay_n \geq B(x)\cdot\omega_n + H(x)\ ,\ n=1,2,\dots,N\ . \end{cases}$$

4.2.1.2 Example: Euclidean Sample Space and Nonnegative Loss Function

When $\Omega = \mathbb{R}^M$ is the Euclidean space and $g(x,\omega) = (f(x,\omega))_+ = \max\{f(x,\omega),0\}$, the constraint

$$\theta + \omega^\top\phi + \langle \omega\omega^\top, \Psi\rangle \geq g(x,\omega)\ ,\ \forall\omega\in\Omega=\mathbb{R}^M\ , \tag{4.5}$$

is equivalent to:

$$\begin{pmatrix}\omega\\1\end{pmatrix}^\top \begin{pmatrix} \Psi & \phi/2 \\ \phi^\top/2 & \theta \end{pmatrix} \begin{pmatrix}\omega\\1\end{pmatrix} \geq 0\ ,\ \forall\omega\in\mathbb{R}^M\ , \tag{4.6a}$$

$$\begin{pmatrix}\omega\\1\end{pmatrix}^\top \begin{pmatrix} \Psi & \phi/2 \\ \phi^\top/2 & \theta \end{pmatrix} \begin{pmatrix}\omega\\1\end{pmatrix} \geq f(x,\omega)\ ,\ \forall\omega\in\mathbb{R}^M\ . \tag{4.6b}$$

We claim that (4.6a) holds if and only if the symmetric matrix

$$Q := \begin{pmatrix} \Psi & \phi/2 \\ \phi^\top/2 & \theta \end{pmatrix} \succeq 0\ .$$

It is clear that $Q \succeq 0$ implies (4.6a). For the opposite direction, assume Q is not positive semidefinite, then there exist $\lambda < 0$, $s \in \mathbb{R}^M$, and $t \in \mathbb{R}$ such that

$$Q\begin{pmatrix}s\\t\end{pmatrix} = \lambda\begin{pmatrix}s\\t\end{pmatrix}\ ,\ \|s\|^2 + t^2 > 0\ ,$$

where the norm is 2-norm. If $t \neq 0$, we can let $\omega = s/t$ and then it contradicts (4.6a). Thus, we only need to consider the case where $t = 0$. Then $\Psi s = \lambda s$ for $\lambda < 0$ and $s \neq 0$. Thus, $\theta + s^\top \phi + \langle ss^\top, \Psi \rangle \leq \theta + \|s\|\|\phi\| + \lambda \|s\|^2 \to -\infty$ as $\|s\| \to +\infty$, which also contradicts (4.6a). Therefore, constraint (4.5) is equivalent to the combination of $Q \succeq 0$ and (4.6b).

4.2.1.3 Example: Quadratic-Constrained Sample Space and a Special Loss Function

Suppose

$$\Omega = \left\{ \omega \in \mathbb{R}^M \;\middle|\; \begin{pmatrix} \omega \\ 1 \end{pmatrix}^\top W_l \begin{pmatrix} \omega \\ 1 \end{pmatrix} \leq 0 \,,\; l = 1, \ldots, L \right\} ,$$

with $W_l \in \mathbb{S}^M, l = 1, \ldots, L$ is a quadratic-constrained sample space. Assume the loss function $g(x, \omega) = (A(x) + B(x)^\top \omega + \omega^\top C(x)\omega)_+ = \max\{A(x) + B(x)^\top \omega + \omega^\top C(x)\omega, 0\}$ with $A(x) \in \mathbb{R}$, $B(x) \in \mathbb{R}^M$, and $C(x) \in \mathbb{S}^M$. Then the constraint

$$\theta + \omega^\top \phi + \langle \omega\omega^\top, \Psi \rangle \geq g(x, \omega) \,,\; \forall \omega \in \Omega \,, \tag{4.7}$$

is equivalent to:

$$\begin{pmatrix} \omega \\ 1 \end{pmatrix}^\top \begin{pmatrix} \Psi & \phi/2 \\ \phi^\top/2 & \theta \end{pmatrix} \begin{pmatrix} \omega \\ 1 \end{pmatrix} \geq 0 \,,\; \forall \omega \,,\; \text{s.t.} \begin{pmatrix} \omega \\ 1 \end{pmatrix}^\top W_l \begin{pmatrix} \omega \\ 1 \end{pmatrix} \leq 0 \,,\; l = 1, \ldots, L \,, \tag{4.8a}$$

$$\begin{aligned} &\begin{pmatrix} \omega \\ 1 \end{pmatrix}^\top \begin{pmatrix} \Psi & \phi/2 \\ \phi^\top/2 & \theta \end{pmatrix} \begin{pmatrix} \omega \\ 1 \end{pmatrix} \geq A(x) + B(x)^\top \omega + \omega^\top C(x)\omega \,, \\ &\qquad \forall \omega \,,\; \text{s.t.} \begin{pmatrix} \omega \\ 1 \end{pmatrix}^\top W_l \begin{pmatrix} \omega \\ 1 \end{pmatrix} \leq 0 \,,\; l = 1, \ldots, L \,. \end{aligned} \tag{4.8b}$$

To reformulate (4.8a), we first introduce a useful lemma.

Lemma 4.1 *Assume $A_i \in \mathbb{S}^M, i = 0, 1, \ldots, P$ and $\tau_i \geq 0, i = 1, \ldots, P$ satisfy that $A_0 + \sum_{p=1}^P \tau_i A_i \succeq 0$, then for any $\omega \in \mathbb{R}^M$ such that $\omega^\top A_i \omega \leq 0, i = 1, \ldots, P$, we have $\omega^\top A_0 \omega \geq 0$.*

Proof By $A_0 + \sum_{p=1}^P \tau_i A_i \succeq 0$, we have $\omega^\top (A_0 + \sum_{p=1}^P \tau_i A_i)\omega \geq 0$. Thus, $\omega^\top A_0 \omega \geq -\sum_{p=1}^P \tau_i \omega^\top A_i \omega \geq 0$. □

According to Lemma 4.1, a conservative approximation of constraint (4.8a) is $\exists \tau_0 \in \mathbb{R}_+^L$, s.t. $Q + \sum_{l=1}^L \tau_{0,l} W_l \succeq 0$. "Conservative approximation" here means that the feasible sets defined by the constraints satisfy:

$$\left\{ Q = \begin{pmatrix} \Psi & \phi/2 \\ \phi^\top/2 & \theta \end{pmatrix} \in \mathbb{S}^{M+1} \;\middle|\; (4.8a) \right\}$$
$$\supset \left\{ Q = \begin{pmatrix} \Psi & \phi/2 \\ \phi^\top/2 & \theta \end{pmatrix} \in \mathbb{S}^{M+1} \;\middle|\; \exists \tau_0 \in \mathbb{R}_+^L \,,\; \text{s.t. } Q + \sum_{l=1}^L \tau_{0,l} W_l \succeq 0 \right\} .$$

Constraint (4.8b) can be reformulated similarly. Finally, constraint (4.7) can be conservatively approximated by the following semidefinite constraints:

$$\exists \tau_0, \tau_1 \in \mathbb{R}_+^L \,,\; \text{s.t. } \begin{pmatrix} \Psi & \phi/2 \\ \phi^\top/2 & \theta \end{pmatrix} + \sum_{l=1}^L \tau_{0,l} W_l \succeq 0 \,,$$
$$\begin{pmatrix} \Psi & \phi/2 \\ \phi^\top/2 & \theta \end{pmatrix} + \sum_{l=1}^L \tau_{1,l} W_l - \begin{pmatrix} C(x) & B(x)/2 \\ B(x)^\top/2 & A(x) \end{pmatrix} \succeq 0 \,.$$

Exercise 4.2 Prove the reformulation of (4.8b).

4.2.2 Ambiguity Set of Ambiguous Moments

Now we turn to the ambiguity set $\mathcal{P}_\mathrm{M}(\Omega, \mu_0, \Sigma_0, \gamma_1, \gamma_2)$. To begin with, we transform the constraint $(\mathbb{E}^\mathbb{P}[\omega] - \mu_0)^\top \Sigma_0^{-1} (\mathbb{E}^\mathbb{P}[\omega] - \mu_0) \le \gamma_1$ in the definition of $\mathcal{P}_\mathrm{M}(\Omega, \mu_0, \Sigma_0, \gamma_1, \gamma_2)$ using the tool named Schur complement.

Definition 4.3 (*Schur complement*) Assume $A \in \mathbb{R}^{M \times M}$ is invertible and

$$Q = \begin{pmatrix} A & B \\ C & D \end{pmatrix} \in \mathbb{R}^{(M+N)\times(M+N)} \,.$$

Then the *Schur complement* of A in Q is defined as:

$$Q/A := D - CA^{-1}B \,. \tag{4.9}$$

Schur complement arises from Gauss elimination. Suppose we want to eliminate C in the second row block (C, D) using the first row block (A, B), then subtract $CA^{-1}(A, B)$ from (C, D) and the result is $(0, D - CA^{-1}B)$. This process can be written as:

$$\begin{pmatrix} I & 0 \\ -CA^{-1} & I \end{pmatrix} Q = \begin{pmatrix} A & B \\ 0 & Q/A \end{pmatrix} .$$

Schur complement can also be defined for complex matrices and it has fruitful properties. Here we characterize positive semidefinite matrices with the help of Schur complement.

Proposition 4.1 *Assume $A \in \mathbb{R}^{M\times M}$ is positive definite and the symmetric matrix*

$$Q = \begin{pmatrix} A & B \\ B^\top & D \end{pmatrix} \in \mathbb{R}^{(M+N)\times(M+N)} .$$

Then $Q \succeq 0$ if and only if the Schur complement $Q/A = D - B^\top A^{-1}B \succeq 0$.

Proof Let

$$N := \begin{pmatrix} I & A^{-1}B \\ 0 & I \end{pmatrix} .$$

It is clear that N is invertible. A direct calculation can verify that

$$N^\top \begin{pmatrix} A & 0 \\ 0 & Q/A \end{pmatrix} N = Q .$$

Then

$$\begin{aligned} Q \succeq 0 &\iff x^\top Qx \geq 0\,,\ \forall x \in \mathbb{R}^{M+N} \\ &\iff (Nx)^\top \begin{pmatrix} A & 0 \\ 0 & Q/A \end{pmatrix} (Nx) \geq 0\,,\ \forall x \in \mathbb{R}^{M+N} \\ &\iff \begin{pmatrix} A & 0 \\ 0 & Q/A \end{pmatrix} \succeq 0 \\ &\iff \begin{pmatrix} y \\ z \end{pmatrix}^\top \begin{pmatrix} A & 0 \\ 0 & Q/A \end{pmatrix} \begin{pmatrix} y \\ z \end{pmatrix} = y^\top Ay + z^\top (Q/A)z \geq 0\,,\ \forall y \in \mathbb{R}^M\,,\ z \in \mathbb{R}^N \\ &\iff z^\top (Q/A)z \geq 0\,,\ \forall z \in \mathbb{R}^N \\ &\iff Q/A \succeq 0\,, \end{aligned}$$

which has used the facts that N is invertible and $A \succ 0$. □

The constraint $(\mathbb{E}^{\mathbb{P}}[\omega] - \mu_0)^\top \Sigma_0^{-1}(\mathbb{E}^{\mathbb{P}}[\omega] - \mu_0) \leq \gamma_1$ can be equivalently transformed into a positive semidefinite cone constraint by regarding $\gamma_1 - (\mathbb{E}^{\mathbb{P}}[\omega] - \mu_0)^\top \Sigma_0^{-1}(\mathbb{E}^{\mathbb{P}}[\omega] - \mu_0)$ as the Schur complement in a matrix.

Corollary 4.1 *When $\Sigma_0 \succ 0$, we have:*

$$\begin{aligned} &(\mathbb{E}^{\mathbb{P}}[\omega] - \mu_0)^\top \Sigma_0^{-1}(\mathbb{E}^{\mathbb{P}}[\omega] - \mu_0) \leq \gamma_1 \\ \iff &\begin{pmatrix} \Sigma_0 & \mathbb{E}^{\mathbb{P}}[\omega] - \mu_0 \\ (\mathbb{E}^{\mathbb{P}}[\omega] - \mu_0)^\top & \gamma_1 \end{pmatrix} \succeq 0 \\ \iff &\int_\Omega \begin{pmatrix} \Sigma_0 & \omega - \mu_0 \\ (\omega - \mu_0)^\top & \gamma_1 \end{pmatrix} \mathrm{d}\mathbb{P}(\omega) \succeq 0 . \end{aligned}$$

The second equivalence uses the fact that the expectation of a constant is itself. For example, $\mathbb{E}^{\mathbb{P}}[\Sigma_0] = \int_\Omega \Sigma_0 \mathrm{d}\mathbb{P}(\omega) = \Sigma_0$.

Now we are ready to transform the worst-case expectation. Based on Definition 4.2 and Corollary 4.1, we have:

$$\sup_{\mathbb{P}\in\mathcal{P}_{\mathrm{M}}(\Omega,\mu_0,\Sigma_0,\gamma_1,\gamma_2)} \mathbb{E}^{\mathbb{P}}[g(x,\omega)] \tag{4.10a}$$

$$= \left\{ \begin{aligned} & \sup_{\mathbb{P}\in C} \int_\Omega g(x,\omega)\mathrm{d}\mathbb{P}(\omega) \\ & \text{s.t. } \int_\Omega \mathrm{d}\mathbb{P}(\omega) = 1 : \theta \ , \\ & \int_\Omega \begin{pmatrix} \Sigma_0 & \omega-\mu_0 \\ (\omega-\mu_0)^\top & \gamma_1 \end{pmatrix} \mathrm{d}\mathbb{P}(\omega) \succeq 0 : -\Phi \ , \\ & \int_\Omega (\omega-\mu_0)(\omega-\mu_0)^\top \mathrm{d}\mathbb{P}(\omega) \preceq \gamma_2\Sigma_0 : \Psi \end{aligned} \right. \tag{4.10b}$$

$$= \left\{ \begin{aligned} & \inf_{\theta,\Phi,\Psi} \ \theta + \langle \gamma_2\Sigma_0, \Psi\rangle \\ & \text{s.t. } \Phi \succeq 0 \ , \ \ \Psi \succeq 0 \ , \\ & \qquad \theta - \left\langle \begin{pmatrix} \Sigma_0 & \omega-\mu_0 \\ (\omega-\mu_0)^\top & \gamma_1 \end{pmatrix}, \Phi \right\rangle + \langle (\omega-\mu_0)(\omega-\mu_0)^\top, \Psi\rangle \\ & \qquad \qquad \geq g(x,\omega) \ , \ \forall \omega \in \Omega \ . \end{aligned} \right. \tag{4.10c}$$

The last equality above follows from the strong duality of conic programs in Proposition 2.45, because the Dirac distribution $\mathbb{P} = \delta_{\mu_0}$ lies in the relative interior of the feasible region. The dual program (4.10c) can be regarded as an RO problem with semidefinite cone constraints. In fact, this problem can be further simplified.

Theorem 4.3 *For any $x \in \mathcal{X}$, the worst-case expectation*

$$\sup_{\mathbb{P}\in\mathcal{P}_{\mathrm{M}}(\Omega,\mu_0,\Sigma_0,\gamma_1,\gamma_2)} \mathbb{E}^{\mathbb{P}}[g(x,\omega)] \ ,$$

equals the optimal value

$$\begin{aligned} & \min_{\Psi\succeq 0,q,r} \ r + \langle \gamma_2\Sigma_0 + \mu_0\mu_0^\top, \Psi\rangle + \mu_0^\top q + \sqrt{\gamma_1}\|\Sigma_0^{1/2}(q+2\Psi\mu_0)\| \\ & \text{s.t. } r \geq g(x,\omega) - \langle \omega\omega^\top, \Psi\rangle - \omega^\top q \ , \ \forall \omega \in \Omega \ , \end{aligned} \tag{4.11}$$

where $\Sigma_0^{1/2}$ is the unique matrix such that $\Sigma_0^{1/2}\Sigma_0^{1/2} = (\Sigma_0^{1/2})^\top\Sigma_0^{1/2} = \Sigma_0$ for positive definite Σ_0.

Proof Let

$$\Phi = \begin{pmatrix} P & p \\ p^\top & s \end{pmatrix} \ , \ r = \theta - \langle \Sigma_0, P \rangle + 2\mu_0^\top p - \gamma_1 s + \langle \mu_0 \mu_0^\top, \Psi \rangle \ ,$$

then problem (4.10c) is equivalently transformed into:

$$\begin{aligned} \min_{\Psi, P, p, s, r} \ & \langle \gamma_2 \Sigma_0 - \mu_0 \mu_0^\top, \Psi \rangle + r + \langle \Sigma_0, P \rangle - 2\mu_0^\top p + \gamma_1 s \\ \text{s.t. } & \Psi \succeq 0 \ , \ \begin{pmatrix} P & p \\ p^\top & s \end{pmatrix} \succeq 0 \ , \\ & \langle \omega \omega^\top, \Psi \rangle - 2\omega^\top (p + \Psi \mu_0) + r - g(x, \omega) \geq 0 \ , \ \forall \omega \in \Omega \ . \end{aligned} \tag{4.12}$$

Now fix (Ψ, r) and optimize the above problem with respect to (P, p, s). First, consider the case where the optimal solution $s^* > 0$. By Schur complement property in Proposition 4.1, the constraint $\Phi \succeq 0$ is equivalent to $P \succeq pp^\top/s$. Note that the last constraint in (4.12) does not contain P and s. In the objective function of (4.12), P only appears in the term of $\langle \Sigma_0, P \rangle$. Since $\Sigma_0 \succ 0$, $P^* = pp^\top/s$. Therefore, $s^* > 0$ should be optimal for the problem $\min_{s>0} p^\top \Sigma_0 p/s + \gamma_1 s$, which means $s^* = \sqrt{p^\top \Sigma_0 p/\gamma_1}$. Let $q = -2(p + \Psi\mu_0)$, then the objective function of program (4.12) is $r + \langle \gamma_2 \Sigma_0 + \mu_0 \mu_0^\top, \Psi \rangle + \mu_0^\top q + \sqrt{\gamma_1} \| \Sigma_0^{1/2} (q + 2\Psi\mu_0) \|$.

Second, assume $s^* = 0$, then by the constraint $\Phi \succeq 0$ and Schur complement we have $p^* = 0$. Then the optimization comes down to $\min_P \langle \Sigma_0, P \rangle$ subject to $P \succeq 0$, and thus the optimal solution $P^* = 0$. Therefore, in this case, the objective function in program (4.12) equals $r + \langle \gamma_2 \Sigma_0 + \mu_0 \mu_0^\top, \Psi \rangle + \mu_0^\top q + \sqrt{\gamma_1} \| \Sigma_0^{1/2} (q + 2\Psi\mu_0) \|$ for $q = -2(p + \Psi\mu_0)$. The transformation of the constraints is straightforward. □

Based on Theorem 4.3, the distributionally robust optimization (DRO) problem

$$\min_{x \in \mathcal{X}} \sup_{\mathbb{P} \in \mathcal{P}_{\mathrm{M}}(\Omega, \mu_0, \Sigma_0, \gamma_1, \gamma_2)} \mathbb{E}^{\mathbb{P}} [g(x, \omega)] \ , \tag{4.13}$$

is equivalent to:

$$\begin{aligned} \min_{x \in \mathcal{X}, \Psi \succeq 0, q, r} \ & r + \langle \gamma_2 \Sigma_0 + \mu_0 \mu_0^\top, \Psi \rangle + \mu_0^\top q + \sqrt{\gamma_1} \| \Sigma_0^{1/2} (q + 2\Psi\mu_0) \| \\ \text{s.t. } & r \geq g(x, \omega) - \langle \omega \omega^\top, \Psi \rangle - \omega^\top q \ , \ \forall \omega \in \Omega \ , \end{aligned}$$

which is also an RO problem with semidefinite cone constraints and can be solved by RO techniques.

According to the complexity analysis in Delage and Ye (2010), if the RO problem

$$\min_{x \in \mathcal{X}} \max_{\omega \in \Omega} g(x, \omega) \ ,$$

is tractable, then the DRO problem (4.13) is also tractable. Moreover, the following conditions are sufficient for tractability:

1. Ω is convex and compact. For any $\omega \in \Omega$, we can either confirm that $\omega \in \Omega$ or find a separating hyperplane in time polynomial in the dimension of ω;
2. $g(x, \omega) = \max_{k=1,\dots,K} g_k(x, \omega)$ for functions $g_k(x, \omega)$ concave in ω. For fixed (x, ω), we can calculate $g_k(x, \omega)$ and find a supergradient of $g_k(x, \omega)$ in ω in polynomial time;
3. $\mathcal{X}$ is convex and compact. For any $x \in \mathcal{X}$, we can either confirm that $x \in \mathcal{X}$ or find a separating hyperplane in time polynomial in the dimension of x;
4. $g(x, \omega)$ is convex in x. For fixed x, we can find a subgradient of $g(x, \omega)$ in x in polynomial time.

Example 4.1 (*Upper bound of probability*) Suppose we want to find an upper bound for $\mathbb{P}_0(\omega \in S)$, where $S \subset \Omega = \mathbb{R}^M$ and the estimated first- and second-order moments of ω are μ_0 and Σ_0, respectively. Using $\mathbb{P}(\omega \in S) = \mathbb{E}^{\mathbb{P}}[1_S(\omega)]$ with the indicator function 1_S of S, the following worst-case expectation is an upper bound estimation for the probability of $\omega \in S$:

$$\sup_{\mathbb{P} \in \mathcal{P}_{\mathrm{M}}(\Omega, \mu_0, \Sigma_0, \gamma_1, \gamma_2)} \mathbb{E}^{\mathbb{P}}[1_S(\omega)] \; .$$

To make the formulation tractable, we need to represent 1_S as the maximum of some concave functions. Assume S is convex. Let $g_1(\omega) = 1$ for $\omega \in S$ and $g_1(\omega) = -\infty$ otherwise. Let $g_2(\omega) = 0, \forall \omega \in \Omega$. Then g_1 and g_2 are concave and $g(\omega) := \max\{g_1(\omega), g_2(\omega)\} = 1_S(\omega), \forall \omega \in \Omega$ and

$$\mathbb{P}_0(\omega \in S) \leq \sup_{\mathbb{P} \in \mathcal{P}_{\mathrm{M}}(\Omega, \mu_0, \Sigma_0, \gamma_1, \gamma_2)} \mathbb{E}^{\mathbb{P}}[g(\omega)] \; ,$$

whenever $\mathbb{P}_0 \in \mathcal{P}_{\mathrm{M}}(\Omega, \mu_0, \Sigma_0, \gamma_1, \gamma_2)$.

4.3 Reformulation of Robust Chance Constraint

In this subsection, we mainly use the moment-based ambiguity set $\mathcal{P}_{\mathrm{M}}(\mathbb{R}^M, \mu_0, \Sigma_0)$ with exact moments in (4.2). In addition, the considered robust chance constraint has the following form:

$$\inf_{\mathbb{P} \in \mathcal{P}_{\mathrm{M}}(\mathbb{R}^M, \mu_0, \Sigma_0)} \mathbb{P}(A(x) + B(x)^\top \omega \leq 0) \geq \beta \; . \tag{4.14}$$

where $\beta \in [0, 1]$ is a probability threshold. Therefore, the random event we want to ensure with probability β is a polyhedron characterized by x. When $A(x) \in \mathbb{R}$, (4.14) is called an individual robust chance constraint; otherwise, it is a joint robust chance constraint. The contents of this section are based on Zymler et al. (2013).

4.3.1 Individual Robust Chance Constraint

Following the steps in Sect. 3.4, the individual robust chance constraint can be reformulated via value-at-risk (VaR) and conditional value-at-risk (CVaR) as follows:

$$
\begin{aligned}
& \inf_{\mathbb{P}\in\mathcal{P}_{\mathrm{M}}(\mathbb{R}^M,\mu_0,\Sigma_0)} \mathbb{P}(A(x)+B(x)^\top\omega\le 0)\ge\beta \\
\Longleftrightarrow\quad & \sup_{\mathbb{P}\in\mathcal{P}_{\mathrm{M}}(\mathbb{R}^M,\mu_0,\Sigma_0)} \beta\text{-VaR}^{\mathbb{P}}(A(x)+B(x)^\top\omega)\le 0 && \text{(4.15a)}\\
\Longleftarrow\quad & \sup_{\mathbb{P}\in\mathcal{P}_{\mathrm{M}}(\mathbb{R}^M,\mu_0,\Sigma_0)} \beta\text{-CVaR}^{\mathbb{P}}(A(x)+B(x)^\top\omega)\le 0 && \text{(4.15b)}\\
\Longleftrightarrow\quad & \exists r\ ,\ \text{s.t. } r+\frac{1}{1-\beta}\sup_{\mathbb{P}\in\mathcal{P}_{\mathrm{M}}(\mathbb{R}^M,\mu_0,\Sigma_0)} \mathbb{E}^{\mathbb{P}}[(A(x)+B(x)^\top\omega-r)_+]\le 0\,. && \text{(4.15c)}
\end{aligned}
$$

According to Sect. 4.2.1.2, the worst-case expectation in (4.15c) satisfies:

$$
\begin{aligned}
&\sup_{\mathbb{P}\in\mathcal{P}_{\mathrm{M}}(\mathbb{R}^M,\mu_0,\Sigma_0)} \mathbb{E}^{\mathbb{P}}[(A(x)+B(x)^\top\omega-r)_+]\\
&=\begin{cases}\inf\limits_{\theta\in\mathbb{R},\phi\in\mathbb{R}^M,\Psi\in\mathbb{S}^M} & \theta+\mu_0^\top\phi+\langle\Sigma_0+\mu_0\mu_0^\top,\Psi\rangle\\ & \text{s.t. } \theta+\omega^\top\phi+\langle\omega\omega^\top,\Psi\rangle\ge(A(x)+B(x)^\top\omega-r)_+\ ,\ \forall\omega\in\mathbb{R}^M\end{cases}\\
&=\begin{cases}\inf\limits_{\theta\in\mathbb{R},\phi\in\mathbb{R}^M,\Psi\in\mathbb{S}^M} & \theta+\mu_0^\top\phi+\langle\Sigma_0+\mu_0\mu_0^\top,\Psi\rangle\\ & \text{s.t. } \begin{pmatrix}\Psi & \phi/2\\ \phi^\top/2 & \theta\end{pmatrix}\succeq 0\,,\\ & \begin{pmatrix}\omega\\1\end{pmatrix}^\top\begin{pmatrix}\Psi & \phi/2\\ \phi^\top/2 & \theta\end{pmatrix}\begin{pmatrix}\omega\\1\end{pmatrix}\ge A(x)+B(x)^\top\omega-r\ ,\ \forall\omega\in\mathbb{R}^M\end{cases}\\
&=\begin{cases}\inf\limits_{\theta\in\mathbb{R},\phi\in\mathbb{R}^M,\Psi\in\mathbb{S}^M} & \left\langle\begin{pmatrix}\Sigma_0+\mu_0\mu_0^\top & \mu_0\\ \mu_0^\top & 1\end{pmatrix},\begin{pmatrix}\Psi & \phi/2\\ \phi^\top/2 & \theta\end{pmatrix}\right\rangle\\ & \text{s.t. } \begin{pmatrix}\Psi & \phi/2\\ \phi^\top/2 & \theta\end{pmatrix}\succeq 0\,,\ \begin{pmatrix}\Psi & \phi/2\\ \phi^\top/2 & \theta\end{pmatrix}-\begin{pmatrix}0 & B(x)/2\\ B(x)^\top/2 & A(x)-r\end{pmatrix}\succeq 0\,.\end{cases}
\end{aligned}
$$

Therefore, we have the following result.

Proposition 4.2 *The distributionally robust CVaR constraint (4.15b) is equivalent to:*

$$
\begin{aligned}
\exists r\in\mathbb{R}\ ,\ Q\in\mathbb{S}^{M+1}\ ,\ \textit{s.t. } r+\frac{1}{1-\beta}\left\langle\begin{pmatrix}\Sigma_0+\mu_0\mu_0^\top & \mu_0\\ \mu_0^\top & 1\end{pmatrix},Q\right\rangle\le 0\,,&\\
Q\succeq 0\,,\ Q-\begin{pmatrix}0 & B(x)/2\\ B(x)^\top/2 & A(x)-r\end{pmatrix}\succeq 0\,.& \qquad \text{(4.16)}
\end{aligned}
$$

In fact, the reformulation of robust chance constraint by CVaR is exact under some conditions.

Theorem 4.4 (Theorem 2.2 in Zymler et al. (2013)) *Let $\mathcal{P}_{\mathrm{M}}(\mathbb{R}^M, \mu_0, \Sigma_0)$ be the ambiguity set defined in (4.2). Suppose the loss function $g : \mathbb{R}^M \to \mathbb{R}$ is continuous and either concave or quadratic with respect to ω, then:*

$$\inf_{\mathbb{P}\in\mathcal{P}_{\mathrm{M}}(\mathbb{R}^M,\mu_0,\Sigma_0)} \mathbb{P}(g(\omega) \le 0) \ge \beta \iff \sup_{\mathbb{P}\in\mathcal{P}_{\mathrm{M}}(\mathbb{R}^M,\mu_0,\Sigma_0)} \beta\text{-}CVaR^{\mathbb{P}}(g(\omega)) \le 0 \, .$$

Exercise 4.3 Find a counterexample for the conclusion of Theorem 4.4 when the loss function $g(\omega)$ is convex in ω.

Since the considered loss function $g(x, \omega) = A(x) + B(x)^\top \omega$ is linear in ω, the assumptions of Theorem 4.4 are satisfied. Hence, we have Corollary 4.2.

Corollary 4.2 *The individual robust chance constraint*

$$\inf_{\mathbb{P}\in\mathcal{P}_{\mathrm{M}}(\mathbb{R}^M,\mu_0,\Sigma_0)} \mathbb{P}(A(x) + B(x)^\top \omega \le 0) \ge \beta \iff (4.15b) \iff (4.16) \, .$$

4.3.2 Joint Robust Chance Constraint

In the robust chance constraint (4.14), if $A(x) \in \mathbb{R}^K$ for $K > 1$, then (4.14) is a joint robust chance constraint. It can also be written as follows for clarity:

$$\inf_{\mathbb{P}\in\mathcal{P}_{\mathrm{M}}(\mathbb{R}^M,\mu_0,\Sigma_0)} \mathbb{P}(A_k(x) + B_k(x)^\top \omega \le 0 \, , \; k = 1, \ldots, K) \ge \beta \, . \tag{4.17}$$

A basic reformulation method is based on Bonferroni inequality in Proposition 2.7.

Proposition 4.3 *For $\beta_k \in [0, 1], k = 1, \ldots, K$ with $\sum_{k=1}^{K}(1 - \beta_k) \le 1 - \beta$, the joint robust chance constraint (4.17) has the following conservative approximation:*

$$\begin{aligned}
&\exists r_k \in \mathbb{R} \, , \; M_k \in \mathbb{S}^{M+1} \, , \; k = 1, \ldots, K \, , \\
&s.t. \; r_k + \frac{1}{1-\beta_k}\left\langle \begin{pmatrix} \Sigma_0 + \mu_0\mu_0^\top & \mu_0 \\ \mu_0^\top & 1 \end{pmatrix}, M_k \right\rangle \le 0 \, , \; k = 1, \ldots, K \, , \\
&M_k \succeq 0 \, , \; M_k - \begin{pmatrix} 0 & B_k(x)/2 \\ B_k(x)^\top/2 & A_k(x) - r_k \end{pmatrix} \succeq 0 \, , \; k = 1, \ldots, K \, .
\end{aligned} \tag{4.18}$$

Proof We need to prove that (4.18) implies (4.17). By Corollary 4.2,

$$\begin{aligned}
(4.18) &\iff \inf_{\mathbb{P}\in\mathcal{P}_{\mathrm{M}}(\mathbb{R}^M,\mu_0,\Sigma_0)} \mathbb{P}(A_k(x) + B_k(x)^\top \omega \le 0) \ge \beta_k \, , \; k = 1, \ldots, K \\
&\iff \sup_{\mathbb{P}\in\mathcal{P}_{\mathrm{M}}(\mathbb{R}^M,\mu_0,\Sigma_0)} \mathbb{P}(A_k(x) + B_k(x)^\top \omega > 0) \le 1 - \beta_k \, , \; k = 1, \ldots, K \, .
\end{aligned}$$

Assume that (4.18) holds. Using Bonferroni inequality, we have:

$$1 - \inf_{\mathbb{P}\in\mathcal{P}_{\mathrm{M}}(\mathbb{R}^M,\mu_0,\Sigma_0)} \mathbb{P}\left(A_k(x) + B_k(x)^\top\omega \le 0\ ,\ k = 1,\ldots,K\right) \tag{4.19a}$$

$$= \sup_{\mathbb{P}\in\mathcal{P}_{\mathrm{M}}(\mathbb{R}^M,\mu_0,\Sigma_0)} \mathbb{P}\left(\exists 1 \le k_0 \le K\ ,\ \text{s.t. } A_{k_0}(x) + B_{k_0}(x)^\top\omega > 0\right) \tag{4.19b}$$

$$= \sup_{\mathbb{P}\in\mathcal{P}_{\mathrm{M}}(\mathbb{R}^M,\mu_0,\Sigma_0)} \mathbb{P}\left(\bigcup_{k=1}^{K}\left\{\omega\in\mathbb{R}^M \mid A_k(x) + B_k(x)^\top\omega > 0\right\}\right) \tag{4.19c}$$

$$\le \sup_{\mathbb{P}\in\mathcal{P}_{\mathrm{M}}(\mathbb{R}^M,\mu_0,\Sigma_0)} \sum_{k=1}^{K}\mathbb{P}\left(\left\{\omega\in\mathbb{R}^M \mid A_k(x) + B_k(x)^\top\omega > 0\right\}\right) \tag{4.19d}$$

$$\le \sum_{k=1}^{K}\sup_{\mathbb{P}\in\mathcal{P}_{\mathrm{M}}(\mathbb{R}^M,\mu_0,\Sigma_0)} \mathbb{P}\left(\left\{\omega\in\mathbb{R}^M \mid A_k(x) + B_k(x)^\top\omega > 0\right\}\right) \tag{4.19e}$$

$$\le \sum_{k=1}^{K}(1-\beta_k) \le 1-\beta\ . \tag{4.19f}$$

Then (4.17) follows. □

A direct choice of $\beta_k, k = 1,\ldots,K$ is $\beta_k = 1-(1-\beta)/K, k = 1,\ldots,K$, which divides the probability threshold equally. However, the approximation by Proposition 4.3 can be rather conservative even under the optimal choice of $\beta_k, k = 1,\ldots,K$ because of the inequalities in (4.19d) and (4.19e), where the former uses Bonferroni inequality and the latter is due to the differences between worst-case probability distributions. For instance, the approximation does not perform well when the inequalities in the chance constraint are positively correlated.

Proposition 4.4 gives another way to reformulate the joint robust chance constraint (Chen et al. 2010).

Proposition 4.4 (Reformulating joint robust chance constraint by new loss function) *For fixed $\gamma_k > 0,\ k = 1,\ldots,K$, the joint robust chance constraint (4.17) has the following conservative approximation:*

$$\sup_{\mathbb{P}\in\mathcal{P}_{\mathrm{M}}(\mathbb{R}^M,\mu_0,\Sigma_0)} \beta\text{-}CVaR^{\mathbb{P}}\left(\max_{1\le k\le K}\left\{\gamma_k\left(A_k(x) + B_k(x)^\top\omega\right)\right\}\right) \le 0\ . \tag{4.20}$$

Proof By Theorems 3.1 and 3.2, we have:

$$\begin{aligned}
(4.20) \implies & \sup_{\mathbb{P}\in\mathcal{P}_{\mathrm{M}}(\mathbb{R}^M,\mu_0,\Sigma_0)} \beta\text{-VaR}^{\mathbb{P}}\left(\max_{1\le k\le K}\left\{\gamma_k\left(A_k(x) + B_k(x)^\top\omega\right)\right\}\right) \le 0 \\
\iff & \inf_{\mathbb{P}\in\mathcal{P}_{\mathrm{M}}(\mathbb{R}^M,\mu_0,\Sigma_0)} \mathbb{P}\left(\max_{1\le k\le K}\left\{\gamma_k\left(A_k(x) + B_k(x)^\top\omega\right)\right\} \le 0\right) \ge \beta \\
\iff & \inf_{\mathbb{P}\in\mathcal{P}_{\mathrm{M}}(\mathbb{R}^M,\mu_0,\Sigma_0)} \mathbb{P}(\gamma_k(A_k(x) + B_k(x)^\top\omega) \le 0\ ,\ k = 1,\ldots,K) \ge \beta \\
\iff & (4.17)\ ,
\end{aligned}$$

which completes the proof. □

In Proposition 4.4, the only introduced conservatism is due to the reformulation of VaR by CVaR. Therefore, it is reasonable to expect that this method may give less conservative reformulations than that in Proposition 4.3. Another important advantage of (4.20) is its tractability, which is elaborated in the Proposition 4.5.

Proposition 4.5 *For fixed $\gamma_k > 0$, $k = 1, \dots, K$, (4.20) is equivalent to:*

$$\exists r \in \mathbb{R}\,,\ Q \in \mathbb{S}^{M+1}\,,\ \text{s.t. } r + \frac{1}{1-\beta}\left\langle \begin{pmatrix} \Sigma_0 + \mu_0\mu_0^\top & \mu_0 \\ \mu_0^\top & 1 \end{pmatrix}, Q \right\rangle \le 0\,,\ Q \succeq 0\,,$$
$$Q - \begin{pmatrix} 0 & \frac{1}{2}\gamma_k B_k(x) \\ \frac{1}{2}\gamma_k B_k(x)^\top & \gamma_k A_k(x) - r \end{pmatrix} \succeq 0\,,\ k = 1, \dots, K\,. \tag{4.21}$$

Proof By the property of CVaR in Theorem 2.6, we know that (4.20) is equivalent to:

$$\exists r, \text{ s.t. } r + \frac{1}{1-\beta} \sup_{\mathbb{P} \in \mathcal{P}_{\mathrm{M}}(\mathbb{R}^M, \mu_0, \Sigma_0)} \mathbb{E}^{\mathbb{P}}\left[\left(\max_{1 \le k \le K}\left\{\gamma_k\left(A_k(x) + B_k(x)^\top \omega\right)\right\} - r\right)_+\right] \le 0.$$

By a similar deduction to that in Proposition 4.2, we have:

$$\begin{aligned}
& \sup_{\mathbb{P} \in \mathcal{P}_{\mathrm{M}}(\mathbb{R}^M, \mu_0, \Sigma_0)} \mathbb{E}^{\mathbb{P}}\left[\left(\max_{1 \le k \le K}\left\{\gamma_k\left(A_k(x) + B_k(x)^\top \omega\right) - r\right\}\right)_+\right] \\
= & \sup_{\mathbb{P} \in \mathcal{P}_{\mathrm{M}}(\mathbb{R}^M, \mu_0, \Sigma_0)} \mathbb{E}^{\mathbb{P}}\left[\max\left\{0, \max_{1 \le k \le K}\left\{\gamma_k A_k(x) + \gamma_k B_k(x)^\top \omega - r\right\}\right\}\right] \\
= & \left\{ \begin{array}{ll} \inf\limits_{\theta \in \mathbb{R}, \phi \in \mathbb{R}^M, \Psi \in \mathbb{S}^M} & \left\langle \begin{pmatrix} \Sigma_0 + \mu_0\mu_0^\top & \mu_0 \\ \mu_0^\top & 1 \end{pmatrix}, \begin{pmatrix} \Psi & \phi/2 \\ \phi^\top/2 & \theta \end{pmatrix} \right\rangle \\
\text{s.t.} & \begin{pmatrix} \Psi & \phi/2 \\ \phi^\top/2 & \theta \end{pmatrix} \succeq 0\,, \\
& \begin{pmatrix} \Psi & \phi/2 \\ \phi^\top/2 & \theta \end{pmatrix} - \begin{pmatrix} 0 & \frac{1}{2}\gamma_k B(x) \\ \frac{1}{2}\gamma_k B(x)^\top & \gamma_k A(x) - r \end{pmatrix} \succeq 0\,,\ k = 1, \dots, K. \end{array} \right.
\end{aligned}$$

It then follows that (4.20) is equivalent to (4.21). □

Exercise 4.4 Complete the proof details of Proposition 4.5.

Corollary 4.3 *For fixed $\gamma_k > 0$, $k = 1, \dots, K$, (4.21) is a conservative approximation of the joint robust chance constraint (4.17).*

The choice of $\gamma_k > 0$, $k = 1, \dots, K$ can be optimized for better approximation performance. Unfortunately, optimizing $\gamma_k > 0, k = 1, \dots, K$ and x jointly results in bilinear matrix inequalities that are generally intractable. Nonetheless, a convergent iterative algorithm can be designed to optimize $\gamma_k > 0$, $k = 1, \dots, K$, although the optimality is not guaranteed (Zymler et al. 2013).

The above reformulation method can be extended to the case of quadratic-constrained sample space.

Proposition 4.6 *Let the sample space be:*

$$\Omega = \left\{ \omega \in \mathbb{R}^M \;\middle|\; \begin{pmatrix} \omega \\ 1 \end{pmatrix}^\top W_l \begin{pmatrix} \omega \\ 1 \end{pmatrix} \le 0 \;,\; l = 1, \ldots, L \right\} ,$$

where $W_l \in \mathbb{S}^{M+1}, l = 1, \ldots, L$. *Then the joint chance constraint*

$$\inf_{\mathbb{P} \in \mathcal{P}_{\mathrm{M}}(\Omega, \mu_0, \Sigma_0)} \mathbb{P}(A_k(x) + B_k(x)^\top \omega \le 0 \;,\; k = 1, \ldots, K) \ge \beta \;, \tag{4.22}$$

has the following conservative approximation:

$$\begin{aligned} & \exists r \in \mathbb{R} \;,\; Q \in \mathbb{S}^{M+1} \;,\; \tau_k \in \mathbb{R}^L \;,\; k = 0, 1, \ldots, K \;, \\ & s.t. \; r + \frac{1}{1-\beta} \left\langle \begin{pmatrix} \Sigma_0 + \mu_0 \mu_0^\top & \mu_0 \\ \mu_0^\top & 1 \end{pmatrix}, Q \right\rangle \le 0 \;,\; \tau_k \ge 0 \;,\; k = 0, 1, \ldots, K \;, \\ & Q + \sum_{l=1}^{L} \tau_{0,l} W_l \succeq 0 \;, \\ & Q + \sum_{l=1}^{L} \tau_{k,l} W_l - \begin{pmatrix} 0 & \frac{1}{2}\gamma_k B_k(x) \\ \frac{1}{2}\gamma_k B_k(x)^\top & \gamma_k A_k(x) - r \end{pmatrix} \succeq 0 \;,\; k = 1, \ldots, K \;. \end{aligned} \tag{4.23}$$

Proof Similar to the proof of Proposition 4.5, the following worst-case expectation should be examined:

$$\sup_{\mathbb{P} \in \mathcal{P}_{\mathrm{M}}(\Omega, \mu_0, \Sigma_0)} \mathbb{E}^{\mathbb{P}} \left[\max \left\{ 0, \max_{1 \le k \le K} \left\{ \gamma_k A_k(x) + \gamma_k B_k(x)^\top \omega - r \right\} \right\} \right] \tag{4.24a}$$

$$= \begin{cases} \inf\limits_{\theta \in \mathbb{R}, \phi \in \mathbb{R}^M, \Psi \in \mathbb{S}^M} \left\langle \begin{pmatrix} \Sigma_0 + \mu_0 \mu_0^\top & \mu_0 \\ \mu_0^\top & 1 \end{pmatrix}, \begin{pmatrix} \Psi & \phi/2 \\ \phi^\top/2 & \theta \end{pmatrix} \right\rangle \\ \text{s.t.} \; \begin{pmatrix} \omega \\ 1 \end{pmatrix}^\top \begin{pmatrix} \Psi & \phi/2 \\ \phi^\top/2 & \theta \end{pmatrix} \begin{pmatrix} \omega \\ 1 \end{pmatrix} \ge \gamma_k A_k(x) + \gamma_k B_k(x)^\top \omega - r \;,\; k = 1, \ldots, K \;, \\ \begin{pmatrix} \omega \\ 1 \end{pmatrix}^\top \begin{pmatrix} \Psi & \phi/2 \\ \phi^\top/2 & \theta \end{pmatrix} \begin{pmatrix} \omega \\ 1 \end{pmatrix} \ge 0 \;,\; \forall \omega \;,\; \text{s.t.} \; \begin{pmatrix} \omega \\ 1 \end{pmatrix}^\top W_l \begin{pmatrix} \omega \\ 1 \end{pmatrix} \le 0 \;,\; l = 1, \ldots, L \;. \end{cases} \tag{4.24b}$$

By Lemma 4.1, the constraints of (4.24b) has the following conservative approximation:

$$\exists \tau_k \in \mathbb{R}_+^L \,, \; k = 0, 1, \ldots, K \,,$$

$$\text{s.t.} \begin{pmatrix} \Psi & \phi/2 \\ \phi^\top/2 & \theta \end{pmatrix} + \sum_{l=1}^{L} \tau_{0,l} W_l \succeq 0 \,,$$

$$\begin{pmatrix} \Psi & \phi/2 \\ \phi^\top/2 & \theta \end{pmatrix} + \sum_{l=1}^{L} \tau_{k,l} W_l - \begin{pmatrix} 0 & \frac{1}{2}\gamma_k B_k(x) \\ \frac{1}{2}\gamma_k B_k(x)^\top & \gamma_k A_k(x) - r \end{pmatrix} \succeq 0 \,, \; k = 1, \ldots, K \,.$$

Then the conclusion of Proposition 4.6 can follow from the proof process of Proposition 4.5. □

Exercise 4.5 Complete the proof details of Proposition 4.6.

4.4 Incorporating Unimodality

As pointed out in Sect. 1.2.3, different probability distributions may have the same first- and second-order moments, which indicates that the moment-based ambiguity sets may be too large and lead to overly conservative results. Incorporating unimodality is a way to reduce the conservatism of the moment-based ambiguity sets.

4.4.1 Moment-Based Ambiguity Set Incorporating Unimodality

Unimodality is about the shape information of a probability distribution. A detailed introduction about unimodality can be found in Dharmadhikari and Joag-dev (1988). For the 1-dimensional case, a continuous probability distribution $\mathbb{P}$ on $\mathbb{R}$ with the probability density function (PDF) f is unimodal about the mode 0 if the function $f(t)$ is nondecreasing on $(-\infty, 0)$ and nonincreasing on $(0, +\infty)$. For multi-dimensional cases, i.e., $\omega \in \mathbb{R}^M$, an extension of the above notion is that the function $f(td)$ of $t \in \mathbb{R}$ is nonincreasing on $(0, +\infty)$ for any $0 \neq d \in \mathbb{R}^M$, where f is the PDF. Many probability distributions have this property, such as a multivariate normal distribution centered at 0. Formally, the definition of unimodality is as follows.

Definition 4.4 (*α-unimodality*) For any $\alpha > 0$ and probability distribution $\mathbb{P}$ on $\mathbb{R}^M$. $\mathbb{P}$ is called *α-unimodal* about mode 0 if the function $G(t) := t^\alpha \mathbb{P}(S/t)$ is nondecreasing on $(0, +\infty)$ for any measurable set S in $\mathbb{R}^M$, where S/t denotes the set $\{s/t \mid s \in S\}$.

When the probability distribution $\mathbb{P}$ is continuous and has a PDF f, it can be proven that $\mathbb{P}$ is α-unimodal about mode 0 if and only if $t^{\alpha - M} f(d/t)$ is nondecreasing as a function of t on $(0, +\infty)$, or equivalently, $t^{M-\alpha} f(td)$ is nonincreasing on $(0, +\infty)$, for any $0 \neq d \in \mathbb{R}^M$. As $\alpha \to +\infty$, the shape requirement of α-unimodality becomes milder. A special case is $\alpha = M$, where $f(td)$ is nonincreasing on $(0, +\infty)$ for any $d \neq 0$. M-unimodality reduces to the star-unimodality defined as follows.

Definition 4.5 (*Star-shaped*) A set $S \subset \mathbb{R}^M$ is called *star-shaped* about 0 if $tS := \{ts \mid s \in S\} \subset S$ for any $t \in [0, 1]$.

Definition 4.6 (*Star-unimodality*) A probability distribution $\mathbb{P}$ on $\mathbb{R}^M$ is called *star-unimodal* about mode 0 if $\mathbb{P}$ is in the closed convex hull of the set of uniform distributions on star-shaped sets in $\mathbb{R}^M$ about 0.

Now we are ready to define the moment-based ambiguity set incorporating unimodality, which adds the requirement of α-unimodality to the definition of moment-based ambiguity set with exact moments in Definition 4.1.

Definition 4.7 (*Moment-based ambiguity set incorporating unimodality*) The moment-based ambiguity set incorporating α-unimodality about mode 0 on sample space $\mathbb{R}^M$ with exact first- and second-order moments μ_0 and Σ_0 is defined as follows:

$$\mathcal{P}_{\mathrm{M}}(\mathbb{R}^M, \mu_0, \Sigma_0, \alpha) := \left\{ \mathbb{P} \in \mathcal{M}_\alpha(\mathbb{R}^M) \;\middle|\; \begin{array}{l} \mathbb{E}^{\mathbb{P}}[\omega] = \mu_0 \, , \\ \mathbb{E}^{\mathbb{P}}[(\omega - \mu_0)(\omega - \mu_0)^\top] = \Sigma_0 \end{array} \right\} , \tag{4.25}$$

where $\mathcal{M}_\alpha(\mathbb{R}^M)$ is the set of α-unimodal probability distributions about mode 0 on $\mathbb{R}^M$.

Although in Definition 4.7, the probability distribution is required to have mode 0. The general case can be handled easily by a translation. For conciseness, we only treat the case of mode 0 in the remaining part of this section.

Note that the first-order moment and the mode can be different. For instance, the PDF $f(t)$ of the exponential distribution of rate parameter $\lambda > 0$ is defined by $f(t) = \lambda \exp(-\lambda t)$ for $t \geq 0$ and $f(t) = 0$ for $t < 0$. It is easy to see that its mode is 0 but the first-order moment is positive.

Exercise 4.6 Calculate the mode and the first-order moment of the exponential distribution of rate parameter $\lambda > 0$.

4.4.2 Reformulation of Individual Robust Chance Constraint

We want to reformulate the following individual robust chance constraint into a tractable form:

$$\inf_{\mathbb{P} \in \mathcal{P}_{\mathrm{M}}(\mathbb{R}^M, \mu_0, \Sigma_0, \alpha)} \mathbb{P}(A(x) + B(x)^\top \omega \leq 0) \geq \beta \, . \tag{4.26}$$

where $A(x) \in \mathbb{R}$, $B(x) \in \mathbb{R}^M$, $\beta \in [0, 1]$.

Here are two mild assumptions that we will always adopt:

Assumption 1: $(\alpha^2 + 2\alpha)\Sigma_0 \succ \mu_0\mu_0^\top$.

Assumption 2: $A(x) \le 0$.

Assumption 1 is adopted to ensure that the ambiguity set $\mathcal{P}_{\mathrm{M}}(\mathbb{R}^M, \mu_0, \Sigma_0, \alpha) \neq \emptyset$. Assumption 2 means that the random event $A(x) + B(x)^\top \omega$ happens at the mode 0. Since in a robust chance constraint, we usually want to guarantee a random event with a high probability, it is reasonable to assume it holds at the mode.

Theorem 4.5 gives an exact reformulation of the individual robust chance constraint (4.26) by an infinite number of second-order cone constraints.

Theorem 4.5 (Theorem 1 in Li et al. (2019)) *Under Assumption 1 and Assumption 2, the individual robust chance constraint (4.26) is equivalent to the following set of second-order cone constraints:*

$$\sqrt{\frac{\beta - \tau^{-\alpha}}{1-\beta}} \|\Lambda B(x)\|_2 \le -\tau A(x) - \frac{\alpha+1}{\alpha}\mu_0^\top B(x)\ ,\ \forall \tau \ge \beta^{-\frac{1}{\alpha}}\ , \tag{4.27}$$

where

$$\Lambda = \frac{1}{\alpha}\left((\alpha^2 + 2\alpha)\Sigma_0 - \mu_0\mu_0^\top\right)^{\frac{1}{2}}\ . \tag{4.28}$$

There are infinite constraints in (4.27), so the form is still intractable. Nonetheless, we can derive a tractable conservative approximation by using a piecewise linear approximation of the left side of the inequalities in (4.27). Let $h(\tau) = \sqrt{(\beta - \tau^{-\alpha})/(1-\beta)}$. Then it can be verified that $h(\tau)$ is concave on $[\beta^{-1/\alpha}, +\infty)$. Therefore, we can define a piecewise linear function $g(\tau) \ge h(\tau)$, whose idea is illustrated in Fig. 4.1. Choose some numbers $\beta^{-1/\alpha} = n_1 < n_2 < \cdots < n_K = +\infty$. Take the minimum of the tangent lines of $h(\tau)$ at points $n_2, \ldots, n_K$, i.e., $\forall \tau \ge \beta^{-\frac{1}{\alpha}}$,

$$g(\tau) = \min_{k=2,\ldots,K}\left\{\sqrt{\frac{1}{(1-\beta)(\beta - n_k^{-\alpha})}}\left(\frac{\alpha n_k^{-\alpha-1}}{2}\tau + \beta - \left(1 + \frac{\alpha}{2}\right)n_k^{-\alpha}\right)\right\}\ . \tag{4.29}$$

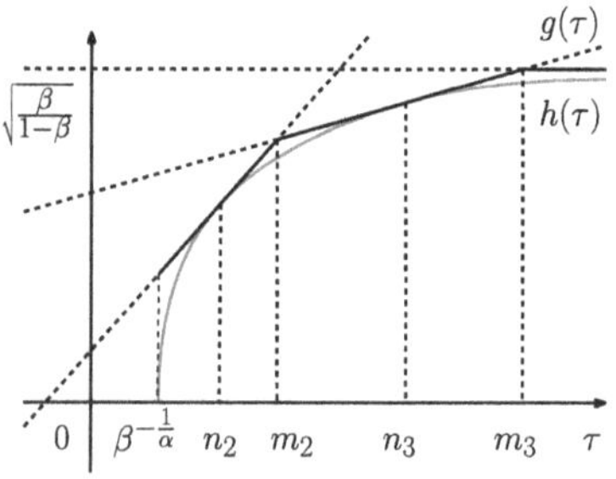

Fig. 4.1 Illustration of approximating the concave function $h(\tau)$ by a piecewise linear function $g(\tau)$ and maintaining $g(\tau) \ge h(\tau), \forall \tau$. $g(\tau)$ is the minimum of some tangent lines of $h(\tau)$

Therefore, a conservative approximation of (4.27) is:

$$g(\tau)\|\Lambda B(x)\|_2 \le -\tau A(x) - \frac{\alpha+1}{\alpha}\mu_0^\top B(x)\ , \ \forall \tau \ge \beta^{-\frac{1}{\alpha}}\ . \tag{4.30}$$

In (4.30), the left side is piecewise linear and the right side is linear, so we only need to check at the breakpoints of $g(\tau)$. Therefore, (4.30) is equivalent to a finite set of second-order cone constraints:

$$g(m_k)\|\Lambda B(x)\|_2 \le -m_k A(x) - \frac{\alpha+1}{\alpha}\mu_0^\top B(x)\ , \ k = 1, \dots, K-1\ , \tag{4.31}$$

where

$$m_k = \frac{\beta\left(1-\sqrt{\frac{\beta-n_k^{-\alpha}}{\beta-n_{k+1}^{-\alpha}}}\right) + \left(1+\frac{\alpha}{2}\right)\left(n_{k+1}^{-\alpha}\sqrt{\frac{\beta-n_k^{-\alpha}}{\beta-n_{k+1}^{-\alpha}}} - n_k^{-\alpha}\right)}{\frac{\alpha}{2}\left(n_{k+1}^{-\alpha-1}\sqrt{\frac{\beta-n_k^{-\alpha}}{\beta-n_{k+1}^{-\alpha}}} - n_k^{-\alpha-1}\right)}\ , \ k = 2, \dots, K-1\ ,$$

$$m_1 = \beta^{-1/\alpha}\ . \tag{4.32}$$

Therefore, we have the following conclusion about the tractable conservative approximation of the individual robust chance constraint (4.26).

Proposition 4.7 *Under Assumption 1 and Assumption 2, assume* $\beta^{-1/\alpha} = n_1 < n_2 < \dots < n_K = +\infty$, *then the individual robust chance constraint (4.26) has the conservative approximation in (4.31), where the function* g *is defined in (4.29) and the points* m_k, $k = 1, \dots, K-1$ *are defined in (4.32).*

4.4.3 Reformulation of Distributionally Robust Conditional Value-at-Risk

We want to reformulate the following distributionally robust CVaR constraint:

$$\sup_{\mathbb{P}\in\mathcal{P}_M(\mathbb{R}^M,\mu_0,\Sigma_0,\alpha)} \beta\text{-CVaR}^{\mathbb{P}}(A(x) + B(x)^\top\omega) \le 0\ , \tag{4.33}$$

where $A(x) \in \mathbb{R}$, $B(x) \in \mathbb{R}^M$, $\beta \in [0, 1]$.

Theorem 4.6 says that the distributionally robust CVaR constraint (4.33) is equivalent to the combination of an infinite number of second-order cone constraints.

Theorem 4.6 (Theorem 3 in Li et al. (2019)) *Under Assumption 1 and Assumption 2, the distributionally robust CVaR constraint (4.33) is equivalent to the following set of second-order cone constraints:*

$$
\begin{aligned}
&\exists r \in \mathbb{R}\ ,\ s.t. \\
&\left\| \begin{pmatrix} (1-\tau^{-\alpha})r - (1-\tau^{-\alpha-1})\mu_0^\top B(x) \\ \frac{\alpha}{\alpha+1}(1-\tau^{-\alpha-1})\Lambda B(x) \end{pmatrix} \right\|_2 \\
&\le -2(1-\beta)A(x) - (1-\tau^{-\alpha-1})\mu_0^\top B(x) + (2\beta - 1 - \tau^{-\alpha})r\ ,\ \forall \tau \ge 1\ , \\
&\left\| \begin{pmatrix} (1-\tau^{-\alpha})r - (1-\tau^{-\alpha-1})\mu_0^\top B(x) \\ \frac{\alpha}{\alpha+1}(1-\tau^{-\alpha-1})\Lambda B(x) \end{pmatrix} \right\|_2 \\
&\le -2(1-\beta)A(x) - (1+\tau^{-\alpha-1})\mu_0^\top B(x) + (2\beta - 1 + \tau^{-\alpha})r\ ,\ \forall \tau \ge 1\ ,
\end{aligned}
\tag{4.34}
$$

where Λ *is defined in (4.28).*

The formulation (4.34) is not tractable because there are infinite constraints. Proposition 4.8 gives a tractable conservative approximation.

Proposition 4.8 (Proposition 10 in Li et al. (2019)) *Suppose* $K \ge 3$ *and* $1 = n_1 < n_2 < \cdots < n_K = +\infty$. *Then a conservative approximation of the distributionally robust CVaR constraint (4.33) is:*

$$
\begin{aligned}
&\exists r \in \mathbb{R}\ ,\ Q \in \mathbb{S}^{M+1}\ ,\ s.t. \\
&r + \frac{1}{1-\beta}\left\langle \begin{pmatrix} \frac{\alpha+2}{\alpha}(\Sigma_0 + \mu_0\mu_0^\top) & \frac{\alpha+1}{\alpha}\mu_0 \\ \frac{\alpha+1}{\alpha}\mu_0^\top & 1 \end{pmatrix}, Q \right\rangle \le -A(x)\ , \\
&Q \succeq 0\ ,\ Q \succeq \begin{pmatrix} 0 & \frac{\alpha}{2(\alpha+1)}B(x) \\ \frac{\alpha}{2(\alpha+1)}B(x)^\top & -r \end{pmatrix}\ , \\
&Q \succeq \begin{pmatrix} 0 & \frac{1}{2(\alpha+1)}\left(\alpha + \frac{n_{k+1}^{-\alpha} - n_k^{-\alpha}}{n_{k+1} - n_k}\right)B(x) \\ \frac{1}{2(\alpha+1)}\left(\alpha + \frac{n_{k+1}^{-\alpha} - n_k^{-\alpha}}{n_{k+1} - n_k}\right)B(x)^\top & \left(\frac{n_{k+1}n_k^{-\alpha} - n_k n_{k+1}^{-\alpha}}{(\alpha+1)(n_{k+1} - n_k)} - 1\right)r \end{pmatrix}\ , \\
&Q \succeq \begin{pmatrix} 0 & \frac{n_k^{-\alpha} - n_{k+1}^{-\alpha}}{2(\alpha+1)(n_{k+1} - n_k)}B(x) \\ \frac{n_k^{-\alpha} - n_{k+1}^{-\alpha}}{2(\alpha+1)(n_{k+1} - n_k)}B(x)^\top & -\frac{n_{k+1}n_k^{-\alpha} - n_k n_{k+1}^{-\alpha}}{(\alpha+1)(n_{k+1} - n_k)}r \end{pmatrix}\ ,\ k = 1, \ldots, K-1\ .
\end{aligned}
$$

References

Chen W, Sim M, Sun J et al (2010) From CVaR to uncertainty set: implications in joint chance-constrained optimization. Oper. Res. 58(2):470–485

Delage E, Ye Y (2010) Distributionally robust optimization under moment uncertainty with application to data-driven problems. Oper. Res. 58(3):595–612

Dharmadhikari S, Joag-dev K (eds) (1988) Unimodality, convexity, and applications. Academic Press Inc, San Diego

Li B, Jiang R, Mathieu JL (2019) Ambiguous risk constraints with moment and unimodality information. Math. Program. 173:151–192

Scarf HE (1958) A min-max solution of an inventory problem. In: Arrow KS, Karlin S, Scarf H (eds) Studies in the mathematical theory of inventory and production. Stanford University Press, Stanford, CA, pp 201–209

Zymler S, Kuhn D, Rustem B (2013) Distributionally robust joint chance constraints with second-order moment information. Math. Program. 137:167–198

Chapter 5
f-Divergence Distributionally Robust Optimization

Abstract This chapter is devoted to f-divergence distributionally robust optimization, where the ambiguity set consists of the probability distributions whose f-divergence to the empirical probability distribution is within a given threshold. An overview of f-divergence is presented first to give a sense of what the f-divergence ambiguity set is, including the definitions, important properties, and the relationships between different types of f-divergences. Then the f-divergence ambiguity set is formally defined, followed by the reformulation methods of the corresponding worst-case expectations and robust chance constraints.

5.1 f-Divergence

f-divergence is a collection of divergences that can measure the difference between two probability distributions, where one of the probability distributions is usually required to be absolutely continuous with respect to the other one.

5.1.1 Definitions

Recall from Definition 2.29 that $\mathbb{P}' \ll \mathbb{P}$ ($\mathbb{P}'$ is absolutely continuous with respect to $\mathbb{P}$) means that for any measurable set S, $\mathbb{P}'(S) > 0$ implies $\mathbb{P}(S) > 0$. When $\mathbb{P}' \ll \mathbb{P}$, there is a density function $\mathrm{d}\mathbb{P}'/\mathrm{d}\mathbb{P}$ so that $\mathbb{P}'(S) = \int_S (\mathrm{d}\mathbb{P}'/\mathrm{d}\mathbb{P})(\omega)\mathrm{d}\mathbb{P}(\omega)$ for any measurable set S. The original definition of f-divergence makes use of the function $\mathrm{d}\mathbb{P}'/\mathrm{d}\mathbb{P}$, which is in Definition 5.1.

Definition 5.1 (*f-divergence $D_f(\mathbb{P}'||\mathbb{P})$ for $\mathbb{P}' \ll \mathbb{P}$*) Assume $\mathbb{P}$ and $\mathbb{P}'$ are two probability distributions on Ω and $\mathbb{P}' \ll \mathbb{P}$. Let $f : [0, +\infty) \to \mathbb{R} \cup \{+\infty\}$ be a function convex on $(0, +\infty)$ and satisfy:

R. Xie and W. Wei, *Distributionally Robust Optimization and its Applications in Power System Energy Storage Sizing*,
https://doi.org/10.1007/978-981-97-2566-3_5

Table 5.1 Various f-divergences

f-divergence	$f(t)$	$l_f = \lim_{t\to+\infty} \frac{f(t)}{t}$	Conjugate $f^*(s)$
Kullback-Leibler divergence	$t\log t - t + 1$	$+\infty$	$\exp(s) - 1$
Total variation distance	$\lvert t-1\rvert/2$	$\frac{1}{2}$	$\begin{cases} -\frac{1}{2}, & s \le -\frac{1}{2}, \\ s, & -\frac{1}{2} \le s \le \frac{1}{2}, \\ +\infty, & s > \frac{1}{2} \end{cases}$
Squared Hellinger distance	$(\sqrt{t}-1)^2$	1	$\begin{cases} \frac{s}{1-s}, & s < 1, \\ +\infty, & s \ge 1 \end{cases}$
Pearson χ^2 divergence	$(t-1)^2$	$+\infty$	$\begin{cases} -1, & s \le -2, \\ \frac{s^2}{4} + s, & s > -2 \end{cases}$

$$\begin{cases} f(t) < +\infty, \ \forall 0 < t < +\infty, \\ f(1) = 0, \\ f(0) = \lim\limits_{t\to 0+} f(t) \in \mathbb{R} \cup \{+\infty\}. \end{cases} \tag{5.1}$$

Then the *f-divergence* $D_f(\mathbb{P}'||\mathbb{P})$ is defined by

$$D_f(\mathbb{P}'||\mathbb{P}) := \int_\Omega f\left(\frac{\mathrm{d}\mathbb{P}'}{\mathrm{d}\mathbb{P}}(\omega)\right) \mathrm{d}\mathbb{P}(\omega). \tag{5.2}$$

We already know that when $\mathbb{P}' \ll \mathbb{P}$, $\int_\Omega (\mathrm{d}\mathbb{P}'/\mathrm{d}\mathbb{P})(\omega)\mathrm{d}\mathbb{P}(\omega) = \mathbb{P}'(\Omega) = 1$. In Definition 5.1, $D_f(\mathbb{P}'||\mathbb{P})$ differs from $\int_\Omega (\mathrm{d}\mathbb{P}'/\mathrm{d}\mathbb{P})(\omega)\mathrm{d}\mathbb{P}(\omega)$ because of the function f. In addition, different functions f lead to different f-divergences as shown in Table 5.1, which will be illustrated in detail later in this section.

f-divergence can be extended to the case when $\mathbb{P}'$ is not absolutely continuous with respect to $\mathbb{P}$. This extension utilizes the property of f that $\lim_{t\to+\infty} f(t)/t$ is finite or $+\infty$ as Lemma 5.1 shows.

Lemma 5.1 *Assume $f : (0, +\infty) \to \mathbb{R}$ is a convex function satisfying $f(1) = 0$, then $\lim_{t\to+\infty} f(t)/t$ is finite or equals $+\infty$.*

Proof Because f is convex and $f(1) = 0$, the slope $(f(t) - f(1))/(t - 1) = f(t)/(t - 1)$ will be nondecreasing as t increases. Hence, the conclusion follows from $\lim_{t\to+\infty} f(t)/t = \lim_{t\to+\infty} f(t)/(t - 1)$. □

For notation brevity, let

$$l_f := \lim_{t\to+\infty} \frac{f(t)}{t},$$

in this chapter. Then Lemma 5.1 indicates that l_f is finite or equals $+\infty$ for convex $f : (0, +\infty) \to \mathbb{R}$.

Definition 5.2 (f-divergence without absolute continuity) Assume $\mathbb{P}$ and $\mathbb{P}'$ are two probability distributions on Ω and $f : [0, +\infty) \to \mathbb{R} \cup \{+\infty\}$ is a function convex on $(0, +\infty)$ satisfying (5.1). Let

$$f\left(\frac{a}{0}\right) \cdot 0 := \lim_{t \to 0+} f\left(\frac{a}{t}\right) \cdot t = a \lim_{t \to +\infty} \frac{f(t)}{t} = a l_f \ , \ \forall a > 0 \ ,$$

and define the f-*divergence* $D_f(\mathbb{P}'||\mathbb{P})$ by (5.2).

Example 5.1 (*f-divergence of discrete probability distributions*) Let $\mathbb{P}$ and $\mathbb{P}'$ be discrete probability distributions on $\Omega = \{\omega_1, \omega_2, \ldots, \omega_n, \ldots\}$. Then the f-divergence

$$\begin{aligned}
D_f(\mathbb{P}', \mathbb{P}) &= \sum_{n=1}^{+\infty} f\left(\frac{\mathbb{P}'(\omega_n)}{\mathbb{P}(\omega_n)}\right) \mathbb{P}(\omega_n) \\
&= \sum_{n:\mathbb{P}(\omega_n)>0} f\left(\frac{\mathbb{P}'(\omega_n)}{\mathbb{P}(\omega_n)}\right) \mathbb{P}(\omega_n) + \left(\lim_{t\to+\infty} \frac{f(t)}{t}\right) \cdot \sum_{n:\mathbb{P}(\omega_n)=0} \mathbb{P}'(\omega_n) \\
&= \sum_{n:\mathbb{P}(\omega_n)>0} f\left(\frac{\mathbb{P}'(\omega_n)}{\mathbb{P}(\omega_n)}\right) \mathbb{P}(\omega_n) + l_f \sum_{n:\mathbb{P}(\omega_n)=0} \mathbb{P}'(\omega_n) \\
&= \sum_{n:\mathbb{P}(\omega_n)>0} f\left(\frac{\mathbb{P}'(\omega_n)}{\mathbb{P}(\omega_n)}\right) \mathbb{P}(\omega_n) + l_f \mathbb{P}'(\{\omega \in \Omega \mid \mathbb{P}(\omega) = 0\}) \ .
\end{aligned}$$

Example 5.2 (*f-divergence of continuous probability distributions*) Let $\mathbb{P}$ and $\mathbb{P}'$ be continuous probability distributions on $\mathbb{R}^M$ with probability density functions (PDFs) p and p', respectively. Then the f-divergence

$$\begin{aligned}
D_f(\mathbb{P}', \mathbb{P}) &= \int_\Omega f\left(\frac{p'(\omega)}{p(\omega)}\right) p(\omega)\mathrm{d}\omega \\
&= \int_{\omega:p(\omega)>0} f\left(\frac{p'(\omega)}{p(\omega)}\right) p(\omega)\mathrm{d}\omega + l_f \int_{\omega:p(\omega)=0} p'(\omega)\mathrm{d}\omega \\
&= \int_{\omega:p(\omega)>0} f\left(\frac{p'(\omega)}{p(\omega)}\right) p(\omega)\mathrm{d}\omega + l_f \mathbb{P}'(\{\omega \in \Omega \mid p(\omega) = 0\}) \ .
\end{aligned}$$

5.1.2 Properties

We introduce the basic properties of general f-divergences, including linearity with respect to f, nonnegativity, the condition for zero f-divergence, the condition for symmetry, convexity, data processing, and conditioning properties. We emphasize that we assume absolute continuity of probability distributions throughout the discussion of properties. In addition, some properties will not hold without absolute continuity.

5.1.2.1 Linearity in f

Proposition 5.1 (Linearity of f-divergence) *Let* $f_1, f_2, \ldots, f_N : [0, +\infty) \to \mathbb{R}$ *be functions convex on* $(0, +\infty)$ *satisfying* (5.1). *Let* $a_1, a_2, \ldots, a_N \in [0, +\infty)$. *Then* $f := \sum_{n=1}^{N} a_n f_n$ *is also convex on* $(0, +\infty)$ *and satisfies* (5.1). *Moreover,*

$$D_f(\mathbb{P}'||\mathbb{P}) = \sum_{n=1}^{N} a_n D_{f_n}(\mathbb{P}'||\mathbb{P}) \ , \ \forall \mathbb{P}' \ll \mathbb{P} \ .$$

Exercise 5.1 Prove Proposition 5.1.

5.1.2.2 Nonnegativity and the Condition for Zero f-Divergence

Theorem 5.1 (Nonnegativity of f-divergence) *The f-divergence in Definition 5.1 is always nonnegative, i.e., for any* $\mathbb{P}' \ll \mathbb{P}$, *we have* $D_f(\mathbb{P}'||\mathbb{P}) \geq 0$. *In addition,* $D_f(\mathbb{P}||\mathbb{P}) = 0$.

Proof By Jensen's inequality in Theorem 2.10 and the convexity of f,

$$\begin{aligned} D_f(\mathbb{P}'||\mathbb{P}) &= \int_\Omega f\left(\frac{\mathrm{d}\mathbb{P}'}{\mathrm{d}\mathbb{P}}(\omega)\right) \mathrm{d}\mathbb{P}(\omega) \\ &\geq f\left(\int_\Omega \frac{\mathrm{d}\mathbb{P}'}{\mathrm{d}\mathbb{P}}(\omega)\mathrm{d}\mathbb{P}(\omega)\right) = f(\mathbb{P}'(\Omega)) = f(1) = 0 \ . \end{aligned}$$

In addition,

$$D_f(\mathbb{P}||\mathbb{P}) = \int_\Omega f\left(\frac{\mathrm{d}\mathbb{P}}{\mathrm{d}\mathbb{P}}(\omega)\right) \mathrm{d}\mathbb{P}(\omega) = \int_\Omega f(1)\mathrm{d}\mathbb{P}(\omega) = 0 \ ,$$

which completes the proof. □

In the above proof, we utilize the fact that $\int_\Omega (\mathrm{d}\mathbb{P}'/\mathrm{d}\mathbb{P})(\omega)\mathrm{d}\mathbb{P}(\omega) = 1$ and $\int_\Omega \mathrm{d}\mathbb{P}'(\omega) = 1$. Therefore, if $f(t) = c(t-1)$ for some constant $c \in \mathbb{R}$, we have:

$$\begin{aligned} D_f(\mathbb{P}'||\mathbb{P}) &= \int_\Omega c\left(\frac{\mathrm{d}\mathbb{P}'}{\mathrm{d}\mathbb{P}}(\omega) - 1\right) \mathrm{d}\mathbb{P}(\omega) \\ &= c\left(\int_\Omega \frac{\mathrm{d}\mathbb{P}'}{\mathrm{d}\mathbb{P}}(\omega)\mathrm{d}\mathbb{P}(\omega) - \int_\Omega \mathrm{d}\mathbb{P}(\omega)\right) = 0 \ , \end{aligned}$$

where $\mathbb{P}$ and $\mathbb{P}'$ are arbitrary. Proposition 5.2 says that $f(t) = c(t-1)$ for some c is also the necessary condition for D_f to be always 0.

Proposition 5.2 (Necessary and sufficient condition for f-divergence being always zero) *In Definition 5.1,* $D_f(\mathbb{P}'||\mathbb{P}) = 0$ *for arbitrary probability distributions* $\mathbb{P}$ *and* $\mathbb{P}'$ *subject to* $\mathbb{P}' \ll \mathbb{P}$ *if and only if there is a constant* $c \in \mathbb{R}$ *such that* $f(t) = c(t-1), \forall t \in [0, +\infty)$.

Proof We already know $f(t) = c(t-1)$ is a sufficient condition and we only need to prove the opposite direction. To this end, suppose $D_f(\mathbb{P}'||\mathbb{P}) = 0, \forall \mathbb{P}' \ll \mathbb{P}$. Let $c := -f(0)$. Fix $N \in \mathbb{N}_+$. Let $\Omega = \{\omega_1, \omega_2, \ldots, \omega_N\}$. Construct probability distributions $\mathbb{P}$ and $\mathbb{P}'$ as follows:

$$\mathbb{P}(\omega_n) = \frac{1}{N}\ , \quad n = 1, 2, \ldots, N\ ,$$
$$\mathbb{P}'(\omega_1) = 1\ ,\ \mathbb{P}'(\omega_n) = 0\ ,\ n = 2, 3, \ldots, N\ .$$

Then $\mathbb{P}' \ll \mathbb{P}$ and

$$0 = D_f(\mathbb{P}'||\mathbb{P}) = \int_\Omega f\left(\frac{\mathrm{d}\mathbb{P}'}{\mathrm{d}\mathbb{P}}(\omega)\right) \mathrm{d}\mathbb{P}(\omega) = \frac{1}{N} f(N) + \frac{N-1}{N} f(0)\ .$$

Thus,

$$f(N) = -(N-1)f(0) = (N-1)c\ ,\ \forall N \in \mathbb{N}_+\ .$$

Noting that $f(0) = -c$, we have $f(N) = (N-1)c, \forall N \in \mathbb{N}$. For any $t \in [0, +\infty)$, there exists $N_t \in \mathbb{N}$ such that $N_t \le t < N_t + 1$. Because f is convex,

$$\begin{aligned} f(t) &= f((N_t + 1 - t)N_t + (t - N_t)(N_t + 1)) \\ &\le (N_t + 1 - t) f(N_t) + (t - N_t) f(N_t + 1) = c(t-1)\ . \end{aligned}$$

In addition, by Proposition 2.19,

$$\frac{f(N_t+1) - f(t)}{N_t + 1 - t} \le \frac{f(N_t+2) - f(N_t+1)}{1} = c \implies f(t) \ge c(t-1)\ .$$

Therefore, $f(t) = c(t-1), \forall t \in [0, +\infty)$. □

Combining the conclusion of Proposition 5.2 with the linearity of f-divergence, we obtain the necessary and sufficient condition for two f-divergences being the same:

Corollary 5.1 (Necessary and sufficient condition for f-divergences being the same) *Assume D_f and D_g are f-divergences. Then $D_f(\mathbb{P}'||\mathbb{P}) = D_g(\mathbb{P}'||\mathbb{P})$ for any probability distributions $\mathbb{P}' \ll \mathbb{P}$ if and only if there exists $c \in \mathbb{R}$ such that $g(t) = f(t) + c(t-1), \forall t \in [0, +\infty)$.*

Thus, there are many functions f that lead to the same f-divergence in Definition 5.1 and they differ by a term $c(t-1)$. In particular, the functions f listed in Table 5.1 are not unique. However, the term $c(t-1)$ does affect $l_f = \lim_{t\to+\infty} f(t)/t$ and therefore has an impact on the f-divergence without absolute continuity in Definition 5.2.

Since f is convex on $(0, +\infty)$ and $f(1) = 0$, we can add a restriction that $f(t) \ge 0, \forall t \ge 0$. In other words, for any f-divergence D_f, there is a function f satisfies the conditions in Definition 5.1 and that $f(t) \geq 0, \forall t \geq 0$. The functions f in Table 5.1 all satisfy this requirement. If f is differentiable at $t = 1$, then such f is unique.

We then turn to the following question: *Does* $D_f(\mathbb{P}'||\mathbb{P}) = 0$ *indicate* $\mathbb{P}' = \mathbb{P}$ *as probability distributions?* We know that the answer is no when $f(x) = c(x-1)$ for some $c \in \mathbb{R}$. Nonetheless, if f is a strictly convex function, the answer is actually yes.

Proposition 5.3 *Let D_f be an f-divergence with a strictly convex f. Assume $\mathbb{P}' \ll \mathbb{P}$ on Ω and $D_f(\mathbb{P}'||\mathbb{P}) = 0$, then $\mathbb{P}' = \mathbb{P}$ as probability distributions.*

Proof According to the proof of Theorem 5.1 and the conclusion of Jensen's inequality for strictly convex function in Theorem 2.10, there is some $c \in \mathbb{R}$ such that

$$\mathbb{P}\left(\left\{\omega \in \Omega \,\middle|\, \frac{\mathrm{d}\mathbb{P}'}{\mathrm{d}\mathbb{P}}(\omega) = c\right\}\right) = 1 .$$

Note that

$$c = \int_\Omega c \cdot \mathrm{d}\mathbb{P}(\omega) = \int_\Omega \frac{\mathrm{d}\mathbb{P}'}{\mathrm{d}\mathbb{P}}(\omega)\mathrm{d}\mathbb{P}(\omega) = \int_\Omega \mathrm{d}\mathbb{P}'(\omega) = 1 ,$$

so $\mathbb{P}' = \mathbb{P}$ as probability distributions. □

Remark 5.1 Proposition 5.3 gives a sufficient but not necessary condition for $D_f(\mathbb{P}'||\mathbb{P}) = 0 \iff \mathbb{P}' = \mathbb{P}, \forall \mathbb{P}' \ll \mathbb{P}$. For example, we will see later in this section that the function f of total variation distance is not strictly convex but the total variation distance equals zero if and only if the two probability distributions are the same.

Exercise 5.2 Use Proposition 5.3 to show that Kullback-Leibler (KL) divergence, squared Hellinger distance, and Pearson χ^2 divergence in Table 5.1 have the property that $D_f(\mathbb{P}'||\mathbb{P}) = 0 \iff \mathbb{P}' = \mathbb{P}, \forall \mathbb{P}' \ll \mathbb{P}$.

5.1.2.3 The Condition for Symmetry

From the definition formula in (5.2), we see that f-divergence is generally not symmetric, i.e., $D_f(\mathbb{P}'||\mathbb{P}) \neq D_f(\mathbb{P}||\mathbb{P}')$, which is one of the burdens prohibiting it from being a distance function. However, when the function f meets some additional requirements, f-divergence can be symmetric. To begin with, we consider a more general question: *Given function f, is there a function g so that* $D_g(\mathbb{P}||\mathbb{P}') = D_f(\mathbb{P}'||\mathbb{P})$?

For simplicity, assume $\mathbb{P}' \ll \mathbb{P}$ and $\mathbb{P} \ll \mathbb{P}'$. Observe that

$$D_f(\mathbb{P}'||\mathbb{P}) = \int_\Omega f\left(\frac{\mathrm{d}\mathbb{P}'}{\mathrm{d}\mathbb{P}}(\omega)\right)\mathrm{d}\mathbb{P}(\omega) = \int_\Omega f\left(\frac{\mathrm{d}\mathbb{P}'}{\mathrm{d}\mathbb{P}}(\omega)\right)\frac{\mathrm{d}\mathbb{P}}{\mathrm{d}\mathbb{P}'}(\omega)\mathrm{d}\mathbb{P}'(\omega) ,$$

prompting us to consider the function $g(t) = tf(1/t)$. Lemma 5.2 says such g satisfy the requirements of the functions in f-divergence.

Lemma 5.2 *Assume $f : (0, +\infty) \to \mathbb{R}$ is a convex function and $f(1) = 0$. Let $g : (0, +\infty) \to \mathbb{R}$ be a function defined by $g(t) = tf(1/t)$. Then g is a convex function, $g(1) = 0$, and*

$$\lim_{t\to 0+} g(t) = \lim_{t\to +\infty} \frac{f(t)}{t} = l_f \ ,$$

is either finite or $+\infty$.

Proof Because f is convex, for any $t_1, t_2 \in (0, +\infty)$ and $\lambda \in [0, 1]$,

$$\begin{aligned}
&g(\lambda t_1 + (1-\lambda)t_2)\\
&= (\lambda t_1 + (1-\lambda)t_2) \cdot f\left(\frac{1}{\lambda t_1 + (1-\lambda)t_2}\right)\\
&\le (\lambda t_1 + (1-\lambda)t_2)\left(\frac{\lambda t_1}{\lambda t_1 + (1-\lambda)t_2} f\left(\frac{1}{t_1}\right) + \frac{(1-\lambda)t_2}{\lambda t_1 + (1-\lambda)t_2} f\left(\frac{1}{t_2}\right)\right)\\
&= \lambda t_1 f\left(\frac{1}{t_1}\right) + (1-\lambda)t_2 f\left(\frac{1}{t_2}\right) = \lambda g(t_1) + (1-\lambda) g(t_2) \ ,
\end{aligned}$$

which shows g is convex. $g(1) = f(1) = 0$. $\lim_{t\to 0+} g(t) = \lim_{t\to 0+} tf(1/t) = \lim_{t\to +\infty} f(t)/t = l_f$ is either finite or $+\infty$ by Lemma 5.1. □

Then according to the previous analysis, we have the following conclusion:

Proposition 5.4 *Let $f : [0, +\infty) \to \mathbb{R} \cup \{+\infty\}$ be a function convex on $(0, +\infty)$ satisfying (5.1). Assume the two probability distributions $\mathbb{P}$ and $\mathbb{P}'$ satisfy $\mathbb{P}' \ll \mathbb{P}$ and $\mathbb{P} \ll \mathbb{P}'$. Let $g : [0, +\infty) \to \mathbb{R} \cup \{+\infty\}$ be $g(t) := tf(1/t), \forall t \in (0, +\infty)$ and $g(0) := \lim_{t\to 0+} g(t)$. Then*

$$D_f(\mathbb{P}'||\mathbb{P}) = D_g(\mathbb{P}||\mathbb{P}') \ . \tag{5.3}$$

Although f-divergence D_f is not symmetric in general, $D_{(f+g)/2}$ is always symmetric with g defined in Proposition 5.4. In other words, f-divergence D_f can be revised to maintain symmetry by using

$$D_{(f+g)/2}(\mathbb{P}'||\mathbb{P}) = \frac{1}{2}(D_f(\mathbb{P}'||\mathbb{P}) + D_f(\mathbb{P}||\mathbb{P}')) \ , \ \forall \mathbb{P}' \ll \mathbb{P} \ , \ \mathbb{P} \ll \mathbb{P}' \ .$$

Recall from Corollary 5.1 the necessary and sufficient condition for f-divergences being the same, we have the following conclusion about when an f-divergence is symmetric:

Corollary 5.2 (Necessary and sufficient condition for f-divergence being symmetric) *Assume D_f is an f-divergence. Define $g : [0, +\infty) \to \mathbb{R} \cup \{+\infty\}$ by $g(t) := tf(1/t), \forall t \in (0, +\infty)$ and $g(0) := \lim_{t\to 0+} g(t)$. Then D_f is symmetric if and only if there is $c \in \mathbb{R}$ so that $g(t) = f(t) + c(t - 1), \forall t \in [0, +\infty)$.*

Exercise 5.3 Prove that the total variation distance and squared Hellinger distance in Table 5.1 are symmetric.

5.1.2.4 Convexity

f-divergence is jointly convex with respect to the two probability distributions.

Theorem 5.2 (Convexity of f-divergence) *Assume D_f is an f-divergence; $\mathbb{P}_1$, $\mathbb{P}_2$, $\mathbb{P}'_1$, and $\mathbb{P}'_2$ are probability distributions on Ω with the properties $\mathbb{P}'_1 \ll \mathbb{P}_1$ and $\mathbb{P}'_2 \ll \mathbb{P}_2$. Then for any $\lambda \in [0, 1]$, $\lambda\mathbb{P}'_1 + (1 - \lambda)\mathbb{P}'_2 \ll \lambda\mathbb{P}_1 + (1 - \lambda)\mathbb{P}_2$ and*

$$\begin{aligned}&D_f(\lambda\mathbb{P}'_1 + (1 - \lambda)\mathbb{P}'_2||\lambda\mathbb{P}_1 + (1 - \lambda)\mathbb{P}_2)\\ &\le \lambda D_f(\mathbb{P}'_1||\mathbb{P}_1) + (1 - \lambda)D_f(\mathbb{P}'_2||\mathbb{P}_2)\ .\end{aligned}$$

Proof We only need to consider the case $\lambda \in (0, 1)$. Let S be a measurable subset of Ω and $(\lambda\mathbb{P}_1 + (1 - \lambda)\mathbb{P}_2)(S) = 0$. Then $\mathbb{P}_1(S) = \mathbb{P}_2(S) = 0$. By $\mathbb{P}'_1 \ll \mathbb{P}_1$ and $\mathbb{P}'_2 \ll \mathbb{P}_2$, we have $\mathbb{P}'_1(S) = \mathbb{P}'_2(S) = 0$. Hence, $(\lambda\mathbb{P}'_1 + (1 - \lambda)\mathbb{P}'_2)(S) = 0$. Since S is arbitrary, we have $\mathbb{P}'_1 \ll \lambda\mathbb{P}_1 + (1 - \lambda)\mathbb{P}_2$, $\mathbb{P}'_2 \ll \lambda\mathbb{P}_1 + (1 - \lambda)\mathbb{P}_2$, and $\lambda\mathbb{P}'_1 + (1 - \lambda)\mathbb{P}'_2 \ll \lambda\mathbb{P}_1 + (1 - \lambda)\mathbb{P}_2$. Denote $\lambda\mathbb{P}_1 + (1 - \lambda)\mathbb{P}_2$ by $\mathbb{P}_3$. According to the convexity of f,

$$\begin{aligned}&D_f(\lambda\mathbb{P}'_1 + (1 - \lambda)\mathbb{P}'_2||\mathbb{P}_3)\\ &= \int_\Omega f\left(\frac{\mathrm{d}(\lambda\mathbb{P}'_1 + (1 - \lambda)\mathbb{P}'_2)}{\mathrm{d}\mathbb{P}_3}(\omega)\right)\mathrm{d}\mathbb{P}_3(\omega)\\ &= \int_\Omega f\left(\frac{\mathrm{d}(\lambda\mathbb{P}_1)}{\mathrm{d}\mathbb{P}_3}(\omega)\cdot\frac{\mathrm{d}\mathbb{P}'_1}{\mathrm{d}\mathbb{P}_1}(\omega) + \frac{\mathrm{d}((1 - \lambda)\mathbb{P}_2)}{\mathrm{d}\mathbb{P}_3}(\omega)\cdot\frac{\mathrm{d}\mathbb{P}'_2}{\mathrm{d}\mathbb{P}_2}(\omega)\right)\mathrm{d}\mathbb{P}_3(\omega)\\ &\le \int_\Omega \left(\frac{\mathrm{d}(\lambda\mathbb{P}_1)}{\mathrm{d}\mathbb{P}_3}(\omega)\cdot f\left(\frac{\mathrm{d}\mathbb{P}'_1}{\mathrm{d}\mathbb{P}_1}(\omega)\right) + \frac{\mathrm{d}((1 - \lambda)\mathbb{P}_2)}{\mathrm{d}\mathbb{P}_3}(\omega)\cdot f\left(\frac{\mathrm{d}\mathbb{P}'_2}{\mathrm{d}\mathbb{P}_2}(\omega)\right)\right)\mathrm{d}\mathbb{P}_3(\omega)\\ &= \lambda\int_\Omega f\left(\frac{\mathrm{d}\mathbb{P}'_1}{\mathrm{d}\mathbb{P}_1}(\omega)\right)\mathrm{d}\mathbb{P}_1(\omega) + (1 - \lambda)\int_\Omega f\left(\frac{\mathrm{d}\mathbb{P}'_2}{\mathrm{d}\mathbb{P}_2}(\omega)\right)\mathrm{d}\mathbb{P}_2(\omega)\\ &= \lambda D_f(\mathbb{P}'_1||\mathbb{P}_1) + (1 - \lambda)D_f(\mathbb{P}'_2||\mathbb{P}_2)\ ,\end{aligned}$$

where we use the fact that

$$\frac{\mathrm{d}(\lambda\mathbb{P}_1)}{\mathrm{d}\mathbb{P}_3}(\omega) + \frac{\mathrm{d}((1 - \lambda)\mathbb{P}_2)}{\mathrm{d}\mathbb{P}_3}(\omega) = \frac{\mathrm{d}\mathbb{P}_3}{\mathrm{d}\mathbb{P}_3}(\omega) = 1\ .$$

This completes the proof. □

A direct consequence of the joint convexity in Theorem 5.2 is that the f-divergence $D_f(\mathbb{P}'||\mathbb{P})$ is both convex as a function of $\mathbb{P}$ and as a function of $\mathbb{P}'$.

5.1.2.5 Data Processing

The data processing property of f-divergence is about how the f-divergence will change after the two probability distributions are both processed by a conditional probability distribution. The answer is that the f-divergence is always nonincreasing after data processing. To illustrate it, we first consider the f-divergence of joint probability distributions. Theorem 5.3 says that the f-divergence of joint probability distributions is no smaller than that of marginal probability distributions.

Theorem 5.3 *Let D_f be an f-divergence. Assume the joint probability distributions $\mathbb{P}'_{\alpha,\beta} \ll \mathbb{P}_{\alpha,\beta}$ on $\Omega_\alpha \times \Omega_\beta$. Assume $\mathbb{P}_{\alpha,\beta} \in \mathcal{J}(\mathbb{P}_\alpha, \mathbb{P}_\beta)$ and $\mathbb{P}'_{\alpha,\beta} \in \mathcal{J}(\mathbb{P}'_\alpha, \mathbb{P}'_\beta)$, which means $\mathbb{P}_{\alpha,\beta}$ and $\mathbb{P}'_{\alpha,\beta}$ have marginal probability distributions $\mathbb{P}_\alpha, \mathbb{P}_\beta$ and $\mathbb{P}'_\alpha, \mathbb{P}'_\beta$, respectively. There are conditional probability distributions $\mathbb{P}_{\beta|\alpha}$ and $\mathbb{P}'_{\beta|\alpha}$ such that $\mathbb{P}_{\alpha,\beta} = \mathbb{P}_{\beta|\alpha}\mathbb{P}_\alpha$ and $\mathbb{P}'_{\alpha,\beta} = \mathbb{P}'_{\beta|\alpha}\mathbb{P}'_\alpha$. Then $\mathbb{P}'_\alpha \ll \mathbb{P}_\alpha$, $\mathbb{P}'_{\beta|\alpha}(\cdot|S_\alpha) \ll \mathbb{P}_{\beta|\alpha}(\cdot|S_\alpha)$ when $\mathbb{P}'_\alpha(S_\alpha) > 0$, and*

$$D_f(\mathbb{P}'_\alpha||\mathbb{P}_\alpha) \le D_f(\mathbb{P}'_{\alpha,\beta}||\mathbb{P}_{\alpha,\beta}) = D_f(\mathbb{P}'_{\beta|\alpha}\mathbb{P}'_\alpha||\mathbb{P}_{\beta|\alpha}\mathbb{P}_\alpha) .$$

Proof Assume $\mathbb{P}_\alpha(S_\alpha) = 0$ for S_α measurable in Ω_α. Then $\mathbb{P}_{\alpha,\beta}(S_\alpha \times \Omega_b) = \mathbb{P}_\alpha(S_\alpha) = 0$. By $\mathbb{P}'_{\alpha,\beta} \ll \mathbb{P}_{\alpha,\beta}$, we have $\mathbb{P}'_\alpha(S_\alpha) = \mathbb{P}'_{\alpha,\beta}(S_\alpha \times \Omega_b) = 0$, which shows $\mathbb{P}'_\alpha \ll \mathbb{P}_\alpha$.

Let S_α be a measurable set in Ω_α with $\mathbb{P}'_\alpha(S_\alpha) > 0$, then $\mathbb{P}_\alpha(S_\alpha) > 0$. Assume $\mathbb{P}_{\beta|\alpha}(S_\beta|S_\alpha) = 0$ for some S_β measurable in Ω_β. Then $\mathbb{P}_{\alpha,\beta}(S_\alpha \times S_\beta) = \mathbb{P}_{\beta|\alpha}(S_\beta|S_\alpha)\mathbb{P}_\alpha(S_\alpha) = 0$. Since $\mathbb{P}'_{\alpha,\beta} \ll \mathbb{P}_{\alpha,\beta}$, $0 = \mathbb{P}'_{\alpha,\beta}(S_\alpha \times S_\beta) = \mathbb{P}'_{\beta|\alpha}(S_\beta|S_\alpha)\mathbb{P}'_\alpha(S_\alpha)$, which implies $\mathbb{P}'_{\beta|\alpha}(S_\beta|S_\alpha) = 0$. This shows $\mathbb{P}'_{\beta|\alpha}(\cdot|S_\alpha) \ll \mathbb{P}_{\beta|\alpha}(\cdot|S_\alpha)$ when $\mathbb{P}'_\alpha(S_\alpha) > 0$.

By Jensen's inequality in Theorem 2.10 and the convexity of f, we have:

$$\begin{aligned}
& D_f(\mathbb{P}'_{\alpha,\beta}||\mathbb{P}_{\alpha,\beta}) \\
&= \int_{\Omega_\alpha \times \Omega_\beta} f\left(\frac{\mathrm{d}\mathbb{P}'_{\alpha,\beta}}{\mathrm{d}\mathbb{P}_{\alpha,\beta}}(\omega_\alpha, \omega_\beta)\right) \mathrm{d}\mathbb{P}_{\alpha,\beta}(\omega_\alpha, \omega_\beta) \\
&= \int_{\Omega_\alpha \times \Omega_\beta} f\left(\frac{\mathrm{d}(\mathbb{P}'_{\beta|\alpha}\mathbb{P}'_\alpha)}{\mathrm{d}(\mathbb{P}_{\beta|\alpha}\mathbb{P}_\alpha)}(\omega_\alpha, \omega_\beta)\right) \mathrm{d}(\mathbb{P}_{\beta|\alpha}\mathbb{P}_\alpha)(\omega_\alpha, \omega_\beta) \\
&= \int_{\Omega_\alpha}\int_{\Omega_\beta} f\left(\frac{\mathrm{d}\mathbb{P}'_\alpha}{\mathrm{d}\mathbb{P}_\alpha}(\omega_\alpha)\frac{\mathrm{d}\mathbb{P}'_{\beta|\alpha}}{\mathrm{d}\mathbb{P}_{\beta|\alpha}}(\omega_\beta|\omega_\alpha)\right) \mathrm{d}\mathbb{P}_{\beta|\alpha}(\omega_\beta|\omega_\alpha)\mathrm{d}\mathbb{P}_\alpha(\omega_\alpha) \\
&\ge \int_{\Omega_\alpha} f\left(\int_{\Omega_\beta} \frac{\mathrm{d}\mathbb{P}'_\alpha}{\mathrm{d}\mathbb{P}_\alpha}(\omega_\alpha)\frac{\mathrm{d}\mathbb{P}'_{\beta|\alpha}}{\mathrm{d}\mathbb{P}_{\beta|\alpha}}(\omega_\beta|\omega_\alpha)\mathrm{d}\mathbb{P}_{\beta|\alpha}(\omega_\beta|\omega_\alpha)\right) \mathrm{d}\mathbb{P}_\alpha(\omega_\alpha) \\
&= \int_{\Omega_\alpha} f\left(\frac{\mathrm{d}\mathbb{P}'_\alpha}{\mathrm{d}\mathbb{P}_\alpha}(\omega_\alpha)\int_{\Omega_\beta} \mathrm{d}\mathbb{P}'_{\beta|\alpha}(\omega_\beta|\omega_\alpha)\right) \mathrm{d}\mathbb{P}_\alpha(\omega_\alpha)
\end{aligned}$$

$$= \int_{\Omega_\alpha} f\left(\frac{\mathrm{d}\mathbb{P}'_\alpha}{\mathrm{d}\mathbb{P}_\alpha}(\omega_\alpha)\right) \mathrm{d}\mathbb{P}_\alpha(\omega_\alpha) = D_f(\mathbb{P}'_\alpha||\mathbb{P}_\alpha) ,$$

which completes the proof. □

Proposition 5.5 gives a condition for the equality in Theorem 5.3: When the conditional probability distributions are the same, the f-divergence of the joint probability distributions equals that of the original probability distributions.

Proposition 5.5 *In Theorem 5.3, if* $\mathbb{P}_{\beta|\alpha} = \mathbb{P}'_{\beta|\alpha}$*, then*

$$D_f(\mathbb{P}'_\alpha||\mathbb{P}_\alpha) = D_f(\mathbb{P}'_{\alpha,\beta}||\mathbb{P}_{\alpha,\beta}) = D_f(\mathbb{P}_{\beta|\alpha}\mathbb{P}'_\alpha||\mathbb{P}_{\beta|\alpha}\mathbb{P}_\alpha) .$$

Proof Similar to the proof of Theorem 5.3, we have:

$$\begin{aligned} &D_f(\mathbb{P}_{\beta|\alpha}\mathbb{P}'_\alpha||\mathbb{P}_{\beta|\alpha}\mathbb{P}_\alpha) \\ &= \int_{\Omega_\alpha\times\Omega_\beta} f\left(\frac{\mathrm{d}(\mathbb{P}_{\beta|\alpha}\mathbb{P}'_\alpha)}{\mathrm{d}(\mathbb{P}_{\beta|\alpha}\mathbb{P}_\alpha)}(\omega_\alpha,\omega_\beta)\right) \mathrm{d}(\mathbb{P}_{\beta|\alpha}\mathbb{P}_\alpha)(\omega_\alpha,\omega_\beta) \\ &= \int_{\Omega_\alpha}\int_{\Omega_\beta} f\left(\frac{\mathrm{d}\mathbb{P}'_\alpha}{\mathrm{d}\mathbb{P}_\alpha}(\omega_\alpha)\frac{\mathrm{d}\mathbb{P}_{\beta|\alpha}}{\mathrm{d}\mathbb{P}_{\beta|\alpha}}(\omega_\beta|\omega_\alpha)\right) \mathrm{d}\mathbb{P}_{\beta|\alpha}(\omega_\beta|\omega_\alpha)\mathrm{d}\mathbb{P}_\alpha(\omega_\alpha) \\ &= \int_{\Omega_\alpha} f\left(\frac{\mathrm{d}\mathbb{P}'_\alpha}{\mathrm{d}\mathbb{P}_\alpha}(\omega_\alpha)\right)\left(\int_{\Omega_\beta} \mathrm{d}\mathbb{P}_{\beta|\alpha}(\omega_\beta|\omega_\alpha)\right) \mathrm{d}\mathbb{P}_\alpha(\omega_\alpha) \\ &= \int_{\Omega_\alpha} f\left(\frac{\mathrm{d}\mathbb{P}'_\alpha}{\mathrm{d}\mathbb{P}_\alpha}(\omega_\alpha)\right) \mathrm{d}\mathbb{P}_\alpha(\omega_\alpha) = D_f(\mathbb{P}'_\alpha||\mathbb{P}_\alpha) , \end{aligned}$$

which completes the proof. □

Now we are ready to introduce the data processing property of f-divergence.

Theorem 5.4 (Data processing property of f-divergence) *Assume* $\mathbb{P}'_\alpha \ll \mathbb{P}_\alpha$ *are probability distributions on* Ω_α. $\mathbb{P}_{\beta|\alpha}$ *is a conditional probability distribution. In addition,* $\mathbb{P}_{\alpha,\beta} := \mathbb{P}_{\beta|\alpha}\mathbb{P}_\alpha \in \mathcal{J}(\mathbb{P}_\alpha, \mathbb{P}_\beta)$ *and* $\mathbb{P}'_{\alpha,\beta} := \mathbb{P}_{\beta|\alpha}\mathbb{P}'_\alpha \in \mathcal{J}(\mathbb{P}'_\alpha, \mathbb{P}'_\beta)$ *for probability distributions* $\mathbb{P}_\beta$ *and* $\mathbb{P}'_\beta$ *on* Ω_β*. Then*

$$D_f(\mathbb{P}'_\beta||\mathbb{P}_\beta) \le D_f(\mathbb{P}'_\alpha||\mathbb{P}_\alpha) .$$

Remark 5.2 Figure 5.1 is a diagram showing the data processing property. The conditional probability distribution $\mathbb{P}_{\beta|\alpha}$ is regarded as a data processing block, through which $\mathbb{P}_\alpha$ and $\mathbb{P}'_\alpha$ are transformed into $\mathbb{P}_\beta$ and $\mathbb{P}'_\beta$.

Proof Combining the conclusions of Theorem 5.3 and Proposition 5.5, we have

$$D_f(\mathbb{P}'_\beta||\mathbb{P}_\beta) \le D_f(\mathbb{P}'_{\alpha,\beta}||\mathbb{P}_{\alpha,\beta}) = D_f(\mathbb{P}_{\beta|\alpha}\mathbb{P}'_\alpha||\mathbb{P}_{\beta|\alpha}\mathbb{P}_\alpha) = D_f(\mathbb{P}'_\alpha||\mathbb{P}_\alpha) ,$$

which is the desired data processing property. □

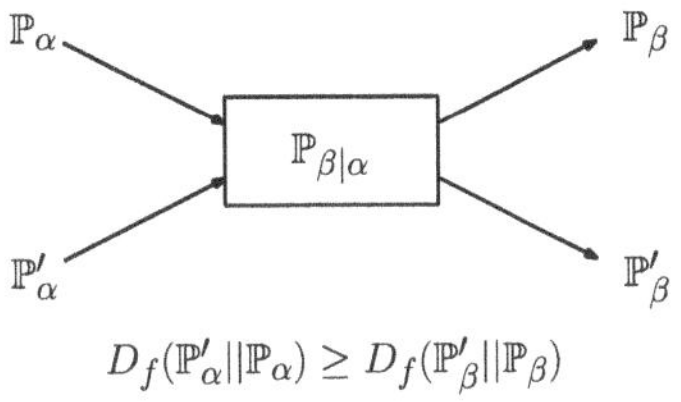

Fig. 5.1 Probability distributions $\mathbb{P}_\alpha$ and $\mathbb{P}'_\alpha$ are processed by the block of conditional probability distribution $\mathbb{P}_{\beta|\alpha}$ and then $\mathbb{P}_\beta$ and $\mathbb{P}'_\beta$ are obtained. The f-divergence will not increase after the data processing

5.1.2.6 Conditioning

For conditional probability distributions, there is a definition of the conditional f-divergence.

Definition 5.3 (*Conditional f-divergence*) Let $\mathbb{P}_\alpha$ be a probability distribution on Ω_α. Assume $\mathbb{P}_{\beta|\alpha}$ and $\mathbb{P}'_{\beta|\alpha}$ are conditional probability distributions. $\mathbb{P}'_{\beta|\alpha}(\cdot|\omega_\alpha) \ll \mathbb{P}_{\beta|\alpha}(\cdot|\omega_\alpha)$ are probability distributions on Ω_β for any $\omega_\alpha \in \Omega_\alpha$. The *conditional f-divergence* is defined by

$$D_f(\mathbb{P}'_{\beta|\alpha}||\mathbb{P}_{\beta|\alpha}|\mathbb{P}_\alpha) := \int_{\Omega_\alpha} D_f(\mathbb{P}'_{\beta|\alpha}(\cdot|\omega_\alpha)||\mathbb{P}_{\beta|\alpha}(\cdot|\omega_\alpha))\mathrm{d}\mathbb{P}_\alpha(\omega_\alpha) .$$

In Definition 5.3, the conditional f-divergence is the expectation of the f-divergence between conditional probability distributions. There is another formula to calculate the conditional f-divergence.

Lemma 5.3 *In Definition 5.3,*

$$D_f(\mathbb{P}'_{\beta|\alpha}||\mathbb{P}_{\beta|\alpha}|\mathbb{P}_\alpha) = D_f(\mathbb{P}'_{\beta|\alpha}\mathbb{P}_\alpha||\mathbb{P}_{\beta|\alpha}\mathbb{P}_\alpha) .$$

Proof Using the definitions and Fubini's theorem in Theorem 2.3, we have:

$$\begin{aligned}
&D_f(\mathbb{P}'_{\beta|\alpha}\mathbb{P}_\alpha||\mathbb{P}_{\beta|\alpha}\mathbb{P}_\alpha)\\
&= \int_{\Omega_\alpha \times \Omega_\beta} f\left(\frac{\mathrm{d}(\mathbb{P}'_{\beta|\alpha}\mathbb{P}_\alpha)}{\mathrm{d}(\mathbb{P}_{\beta|\alpha}\mathbb{P}_\alpha)}(\omega_\alpha, \omega_\beta)\right) \mathrm{d}(\mathbb{P}_{\beta|\alpha}\mathbb{P}_\alpha)(\omega_\alpha, \omega_\beta)\\
&= \int_{\Omega_\alpha}\int_{\Omega_\beta} f\left(\frac{\mathrm{d}\mathbb{P}'_{\beta|\alpha}}{\mathrm{d}\mathbb{P}_{\beta|\alpha}}(\omega_\beta|\omega_\alpha)\right) \mathrm{d}\mathbb{P}_{\beta|\alpha}(\omega_\beta|\omega_\alpha)\mathrm{d}\mathbb{P}_\alpha(\omega_\alpha)\\
&= \int_{\Omega_\alpha} D_f(\mathbb{P}'_{\beta|\alpha}(\cdot|\omega_\alpha)||\mathbb{P}_{\beta|\alpha}(\cdot|\omega_\alpha))\mathrm{d}\mathbb{P}_\alpha(\omega_\alpha) ,
\end{aligned}$$

which is the definition of $D_f(\mathbb{P}'_{\beta|\alpha}||\mathbb{P}_{\beta|\alpha}|\mathbb{P}_\alpha)$. □

Theorem 5.5 says that f-divergence never decreases after conditioning.

Theorem 5.5 (f-divergence is nondecreasing after conditioning) *In Definition 5.3, assume* $\mathbb{P}_{\beta|\alpha}\mathbb{P}_\alpha \in \mathcal{J}(\mathbb{P}_\alpha, \mathbb{P}_\beta)$ *and* $\mathbb{P}'_{\beta|\alpha}\mathbb{P}_\alpha \in \mathcal{J}(\mathbb{P}_\alpha, \mathbb{P}'_\beta)$ *for probability distributions* $\mathbb{P}_\beta$ *and* $\mathbb{P}'_\beta$ *on* Ω_β*. Then* $\mathbb{P}'_\beta \ll \mathbb{P}_\beta$ *and*

$$D_f(\mathbb{P}'_{\beta|\alpha}||\mathbb{P}_{\beta|\alpha}|\mathbb{P}_\alpha) \geq D_f(\mathbb{P}'_\beta||\mathbb{P}_\beta) . \tag{5.4}$$

Proof If $\mathbb{P}_\beta(S_\beta) = 0$ for some measurable S_β in Ω_β, we have:

$$\int_{\Omega_\alpha}\int_{S_\beta} \mathrm{d}\mathbb{P}_{\beta|\alpha}(\omega_\beta|\omega_\alpha)\mathrm{d}\mathbb{P}_\alpha(\omega_\alpha) = 0 ,$$

which indicates that $\mathbb{P}_\alpha(\{\omega_\alpha \in \Omega_\alpha \mid \mathbb{P}_{\beta|\alpha}(S_\beta|\omega_\alpha) = 0\}) = 1$. Since $\mathbb{P}'_{\beta|\alpha}(\cdot|\omega_\alpha) \ll \mathbb{P}_{\beta|\alpha}(\cdot|\omega_\alpha)$, $\forall\alpha \in \Omega_\alpha$, we have $\mathbb{P}_\alpha(\{\omega_\alpha \in \Omega_\alpha \mid \mathbb{P}'_{\beta|\alpha}(S_\beta|\omega_\alpha) = 0\}) = 1$ and hence

$$\mathbb{P}'_\beta(S_\beta) = \int_{\Omega_\alpha}\int_{S_\beta} \mathrm{d}\mathbb{P}'_{\beta|\alpha}(\omega_\beta|\omega_\alpha)\mathrm{d}\mathbb{P}_\alpha(\omega_\alpha) = 0 ,$$

which shows $\mathbb{P}'_\beta \ll \mathbb{P}_\beta$. By Theorem 5.3 and Lemma 5.3, we have

$$D_f(\mathbb{P}'_\beta||\mathbb{P}_\beta) \leq D_f(\mathbb{P}'_{\beta|\alpha}\mathbb{P}_\alpha||\mathbb{P}_{\beta|\alpha}\mathbb{P}_\alpha) = D_f(\mathbb{P}'_{\beta|\alpha}||\mathbb{P}_{\beta|\alpha}|\mathbb{P}_\alpha) ,$$

which completes the proof. □

5.1.3 *Kullback-Leibler Divergence*

KL divergence is among the most popular f-divergences. Therefore, we mainly focus on KL divergence in this subsection. Nonetheless, other commonly used f-divergence will also be introduced in this section later. KL divergence was named after Solomon Kullback and Richard Leibler (1951), and is also called relative entropy sometimes. KL divergence has a strong motivation and many applications in information theory, for which readers can refer to textbooks such as MacKay (2003). Besides information theory, KL divergence is widely used in random processes, statistical inference, machine learning, and other scientific and engineering problems.

5.1.3.1 Definition

KL divergence is a kind of f-divergences with $f(t) = t\log t - t + 1, \forall t \in (0, +\infty)$ and $f(0) = 1$. This function f has some properties, which are stated in Exercise 5.4.

Exercise 5.4 Let $f(t) = t\log t - t + 1, \forall t \in (0, +\infty)$ and $f(0) = 1$. Prove that f is strictly convex on $(0, +\infty)$, $f(1) = 0$, and $f(t) \geq 0, \forall t \in [0, +\infty)$.

The definition of KL divergence is as follows:

Definition 5.4 (*KL divergence*) Let $\mathbb{P}$ and $\mathbb{P}'$ be probability distributions on Ω, then the *KL divergence* $D_{\mathrm{KL}}(\mathbb{P}'||\mathbb{P})$ from $\mathbb{P}'$ to $\mathbb{P}$ is defined as:

$$D_{\mathrm{KL}}(\mathbb{P}'||\mathbb{P}) = \int_{\Omega} \log\left(\frac{\mathrm{d}\mathbb{P}'}{\mathrm{d}\mathbb{P}}(\omega)\right) \mathrm{d}\mathbb{P}'(\omega) .$$

Remark 5.3 When $\mathbb{P}'$ is not absolutely continuous with respect to $\mathbb{P}$, there is a measurable set $S \subset \Omega$ such that $\mathbb{P}'(S) > 0$ and $\mathbb{P}(S) = 0$. Thus, $D_{\mathrm{KL}}(\mathbb{P}'||\mathbb{P}) = +\infty$ according to Definition 5.4. Therefore, KL divergence is only useful when $\mathbb{P}' \ll \mathbb{P}$ and we assume absolute continuity in the analysis of KL divergence.

Example 5.3 (*KL divergence of discrete probability distributions*) Let $\mathbb{P}$ and $\mathbb{P}'$ be discrete probability distributions on $\{\omega_1, \omega_2, \ldots, \omega_n, \ldots\}$ with $\mathbb{P}(\omega_n) > 0, \forall n \in \mathbb{N}_+$. Then

$$\begin{aligned} D_{\mathrm{KL}}(\mathbb{P}'||\mathbb{P}) &= \sum_{n=1}^{+\infty} \mathbb{P}'(\omega_n) \log\left(\frac{\mathbb{P}'(\omega_n)}{\mathbb{P}(\omega_n)}\right) \\ &= \sum_{n:\mathbb{P}'(\omega_n)>0} \mathbb{P}'(\omega_n) \log\left(\frac{\mathbb{P}'(\omega_n)}{\mathbb{P}(\omega_n)}\right) . \end{aligned} \tag{5.5}$$

Example 5.4 (*KL divergence of continuous probability distributions*) Let $\mathbb{P}$ and $\mathbb{P}'$ be continuous probability distributions on $\mathbb{R}^M$ with PDFs p and p', respectively. Moreover, $p(\omega) = 0 \Rightarrow p'(\omega) = 0$ for any $\omega \in \mathbb{R}^M$. Then

$$\begin{aligned} D_{\mathrm{KL}}(\mathbb{P}'||\mathbb{P}) &= \int_{\mathbb{R}^M} \log\left(\frac{p'(\omega)}{p(\omega)}\right) \cdot p'(\omega)\mathrm{d}\omega \\ &= \int_{\omega\subset\mathbb{R}^M:p'(\omega)>0} \log\left(\frac{p'(\omega)}{p(\omega)}\right) \cdot p'(\omega)\mathrm{d}\omega . \end{aligned} \tag{5.6}$$

5.1.3.2 Properties

Because KL divergence belongs to f-divergences, it naturally has the properties of general f-divergence, which are listed in Proposition 5.6.

Proposition 5.6 *KL divergence D_{KL} has the following properties:*

1. *(Nonnegativity and point-separating) $D_{\mathrm{KL}}(\mathbb{P}'||\mathbb{P}) \geq 0, \forall \mathbb{P}' \ll \mathbb{P}$. In addition, $D_{\mathrm{KL}}(\mathbb{P}'||\mathbb{P}) = 0$ if and only if $\mathbb{P}' = \mathbb{P}$ as probability distributions.*
2. *(Convexity) For any $\lambda \in [0, 1]$ and probability distributions $\mathbb{P}'_1 \ll \mathbb{P}_1$ and $\mathbb{P}'_2 \ll \mathbb{P}_2$,*

$$\begin{aligned} &D_{\mathrm{KL}}(\lambda\mathbb{P}'_1 + (1-\lambda)\mathbb{P}'_2||\lambda\mathbb{P}_1 + (1-\lambda)\mathbb{P}_2) \\ &\leq \lambda D_{\mathrm{KL}}(\mathbb{P}'_1||\mathbb{P}_1) + (1-\lambda)D_{\mathrm{KL}}(\mathbb{P}'_2||\mathbb{P}_2) . \end{aligned}$$

3. *(Data processing) Assume $\mathbb{P}'_\alpha \ll \mathbb{P}_\alpha$ are probability distributions; $\mathbb{P}_{\beta|\alpha}$ is a conditional probability distribution; $\mathbb{P}_{\beta|\alpha}\mathbb{P}_\alpha \in \mathcal{J}(\mathbb{P}_\alpha, \mathbb{P}_\beta)$ and $\mathbb{P}_{\beta|\alpha}\mathbb{P}'_\alpha \in \mathcal{J}(\mathbb{P}'_\alpha, \mathbb{P}'_\beta)$. Then*

$$D_{\mathrm{KL}}(\mathbb{P}'_\beta||\mathbb{P}_\beta) \le D_{\mathrm{KL}}(\mathbb{P}'_\alpha||\mathbb{P}_\alpha) . \tag{5.7}$$

4. *(Conditioning) Assume $\mathbb{P}_\alpha$ is a probability distribution on Ω_α; $\mathbb{P}_{\beta|\alpha}$ and $\mathbb{P}'_{\beta|\alpha}$ are conditional probability distributions, where $\mathbb{P}'_{\beta|\alpha}(\cdot|\omega_\alpha) \ll \mathbb{P}_{\beta|\alpha}(\cdot|\omega_\alpha)$ for any $\omega_\alpha \in \Omega_\alpha$; $\mathbb{P}_{\beta|\alpha}\mathbb{P}_\alpha \in \mathcal{J}(\mathbb{P}_\alpha, \mathbb{P}_\beta)$ and $\mathbb{P}'_{\beta|\alpha}\mathbb{P}_\alpha \in \mathcal{J}(\mathbb{P}_\alpha, \mathbb{P}'_\beta)$, where $\mathbb{P}_\beta$ and $\mathbb{P}'_\beta$ are probability distributions on Ω_β. Then*

$$D_{\mathrm{KL}}(\mathbb{P}'_\beta||\mathbb{P}_\beta) \le D_{\mathrm{KL}}(\mathbb{P}'_{\beta|\alpha}||\mathbb{P}_{\beta|\alpha}|\mathbb{P}_\alpha) . \tag{5.8}$$

Remark 5.4 The discrete version of nonnegativity and point-separating properties of KL divergence is called *Gibbs' inequality* in information theory: Let $\mathbb{P}$ and $\mathbb{P}'$ be discrete probability distributions on $\{\omega_1, \omega_2, \ldots, \omega_n, \ldots,\}$ with $\mathbb{P}(\omega_n) > 0$, $\mathbb{P}'(\omega_n) > 0$, $\forall n \in \mathbb{N}_+$, then

$$-\sum_{n\in\mathbb{N}_+} \mathbb{P}'(\omega_n)\log(\mathbb{P}'(\omega_n)) \le -\sum_{n\in\mathbb{N}_+} \mathbb{P}'(\omega_n)\log(\mathbb{P}(\omega_n)) , \tag{5.9}$$

and the equation holds if and only if $\mathbb{P}'(\omega_n) = \mathbb{P}(\omega_n)$, $\forall n \in \mathbb{N}_+$. In (5.9), the left side is the information entropy of $\mathbb{P}'$ and the right side is the cross entropy of $\mathbb{P}'$ with $\mathbb{P}$.

Remark 5.5 The data processing and conditioning properties will be revisited in Proposition 5.7, which gives the necessary and sufficient conditions for the equalities.

Example 5.5 shows that KL divergence is not symmetric and has no upper bounds, where the sample space only contains two points.

Example 5.5 (*KL divergence is asymmetric and unbounded from above*) Let $\Omega = \{a, b\}$. Define probability distributions $\mathbb{P}$ and $\mathbb{P}'$ as follows:

$$\mathbb{P}(a) = p , \ \mathbb{P}(b) = 1 - p , \ \mathbb{P}'(a) = \mathbb{P}'(b) = \frac{1}{2} ,$$

where $p \in (0, 1)$. The KL divergence can be calculated by

$$\begin{aligned} D_{\mathrm{KL}}(\mathbb{P}'||\mathbb{P}) &= \mathbb{P}'(a)\log\left(\frac{\mathbb{P}'(a)}{\mathbb{P}(a)}\right) + \mathbb{P}'(b)\log\left(\frac{\mathbb{P}'(b)}{\mathbb{P}(b)}\right) \\ &= \frac{1}{2}\log\left(\frac{1/2}{p}\right) + \frac{1}{2}\log\left(\frac{1/2}{1-p}\right) = -\frac{1}{2}\log(4p(1-p)) . \end{aligned}$$

Thus, $D_{\mathrm{KL}}(\mathbb{P}'||\mathbb{P}) \to +\infty$ as $p \to 0+$ or $p \to 1-$, which shows KL divergence is unbounded from above. We can also calculate $D_{\mathrm{KL}}(\mathbb{P}||\mathbb{P}')$:

$$D_{\mathrm{KL}}(\mathbb{P}||\mathbb{P}') = p\log(2p) + (1-p)\log(2(1-p)) ,$$

so KL divergence is not symmetric.

Example 5.6 (*Symmetric divergences related to KL divergence*) KL divergence is asymmetric and it has two related symmetric f-divergences, i.e., Jeffreys divergence and Jensen-Shannon divergence. Assume that probability distributions $\mathbb{P}$ and $\mathbb{P}'$ satisfy $\mathbb{P}' \ll \mathbb{P}$ and $\mathbb{P} \ll \mathbb{P}'$. *Jeffreys divergence* is:

$$D_{\mathrm{Jeffreys}}(\mathbb{P}'||\mathbb{P}) := \frac{1}{2}D_{\mathrm{KL}}(\mathbb{P}'||\mathbb{P}) + \frac{1}{2}D_{\mathrm{KL}}(\mathbb{P}||\mathbb{P}') .$$

On the contrary, the definition of *Jensen-Shannon divergence* is:

$$D_{\mathrm{JS}}(\mathbb{P}'||\mathbb{P}) := \frac{1}{2}D_{\mathrm{KL}}(\mathbb{P}'||(\mathbb{P}+\mathbb{P}')/2) + \frac{1}{2}D_{\mathrm{KL}}(\mathbb{P}||(\mathbb{P}+\mathbb{P}')/2) .$$

Exercise 5.5 Let $\mathbb{P}$ and $\mathbb{P}'$ be probability distributions defined on Ω. Prove that $(\mathbb{P}+\mathbb{P}')/2$ is also a probability distribution, $\mathbb{P} \ll (\mathbb{P}+\mathbb{P}')/2$, and $\mathbb{P}' \ll (\mathbb{P}+\mathbb{P}')/2$.

According to the conclusion of Exercise 5.5, Jensen-Shannon divergence does not need the absolute continuity assumption. By the definitions, it is clear that both Jeffreys divergence and Jensen-Shannon divergence are symmetric divergences. However, they do not satisfy the triangle inequality, which prevents them from being distance functions.

KL divergence also has properties that general f-divergences do not have.

Theorem 5.6 (Chain rule of KL divergence) *Assume $\mathbb{P}'_{\alpha,\beta} \ll \mathbb{P}_{\alpha,\beta}$ on $\Omega_\alpha \times \Omega_\beta$, where $\mathbb{P}_{\alpha,\beta} = \mathbb{P}_{\beta|\alpha}\mathbb{P}_\alpha \in \mathcal{J}(\mathbb{P}_\alpha, \mathbb{P}_\beta)$ and $\mathbb{P}'_{\alpha,\beta} = \mathbb{P}'_{\beta|\alpha}\mathbb{P}'_\alpha \in \mathcal{J}(\mathbb{P}'_\alpha, \mathbb{P}'_\beta)$ with probability distributions $\mathbb{P}_\alpha$ and $\mathbb{P}'_\alpha$ defined on Ω_α. $\mathbb{P}_{\beta|\alpha}$ and $\mathbb{P}'_{\beta|\alpha}$ are conditional probability distributions. Then*

$$D_{\mathrm{KL}}(\mathbb{P}'_{\alpha,\beta}||\mathbb{P}_{\alpha,\beta}) = D_{\mathrm{KL}}(\mathbb{P}'_{\beta|\alpha}||\mathbb{P}_{\beta|\alpha}|\mathbb{P}'_\alpha) + D_{\mathrm{KL}}(\mathbb{P}'_\alpha||\mathbb{P}_\alpha) .$$

Proof Since $\mathbb{P}'_{\alpha,\beta} \ll \mathbb{P}_{\alpha,\beta}$, we have $\mathbb{P}'_\alpha \ll \mathbb{P}_\alpha$ and $\mathbb{P}'_{\beta|\alpha}(\cdot|S_\alpha) \ll \mathbb{P}_{\beta|\alpha}(\cdot|S_\alpha)$ for any $\mathbb{P}'_\alpha(S_\alpha) > 0$. By direct calculation and Fubini's theorem in Theorem 2.3,

$$
\begin{aligned}
& D_{\mathrm{KL}}(\mathbb{P}'_{\alpha,\beta}||\mathbb{P}_{\alpha,\beta}) \\
&= \int_{\Omega_\alpha \times \Omega_\beta} \log\left(\frac{\mathrm{d}\mathbb{P}'_{\alpha,\beta}}{\mathrm{d}\mathbb{P}_{\alpha,\beta}}(\omega_\alpha, \omega_\beta)\right) \mathrm{d}\mathbb{P}'_{\alpha,\beta}(\omega_\alpha, \omega_\beta) \\
&= \int_{\Omega_\alpha}\int_{\Omega_\beta} \log\left(\frac{\mathrm{d}\mathbb{P}'_{\beta|\alpha}}{\mathrm{d}\mathbb{P}_{\beta|\alpha}}(\omega_\beta|\omega_\alpha)\frac{\mathrm{d}\mathbb{P}'_{\alpha}}{\mathrm{d}\mathbb{P}_{\alpha}}(\omega_\alpha)\right) \mathrm{d}\mathbb{P}'_{\beta|\alpha}(\omega_\beta|\omega_\alpha)\mathrm{d}\mathbb{P}'_{\alpha}(\omega_\alpha) \\
&= \int_{\Omega_\alpha}\int_{\Omega_\beta} \left\{\log\left(\frac{\mathrm{d}\mathbb{P}'_{\beta|\alpha}}{\mathrm{d}\mathbb{P}_{\beta|\alpha}}(\omega_\beta|\omega_\alpha)\right) + \log\left(\frac{\mathrm{d}\mathbb{P}'_{\alpha}}{\mathrm{d}\mathbb{P}_{\alpha}}(\omega_\alpha)\right)\right\} \mathrm{d}\mathbb{P}'_{\beta|\alpha}(\omega_\beta|\omega_\alpha)\mathrm{d}\mathbb{P}'_{\alpha}(\omega_\alpha) \\
&= \int_{\Omega_\alpha}\int_{\Omega_\beta} \log\left(\frac{\mathrm{d}\mathbb{P}'_{\beta|\alpha}}{\mathrm{d}\mathbb{P}_{\beta|\alpha}}(\omega_\beta|\omega_\alpha)\right) \mathrm{d}\mathbb{P}'_{\beta|\alpha}(\omega_\beta|\omega_\alpha)\mathrm{d}\mathbb{P}'_{\alpha}(\omega_\alpha) \\
&\quad + \int_{\Omega_\alpha} \log\left(\frac{\mathrm{d}\mathbb{P}'_{\alpha}}{\mathrm{d}\mathbb{P}_{\alpha}}(\omega_\alpha)\right)\left(\int_{\Omega_\beta} \mathrm{d}\mathbb{P}'_{\beta|\alpha}(\omega_\beta|\omega_\alpha)\right) \mathrm{d}\mathbb{P}'_{\alpha}(\omega_\alpha) \\
&= D_{\mathrm{KL}}(\mathbb{P}'_{\beta|\alpha}||\mathbb{P}_{\beta|\alpha}|\mathbb{P}'_{\alpha}) + D_{\mathrm{KL}}(\mathbb{P}'_{\alpha}||\mathbb{P}_{\alpha}) \,,
\end{aligned}
$$

which completes the proof. □

A special case of the chain rule is for product distributions, which can be comprehended as that KL divergence is additive for independent probability distributions.

Corollary 5.3 (KL divergence of product distribution) *Assume* $\mathbb{P}'_\alpha \ll \mathbb{P}_\alpha$ *on* Ω_α *and* $\mathbb{P}'_\beta \ll \mathbb{P}_\beta$ *on* Ω_β. *Then*

$$
D_{\mathrm{KL}}(\mathbb{P}'_\alpha \times \mathbb{P}'_\beta||\mathbb{P}_\alpha \times \mathbb{P}_\beta) = D_{\mathrm{KL}}(\mathbb{P}'_\alpha||\mathbb{P}_\alpha) + D_{\mathrm{KL}}(\mathbb{P}'_\beta||\mathbb{P}_\beta) \,.
$$

Exercise 5.6 Prove Corollary 5.3 using Theorem 5.6.

Utilizing the chain rule, we can find necessary and sufficient conditions for the equalities in the data processing and conditioning properties.

Proposition 5.7 *1. (Data processing revisited) In* (5.7), *assume* $\mathbb{P}_{\beta|\alpha}\mathbb{P}_\alpha = \mathbb{P}_{\alpha|\beta}\mathbb{P}_\beta$ *and* $\mathbb{P}_{\beta|\alpha}\mathbb{P}'_\alpha = \mathbb{P}'_{\alpha|\beta}\mathbb{P}'_\beta$ *with* $\mathbb{P}_{\alpha|\beta}$ *and* $\mathbb{P}'_{\alpha|\beta}$ *conditional probability distributions, then*

$$
D_{\mathrm{KL}}(\mathbb{P}'_\beta||\mathbb{P}_\beta) = D_{\mathrm{KL}}(\mathbb{P}'_\alpha||\mathbb{P}_\alpha) \iff D_{\mathrm{KL}}(\mathbb{P}'_{\alpha|\beta}||\mathbb{P}_{\alpha|\beta}|\mathbb{P}'_\beta) = 0 \,.
$$

2. *(Conditioning revisited) In* (5.8), *assume* $\mathbb{P}_{\beta|\alpha}\mathbb{P}_\alpha = \mathbb{P}_{\alpha|\beta}\mathbb{P}_\beta$ *and* $\mathbb{P}'_{\beta|\alpha}\mathbb{P}_\alpha = \mathbb{P}'_{\alpha|\beta}\mathbb{P}'_\beta$ *with* $\mathbb{P}_{\alpha|\beta}$ *and* $\mathbb{P}'_{\alpha|\beta}$ *conditional probability distributions, then*

$$
D_{\mathrm{KL}}(\mathbb{P}'_\beta||\mathbb{P}_\beta) = D_{\mathrm{KL}}(\mathbb{P}'_{\beta|\alpha}||\mathbb{P}_{\beta|\alpha}|\mathbb{P}_\alpha) \iff D_{\mathrm{KL}}(\mathbb{P}'_{\alpha|\beta}||\mathbb{P}_{\alpha|\beta}|\mathbb{P}'_\beta) = 0 \,.
$$

Proof 1. By the chain rule of KL divergence, we have:

$$D_{\text{KL}}(\mathbb{P}'_{\alpha|\beta}\mathbb{P}'_{\beta}||\mathbb{P}_{\alpha|\beta}\mathbb{P}_{\beta}) = D_{\text{KL}}(\mathbb{P}'_{\alpha|\beta}||\mathbb{P}_{\alpha|\beta}|\mathbb{P}'_{\beta}) + D_{\text{KL}}(\mathbb{P}'_{\beta}||\mathbb{P}_{\beta}) ,$$

and

$$\begin{aligned} D_{\text{KL}}(\mathbb{P}_{\beta|\alpha}\mathbb{P}'_{\alpha}||\mathbb{P}_{\beta|\alpha}\mathbb{P}_{\alpha}) &= D_{\text{KL}}(\mathbb{P}_{\beta|\alpha}||\mathbb{P}_{\beta|\alpha}|\mathbb{P}'_{\alpha}) + D_{\text{KL}}(\mathbb{P}'_{\alpha}||\mathbb{P}_{\alpha}) \\ &= D_{\text{KL}}(\mathbb{P}'_{\alpha}||\mathbb{P}_{\alpha}) . \end{aligned}$$

Then the conclusion follows from $\mathbb{P}_{\beta|\alpha}\mathbb{P}_{\alpha} = \mathbb{P}_{\alpha|\beta}\mathbb{P}_{\beta}$ and $\mathbb{P}_{\beta|\alpha}\mathbb{P}'_{\alpha} = \mathbb{P}'_{\alpha|\beta}\mathbb{P}'_{\beta}$.

2. Again by the chain rule,

$$\begin{aligned} & D_{\text{KL}}(\mathbb{P}'_{\alpha|\beta}||\mathbb{P}_{\alpha|\beta}|\mathbb{P}'_{\beta}) + D_{\text{KL}}(\mathbb{P}'_{\beta}||\mathbb{P}_{\beta}) \\ = {} & D_{\text{KL}}(\mathbb{P}'_{\alpha|\beta}\mathbb{P}'_{\beta}||\mathbb{P}_{\alpha|\beta}\mathbb{P}_{\beta}) \\ = {} & D_{\text{KL}}(\mathbb{P}'_{\beta|\alpha}\mathbb{P}_{\alpha}||\mathbb{P}_{\beta|\alpha}\mathbb{P}_{\alpha}) \\ = {} & D_{\text{KL}}(\mathbb{P}'_{\beta|\alpha}||\mathbb{P}_{\beta|\alpha}|\mathbb{P}_{\alpha}) + D_{\text{KL}}(\mathbb{P}_{\alpha}||\mathbb{P}_{\alpha}) \\ = {} & D_{\text{KL}}(\mathbb{P}'_{\beta|\alpha}||\mathbb{P}_{\beta|\alpha}|\mathbb{P}_{\alpha}) , \end{aligned}$$

which implies the desired conclusion.

□

In the following subsections, we calculate the KL divergences of uniform distributions and normal distributions as examples.

5.1.3.3 Example: Uniform Distributions

Consider two uniform distributions on rectangular cuboids in $\mathbb{R}^M$. The two rectangular cuboids are

$$\begin{aligned} S &:= \{(x_1, \ldots, x_M) \in \mathbb{R}^M \mid a_m \leq x_m \leq b_m , \ \forall 1 \leq m \leq M\} , \\ S' &:= \{(x_1, \ldots, x_M) \in \mathbb{R}^M \mid a'_m \leq x_m \leq b'_m , \ \forall 1 \leq m \leq M\} , \end{aligned}$$

where the parameters $a_m \leq a'_m < b'_m \leq b_m$, $\forall 1 \leq m \leq M$. Thus, $S' \subset S$ and their volumes satisfy

$$\text{Volume}(S) = \prod_{m=1}^{M}(b_m - a_m) \geq \prod_{m=1}^{M}(b'_m - a'_m) = \text{Volume}(S') .$$

Denote the uniform distributions on S and S' by $\mathbb{P}$ and $\mathbb{P}'$, respectively. Their PDFs are:

$$p(x_1, \ldots, x_M) = \begin{cases} 1/\text{Volume}(S) \ , & \text{if } x \in S \ , \\ 0 \ , & \text{if } x \in \mathbb{R}^M \setminus S \ , \end{cases}$$

$$p'(x_1, \ldots, x_M) = \begin{cases} 1/\text{Volume}(S') \ , & \text{if } x \in S' \ , \\ 0 \ , & \text{if } x \in \mathbb{R}^M \setminus S' \ . \end{cases}$$

Since $S' \subset S$, we have $\mathbb{P}' \ll \mathbb{P}$, and the KL divergence is as follows by (5.6):

$$\begin{aligned} D_{\text{KL}}(\mathbb{P}'||\mathbb{P}) &= \int_{S'} \log\left(\frac{p'(x)}{p(x)}\right) \cdot p'(x)\mathrm{d}x \\ &= \int_{S'} \log\left(\frac{\text{Volume}(S)}{\text{Volume}(S')}\right) \cdot \frac{\mathrm{d}x}{\text{Volume}(S')} \\ &= \log\left(\frac{\text{Volume}(S)}{\text{Volume}(S')}\right) = \log\left(\frac{\prod_{m=1}^M (b_m - a_m)}{\prod_{m=1}^M (b'_m - a'_m)}\right) . \end{aligned}$$

5.1.3.4 Example: Normal Distributions

Let $\mathcal{N}(\mu, \sigma^2)$ and $\mathcal{N}(\mu', (\sigma')^2)$ be two normal distributions on $\mathbb{R}$. Then their PDFs are:

$$p(t) = \frac{1}{\sqrt{2\pi}\sigma} \exp\left(-\frac{1}{2}\left(\frac{t-\mu}{\sigma}\right)^2\right) , \quad \forall t \in \mathbb{R} ,$$

$$p'(t) = \frac{1}{\sqrt{2\pi}\sigma'} \exp\left(-\frac{1}{2}\left(\frac{t-\mu'}{\sigma'}\right)^2\right) , \quad \forall t \in \mathbb{R} .$$

According to (5.6),

$$\begin{aligned} &D_{\text{KL}}(\mathcal{N}(\mu', (\sigma')^2)||\mathcal{N}(\mu, \sigma^2)) \\ &= \int_{\mathbb{R}} \log\left(\frac{p'(t)}{p(t)}\right) \cdot p'(t)\mathrm{d}t \\ &= \int_{\mathbb{R}} \left(\log\frac{\sigma}{\sigma'} - \frac{1}{2}\left(\frac{t-\mu'}{\sigma'}\right)^2 + \frac{1}{2}\left(\frac{t-\mu}{\sigma}\right)^2\right) p'(t)\mathrm{d}t . \end{aligned}$$

Because $p'(t)$ is the PDF of $\mathcal{N}(\mu', (\sigma')^2)$, we have:

$$\int_{\mathbb{R}} p'(t)\mathrm{d}t = 1 , \quad \int_{\mathbb{R}} tp'(t)\mathrm{d}t = \mu' , \quad \int_{\mathbb{R}} (t-\mu')^2 p'(t)\mathrm{d}t = (\sigma')^2 .$$

Thus,

$$
\begin{aligned}
& D_{\mathrm{KL}}(\mathcal{N}(\mu', (\sigma')^2)||\mathcal{N}(\mu, \sigma^2)) \\
&= \left(\log \frac{\sigma}{\sigma'} + \frac{\mu^2 - (\mu')^2}{2\sigma^2}\right) \int_{\mathbb{R}} p'(t)\mathrm{d}t + \frac{\mu' - \mu}{\sigma^2} \int_{\mathbb{R}} tp'(t)\mathrm{d}t \\
&\quad + \left(-\frac{1}{2(\sigma')^2} + \frac{1}{2\sigma^2}\right) \int_{\mathbb{R}} (t - \mu')^2 p'(t)\mathrm{d}t \\
&= \log \frac{\sigma}{\sigma'} + \frac{\mu^2 - (\mu')^2}{2\sigma^2} + \frac{\mu' - \mu}{\sigma^2} \cdot \mu' + \left(-\frac{1}{2(\sigma')^2} + \frac{1}{2\sigma^2}\right) (\sigma')^2 \\
&= \log \frac{\sigma}{\sigma'} - \frac{1}{2} + \frac{(\mu - \mu')^2 + (\sigma')^2}{2\sigma^2} .
\end{aligned}
$$

When $\sigma = \sigma'$,

$$
D_{\mathrm{KL}}(\mathcal{N}(\mu', \sigma^2)||\mathcal{N}(\mu, \sigma^2)) = \frac{(\mu - \mu')^2}{2\sigma^2} = D_{\mathrm{KL}}(\mathcal{N}(\mu, \sigma^2)||\mathcal{N}(\mu', \sigma^2)) ,
$$

is symmetric.

5.1.4 Other Types of f-Divergences

We introduce some other types of commonly used f-divergences, namely, total variation distance, squared Hellinger distance, and Pearson χ^2 divergence. The relationships between KL divergence and these popular f-divergences will be discussed later.

5.1.4.1 Total Variation Distance

Total variation distance is one of the most popular distance functions for probability distributions. It has a direct definition and several equivalent definitions. We first introduce the direct definition and then show that total variation distance belongs to f-divergence. In fact, total variation distance has a connection with Wasserstein distance, so we will also come back to it in Chap. 6.

Definition 5.5 (*Total variation distance*) Let $\mathbb{P}$ and $\mathbb{P}'$ be two probability distributions defined on the sample space $(\Omega, \mathcal{F})$. Then the *total variation distance* between $\mathbb{P}$ and $\mathbb{P}'$ is defined as follows:

$$
d_{\mathrm{TV}}(\mathbb{P}, \mathbb{P}') := \sup_{S \in \mathcal{F}} \left\{\mathbb{P}(S) - \mathbb{P}'(S)\right\} . \tag{5.10}
$$

As (5.10) shows, total variation distance $d_{\mathrm{TV}}(\mathbb{P}, \mathbb{P}')$ is the maximum difference of probability on a measurable subset. Therefore, $d_{\mathrm{TV}}(\mathbb{P}, \mathbb{P}') \in [0, 1]$ always holds and no absolute continuity assumption is needed. We check that total variation distance is really a distance function.

Theorem 5.7 *Total variation distance is a distance function.*

Proof Assume $\mathbb{P}, \mathbb{P}'$, and $\mathbb{P}''$ are probability distributions on the sample space $(\Omega, \mathcal{F})$.

1. Nonnegativity: $d_{\mathrm{TV}}(\mathbb{P}, \mathbb{P}') = \sup_{S\in\mathcal{F}}\{\mathbb{P}(S) - \mathbb{P}'(S)\} \geq \mathbb{P}(\varnothing) - \mathbb{P}'(\varnothing) = 0$.
2. Point-separating: Assume $d_{\mathrm{TV}}(\mathbb{P}, \mathbb{P}') = 0$, then for any $S \in \mathcal{F}, \mathbb{P}(S) - \mathbb{P}'(S) \leq 0$ and $\mathbb{P}(\Omega \setminus S) - \mathbb{P}'(\Omega \setminus S) \leq 0$. Then by

$$\mathbb{P}(S) - \mathbb{P}'(S) = (1 - \mathbb{P}(\Omega \setminus S)) - (1 - \mathbb{P}'(\Omega \setminus S)) = \mathbb{P}'(\Omega \setminus S) - \mathbb{P}(\Omega \setminus S) ,$$

   we have $\mathbb{P}(S) = \mathbb{P}'(S)$. Therefore, $\mathbb{P} = \mathbb{P}'$ as probability distributions.
3. Symmetry: For $S \in \mathcal{F}, \mathbb{P}(S) - \mathbb{P}'(S) = \mathbb{P}'(\Omega \setminus S) - \mathbb{P}(\Omega \setminus S)$. Note that by the definition of σ-algebra, $\mathcal{F} = \{S \mid S \in \mathcal{F}\} = \{S \mid \Omega \setminus S \in \mathcal{F}\}$. Thus,

$$\begin{aligned} d_{\mathrm{TV}}(\mathbb{P}, \mathbb{P}') &= \sup_{S\in\mathcal{F}}\{\mathbb{P}(S) - \mathbb{P}'(S)\} \\ &= \sup_{S\in\mathcal{F}}\{\mathbb{P}'(\Omega \setminus S) - \mathbb{P}(\Omega \setminus S)\} \\ &= \sup_{\Omega\setminus S\in\mathcal{F}}\{\mathbb{P}'(\Omega \setminus S) - \mathbb{P}(\Omega \setminus S)\} = d_{\mathrm{TV}}(\mathbb{P}', \mathbb{P}) . \end{aligned}$$

4. Triangle inequality:

$$\begin{aligned} d_{\mathrm{TV}}(\mathbb{P}, \mathbb{P}'') &= \sup_{S\in\mathcal{F}}\{\mathbb{P}(S) - \mathbb{P}''(S)\} \\ &= \sup_{S\in\mathcal{F}}\{(\mathbb{P}(S) - \mathbb{P}'(S)) + (\mathbb{P}'(S) - \mathbb{P}''(S))\} \\ &\leq \sup_{S\in\mathcal{F}}\{\mathbb{P}(S) - \mathbb{P}'(S)\} + \sup_{S\in\mathcal{F}}\{\mathbb{P}'(S) - \mathbb{P}''(S)\} \\ & d_{\mathrm{TV}}(\mathbb{P}, \mathbb{P}') + d_{\mathrm{TV}}(\mathbb{P}', \mathbb{P}'') . \end{aligned}$$

In conclusion, $d_{\mathrm{TV}}(\cdot, \cdot)$ is a distance function. □

Theorem 5.8 *The total variation distance $d_{\mathrm{TV}}(\mathbb{P}, \mathbb{P}')$ defined in Definition 5.5 satisfies:*

$$d_{\mathrm{TV}}(\mathbb{P}, \mathbb{P}') = \frac{1}{2}\sup_{|f|\leq 1}\left\{\int_\Omega f(\omega)\mathrm{d}\mathbb{P}(\omega) - \int_\Omega f(\omega)\mathrm{d}\mathbb{P}'(\omega)\right\} . \tag{5.11}$$

Proof We start from the right side:

$$
\begin{aligned}
&\sup_{|f|\le 1}\left\{\int_\Omega f(\omega)\mathrm{d}\mathbb{P}(\omega)-\int_\Omega f(\omega)\mathrm{d}\mathbb{P}'(\omega)\right\}\\
&=\sup_{|f|\le 1}\int_\Omega f(\omega)(\mathrm{d}\mathbb{P}-\mathrm{d}\mathbb{P}')(\omega)\\
&=\int_\Omega |\mathrm{d}\mathbb{P}-\mathrm{d}\mathbb{P}'|(\omega)\\
&=\int_\Omega (\mathrm{d}\mathbb{P}-\mathrm{d}\mathbb{P}')_+(\omega)+\int_\Omega (\mathrm{d}\mathbb{P}-\mathrm{d}\mathbb{P}')_-(\omega)\\
&=\sup_{S\in\mathcal{F}}\left\{\mathbb{P}(S)-\mathbb{P}'(S)\right\}+\sup_{S\in\mathcal{F}}\left\{\mathbb{P}'(S)-\mathbb{P}(S)\right\}\\
&=d_{\mathrm{TV}}(\mathbb{P},\mathbb{P}')+d_{\mathrm{TV}}(\mathbb{P}',\mathbb{P})=2d_{\mathrm{TV}}(\mathbb{P},\mathbb{P}')\ ,
\end{aligned}
$$

where $(\cdot)_+$ and $(\cdot)_-$ denote the positive part and negative part, respectively. □

Remark 5.6 From the proof of Theorem 5.8, we also know that

$$
d_{\mathrm{TV}}(\mathbb{P},\mathbb{P}')=\frac{1}{2}\int_\Omega |\mathrm{d}\mathbb{P}-\mathrm{d}\mathbb{P}'|(\omega)\ . \tag{5.13}
$$

As listed in Table 5.1, the function f for total variation distance is $f(t)=|t-1|/2, \forall t\in[0,+\infty)$. In addition, $tf(1/t)=t|(1/t)-1|/2=|1-t|/2=f(t)$, $\forall t\in(0,+\infty)$. Thus, total variation distance is symmetric and its formula is:

$$
d_{\mathrm{TV}}(\mathbb{P}',\mathbb{P})=\frac{1}{2}\int\left|\frac{\mathrm{d}\mathbb{P}'}{\mathrm{d}\mathbb{P}}-1\right|\cdot\mathrm{d}\mathbb{P}=\frac{1}{2}\int|\mathrm{d}\mathbb{P}'-\mathrm{d}\mathbb{P}|\ ,
$$

which is consistent with (5.13). Therefore, total variation distance is a kind of f-divergences and possesses the properties of general f-divergences.

Theorem 5.9 (Scheffé theorem) *Assume $\mathbb{P}$ and $\mathbb{P}'$ are probability distributions on Ω. Then*

$$
d_{\mathrm{TV}}(\mathbb{P},\mathbb{P}')=1-\int_\Omega \min\{d\mathbb{P},d\mathbb{P}'\}\ . \tag{5.14}
$$

Proof We observe that for any $t_1,t_2\ge 0$, $|t_1-t_2|=\max\{t_1,t_2\}-\min\{t_1,t_2\}$ and $\max\{t_1,t_2\}+\min\{t_1,t_2\}=t_1+t_2$, so $|t_1-t_2|=t_1+t_2-2\min\{t_1,t_2\}$. Then

$$
\begin{aligned}
d_{\mathrm{TV}}(\mathbb{P},\mathbb{P}')&=\frac{1}{2}\int_\Omega |\mathrm{d}\mathbb{P}-\mathrm{d}\mathbb{P}'|(\omega)\\
&=\frac{1}{2}\int_\Omega \left(\mathrm{d}\mathbb{P}+\mathrm{d}\mathbb{P}'-2\min\{\mathrm{d}\mathbb{P},\mathrm{d}\mathbb{P}'\}\right)\\
&=1-\int_\Omega \min\{\mathrm{d}\mathbb{P},\mathrm{d}\mathbb{P}'\}\ ,
\end{aligned}
$$

which completes the proof. □

Example 5.7 (*Total variation distance of discrete probability distributions*) When $\mathbb{P}$ and $\mathbb{P}'$ are two discrete probability distributions on Ω and the countable set $\{\omega_1, \omega_2, \ldots, \omega_n, \ldots\} \subset \Omega$ satisfies $\sum_{n=1}^{+\infty} \mathbb{P}(\omega_n) = \sum_{n=1}^{+\infty} \mathbb{P}'(\omega_n) = 1$, the total variation distance becomes:

$$d_{\mathrm{TV}}(\mathbb{P}, \mathbb{P}') = \frac{1}{2} \sum_{n=1}^{+\infty} |\mathbb{P}(\omega_n) - \mathbb{P}'(\omega_n)| .$$

Example 5.8 (*Total variation distance of continuous probability distributions*) When $\mathbb{P}$ and $\mathbb{P}'$ are two continuous probability distributions on $\mathbb{R}^M$ with PDFs $p(x)$ and $p'(x)$, respectively. The total variation distance between them is:

$$d_{\mathrm{TV}}(\mathbb{P}', \mathbb{P}) = \frac{1}{2} \int_{\mathbb{R}^M} |p'(\omega) - p(\omega)| \mathrm{d}\omega . \tag{5.15}$$

5.1.4.2 Squared Hellinger Distance

Definition 5.6 (*Hellinger distance*) Let $\mathbb{P}$ and $\mathbb{P}'$ be two probability distributions on Ω, the *Hellinger distance* between them is defined by:

$$d_{\mathrm{H}}(\mathbb{P}, \mathbb{P}') = \left(\int_{\Omega} \left(\sqrt{\mathrm{d}\mathbb{P}(\omega)} - \sqrt{\mathrm{d}\mathbb{P}'(\omega)} \right)^2 \right)^{\frac{1}{2}} . \tag{5.16}$$

It is clear from the definition that Hellinger distance is nonnegative, point-separating, and symmetric. Its triangle inequality follows from Minkowski inequality. Therefore, Hellinger distance is a distance function of probability distributions. Moreover, Lemma 5.4 directly follows from the definition.

Lemma 5.4 *For any probability distributions $\mathbb{P}$ and $\mathbb{P}'$ on Ω, the Hellinger distance satisfies:*

$$d_{\mathrm{H}}(\mathbb{P}, \mathbb{P}') = \sqrt{2} \left(1 - \int_{\Omega} \sqrt{d\mathbb{P}(\omega) d\mathbb{P}'(\omega)} \right)^{\frac{1}{2}} \in [0, \sqrt{2}] .$$

Using Lemma 5.4, we can easily obtain an important property of Hellinger distance about product distributions.

Corollary 5.4 *Assume $\mathbb{P}_\alpha$ and $\mathbb{P}'_\alpha$ are probability distributions on Ω_α. $\mathbb{P}_\beta$ and $\mathbb{P}'_\beta$ are probability distributions on Ω_β. Then squared Hellinger distance $d_{\mathrm{H}}^2(\mathbb{P}_\alpha \times \mathbb{P}_\beta, \mathbb{P}'_\alpha \times \mathbb{P}'_\beta)$ between the product distributions satisfy:*

$$1 - \frac{1}{2} d_{\mathrm{H}}^2(\mathbb{P}_\alpha \times \mathbb{P}_\beta, \mathbb{P}'_\alpha \times \mathbb{P}'_\beta) = \left(1 - \frac{1}{2} d_{\mathrm{H}}^2(\mathbb{P}_\alpha, \mathbb{P}'_\alpha) \right) \left(1 - \frac{1}{2} d_{\mathrm{H}}^2(\mathbb{P}_\beta, \mathbb{P}'_\beta) \right) .$$

Exercise 5.7 Prove Corollary 5.4 using Lemma 5.4.

According to (5.16), the squared Hellinger distance

$$d_{\mathrm{H}}^2(\mathbb{P},\mathbb{P}') = \int_\Omega \left(\sqrt{\mathrm{d}\mathbb{P}(\omega)} - \sqrt{\mathrm{d}\mathbb{P}'(\omega)}\right)^2 = \int_\Omega \left(\sqrt{\frac{\mathrm{d}\mathbb{P}'}{\mathrm{d}\mathbb{P}}(\omega)} - 1\right)^2 \mathrm{d}\mathbb{P}(\omega)\ ,$$

which shows that it is the f-divergence with function $f(t) = (\sqrt{t} - 1)^2$. Note that squared Hellinger distance is symmetric and takes value in $[0, 2]$ for all probability distributions on the same sample spaces, where no absolute continuity assumption is needed.

5.1.4.3 Pearson χ^2 Divergence

In *Pearson χ^2 divergence*, the function $f(t) = (t-1)^2$. Hence,

$$D_{\chi^2}(\mathbb{P}'||\mathbb{P}) = \int_\Omega \left(\frac{\mathrm{d}\mathbb{P}'}{\mathrm{d}\mathbb{P}}(\omega) - 1\right)^2 \mathrm{d}\mathbb{P}(\omega) = \int_\Omega \left(\frac{\mathrm{d}\mathbb{P}'}{\mathrm{d}\mathbb{P}}(\omega)\right)^2 \mathrm{d}\mathbb{P}(\omega) - 1\ ,$$

for probability distributions $\mathbb{P}$ and $\mathbb{P}'$ on Ω. This divergence is not symmetric and equals $+\infty$ if $\mathbb{P}'$ is not absolutely continuous with respect to $\mathbb{P}$.

Example 5.9 (*Pearson χ^2 divergence of continuous probability distributions*) When $\mathbb{P}$ and $\mathbb{P}'$ are continuous probability distributions on $\mathbb{R}^M$ with PDFs $p(x)$ and $p'(x)$, respectively, their Pearson χ^2 divergence is:

$$D_{\chi^2}(\mathbb{P}'||\mathbb{P}) = \int_{\mathbb{R}^M} \left(\frac{p'(t)}{p(t)} - 1\right)^2 p(t)\mathrm{d}t = \int_{\mathbb{R}^M} \frac{(p'(t) - p(t))^2}{p(t)}\mathrm{d}t\ .$$

Example 5.10 (*Pearson χ^2 divergence of discrete probability distributions*) For discrete probability distributions $\mathbb{P}'$ and $\mathbb{P}$ on $\{\omega_1, \omega_2, \ldots, \omega_n, \ldots\}$, the Pearson χ^2 divergence becomes:

$$D_{\chi^2}(\mathbb{P}'||\mathbb{P}) = \sum_{n=1}^{\infty} \left(\frac{\mathbb{P}'(\omega_n)}{\mathbb{P}(\omega_n)} - 1\right)^2 \mathbb{P}(\omega_n) = \sum_{n=1}^{\infty} \frac{(\mathbb{P}'(\omega_n) - \mathbb{P}(\omega_n))^2}{\mathbb{P}(\omega_n)}\ .$$

Proposition 5.8 (Pearson χ^2 divergence of product distributions) *Assume $\mathbb{P}_\alpha$ and $\mathbb{P}'_\alpha$ are probability distributions on Ω_α. $\mathbb{P}_\beta$ and $\mathbb{P}'_\beta$ are probability distributions on Ω_β. Then*

$$D_{\chi^2}(\mathbb{P}'_\alpha \times \mathbb{P}'_\beta||\mathbb{P}_\alpha \times \mathbb{P}_\beta) + 1 = (D_{\chi^2}(\mathbb{P}'_\alpha||\mathbb{P}_\alpha) + 1)(D_{\chi^2}(\mathbb{P}'_\beta||\mathbb{P}_\beta) + 1)\ .$$

Exercise 5.8 Prove Proposition 5.8 by direct calculation.

5.1.5 Relationships Between Different f-Divergences

We introduce some inequalities between the aforementioned different f-divergences. For simplicity, we always assume $\mathbb{P}' \ll \mathbb{P}$ are probability distributions defined on Ω. However, these relationships also apply to general cases.

Theorem 5.10 introduces Pinsker's inequality, which is about how KL divergence bounds total variation distance.

Theorem 5.10 (Pinsker's inequality)

$$d_{\mathrm{TV}}(\mathbb{P}', \mathbb{P}) \le \sqrt{D_{\mathrm{KL}}(\mathbb{P}'||\mathbb{P})/2}. \tag{5.17}$$

Proof The following proof is from Tsybakov (2009). In KL divergence, the f function satisfies $f(t) = t\log t - t + 1, t \in (0, +\infty)$ and $f(0) = 1$. We claim that

$$\left(\frac{4}{3} + \frac{2}{3}t\right) f(t) \ge (t-1)^2 \ , \ \forall t \ge 0 \ , \tag{5.18}$$

where the proof is left as an exercise. Then

$$\begin{aligned}
d_{\mathrm{TV}}(\mathbb{P}', \mathbb{P}) &= \frac{1}{2}\int_\Omega |\mathrm{d}\mathbb{P}'(\omega) - \mathrm{d}\mathbb{P}(\omega)| && (5.19a)\\
&= \frac{1}{2}\int_\Omega \left|\frac{\mathrm{d}\mathbb{P}'}{\mathrm{d}\mathbb{P}}(\omega) - 1\right| \mathrm{d}\mathbb{P}(\omega) && (5.19b)\\
&\le \frac{1}{2}\int_\Omega \sqrt{\left(\frac{4}{3} + \frac{2}{3}\frac{\mathrm{d}\mathbb{P}'}{\mathrm{d}\mathbb{P}}(\omega)\right) f\left(\frac{\mathrm{d}\mathbb{P}'}{\mathrm{d}\mathbb{P}}(\omega)\right)} \cdot \mathrm{d}\mathbb{P}(\omega) && (5.19c)\\
&\le \frac{1}{2}\sqrt{\int_\Omega \left(\frac{4}{3} + \frac{2}{3}\frac{\mathrm{d}\mathbb{P}'}{\mathrm{d}\mathbb{P}}(\omega)\right) \mathrm{d}\mathbb{P}(\omega)}\sqrt{\int_\Omega f\left(\frac{\mathrm{d}\mathbb{P}'}{\mathrm{d}\mathbb{P}}(\omega)\right) \mathrm{d}\mathbb{P}(\omega)} && (5.19d)\\
&= \frac{1}{2}\cdot\sqrt{2}\cdot\sqrt{D_{\mathrm{KL}}(\mathbb{P}'||\mathbb{P})} = \sqrt{D_{\mathrm{KL}}(\mathbb{P}'||\mathbb{P})/2}\ , && (5.19e)
\end{aligned}$$

where Cauchy-Schwarz inequality is used to get (5.19d) from (5.19c). □

Exercise 5.9 Prove (5.18).

We know from Sect. 5.1.4.1 that total variation distance $d_{\mathrm{TV}}(\mathbb{P}', \mathbb{P}) \in [0, 1]$. In contrast, KL divergence $D_{\mathrm{KL}}(\mathbb{P}'||\mathbb{P})$ has no upper bounds, which has been discussed in Sect. 5.1.3.2. Therefore, Pinsker's inequality in (5.17) does not give an actual bound on $d_{\mathrm{TV}}(\mathbb{P}', \mathbb{P})$ when $D_{\mathrm{KL}}(\mathbb{P}'||\mathbb{P}) \ge 2$. Another inequality is still effective under this circumstance:

Proposition 5.9

$$d_{\mathrm{TV}}(\mathbb{P}', \mathbb{P}) \leq 1 - \frac{1}{2}\exp(-D_{\mathrm{KL}}(\mathbb{P}'||\mathbb{P})) \ . \tag{5.20}$$

In order to prove Proposition 5.9, we first introduce Lemma 5.5, which will also be used later for bounds of Hellinger distance.

Lemma 5.5

$$\begin{aligned}&\left(\int_\Omega \sqrt{\mathrm{d}\mathbb{P}'(\omega)\mathrm{d}\mathbb{P}(\omega)}\right)^2 \\ &\leq \left(\int_\Omega \min\{\mathrm{d}\mathbb{P}'(\omega), \mathrm{d}\mathbb{P}(\omega)\}\right)\left(2 - \int_\Omega \min\{\mathrm{d}\mathbb{P}'(\omega), \mathrm{d}\mathbb{P}(\omega)\}\right) \ .\end{aligned}$$

Proof

$$\left(\int_\Omega \sqrt{\mathrm{d}\mathbb{P}'(\omega)\mathrm{d}\mathbb{P}(\omega)}\right)^2 \tag{5.21a}$$

$$= \left(\int_\Omega \sqrt{\min\{\mathrm{d}\mathbb{P}'(\omega), \mathrm{d}\mathbb{P}(\omega)\}\max\{\mathrm{d}\mathbb{P}'(\omega), \mathrm{d}\mathbb{P}(\omega)\}}\right)^2 \tag{5.21b}$$

$$\leq \left(\int_\Omega \min\{\mathrm{d}\mathbb{P}'(\omega), \mathrm{d}\mathbb{P}(\omega)\}\right)\left(\int_\Omega \max\{\mathrm{d}\mathbb{P}'(\omega), \mathrm{d}\mathbb{P}(\omega)\}\right) \tag{5.21c}$$

$$= \left(\int_\Omega \min\{\mathrm{d}\mathbb{P}'(\omega), \mathrm{d}\mathbb{P}(\omega)\}\right)\left(2 - \int_\Omega \min\{\mathrm{d}\mathbb{P}'(\omega), \mathrm{d}\mathbb{P}(\omega)\}\right) , \tag{5.21d}$$

where Cauchy-Schwarz inequality is used to get (5.21c) from (5.21b); from (5.21c) to (5.21d) the following fact is utilized:

$$\int_\Omega \left(\min\{\mathrm{d}\mathbb{P}'(\omega), \mathrm{d}\mathbb{P}(\omega)\} + \max\{\mathrm{d}\mathbb{P}'(\omega), \mathrm{d}\mathbb{P}(\omega)\}\right) = \int_\Omega \mathrm{d}\mathbb{P}'(\omega) + \int_\Omega \mathrm{d}\mathbb{P}(\omega) = 2 \ .$$

This completes the proof of Lemma 5.5. □

Proof (*Proof of Proposition* 5.9) From Lemma 5.5, we have:

$$\left(\int_\Omega \sqrt{\mathrm{d}\mathbb{P}'(\omega)\mathrm{d}\mathbb{P}(\omega)}\right)^2 \leq 2\int_\Omega \min\{\mathrm{d}\mathbb{P}'(\omega), \mathrm{d}\mathbb{P}(\omega)\} \ .$$

Then by Theorem 5.9,

$$
\begin{aligned}
d_{\mathrm{TV}}(\mathbb{P}', \mathbb{P}) &= 1 - \int_\Omega \min\{\mathrm{d}\mathbb{P}'(\omega), \mathrm{d}\mathbb{P}(\omega)\} && \text{(5.22a)}\\
&\leq 1 - \frac{1}{2}\left(\int_\Omega \sqrt{\mathrm{d}\mathbb{P}'(\omega)\mathrm{d}\mathbb{P}(\omega)}\right)^2 && \text{(5.22b)}\\
&= 1 - \frac{1}{2}\exp\left(2\log\left(\int_\Omega \sqrt{\mathrm{d}\mathbb{P}'(\omega)\mathrm{d}\mathbb{P}(\omega)}\right)\right) && \text{(5.22c)}\\
&= 1 - \frac{1}{2}\exp\left(2\log\left(\int_\Omega \sqrt{\frac{\mathrm{d}\mathbb{P}(\omega)}{\mathrm{d}\mathbb{P}'(\omega)}}\cdot d\mathbb{P}'(\omega)\right)\right) && \text{(5.22d)}\\
&\leq 1 - \frac{1}{2}\exp\left(\int_\Omega 2\log\left(\sqrt{\frac{\mathrm{d}\mathbb{P}(\omega)}{\mathrm{d}\mathbb{P}'(\omega)}}\right) d\mathbb{P}'(\omega)\right) && \text{(5.22e)}\\
&= 1 - \frac{1}{2}\exp\left(-\int_\Omega \log\left(\frac{\mathrm{d}\mathbb{P}'(\omega)}{\mathrm{d}\mathbb{P}(\omega)}\right) \mathrm{d}\mathbb{P}'(\omega)\right) && \text{(5.22f)}\\
&= 1 - \frac{1}{2}\exp(-D_{\mathrm{KL}}(\mathbb{P}'||\mathbb{P}))\,, && \text{(5.22g)}
\end{aligned}
$$

where Jensen's inequality in Theorem 2.10 is used with respect to the convex function $-\log$ to get (5.22e) from (5.22d). □

Lemma 5.5 also helps to connect Hellinger distance and total variation distance, as stated in Proposition 5.10.

Proposition 5.10

$$
\frac{1}{2}d_{\mathrm{H}}^2(\mathbb{P}', \mathbb{P}) \leq d_{\mathrm{TV}}(\mathbb{P}', \mathbb{P}) \leq \sqrt{d_{\mathrm{H}}^2(\mathbb{P}', \mathbb{P})\left(1 - \frac{1}{4}d_{\mathrm{H}}^2(\mathbb{P}', \mathbb{P})\right)} \leq d_{\mathrm{H}}(\mathbb{P}', \mathbb{P})\,.
$$

Proof Combining the formula of total variation distance in Theorem 5.9 with the formula of Hellinger distance in Lemma 5.4, we have:

$$
\begin{aligned}
d_{\mathrm{TV}}(\mathbb{P}', \mathbb{P}) &= 1 - \int_\Omega \min\{\mathrm{d}\mathbb{P}'(\omega), \mathrm{d}\mathbb{P}(\omega)\}\\
&\geq 1 - \int_\Omega \sqrt{\mathrm{d}\mathbb{P}'(\omega)\mathrm{d}\mathbb{P}(\omega)} = \frac{1}{2}d_{\mathrm{H}}^2(\mathbb{P}', \mathbb{P})\,,
\end{aligned}
$$

where the fact that $\min\{t_1, t_2\} \leq \sqrt{t_1 t_2}$, $\forall t_1, t_2 \geq 0$ is used.

By Lemma 5.5, we have:

$$
\begin{aligned}
\left(1 - \frac{1}{2}d_{\mathrm{H}}^2(\mathbb{P}', \mathbb{P})\right)^2 &= \left(\int_\Omega \sqrt{\mathrm{d}\mathbb{P}'(\omega)\mathrm{d}\mathbb{P}(\omega)}\right)^2\\
&\leq \left(\int_\Omega \min\{\mathrm{d}\mathbb{P}'(\omega), \mathrm{d}\mathbb{P}(\omega)\}\right)\left(2 - \int_\Omega \min\{\mathrm{d}\mathbb{P}'(\omega), \mathrm{d}\mathbb{P}(\omega)\}\right)\\
&= \left(1 - d_{\mathrm{TV}}(\mathbb{P}', \mathbb{P})\right)\left(1 + d_{\mathrm{TV}}(\mathbb{P}', \mathbb{P})\right)\,,
\end{aligned}
$$

which can be simplified into:

$$d_{\mathrm{TV}}(\mathbb{P}', \mathbb{P}) \le \sqrt{d_{\mathrm{H}}^2(\mathbb{P}', \mathbb{P})\left(1 - \frac{1}{4} d_{\mathrm{H}}^2(\mathbb{P}', \mathbb{P})\right)} \le d_{\mathrm{H}}(\mathbb{P}', \mathbb{P}) \; .$$

This completes the proof. □

Proposition 5.10 shows that Hellinger distance is closely associated with total variation distance. Since total variation distance can be bounded by KL divergence as shown in Theorem 5.10 and Proposition 5.9, it is not a surprise that Hellinger distance can also be bounded by KL divergence.

Proposition 5.11

$$d_{\mathrm{H}}^2(\mathbb{P}', \mathbb{P}) \le D_{\mathrm{KL}}(\mathbb{P}' || \mathbb{P}) \; .$$

Proof Using the inequality $\log t \le t - 1, \forall t > 0$, we have:

$$\begin{aligned} D_{\mathrm{KL}}(\mathbb{P}' || \mathbb{P}) &= \int_{\Omega} \log\left(\frac{\mathrm{d}\mathbb{P}'}{\mathrm{d}\mathbb{P}}(\omega)\right) \cdot \mathrm{d}\mathbb{P}'(\omega) \\ &= -2 \int_{\Omega} \log \sqrt{\frac{\mathrm{d}\mathbb{P}}{\mathrm{d}\mathbb{P}'}(\omega)} \cdot \mathrm{d}\mathbb{P}'(\omega) \\ &\ge -2 \int_{\Omega} \left(\sqrt{\frac{\mathrm{d}\mathbb{P}}{\mathrm{d}\mathbb{P}'}(\omega)} - 1\right) \mathrm{d}\mathbb{P}'(\omega) \\ &= -2 \left(\int_{\Omega} \sqrt{\frac{\mathrm{d}\mathbb{P}}{\mathrm{d}\mathbb{P}'}(\omega)} \cdot \mathrm{d}\mathbb{P}'(\omega) - 1\right) = d_{\mathrm{H}}^2(\mathbb{P}', \mathbb{P}) \; , \end{aligned}$$

which completes the proof. □

Proposition 5.12 bounds KL divergence from above by Pearson χ^2 divergence.

Proposition 5.12

$$D_{\mathrm{KL}}(\mathbb{P}' || \mathbb{P}) \le \log(1 + D_{\chi^2}(\mathbb{P}' || \mathbb{P})) \le D_{\chi^2}(\mathbb{P}' || \mathbb{P}) \; .$$

Proof

$$\begin{aligned} D_{\mathrm{KL}}(\mathbb{P}' || \mathbb{P}) &= \int_{\Omega} \log\left(\frac{\mathrm{d}\mathbb{P}'}{\mathrm{d}\mathbb{P}}(\omega)\right) \cdot \mathrm{d}\mathbb{P}'(\omega) && \text{(5.23a)} \\ &\le \log\left(\int_{\Omega} \frac{\mathrm{d}\mathbb{P}'}{\mathrm{d}\mathbb{P}}(\omega) \mathrm{d}\mathbb{P}'(\omega)\right) && \text{(5.23b)} \\ &= \log(1 + D_{\chi^2}(\mathbb{P}' || \mathbb{P})) && \text{(5.23c)} \\ &\le D_{\chi^2}(\mathbb{P}' || \mathbb{P}) \; , && \text{(5.23d)} \end{aligned}$$

where (5.23b) follows from applying Jensen's inequality in Theorem 2.10 to the convex function $-\log$ and (5.23d) is because $\log(1+t) \le t, \forall t > -1$. □

Combining Propositions 5.10, 5.11, and 5.12, we have:

$$\frac{1}{2}d_{\mathrm{H}}^2(\mathbb{P}', \mathbb{P}) \le d_{\mathrm{TV}}(\mathbb{P}', \mathbb{P}) \le d_{\mathrm{H}}(\mathbb{P}', \mathbb{P}) \le \sqrt{D_{\mathrm{KL}}(\mathbb{P}'||\mathbb{P})} \le \sqrt{D_{\chi^2}(\mathbb{P}'||\mathbb{P})} \,,$$

which provides guidance on the relationship between these f divergences.

5.2 f-Divergence Ambiguity Set

In this section, we first introduce the definition of f-divergence ambiguity set, then discuss how to choose the divergence threshold parameter.

5.2.1 *Definition*

The definition of f-divergence ambiguity set is:

Definition 5.7 (f*-divergence ambiguity set*) Let $\mathbb{P}_0$ be a probability distribution on Ω. Suppose D_f is a f-divergence and $d_0 \in \mathbb{R}_+$ is an upper threshold of the f-divergence. Then the f*-divergence ambiguity set* is defined as:

$$\mathcal{P}_f(\Omega, \mathbb{P}_0, d_0) := \{\mathbb{P} \in \mathcal{M}(\Omega) \mid D_f(\mathbb{P}||\mathbb{P}_0) \le d_0\} \,.$$

5.2.1.1 Example: Kullback-Leibler-Divergence Ambiguity Set and Discrete Empirical Probability Distribution

Suppose $\omega_1, \omega_2, \ldots, \omega_N$ are the samples of random variable ω on $\mathbb{R}^M$. Following the idea in (1.3), the empirical probability distribution is formed as:

$$\mathbb{P}_0 := \frac{1}{N}\sum_{n=1}^{N} \delta_{\omega_n} \,, \tag{5.24}$$

where δ_{ω_n} represents the Dirac distribution at ω_n. Then the KL-divergence ambiguity set is:

$$\mathcal{P}_{\mathrm{KL}}(\mathbb{R}^M, \mathbb{P}_0, d_0) = \{\mathbb{P} \in \mathcal{M}(\mathbb{R}^M) \mid D_{\mathrm{KL}}(\mathbb{P}||\mathbb{P}_0) \le d_0\} \,. \tag{5.25}$$

Recall from Sect. 5.1.3.2 that KL divergence $D_{\mathrm{KL}}(\mathbb{P}||\mathbb{P}_0)$ equals $+\infty$ if $\mathbb{P} \ll \mathbb{P}_0$ does not hold. Therefore, for any $\mathbb{P} \in \mathcal{P}_{\mathrm{KL}}(\mathbb{R}^M, \mathbb{P}_0, d_0)$, we have $\mathbb{P} \ll \mathbb{P}_0$. Let $\Omega_N :=$

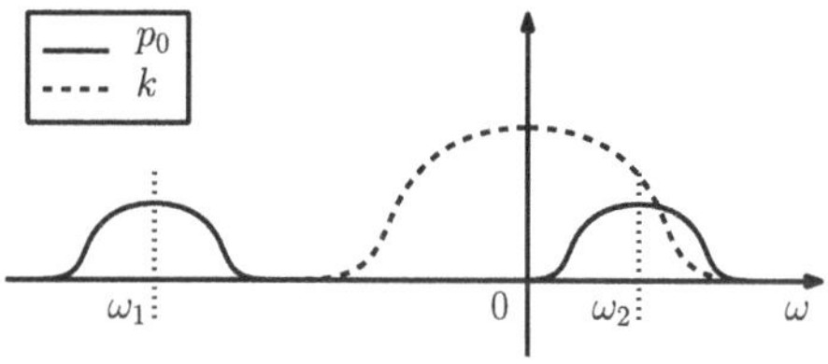

Fig. 5.2 The empirical probability density function (PDF) p_0 is constructed by kernel density function k

$\{\omega \in \mathbb{R}^M \mid \exists n\ ,\ \text{s.t. } \omega = \omega_n\}$. Then $\mathbb{P}_0(\mathbb{R}^M \setminus \Omega_N) = 0$ indicates $\mathbb{P}(\mathbb{R}^M \setminus \Omega_N) = 0$, which means $\mathbb{P}$ should be a discrete probability distribution on Ω_N. Therefore,

$$\begin{aligned}\mathcal{P}_{\mathrm{KL}}(\mathbb{R}^M, \mathbb{P}_0, d_0) &= \mathcal{P}_{\mathrm{KL}}(\Omega_N, \mathbb{P}_0, d_0) \\ &= \left\{ \mathbb{P} \;\middle|\; \begin{array}{l} \sum_{\omega \in \Omega_N} \mathbb{P}(\omega) = 1\ ,\ \mathbb{P}(\omega) \geq 0\ ,\ \forall \omega \in \Omega_N\ , \\ \sum_{\omega \in \Omega_N} \mathbb{P}(\omega) \log \left(\dfrac{\mathbb{P}(\omega)}{\mathbb{P}_0(\omega)} \right) \leq d_0 \end{array} \right\} ,\end{aligned}$$

which only contains some discrete probability distributions on the set of data samples.

5.2.1.2 Example: Kullback-Leibler-Divergence Ambiguity Set and Continuous Empirical Probability Distribution

Some methods can form the empirical probability distribution as a continuous probability distribution, such as the kernel density estimator approach (Parzen 1962): Again, suppose we have samples $\omega_1, \ldots, \omega_N$ of random variable ω on $\mathbb{R}^M$. Let $\mathbb{K}$ be a continuous probability distribution on $\mathbb{R}_M$ so that its mean is 0. Suppose the PDF of $\mathbb{K}$ is k. We will use $\mathbb{K}$ as a kernel, i.e., the empirical probability distribution $\mathbb{P}_0$ is defined as the continuous probability distribution with the density:

$$p_0(\omega) := \frac{1}{N} \sum_{n=1}^{N} k(\omega - \omega_N)\ ,$$

which means we assign a density with the shape of k and $1/N$ the mass to each sample, as illustrated in Fig. 5.2.

The KL-divergence ambiguity set is defined in (5.25). For any $\mathbb{P} \in \mathcal{P}_{\mathrm{KL}}(\Omega, \mathbb{P}_0, d_0)$, the absolute continuity $\mathbb{P} \ll \mathbb{P}_0$ should hold. Now $\mathbb{P}_0$ is a continuous probability distribution with PDF p_0, so $\mathbb{P}$ should also be continuous and its density p should satisfy $p_0(\omega) = 0 \Rightarrow p(\omega) = 0$ for any $\omega \in \mathbb{R}^M$. In conclusion,

$$\mathcal{P}_{\mathrm{KL}}(\mathbb{R}^M, \mathbb{P}_0, d_0) = \left\{ \mathbb{P} \in \mathbb{R}^M \;\middle|\; \begin{array}{l} \mathbb{P} \text{ is continuous with density } p\,, \\ \{\omega \mid p_0(\omega) = 0\} \subset \{\omega \mid p(\omega) = 0\}\,, \\ \displaystyle\int_{\omega: p(\omega)>0} p(\omega) \log\left(\frac{p(\omega)}{p_0(\omega)}\right) \mathrm{d}\omega \le d_0 \end{array} \right\},$$

which only contains some continuous probability distributions that are absolutely continuous with respect to the empirical probability distribution $\mathbb{P}_0$.

5.2.2 *The Choice of Parameter*

Assume Ω is a finite sample space. The empirical probability distribution $\mathbb{P}_0$ is constructed by (5.24) based on the N independent samples drawn from the real probability distribution $\mathbb{P}^*$. Suppose the f-divergence D_f satisfies that $\mathrm{d}^2 f/\mathrm{d}t^2$ exists at $t = 1$, then as $N \to \infty$,

$$\frac{2N}{\frac{\mathrm{d}^2 f}{\mathrm{d}t^2}(1)} D_f(\mathbb{P}^* || \mathbb{P}_0)\,,$$

asymptotically follows the χ^2 distribution with $|\Omega| - 1$ degrees of freedom (Pardo 2005), where $|\Omega|$ denotes the number of elements in Ω. Therefore, it was proposed in Ben-Tal et al. (2013) to choose the divergence threshold d_0 as follows:

$$d_0 := \frac{\frac{\mathrm{d}^2 f}{\mathrm{d}t^2}(1)}{2N} \chi^2_{|\Omega|-1,\beta}\,. \tag{5.26}$$

In (5.26), β is the desired confidence to believe $D_f(\mathbb{P}^* || \mathbb{P}) \le d_0$. $\chi^2_{|\Omega|-1,\beta}$ denotes the β percentile of $\chi^2_{|\Omega|-1}$, the χ^2 distribution with $|\Omega| - 1$ degrees of freedom. In other words, the probability of random variable X being smaller than $\chi^2_{|\Omega|-1,\beta}$ is β, where X follows the $\chi^2_{|\Omega|-1}$ distribution.

5.3 Reformulation of Worst-Case Expectation

In this section, we investigate how to reformulate the worst-case expectation under the f-divergence ambiguity set:

$$\sup_{\mathbb{P} \in \mathcal{P}_f(\Omega, \mathbb{P}_0, d_0)} \mathbb{E}^{\mathbb{P}}[g(x, \omega)]\,. \tag{5.27}$$

We assume $d_0 > 0$ and $g(x, \omega)$ is bounded. For presentation simplicity, we focus on the case of finite sample space Ω. This is a common assumption in most applications of f-divergence distributionally robust optimization (DRO). Nonetheless, the conclusions in this section can be extended to the general case. We first derive under general f-divergence and then focus on KL divergence.

5.3.1 General f-Divergence

According to the definition of f-divergence ambiguity set in Definition 5.7, the worst-case expectation (5.27) equals the optimal value of the following optimization problem:

$$\sup_{p} \sum_{\omega\in\Omega} g(x,\omega)p_\omega \tag{5.28a}$$

$$\text{s.t.} \sum_{\omega\in\Omega} f\left(\frac{p_\omega}{\mathbb{P}_0(\omega)}\right)\mathbb{P}_0(\omega) \le d_0 : \lambda\ , \tag{5.28b}$$

$$\sum_{\omega\in\Omega} p_\omega = 1 : \mu\ , \tag{5.28c}$$

$$p_w \ge 0\ ,\ \forall\omega\in\Omega\ , \tag{5.28d}$$

where $p_\omega := \mathbb{P}(\omega)$, $\forall\omega\in\Omega$. Because f is convex, (5.28) is a convex optimization problem. Because we assume $d_0 > 0$, (5.28) is strictly feasible, and thus Slater's condition is satisfied. Therefore, strong duality holds according to Proposition 2.14, which means:

$$\sup_{\mathbb{P}\in\mathcal{P}_f(\Omega,\mathbb{P}_0,d_0)} \mathbb{E}^{\mathbb{P}}[g(x,\omega)] = \inf_{\lambda\ge 0,\mu\in\mathbb{R}}\ \sup_{p:p_\omega\ge 0,\forall\omega\in\Omega} L(p,\lambda,\mu)\ , \tag{5.29}$$

where

$$\begin{aligned} L(p,\lambda,\mu) &= \sum_{\omega\in\Omega} g(x,\omega)p_\omega + \lambda\left(d_0 - \sum_{\omega\in\Omega} f\left(\frac{p_\omega}{\mathbb{P}_0(\omega)}\right)\mathbb{P}_0(\omega)\right) + \mu\left(1-\sum_{\omega\in\Omega} p_\omega\right) \\ &= d_0\lambda + \mu + \sum_{\omega\in\Omega}\left((g(x,\omega)-\mu)p_\omega - \lambda\mathbb{P}_0(\omega) f\left(\frac{p_\omega}{\mathbb{P}_0(\omega)}\right)\right). \end{aligned}$$

$L(p,\lambda,\mu)$ is linear in λ and μ. Because $h(\lambda,\mu) := \sup_{p:p_\omega\ge 0,\forall\omega\in\Omega} L(p,\lambda,\mu)$ is the supremum of a set of linear functions, it is convex in λ and μ.

We first prove that we can assume $\lambda > 0$ without loss of generality, i.e., for any $\mu_0\in\mathbb{R}$, $\inf_{\lambda\ge 0} h(\lambda,\mu_0) = \inf_{\lambda>0} h(\lambda,\mu_0)$. To this end, we only need to prove that $h(0,\mu_0) \ge \inf_{\lambda>0} h(\lambda,\mu_0)$. If $h(0,\mu_0) < +\infty$, then $h(\lambda,\mu_0) < +\infty$, $\forall\lambda > 0$ because $p_\omega \ge 0$, $\forall\omega\in\Omega$ and $f(t)\ge 0$, $\forall t\in[0,+\infty)$. Therefore, in the following, we suppose $h(0,\mu_0) < \inf_{\lambda>0} h(\lambda,\mu_0) < +\infty$ and prove by finding a contradiction. Then for any $\lambda_0 > 0$, there is $a\in(0,1)$ such that $\inf_{\lambda>0} h(\lambda,\mu_0) > ah(0,\mu_0) +$

$(1-a)h(\lambda_0, \mu_0) \geq h((1-a)\lambda_0, \mu_0)$, where the second inequality follows from the convexity of h in λ. However, this leads to the contradiction $\inf_{\lambda>0} h(\lambda, \mu_0) > h((1-a)\lambda_0, \mu_0)$. Therefore, $h(0, \mu_0) \geq \inf_{\lambda>0} h(\lambda, \mu_0)$, $\forall \mu_0 \in \mathbb{R}$. Then we have $\inf_\mu h(0, \mu) \geq \inf_{\lambda>0, \mu\in\mathbb{R}} h(\lambda, \mu)$ and (5.27) $= \inf_{\lambda>0, \mu\in\mathbb{R}} h(\lambda, \mu)$. Therefore, we only consider the case $\lambda > 0$ in the following.

Recall that $f : [0, +\infty) \to \mathbb{R}$ has the following conjugate function:

$$f^*(s) := \sup_{t\geq 0}\{st - f(t)\}\ .$$

The conjugate functions of some common f-divergences are also shown in Table 5.1. These conjugate functions have some properties, which will be used later.

Lemma 5.6 *Assume that the function $f : [0, +\infty) \to \mathbb{R}$ satisfies the conditions for f-divergence in Definition 5.1. Then its conjugate function f^* satisfies:*

1. *f^* is nondecreasing;*
2. *$f^*(s) \geq s$, $\forall s \in \mathbb{R}$;*
3. *$f^*(s) = +\infty$, $\forall s > l_f := \lim_{t\to+\infty} f(t)/t$.*

Proof 1. For $s_1 < s_2$ in $\mathbb{R}$, $f^*(s_1) = \sup_{t\geq 0}\{s_1 t - f(t)\} \leq \sup_{t\geq 0}\{s_2 t - f(t)\} = f^*(s_2)$, so f^* is nondecreasing on $\mathbb{R}$.

2. $f^*(s) = \sup_{t\geq 0}\{st - f(t)\} \geq s - f(1) = s$, $\forall s \in \mathbb{R}$.

3. By Lemma 5.1, l_f is finite or $+\infty$. We only need to consider the case where l_f is finite. For any $s > l_f$, we have:

$$f^*(s) = \sup_{t\geq 0}\{st - f(t)\} \geq \lim_{t\to+\infty}(st - f(t)) = +\infty\ ,$$

which completes the proof. □

When $\lambda > 0$, we have:

$$\sup_{p:p_\omega\geq 0,\forall\omega\in\Omega} L(p, \lambda, \mu) \tag{5.30a}$$

$$= \sup_{p:p_\omega\geq 0,\forall\omega\in\Omega} \left\{ d_0\lambda + \mu + \sum_{\omega\in\Omega}\left((g(x,\omega) - \mu)p_\omega - \lambda\mathbb{P}_0(\omega) f\left(\frac{p_\omega}{\mathbb{P}_0(\omega)}\right)\right)\right\} \tag{5.30b}$$

$$= d_0\lambda + \mu + \sum_{\omega\in\Omega}\sup_{p_\omega\geq 0}\left\{ (g(x,\omega) - \mu)p_\omega - \lambda\mathbb{P}_0(\omega) f\left(\frac{p_\omega}{\mathbb{P}_0(\omega)}\right)\right\} \tag{5.30c}$$

$$\begin{aligned}= d_0\lambda + \mu + \sum_{\omega:\mathbb{P}_0(\omega)>0} \lambda\mathbb{P}_0(\omega)\cdot\sup_{p_\omega\geq 0}\left\{\left(\frac{g(x,\omega)-\mu}{\lambda}\frac{p_\omega}{\mathbb{P}_0(\omega)} - f\left(\frac{p_\omega}{\mathbb{P}_0(\omega)}\right)\right)\right\} \\ + \sum_{\omega:\mathbb{P}_0(\omega)=0}\sup_{p_\omega\geq 0}\left\{(g(x,\omega)-\mu)p_\omega - \lambda p_\omega l_f\right\}\end{aligned} \tag{5.30d}$$

$$
= \begin{cases} d_0\lambda + \mu + \lambda \sum_{\omega:\mathbb{P}_0(\omega)>0} \mathbb{P}_0(\omega) f^*\left(\frac{g(x,\omega)-\mu}{\lambda}\right), & \text{if } \mu \geq U(x) - \lambda l_f , \\ +\infty , & \text{otherwise} . \end{cases} \tag{5.30e}
$$

From (5.30d) to (5.30e), we use the definition of the conjugate function f^*. For $\omega \in \Omega$ such that $\mathbb{P}_0(\omega) > 0$, by Lemma 5.6, $f^*((g(x,\omega) - \mu)/\lambda)) < +\infty$ implies $\mu \geq g(x,\omega) - \lambda l_f$. For $\omega \in \Omega$ such that $\mathbb{P}_0(\omega) = 0$, $\sup_{p_\omega \geq 0}(g(x,\omega) - \mu - \lambda l_f) p_\omega$ equals 0 if $g(x,\omega) - \mu - \lambda l_f \leq 0$ and it equals $+\infty$ otherwise. Thus, we have the equation from (5.30d) to (5.30e), where $U(x) := \sup\{g(x,\omega) \mid \omega \in \Omega\}$.

Combining (5.29) and (5.30), we have:

$$
\begin{aligned}
&\sup_{\mathbb{P}\in\mathcal{P}_f(\Omega,\mathbb{P}_0,d_0)} \mathbb{E}^{\mathbb{P}}[g(x,\omega)] \\
&= \inf_{\lambda>0,\mu\geq\sup_{\omega\in\Omega} g(x,\omega)-\lambda l_f} \left\{ d_0\lambda + \mu + \lambda \sum_{\omega:\mathbb{P}_0(\omega)>0} \mathbb{P}_0(\omega) f^*\left(\frac{g(x,\omega)-\mu}{\lambda}\right) \right\} \\
&= \inf_{\lambda>0,\mu\geq U(x)-\lambda l_f} \left\{ d_0\lambda + \mu + \sum_{\omega\in\Omega} \mathbb{P}_0(\omega) \lambda f^*\left(\frac{g(x,\omega)-\mu}{\lambda}\right) \right\}, \\
&= \inf_{\lambda>0,\mu\geq U(x)-\lambda l_f} \left\{ d_0\lambda + \mu + \mathbb{E}^{\mathbb{P}_0}\left[\lambda f^*\left(\frac{g(x,\omega)-\mu}{\lambda}\right) \right] \right\},
\end{aligned}
$$

which minimizes a convex function of λ and μ over a polyhedral region and thus is a convex optimization problem. This is consistent with the convexity of the Lagrange dual problem.

Exercise 5.10 Prove that the function $\lambda f^*((g(x,\omega) - \mu)/\lambda)$ is convex in (λ, μ).

Theorem 5.11 sums up the above results (Bayraksan and Love 2015):

Theorem 5.11 *Assume $d_0 > 0$ and $g(x,\omega)$ is bounded. The worst-case expectation under f-divergence ambiguity set can be calculated by solving a convex optimization problem:*

$$
\sup_{\mathbb{P}\in\mathcal{P}_f(\Omega,\mathbb{P}_0,d_0)} \mathbb{E}^{\mathbb{P}}[g(x,\omega)] = \begin{cases} \min_{\lambda,\mu} d_0\lambda + \mu + \mathbb{E}^{\mathbb{P}_0}\left[\lambda f^*\left(\frac{g(x,\omega)-\mu}{\lambda}\right) \right] \\ s.t.\ \lambda > 0 ,\ \mu \geq U(x) - \lambda l_f , \end{cases}
$$

where $U(x) := \sup\{g(x,\omega) \mid \omega \in \Omega\}$ and $l_f := \lim_{t\to+\infty} f(t)/t$.

Although we derive it on finite sample space Ω, Theorem 5.11 holds for any sample space Ω, any f-divergence, and any bounded measurable function $g(x,\omega)$, so it is very general.

5.3.2 Kullback-Leibler Divergence

Recall that the KL divergence $D_{\mathrm{KL}}(\mathbb{P}||\mathbb{P}_0)$ from $\mathbb{P}$ to $\mathbb{P}_0$ on finite sample space Ω is:

$$\begin{cases} \sum_{\omega:\mathbb{P}_0(\omega)>0} \mathbb{P}(\omega)\log\left(\frac{\mathbb{P}(\omega)}{\mathbb{P}_0(\omega)}\right), & \text{if } \mathbb{P}_0(\omega)=0 \Rightarrow \mathbb{P}(\omega)=0\ ,\ \forall\omega\in\Omega\ , \\ +\infty\ , & \text{otherwise}\ . \end{cases}$$

Therefore, for any $d_0 \in \mathbb{R}_+$, so long as the probability distribution $\mathbb{P}$ is in the ambiguity set $\mathcal{P}_{\mathrm{KL}}(\Omega, \mathbb{P}_0, d_0)$, it satisfies $\mathbb{P} \ll \mathbb{P}_0$. Hence, we can replace Ω by $\mathrm{supp}(\mathbb{P}_0) := \{\omega \in \Omega \mid \mathbb{P}_0(\omega) > 0\}$, the support of $\mathbb{P}_0$. In other words, $\mathcal{P}_{\mathrm{KL}}(\Omega, \mathbb{P}_0, d_0) = \mathcal{P}_{\mathrm{KL}}(\mathrm{supp}(\mathbb{P}_0), \mathbb{P}_0, d_0)$. Therefore, in the following, we assume $\Omega = \mathrm{supp}(\mathbb{P}_0)$ without loss of generality. Then $\mathbb{P}_0(\omega) > 0$, $\forall\omega \in \Omega$ and

$$\begin{aligned} D_{\mathrm{KL}}(\mathbb{P}||\mathbb{P}_0) &= \sum_{\omega\in\Omega} \mathbb{P}(\omega)\log\left(\frac{\mathbb{P}(\omega)}{\mathbb{P}_0(\omega)}\right) \\ &= \sum_{\omega\in\Omega} \mathbb{P}_0(\omega)\frac{\mathbb{P}(\omega)}{\mathbb{P}_0(\omega)}\log\left(\frac{\mathbb{P}(\omega)}{\mathbb{P}_0(\omega)}\right) \\ &= \sum_{\omega\in\Omega} \mathbb{P}_0(\omega)Q(\omega)\log(Q(\omega))\ , \end{aligned} \tag{5.31}$$

where $Q(\omega) := \mathbb{P}(\omega)/\mathbb{P}_0(\omega)$, $\forall\omega \in \Omega$. Similarly, the expectation

$$\mathbb{E}^{\mathbb{P}}[g(x,\omega)] = \sum_{\omega\in\Omega} \mathbb{P}(\omega)g(x,\omega) = \sum_{\omega\in\Omega} \mathbb{P}_0(\omega)Q(\omega)g(x,\omega)\ . \tag{5.32}$$

Given $\mathbb{P}_0$ and Q on Ω, we can restore $\mathbb{P}$ by $\mathbb{P}(\omega) = \mathbb{P}_0(\omega)Q(\omega)$, $\forall\omega \in \Omega$. $\mathbb{P}$ is a probability distribution, so it must satisfy that $\mathbb{P}(\omega) \geq 0$, $\forall\omega \in \Omega$ and $\sum_{\omega\in\Omega} \mathbb{P}(\omega) = 1$, which is equivalent to:

$$Q(\omega) \geq 0\ ,\ \forall\omega \in \Omega\ ,\ \sum_{\omega\in\Omega} \mathbb{P}_0(\omega)Q(\omega) = 1\ . \tag{5.33}$$

Using (5.31)–(5.33), we know that the worst-case expectation under KL-divergence ambiguity set is:

$$\sup_{\mathbb{P}\in\mathcal{P}_{\mathrm{KL}}(\Omega,\mathbb{P}_0,d_0)} \mathbb{E}^{\mathbb{P}}[g(x,\omega)] = \begin{cases} \max\limits_{Q} \sum_{\omega\in\Omega} \mathbb{P}_0(\omega)Q(\omega)g(x,\omega) \\ \text{s.t.} \sum_{\omega\in\Omega} \mathbb{P}_0(\omega)Q(\omega)\log(Q(\omega)) \leq d_0\ , \\ \quad \sum_{\omega\in\Omega} \mathbb{P}_0(\omega)Q(\omega) = 1\ , \\ \quad Q(\omega) \geq 0\ ,\ \forall\omega \in \Omega\ . \end{cases} \tag{5.34}$$

Exercise 5.11 Show that (5.34) is a convex optimization problem and it is strictly feasible when $d_0 > 0$.

The Lagrangian function of problem (5.34) is:

$$L(Q, \lambda) := \sum_{\omega \in \Omega} \mathbb{P}_0(\omega) Q(\omega) g(x, \omega) - \lambda \left(\sum_{\omega \in \Omega} \mathbb{P}_0(\omega) Q(\omega) \log(Q(\omega)) - d_0 \right) ,$$

for any $\lambda \geq 0$ and $Q \in \mathcal{Q}$, where

$$\mathcal{Q} := \{Q : \Omega \to \mathbb{R} \mid (5.33)\} .$$

By Proposition 5.3, we know that $D_{\mathrm{KL}}(\mathbb{P}||\mathbb{P}_0) = 0$ implies $\mathbb{P} = \mathbb{P}_0$ as probability distributions. Therefore, we focus on the case $d_0 > 0$, and then problem (5.34) has strong duality because it is a convex optimization problem that satisfies Slater's condition. Hence,

$$(5.34) = \min_{\lambda \geq 0} \max_{Q \in \mathcal{Q}} L(Q, \lambda) .$$

The inner maximization problem

$$\max_{Q \in \mathcal{Q}} L(Q, \lambda)$$
$$= \begin{cases} \max\limits_{Q} \; d_0 \lambda + \sum\limits_{\omega \in \Omega} \mathbb{P}_0(\omega)(Q(\omega) g(x, \omega) - \lambda Q(\omega) \log(Q(\omega))) \\ \text{s.t.} \; \sum\limits_{\omega \in \Omega} \mathbb{P}_0(\omega) Q(\omega) = 1 : \mu , \\ Q(\omega) \geq 0 : \nu(\omega) , \; \forall \omega \in \Omega , \end{cases} \tag{5.35}$$

maximizes a concave function over a polyhedral region. Problem (5.35) is convex and strictly feasible. Its objective and constraint functions are C^1. Therefore, we can use the following Karush-Kuhn-Tucker (KKT) conditions to find its optimal solution according to Theorem 2.18:

$$\begin{cases} (5.33) \; , \; \nu(\omega) \geq 0 \; , \; \forall \omega \in \Omega \; , \\ Q(\omega)\nu(\omega) = 0 \; , \; \forall \omega \in \Omega \; , \\ \mathbb{P}_0(\omega)(g(x, \omega) - \lambda \log(Q(\omega)) - \lambda - \mu) + \nu(\omega) = 0 \; , \; \forall \omega \in \Omega \; . \end{cases} \tag{5.36}$$

Exercise 5.12 Show that $\max_{Q \in \mathcal{Q}} L(Q, 0) = U(x) := \sup_{\omega \in \Omega} g(x, \omega)$.

We focus on the case $\lambda > 0$. Then according to the last constraint in (5.36), at the optimal point, we have $Q(\omega) > 0, \forall \omega \in \Omega$. Then $\nu(\omega) = 0, \forall \omega \in \Omega$ and the optimal solution can be obtained by solving

$$\begin{cases} \sum_{\omega \in \Omega} \mathbb{P}_0(\omega) Q(\omega) = 1 \ , \\ g(x,\omega) - \lambda \log(Q(\omega)) - \lambda - \mu = 0 \ , \ \forall \omega \in \Omega \ , \end{cases}$$

which is:

$$Q(\tilde{\omega}) = \frac{\exp(g(x,\tilde{\omega})/\lambda)}{\sum_{\omega \in \Omega} \mathbb{P}_0(\omega) \exp(g(x,\omega)/\lambda)} = \frac{\exp(g(x,\tilde{\omega})/\lambda)}{\mathbb{E}^{\mathbb{P}_0}[\exp(g(x,\omega)/\lambda)]} \ , \ \forall \tilde{\omega} \in \Omega \ . \tag{5.37}$$

Exercise 5.13 Prove (5.37).

Using (5.37), we have:

$$\begin{aligned} &\max_{Q \in Q} L(Q,\lambda) \\ &= d_0\lambda + \mathbb{E}^{\mathbb{P}_0}[Q(\omega)(g(x,\omega) - \lambda \log(Q(\omega)))] \\ &= d_0\lambda + \mathbb{E}^{\mathbb{P}_0}\left[\frac{\exp(g(x,\omega)/\lambda)}{\mathbb{E}^{\mathbb{P}_0}[\exp(g(x,\omega)/\lambda)]}\right. \\ &\quad \left.\cdot \left(g(x,\omega) - \lambda \log\left(\frac{\exp(g(x,\omega)/\lambda)}{\mathbb{E}^{\mathbb{P}_0}[\exp(g(x,\omega)/\lambda)]}\right)\right)\right] \\ &= d_0\lambda + \mathbb{E}^{\mathbb{P}_0}\left[\frac{\exp(g(x,\omega)/\lambda)}{\mathbb{E}^{\mathbb{P}_0}[\exp(g(x,\omega)/\lambda)]}\left(\lambda \log\left(\mathbb{E}^{\mathbb{P}_0}[\exp(g(x,\omega)/\lambda)]\right)\right)\right] \\ &= d_0\lambda + \lambda \log\left(\mathbb{E}^{\mathbb{P}_0}[\exp(g(x,\omega)/\lambda)]\right) \ . \end{aligned}$$

Exercise 5.14 Show that $\max_{Q \in Q} L(Q,\lambda)$ is convex in λ and

$$\lim_{\lambda \to 0+} \left\{d_0\lambda + \lambda \log\left(\mathbb{E}^{\mathbb{P}_0}[\exp(g(x,\omega)/\lambda)]\right)\right\} = U(x) := \sup_{\omega \in \Omega} g(x,\omega) \ .$$

Combining the above results with the strong duality of problem (5.34), we have the following conclusion:

Theorem 5.12 (Theorem 1 in Hu and Hong 2012) *Assume $g(x,\omega)$ is bounded and measurable. Then the worst-case expectation under the KL-divergence ambiguity set satisfies:*

$$\sup_{\mathbb{P} \in \mathcal{P}_{\mathrm{KL}}(\Omega,\mathbb{P}_0,d_0)} \mathbb{E}^{\mathbb{P}}[g(x,\omega)] = \min_{\lambda > 0} \left\{d_0\lambda + \lambda \log\left(\mathbb{E}^{\mathbb{P}_0}[\exp(g(x,\omega)/\lambda)]\right)\right\} \ .$$

5.4 Reformulation of Robust Chance Constraint

In this section, we discuss the reformulation of robust chance constraint under f-divergence ambiguity set, i.e.,

$$\inf_{\mathbb{P}\in\mathcal{P}_f(\Omega,\mathbb{P}_0,d_0)} \mathbb{P}(S(x)) > \beta\,, \tag{5.38}$$

where $\varnothing \neq S(x) \subsetneq \Omega$ and $\beta \in (0, 1)$.

5.4.1 General f-Divergence

We would like to use the reformulation of the worst-case expectation in Sect. 5.3.1. To this end, we rewrite the robust chance constraint (5.38) using worst-case expectation as follows:

$$\begin{aligned}(5.38) \iff & \sup_{\mathbb{P}\in\mathcal{P}_f(\Omega,\mathbb{P}_0,d_0)} \mathbb{P}(\omega \in \Omega \setminus S(x)) \le 1-\beta \\ \iff & \sup_{\mathbb{P}\in\mathcal{P}_f(\Omega,\mathbb{P}_0,d_0)} \mathbb{E}^{\mathbb{P}}[1_{\Omega\setminus S(x)}(\omega)] \le 1-\beta\,,\end{aligned} \tag{5.39}$$

where $1_{\Omega\setminus S(x)}(\omega)$ is the indicator function of the set $\Omega \setminus S(x)$, i.e., $1_{\Omega\setminus S(x)}(\omega) = 1$ for $\omega \in \Omega \setminus S(x)$ and $1_{\Omega\setminus S(x)}(\omega) = 0$ for $\omega \in S(x)$.

According to Theorem 5.11,

$$\begin{aligned}&\sup_{\mathbb{P}\in\mathcal{P}_f(\Omega,\mathbb{P}_0,d_0)} \mathbb{E}^{\mathbb{P}}[1_{\Omega\setminus S(x)}(\omega)] \\ &= \begin{cases} \min\limits_{\lambda,\mu} d_0\lambda + \mu + \mathbb{E}^{\mathbb{P}_0}\left[\lambda f^*\left(\dfrac{1_{\Omega\setminus S(x)}(\omega)-\mu}{\lambda}\right)\right] \\ \text{s.t. } \lambda > 0\,,\ \mu + \lambda l_f \ge \sup\{1_{\Omega\setminus S(x)}(\omega) \mid \omega \in \Omega\} \end{cases} \\ &= \min_{\lambda>0,\mu+\lambda l_f\ge 1} \left\{d_0\lambda + \mu + \lambda \sum_{\omega\in\Omega} \mathbb{P}_0(\omega) f^*\left(\frac{1_{\Omega\setminus S(x)}(\omega)-\mu}{\lambda}\right)\right\}.\end{aligned} \tag{5.40}$$

By the definition of an indicator function,

$$\begin{aligned}&\sum_{\omega\in\Omega} \mathbb{P}_0(\omega) f^*\left(\frac{1_{\Omega\setminus S(x)}(\omega)-\mu}{\lambda}\right) \\ &= \sum_{\omega\in S(x)} \mathbb{P}_0(\omega) f^*\left(-\frac{\mu}{\lambda}\right) + \sum_{\omega\in\Omega\setminus S(x)} \mathbb{P}_0(\omega) f^*\left(\frac{1-\mu}{\lambda}\right) \\ &= \mathbb{P}_0(S(x)) f^*\left(-\frac{\mu}{\lambda}\right) + (1-\mathbb{P}_0(S(x))) f^*\left(\frac{1-\mu}{\lambda}\right) \\ &= \mathbb{P}_0(S(x))\left(f^*\left(-\frac{\mu}{\lambda}\right) - f^*\left(\frac{1-\mu}{\lambda}\right)\right) + f^*\left(\frac{1-\mu}{\lambda}\right).\end{aligned} \tag{5.41}$$

Combining the results in (5.39)–(5.41), we want to find $\lambda > 0$ and $\mu + \lambda l_f \geq 1$ such that

$$d_0\lambda + \mu + \lambda\mathbb{P}_0(S(x))\left(f^*\left(-\frac{\mu}{\lambda}\right) - f^*\left(\frac{1-\mu}{\lambda}\right)\right) + \lambda f^*\left(\frac{1-\mu}{\lambda}\right) \leq 1 - \beta\,. \tag{5.42}$$

The left side of constraint (5.42) is convex in (λ, μ) according to the discussion in Sect. 5.3.1.

Recall from Lemma 5.6 that f^* is nondecreasing and $f^*(s) \geq s, \forall s \in \mathbb{R}$. Thus, $f^*(-\mu/\lambda) - f^*((1-\mu)/\lambda) \leq 0$. We exclude the case $f^*(-\mu/\lambda) - f^*((1-\mu)/\lambda) = 0$: Suppose $f^*(-\mu/\lambda) = f^*((1-\mu)/\lambda)$, then the left side of (5.42) is changed into:

$$d_0\lambda + \mu + \lambda f^*\left(\frac{1-\mu}{\lambda}\right) \geq d_0\lambda + \mu + \lambda \cdot \frac{1-\mu}{\lambda} = 1 + d_0\lambda > 1 - \beta\,,$$

so (5.42) cannot hold if $f^*(-\mu/\lambda) = f^*((1-\mu)/\lambda)$. Therefore, $f^*(-\mu/\lambda) - f^*((1-\mu)/\lambda) < 0$ and (5.42) can be organized into Theorem 5.13:

Theorem 5.13 (Theorem 1 in Jiang and Guan 2016) *The robust chance constraint* (5.38) *can be equivalently reformulated as:*

$$\mathbb{P}_0(S(x)) \geq \min_{\lambda>0,\mu+\lambda l_f\geq 1} \frac{d_0\lambda + \mu - 1 + \beta + \lambda f^*\left(\frac{1-\mu}{\lambda}\right)}{\lambda f^*\left(\frac{1-\mu}{\lambda}\right) - \lambda f^*\left(-\frac{\mu}{\lambda}\right)}\,, \tag{5.43}$$

where $l_f := \lim_{t\to+\infty} f(t)/t$.

Theorem 5.13 is general in the sense that it holds for any sample space Ω, any type of f-divergence, and any random event $\varnothing \neq S(x) \subsetneq \Omega$. Theorem 5.13 indicates that the robust chance constraint under f-divergence ambiguity set is equivalent to an ordinary chance constraint under the empirical probability distribution $\mathbb{P}_0$, where the probability threshold is changed. We emphasize that the new threshold is no smaller than the original one. In fact, for any $\lambda > 0$ and $\mu + \lambda l_f \geq 1$, we have:

$$\frac{d_0\lambda + \mu - 1 + \beta + \lambda f^*\left(\frac{1-\mu}{\lambda}\right)}{\lambda f^*\left(\frac{1-\mu}{\lambda}\right) - \lambda f^*\left(-\frac{\mu}{\lambda}\right)} \geq \beta\,, \tag{5.44}$$

which follows from

$$\lambda(1-\beta)f^*\left(\frac{1-\mu}{\lambda}\right)+\beta\lambda f^*\left(-\frac{\mu}{\lambda}\right)\geq\lambda(1-\beta)\frac{1-\mu}{\lambda}+\beta\lambda\left(-\frac{\mu}{\lambda}\right)=1-\beta-\mu\,.$$

Inequality (5.44) means that the robust chance constraint is more conservative than the ordinary chance constraint.

We adopt variable transformation to simplify the right side of (5.43): Let $a := 1/\lambda$ and $b := \mu/\lambda$. The the right side of (5.43) equals:

$$\min_{a,b:0<a\leq b+l_f}\frac{d_0+b-a+\beta a+f^*(a-b)}{f^*(a-b)-f^*(-b)}\,. \tag{5.45}$$

For a specific type of f-divergence, we can try to solve the optimization problem (5.45) to get the desired probability threshold. Its optimal value must be no smaller than β as we discussed earlier.

5.4.2 Kullback-Leibler Divergence

Recall that for KL divergence, $l_f = +\infty$, $f^*(s) = \exp(s) - 1, \forall s \in \mathbb{R}$. Thus, problem (5.45) can be written as:

$$\begin{aligned}
&\min_{a,b:a>0}\frac{d_0+b-a+\beta a+\exp(a-b)-1}{\exp(a-b)-\exp(-b)}\\
=\;&\min_{a,b:a>0}\frac{\exp(a)+(d_0+b-a+\beta a-1)\exp(b)}{\exp(a)-1}\\
=\;&\min_{a>0}\frac{\exp(a)+\min_b(d_0+b-a+\beta a-1)\exp(b)}{\exp(a)-1}\\
=\;&\min_{a>0}\frac{\exp(a)-\exp(-d_0+a-\beta a)}{\exp(a)-1}\\
=\;&\min_{a>0}\frac{1-\exp(-d_0)(\exp(-a))^{\beta}}{1-\exp(-a)}\\
=\;&\min_{0<t<1}\frac{1-\exp(-d_0)t^{\beta}}{1-t}\,,
\end{aligned}$$

where $t = \exp(-a)$. Define function $h(t) := (1 - \exp(-d_0)t^{\beta})/(1 - t), 0 < t < 1$, then

$$\frac{\mathrm{d}h}{\mathrm{d}t}(t)=\frac{-\exp(-d_0)\beta t^{\beta-1}+\exp(-d_0)\beta t^{\beta}+1-\exp(-d_0)t^{\beta}}{(1-t)^2}\,.$$

Let $k(t) := -\exp(-d_0)\beta t^{\beta-1} + \exp(-d_0)\beta t^{\beta} + 1 - \exp(-d_0)t^{\beta}$, $0 < t < 1$, then

$$\frac{\mathrm{d}k}{\mathrm{d}t}(t) = \exp(-d_0)\beta t^{\beta-2}(1-\beta)(1-t) > 0\ ,\ 0 < t < 1\ ,$$

which indicates $k(t)$ is increasing and $k(t) \geq \lim_{t\to 0+} k(t) = 1$. Since $(1-t)^2$ is decreasing for $0 < t < 1$, $\mathrm{d}h/\mathrm{d}t$ is increasing. Therefore, h is convex. Because

$$\lim_{t\to 0+} \frac{\mathrm{d}h}{\mathrm{d}t}(t) = -\infty\ ,\ \lim_{t\to 1-} \frac{\mathrm{d}h}{\mathrm{d}t}(t) = +\infty\ ,$$

h has a unique minimum and it can be calculated by finding the zero point of the increasing function $k(t), 0 < t < 1$ using a bisection line search. Therefore, we have the following conclusion about the robust chance constraint under KL divergence:

Theorem 5.14 (Proposition 4 in Jiang and Guan 2016) *Assume $\emptyset \neq S(x) \subsetneq \Omega$ and $\beta \in (0, 1)$. The robust chance constraint under KL divergence:*

$$\inf_{\mathbb{P}\in\mathcal{P}_{\mathrm{KL}}(\Omega,\mathbb{P}_0,d_0)} \mathbb{P}(S(x)) > \beta\ ,$$

is equivalent to:

$$\mathbb{P}_0(S(x)) \geq \min_{0<t<1} \frac{1-\exp(-d_0)t^{\beta}}{1-t}\ ,$$

where the right side is to minimize a 1-dimensional convex function and this optimization problem can be solved using bisection line search.

5.4.3 Total Variation Distance

For total variation distance, $l_f = 1/2$, and the conjugate function

$$f^*(s) = \begin{cases} -\frac{1}{2}\ , & \text{if } s \leq -\frac{1}{2}\ , \\ s\ , & \text{if } -\frac{1}{2} \leq s \leq \frac{1}{2}\ , \\ +\infty\ , & \text{if } s > \frac{1}{2}\ . \end{cases}$$

In this case, problem (5.45) can be written as:

$$\min_{a,b:0<a\leq b+\frac{1}{2}} \frac{d_0 + b - a + \beta a + f^*(a-b)}{f^*(a-b) - f^*(-b)}\ . \tag{5.46}$$

Since $0 < a \leq b + 1/2$ and f^* is nondecreasing, $f^*(-b) \leq f^*(a-b) \leq f^*(1/2) = 1/2$. According to the discussion before Theorem 5.13, $f^*(a-b) > f^*(-b)$.

Then there are only two possible cases: 1. $f^*(-b) = -b < a - b = f^*(a-b)$; 2. $f^*(-b) = -1/2 < a - b = f^*(a-b)$. We check them one by one.

1. If $f^*(-b) = -b < a - b = f^*(a-b)$, then $-1/2 \le -b < a - b \le 1/2$ and

$$(5.46) = \min_{b \le \frac{1}{2}, 0 < a \le b + \frac{1}{2}} \frac{d_0 + \beta a}{a} = \min_{0 < a \le 1} \left\{ \beta + \frac{d_0}{a} \right\} = \beta + d_0 \; .$$

2. If $f^*(-b) = -1/2 < a - b = f^*(a-b)$, then $-b \le -1/2 < a - b \le 1/2$ and

$$\begin{aligned}(5.46) &= \min_{b \ge \frac{1}{2}, b - \frac{1}{2} < a \le b + \frac{1}{2}} \frac{d_0 + \beta a}{a - b + \frac{1}{2}} \\ &= \min_{c \ge 0, 0 < a - c \le 1} \left\{ \beta + \frac{d_0 + \beta c}{a - c} \right\} \\ &= \min_{c \ge 0} \{ \beta + d_0 + \beta c \} = \beta + d_0 \; ,\end{aligned}$$

where $c := b - 1/2$.

Combining the above two cases, we have the following conclusion:

Theorem 5.15 (Proposition 3 in Jiang and Guan 2016) *Assume* $\emptyset \ne S(x) \subsetneq \Omega$ *and* $\beta \in (0,1)$. *The robust chance constraint under total variation distance*

$$\inf_{\mathbb{P} \in \mathcal{P}_{\mathrm{TV}}(\Omega, \mathbb{P}_0, d_0)} \mathbb{P}(S(x)) > \beta \iff \mathbb{P}_0(S(x)) \ge \beta + d_0 \; .$$

References

Bayraksan G, Love DK (2015) Data-driven stochastic programming using phi-divergences. INFORMS Tutor Oper Res 1–19

Ben-Tal A, den Hertog D, De Waegenaere A et al (2013) Robust solutions of optimization problems affected by uncertain probabilities. Manag Sci 59(2):341–357

Hu Z, Hong LJ (2012) Kullback-Leibler divergence constrained distributionally robust optimization. Available at Optimization Online. https://optimization-online.org/wp-content/uploads/2012/11/3677.pdf[2023-11-12]

Jiang R, Guan Y (2016) Data-driven chance constrained stochastic program. Math Program 158:291–327

Kullback S, Leibler RA (1951) On information and sufficiency. Ann Math Stat 22(1):79–86

MacKay DJ (2003) Information theory, inference, and learning algorithms. Cambridge University Press, Cambridge

Pardo L (2005) Statistical inference based on divergence measures. Chapman and Hall/CRC, New York

Parzen E (1962) On estimation of a probability density function and mode. Ann Math Stat 33(3):1065–1076

Tsybakov AB (2009) Introduction to nonparametric estimation. Springer Series in Statistics, Springer, New York, NY

Chapter 6
Wasserstein-Distance Distributionally Robust Optimization

Abstract Wasserstein distance can measure the distance between probability distributions and it is closely related to the optimal transport problem. This chapter focuses on Wasserstein-distance distributionally robust optimization. We first provide an overview of Wasserstein distance to establish an intuition about what a Wasserstein-distance ambiguity set should be like. Subsequently, Wasserstein-distance ambiguity sets of discrete and general probability distributions are both defined, with discussions about the parameter choice. Finally, the reformulation methods of worst-case expectations and robust chance constraints are introduced.

6.1 Wasserstein Distance

Wasserstein distance was first introduced by Kantorovich (1960), who was awarded the Nobel Memorial Prize in Economic Sciences in 1975. Wasserstein distance is also called the Kantorovich–Rubinstein metric. In this section, we introduce the definition of Wasserstein distance, its origin from the optimal transport problem, Kantorovich duality, its relation to f-divergence, convergence results, and Wasserstein distance between 1-dimensional probability distributions.

6.1.1 Definition

Let ω be an M-dimensional random variable on sample space $(\Omega, \mathcal{F})$, where $\Omega \subset \mathbb{R}^M$. In practical problems, Ω is usually bounded. Denote the set of probability distributions that has a finite pth order moment on Ω by $\mathcal{M}_p(\Omega)$, where $p \in [1, +\infty)$. Suppose $\mathbb{P}_1, \mathbb{P}_2 \in \mathcal{M}_p(\Omega)$. Recall the definition of joint probability distribution in Definition 2.34. Let $\mathcal{J}(\mathbb{P}_1, \mathbb{P}_2)$ be the set of joint probability distributions with marginal probability distributions $\mathbb{P}_1$ and $\mathbb{P}_2$, i.e.,

$$\mathcal{J}(\mathbb{P}_1, \mathbb{P}_2) := \left\{ \mathbb{P}_{1,2} \in \mathcal{M}(\Omega \times \Omega) \mid \begin{array}{l} \mathbb{P}_{1,2}(S, \Omega) = \mathbb{P}_1(S) \ , \ \forall S \in \mathcal{F} \ , \\ \mathbb{P}_{1,2}(\Omega, S) = \mathbb{P}_2(S) \ , \ \forall S \in \mathcal{F} \end{array} \right\} , \tag{6.1}$$

R. Xie and W. Wei, *Distributionally Robust Optimization and its Applications in Power System Energy Storage Sizing*,
https://doi.org/10.1007/978-981-97-2566-3_6

where $\mathcal{M}(\Omega \times \Omega)$ is the set of probability distributions on $\Omega \times \Omega$.

Now we are ready to define Wasserstein distance.

Definition 6.1 Assume $\mathbb{P}_1, \mathbb{P}_2 \in \mathcal{M}_p(\Omega)$ and $p \in [1, +\infty)$, then the *Wasserstein p-distance* $d_{\mathrm{W},p}(\mathbb{P}_1, \mathbb{P}_2)$ between $\mathbb{P}_1$ and $\mathbb{P}_2$ with norm $\|\cdot\|$ is defined as follows:

$$d_{\mathrm{W},p}(\mathbb{P}_1, \mathbb{P}_2) := \left(\inf_{\mathbb{P}_{1,2} \in \mathcal{J}(\mathbb{P}_1, \mathbb{P}_2)} \int_{\Omega \times \Omega} \|\omega' - \omega''\|^p \cdot \mathrm{d}\mathbb{P}_{1,2}(\omega', \omega'') \right)^{\frac{1}{p}} . \tag{6.2}$$

According to Definition 6.1, $d_{\mathrm{W},p}(\mathbb{P}_1, \mathbb{P}_2)$ is the $1/p$ power of the optimal value of an optimization problem, where the decision variable is the joint probability distribution $\mathbb{P}_{1,2}$ and the goal is to minimize an integral. Actually, the integral can be written as an expectation under the joint probability distribution, i.e., $\mathbb{E}^{\mathbb{P}_{1,2}}[\|\omega' - \omega''\|^p]$ with (ω', ω'') following the probability distribution $\mathbb{P}_{1,2}$. Lemma 6.1 easily follows from Definition 6.1.

Lemma 6.1 *Wasserstein p-distance in Definition 6.1 has a nonnegative and finite value.*

Proof Because the value of a norm is nonnegative, $d_{\mathrm{W},p}(\mathbb{P}_1, \mathbb{P}_2) \geq 0$. The product distribution $\mathbb{P}_1 \times \mathbb{P}_2 \in \mathcal{J}(\mathbb{P}_1, \mathbb{P}_2)$, so by Minkowski inequality,

$$\begin{aligned}
d_{\mathrm{W},p}(\mathbb{P}_1, \mathbb{P}_2) &\leq \left(\int_{\Omega \times \Omega} \|\omega' - \omega''\|^p \cdot \mathrm{d}(\mathbb{P}_1 \times \mathbb{P}_2)(\omega', \omega'') \right)^{\frac{1}{p}} \\
&\leq \left(\int_{\Omega \times \Omega} \|\omega'\|^p \mathrm{d}(\mathbb{P}_1 \times \mathbb{P}_2)(\omega', \omega'') \right)^{\frac{1}{p}} \\
&\quad + \left(\int_{\Omega \times \Omega} \|\omega''\|^p \mathrm{d}(\mathbb{P}_1 \times \mathbb{P}_2)(\omega', \omega'') \right)^{\frac{1}{p}} \\
&= \left(\int_{\Omega} \|\omega'\|^p \mathrm{d}\mathbb{P}_1(\omega') \right)^{\frac{1}{p}} + \left(\int_{\Omega} \|\omega''\|^p \mathrm{d}\mathbb{P}_2(\omega'') \right)^{\frac{1}{p}} < +\infty .
\end{aligned}$$

Therefore, $d_{\mathrm{W},p}(\mathbb{P}_1, \mathbb{P}_2)$ is nonnegative and finite. □

Remark 6.1 In fact, the infimum in (6.2) can always be attained by some $\mathbb{P}_{1,2} \in \mathcal{J}(\mathbb{P}_1, \mathbb{P}_2)$ using theorems from functional analysis, even if Ω is not bounded. Readers can refer to Chap. 1 in Santambrogio (2015) for the proof.

Example 6.1 (*Wasserstein distance between Dirac distributions*) Let $\mathbb{P}_1 = \delta_a$ and $\mathbb{P}_2 = \delta_b$ be Dirac distributions on $\mathbb{R}^M$ at a and b, respectively. Assume that $\mathbb{P}_{1,2} \in \mathcal{J}(\delta_a, \delta_b)$ is a joint probability distribution, then $\mathbb{P}_{1,2}(\{a\} \times \mathbb{R}^M) = \mathbb{P}_{1,2}(\mathbb{R}^M \times \{b\}) = 1$, which implies $\mathbb{P}_{1,2}(a, b) = 1$. Thus, the Wasserstein p-distance between δ_a and δ_b is:

$$d_{\mathrm{W},p}(\delta_a, \delta_b) = \left(\inf_{\mathbb{P}_{1,2} \in \mathcal{J}(\delta_a, \delta_b)} \int_{\mathbb{R}^M \times \mathbb{R}^M} \|\omega' - \omega''\|^p \mathrm{d}\mathbb{P}_{1,2}(\omega', \omega'') \right)^{\frac{1}{p}} = \|a - b\| .$$

6.1.2 Wasserstein Distance and Optimal Transport Problem

We focus on the Wasserstein 1-distance, which is:

$$d_W(\mathbb{P}_1, \mathbb{P}_2) := \inf_{\mathbb{P}_{1,2} \in \mathcal{J}(\mathbb{P}_1, \mathbb{P}_2)} \int_{\Omega \times \Omega} \|\omega' - \omega''\| \cdot d\mathbb{P}_{1,2}(\omega', \omega'') , \tag{6.3}$$

for $\mathbb{P}_1, \mathbb{P}_2 \in \mathcal{M}_1(\Omega)$. We refer to it as *Wasserstein distance* hereinafter. In (6.3), the value of Wasserstein distance is defined by the optimal value of minimizing an integral by adjusting the joint probability distribution $\mathbb{P}_{1,2}$. In fact, Wasserstein distance has a close relationship to the optimal transport problem. The probability distributions $\mathbb{P}_1$ and $\mathbb{P}_2$ can be regarded as mass distributions. Then the joint probability distribution $\mathbb{P}_{1,2}$ can be interpreted as a way to transform the mass distribution $\mathbb{P}_1$ into $\mathbb{P}_2$. The larger the distance, the higher the cost, and the norm $\|\cdot\|$ can measure the transportation cost of unit mass. Therefore, the integral in (6.3) represents the total transportation cost and Wasserstein distance is the minimum transportation cost between mass distributions $\mathbb{P}_1$ and $\mathbb{P}_2$.

Example 6.2 (*Optimal transport of boxes*) Suppose there are 10 boxes at the positions $1, 2, 3$, and 4. All the boxes are the same. The current distribution of boxes is $\mathbb{P}_1$ shown in Fig. 6.1. We want to transport some boxes so that the distribution becomes $\mathbb{P}_2$, which is also depicted in Fig. 6.1. A transport plan $\mathbb{P}_{1,2}$ is shown in Fig. 6.2. The boxes at the grid point are those that will be transported to the position represented by the column from the position represented by the row. For instance, the 3 boxes at the grid point at row 2 column 1 mean that those 3 boxes are transported from position 2 to position 1. Note that the boxes on the diagonal do not move and incur no costs. The box distribution $\mathbb{P}_{1,2}$ at the grid points can be regarded as a joint probability distribution of $\mathbb{P}_1$ and $\mathbb{P}_2$. If transporting a box from position x to y costs $|x - y|$, then the transport plan in Fig. 6.2 costs 7 in total. The optimal transport problem tries to find a way to minimize the total transport cost and Wasserstein distance is defined as the minimum total transport cost.

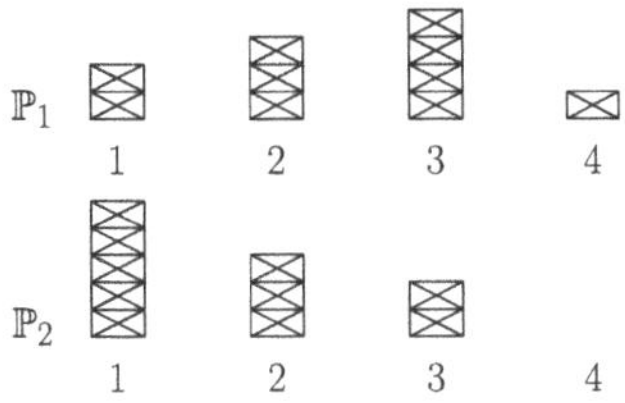

Fig. 6.1 Distributions of 10 identical boxes. $\mathbb{P}_1$ is the current distribution and $\mathbb{P}_2$ is the goal distribution

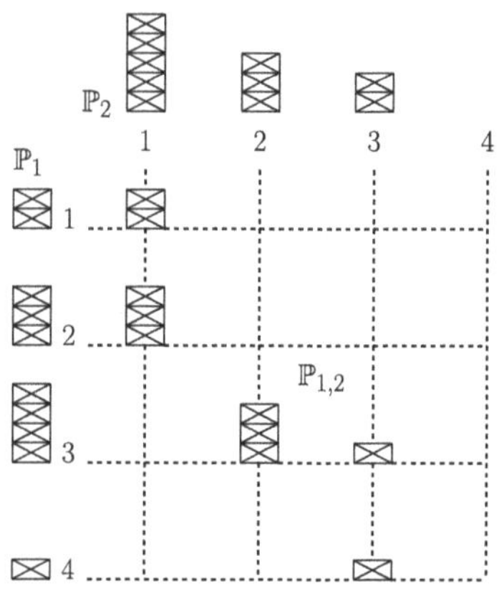

Fig. 6.2 A transport plan for boxes: 3 boxes from position 2 are moved to position 1; 3 boxes from position 3 are moved to position 2; 1 box from position 4 is moved to position 3; 2 boxes at position 1 and 1 box at position 3 do not move

6.1.3 *Kantorovich Duality*

Kantorovich duality (Kantorovich and Rubinshtein 1958) is a kind of optimization problem duality about the problems like that in (6.3). To introduce its idea, we first investigate the discrete probability distributions with a finite number of values.

6.1.3.1 Kantorovich Duality for Discrete Probability Distributions

Suppose $\mathbb{P}_1$ and $\mathbb{P}_2$ are discrete probability distributions on sample space Ω and take values in $S := \{\omega_1, \omega_2, \ldots, \omega_N\} \subset \Omega$. For any joint probability distribution $\mathbb{P}_{1,2} \in \mathcal{J}(\mathbb{P}_1, \mathbb{P}_2)$, we have $\mathbb{P}_{1,2}((\Omega \setminus S) \times \Omega) = \mathbb{P}_1(\Omega \setminus S) = 0$ and $\mathbb{P}_{1,2}(\Omega \times (\Omega \setminus S)) = \mathbb{P}_2(\Omega \setminus S) = 0$, which implies:

$$\begin{aligned} \mathbb{P}_{1,2}(S \times S) &= 1 - \mathbb{P}_{1,2}((\Omega \times \Omega) \setminus (S \times S)) \\ &= 1 - \mathbb{P}_{1,2}(((\Omega \setminus S) \times \Omega) \cup (\Omega \times (\Omega \setminus S))) = 1 \,. \end{aligned}$$

Note that $S \times S = \{(\omega_i, \omega_j) \mid i, j = 1, 2, \ldots, N\}$ is also a finite set, so the joint probability distribution $\mathbb{P}_{1,2} \in \mathcal{J}(\mathbb{P}_1, \mathbb{P}_2)$ must be discrete and can be represented by its probability values at the points in $S \times S$. The optimization problem in (6.3) then becomes:

$$\inf_{\mathbb{P}_{1,2}\in\mathcal{J}(\mathbb{P}_1,\mathbb{P}_2)} \int_{\Omega\times\Omega} \|\omega' - \omega''\| \cdot \mathrm{d}\mathbb{P}_{1,2}(\omega', \omega'') \tag{6.5a}$$

$$= \begin{cases} \min\limits_{\mathbb{P}_{1,2}} \sum_{i=1}^{N}\sum_{j=1}^{N} \|\omega_i - \omega_j\| \cdot \mathbb{P}_{1,2}(\omega_i, \omega_j) \\ \text{s.t.} \sum_{j=1}^{N} \mathbb{P}_{1,2}(\omega_i, \omega_j) = \mathbb{P}_1(\omega_i) : \lambda_i \ , \ i = 1, 2, \ldots, N \ , \\ \sum_{i=1}^{N} \mathbb{P}_{1,2}(\omega_i, \omega_j) = \mathbb{P}_2(\omega_j) : \mu_j \ , \ j = 1, 2, \ldots, N \ , \\ \mathbb{P}_{1,2}(\omega_i, \omega_j) \geq 0 \ , \ i, j = 1, 2, \ldots, N \ , \end{cases} \tag{6.5b}$$

$$= \begin{cases} \max\limits_{\lambda,\mu} \sum_{i=1}^{N} \mathbb{P}_1(\omega_i) \cdot \lambda_i + \sum_{j=1}^{N} \mathbb{P}_2(\omega_j) \cdot \mu_j \\ \text{s.t. } \lambda_i + \mu_j \leq \|\omega_i - \omega_j\| : \mathbb{P}_{1,2}(\omega_i, \omega_j) \ , \ i, j = 1, 2, \ldots, N \ . \end{cases} \tag{6.5c}$$

In (6.5b), the decision variables are $\mathbb{P}_{1,2}(\omega_i, \omega_j)$, $i, j = 1, 2, \ldots, N$. The constraints include the nonnegativity and the requirements for marginal probability distributions. λ_i and μ_j are dual variables. The constraint $\sum_{i=1}^{N}\sum_{j=1}^{N} \mathbb{P}_{1,2}(\omega_i, \omega_j) = 1$ is omitted because it follows from the first constraint and the fact that $\sum_{i=1}^{N} \mathbb{P}_1(\omega_i) = 1$.

The optimization problem in (6.5b) is a linear programming (LP) problem. It is bounded below since $\|\omega_i - \omega_j\| \geq 0$ and $\mathbb{P}_{1,2}(\omega_i, \omega_j) \geq 0$ for any $i, j = 1, 2, \ldots, N$. It is feasible because the joint probability distribution $\mathbb{P}_{1,2} := \mathbb{P}_1 \times \mathbb{P}_2$, i.e., $\mathbb{P}_{1,2}(\omega_i, \omega_j) = \mathbb{P}_1(\omega_1) \cdot \mathbb{P}_2(\omega_2)$, $i, j = 1, 2, \ldots, N$, satisfies all the constraints. Then by the strong duality of LP, (6.5b) equals the optimal value of its dual LP problem, which is in (6.5c). In addition, the optimal values are both attained. The equation between (6.5b) and (6.5c) is the Kantorovich duality in this example.

From the perspective of optimal transport problems, (6.5b) and (6.5c) both have physical meanings (Villani 2003). In (6.5b), the user tries to find the optimal transport plan from the mass distribution $\mathbb{P}_1$ to $\mathbb{P}_2$, where the unit transport cost between ω_i and ω_j is measured by the norm $\|\omega_i - \omega_j\|$. Imagine that a service operator offers the user a contract, where the user knows nothing about how the service operator transports the mass but has to pay λ_i for the departure of mass at ω_i and μ_j for the arrival of mass at ω_j. The service operator decides the prices λ_i and μ_j (not necessarily nonnegative) so that $\lambda_i + \mu_j \leq \|\omega_i - \omega_j\|$ for all $i, j = 1, 2, \ldots, N$. Then the offer is beneficial for the user because the service always costs no more than the original transport plan. Therefore, (6.5b) $\geq$ (6.5c) naturally. The opposite direction means that the service operator has a way to determine the prices λ_i and μ_j so that the total income equals the cost of the optimal transport plan of the user.

Let us take a closer look at λ_i and μ_j. According to the objective function of (6.5c) and the fact that probability values are nonnegative, larger λ_i and μ_j are better. If we fix λ_i, $i = 1, 2, \ldots, N$, then by the constraint $\lambda_i + \mu_j \leq \|\omega_i - \omega_j\|$, $\mu_j = \min\{\|\omega_k - \omega_j\| - \lambda_k \mid k = 1, 2, \ldots, N\}$ is optimal. If we take

$i = j$, then $\mu_j \le -\lambda_j, j = 1, 2, \ldots, N$. Similarly, $\lambda_i = \min\{\|\omega_i - \omega_k\| - \mu_k \mid k = 1, 2, \ldots, N\}$. Assume $\lambda_j = \|\omega_j - \omega_{j_0}\| - \mu_{j_0}$. Then

$$\begin{aligned}\lambda_i - \lambda_j &= \min\{\|\omega_i - \omega_k\| - \mu_k \mid k = 1, 2, \ldots, N\} - (\|\omega_j - \omega_{j_0}\| - \mu_{j_0}) \\ &\le \|\omega_i - \omega_{j_0}\| - \mu_{j_0} - \|\omega_j - \omega_{j_0}\| + \mu_{j_0} \\ &= \|\omega_i - \omega_{j_0}\| - \|\omega_j - \omega_{j_0}\| \le \|\omega_i - \omega_j\| ,\end{aligned}$$

using the triangle inequality property of norm. Combining $\mu_j \le -\lambda_j$ and $\lambda_i - \lambda_j \le \|\omega_i - \omega_j\|$, we see that $\mu_j = -\lambda_j,\ j = 1, 2, \ldots, N$ belongs to the best feasible choice in (6.5c). In this regard,

$$\begin{aligned}&\inf_{\mathbb{P}_{1,2} \in \mathcal{J}(\mathbb{P}_1, \mathbb{P}_2)} \int_{\Omega \times \Omega} \|\omega' - \omega''\| \cdot \mathrm{d}\mathbb{P}_{1,2}(\omega', \omega'') = (6.5\mathrm{c}) \\ &= \begin{cases} \max\limits_{\lambda} \sum\limits_{i=1}^{N} \mathbb{P}_1(\omega_i) \cdot \lambda_i - \sum\limits_{j=1}^{N} \mathbb{P}_2(\omega_j) \cdot \lambda_j \\ \text{s.t. } \lambda_i - \lambda_j \le \|\omega_i - \omega_j\| ,\ i, j = 1, 2, \ldots, N . \end{cases}\end{aligned}$$

Exercise 6.1 Find the primal- and dual-optimal solutions of the example optimal transport problem in Fig. 6.2 and then explain the physical meaning.

6.1.3.2 Kantorovich Duality for General Probability Distributions

Kantorovich duality applies to a wide range of cases (Villani 2003). Here we introduce one of the cases.

Theorem 6.1 (Kantorovich duality) *Assume $\mathbb{P}_1, \mathbb{P}_2 \in \mathcal{M}_1(\Omega)$ with measurable $\Omega \subset \mathbb{R}^M$ and $\|\cdot\|$ is a norm on $\mathbb{R}^M$. Then*

$$d_{\mathrm{W}}(\mathbb{P}_1, \mathbb{P}_2)$$

$$= \inf_{\mathbb{P}_{1,2} \in \mathcal{J}(\mathbb{P}_1, \mathbb{P}_2)} \int_{\Omega \times \Omega} \|\omega' - \omega''\| \cdot \mathrm{d}\mathbb{P}_{1,2}(\omega', \omega'') \tag{6.6a}$$

$$= \begin{cases} \sup\limits_{f,g} \int_{\Omega} f(\omega')\mathrm{d}\mathbb{P}_1(\omega') + \int_{\Omega} g(\omega'')\mathrm{d}\mathbb{P}_2(\omega'') \\ s.t.\ f(\omega') + g(\omega'') \le \|\omega' - \omega''\| ,\ \forall \omega', \omega'' \in \Omega , \\ \qquad f : \Omega \to \mathbb{R} \text{ and } g : \Omega \to \mathbb{R} \text{ are continuous and bounded} , \end{cases} \tag{6.6b}$$

$$= \begin{cases} \sup\limits_{f} \int_{\Omega} f(\omega')\mathrm{d}\mathbb{P}_1(\omega') - \int_{\Omega} f(\omega'')\mathrm{d}\mathbb{P}_2(\omega'') \\ s.t.\ f(\omega') - f(\omega'') \le \|\omega' - \omega''\| ,\ \forall \omega', \omega'' \in \Omega . \end{cases} \tag{6.6c}$$

Remark 6.2 The constraint $f(\omega') - f(\omega'') \le \|\omega' - \omega''\|, \forall \omega', \omega'' \in \Omega$ is equivalent to $|f(\omega') - f(\omega'')| \le \|\omega' - \omega''\|, \forall \omega', \omega'' \in \Omega$, which exactly means that the function f is 1-Lipschitz continuous. When Ω is assumed to be bounded, Lipschitz continuous functions on Ω must be bounded. Therefore, the optimization problem in (6.6c) tries to find the supremum of the difference between the expectations of a 1-Lipschitz continuous function under probability distributions $\mathbb{P}_1$ and $\mathbb{P}_2$.

The rigorous proof of (6.6a) = (6.6b) in Theorem 6.1 uses conclusions from conic duality and functional analysis. Here we explain the basic idea under the assumption that Ω is bounded. Relation (6.6) still holds when Ω is not bounded (Villani 2003). Readers can refer to Santambrogio (2015) or Villani (2003) for proof details.

Recall from Remark 2.1 of Lemma 2.5 in Sect. 2.2.3 that $\mathbb{P}_{1,2} \in \mathcal{J}(\mathbb{P}_1, \mathbb{P}_2)$ if and only if for all bounded and continuous functions $f, g : \Omega \to \mathbb{R}$, we have:

$$\int_{\Omega\times\Omega} (f(\omega) + g(\omega'))\mathrm{d}\mathbb{P}_{1,2}(\omega', \omega'') = \int_{\Omega} f(\omega')\mathrm{d}\mathbb{P}_1(\omega') + \int_{\Omega} g(\omega'')\mathrm{d}\mathbb{P}_2(\omega'') .$$

It follows that when $\mathbb{P}_{1,2}$ is a measure (not necessarily a probability distribution) defined on $\Omega \times \Omega$,

$$\begin{aligned} \sup_{f,g\in C_b(\Omega)} &\left\{ \int_{\Omega} f(\omega')\mathrm{d}\mathbb{P}_1(\omega') + \int_{\Omega} g(\omega'')\mathrm{d}\mathbb{P}_2(\omega'') \right. \\ &\left. - \int_{\Omega\times\Omega} (f(\omega') + g(\omega''))\mathrm{d}\mathbb{P}_{1,2}(\omega', \omega'') \right\} = \begin{cases} 0 , & \text{if } \mathbb{P}_{1,2} \in \mathcal{J}(\mathbb{P}_1, \mathbb{P}_2) , \\ +\infty , & \text{otherwise} , \end{cases} \end{aligned} \tag{6.7}$$

where $C_b(\Omega)$ is the set of bounded and continuous functions defined on Ω. Note that (6.7) is a characteristic function regarded as a function of $\mathbb{P}_{1,2}$. Then we can equivalently transform the optimal transport problem (6.6a) into:

$$\begin{aligned} \inf_{\mathbb{P}_{1,2}} &\left\{ \int_{\Omega\times\Omega} \|\omega' - \omega''\| \cdot \mathrm{d}\mathbb{P}_{1,2}(\omega', \omega'') + \sup_{f,g\in C_b(\Omega)} \left\{ \int_{\Omega} f(\omega')\mathrm{d}\mathbb{P}_1(\omega') \right. \right. \\ &\left. \left. + \int_{\Omega} g(\omega'')\mathrm{d}\mathbb{P}_2(\omega'') - \int_{\Omega\times\Omega} (f(\omega') + g(\omega''))\mathrm{d}\mathbb{P}_{1,2}(\omega', \omega'') \right\} \right\} , \end{aligned} \tag{6.8}$$

where the constraint $\mathbb{P}_{1,2} \in \mathcal{J}(\mathbb{P}_1, \mathbb{P}_2)$ is eliminated by taking advantage of (6.7). Bringing forward $\sup_{f,g\in C_b(\Omega)}$ and merging the integrals under the measure $\mathbb{P}_{1,2}$, (6.8) is then equivalent to:

$$\begin{aligned} \inf_{\mathbb{P}_{1,2}} \sup_{f,g\in C_b(\Omega)} &\left\{ \int_{\Omega} f(\omega')\mathrm{d}\mathbb{P}_1(\omega') + \int_{\Omega} g(\omega'')\mathrm{d}\mathbb{P}_2(\omega'') \right. \\ &\left. + \int_{\Omega\times\Omega} \left(\|\omega' - \omega''\| - f(\omega') - g(\omega'')\right) \mathrm{d}\mathbb{P}_{1,2}(\omega', \omega'') \right\} . \end{aligned} \tag{6.9}$$

Problem (6.9) is an "inf sup" problem. We want to exchange the order so that it becomes a "sup inf" problem. Equality does not always hold for such an exchange, but for the case we considered, it indeed holds and that is where conic duality and functional analysis are needed. Here we take it for granted and then

$$
\begin{aligned}
(6.9) = \sup_{f,g \in C_b(\Omega)} \inf_{\mathbb{P}_{1,2}} & \left\{ \int_{\Omega} f(\omega')\mathrm{d}\mathbb{P}_1(\omega') + \int_{\Omega} g(\omega'')\mathrm{d}\mathbb{P}_2(\omega'') \right. \\
& \left. + \int_{\Omega \times \Omega} \left(\|\omega' - \omega''\| - f(\omega') - g(\omega'') \right) \mathrm{d}\mathbb{P}_{1,2}(\omega', \omega'') \right\} \qquad (6.10a) \\
= \sup_{f,g \in C_b(\Omega)} & \left\{ \int_{\Omega} f(\omega')\mathrm{d}\mathbb{P}_1(\omega') + \int_{\Omega} g(\omega'')\mathrm{d}\mathbb{P}_2(\omega'') \right. \\
& \left. + \inf_{\mathbb{P}_{1,2}} \int_{\Omega \times \Omega} \left(\|\omega' - \omega''\| - f(\omega') - g(\omega'') \right) \mathrm{d}\mathbb{P}_{1,2}(\omega', \omega'') \right\} . \qquad (6.10b)
\end{aligned}
$$

We observe that

$$
\begin{aligned}
& \inf_{\mathbb{P}_{1,2}} \int_{\Omega \times \Omega} \left(\|\omega' - \omega''\| - f(\omega') - g(\omega'') \right) \mathrm{d}\mathbb{P}_{1,2}(\omega', \omega'') \\
&= \begin{cases} 0 , & \text{if } \|\omega' - \omega''\| - f(\omega') - g(\omega'') \geq 0 , \ \forall (\omega', \omega'') \in \Omega \times \Omega , \\ -\infty , & \text{otherwise} . \end{cases}
\end{aligned}
$$

Therefore, f and g in the optimal solution of problem (6.10b) must satisfy $\|\omega' - \omega''\| - f(\omega') - g(\omega'') \geq 0, \forall (\omega', \omega'') \in \Omega \times \Omega$. In other words, $f(\omega') + g(\omega'') \leq \|\omega' - \omega''\|, \forall \omega', \omega'' \in \Omega$. Use this constraint and eliminate the infimum term in (6.10b), then we arrive at the optimization problem in (6.6b). This is how (6.6a) = (6.6b) can be proved in Theorem 6.1.

***Proof of* (6.6b) = (6.6c) *in Theorem* 6.1 *with bounded* $\boldsymbol{\Omega}$** We prove (6.6b) $\geq$ (6.6c) and (6.6b) $\leq$ (6.6c), respectively.

1. Prove that (6.6b) $\geq$ (6.6c): Let $f : \Omega \to \mathbb{R}$ be a feasible solution of problem (6.6c), which means that f satisfies $f(\omega') - f(\omega'') \leq \|\omega' - \omega''\|, \forall \omega', \omega'' \in \Omega$. Then $f(\omega'') - f(\omega') \leq \|\omega'' - \omega'\| = \|\omega' - \omega''\|$ and thus $|f(\omega') - f(\omega'')| \leq \|\omega' - \omega''\|, \forall \omega', \omega'' \in \Omega$, which is the definition of the function $f : \Omega \to \mathbb{R}$ being 1-Lipschitz continuous. By Propositions 2.5 and 2.6, f is continuous and bounded. Define function $g : \Omega \to \mathbb{R}$ by $g(\omega) := -f(\omega), \forall \omega \in \Omega$. Then g is also continuous and bounded and additionally, we have:

$$
f(\omega') + g(\omega'') = f(\omega') - f(\omega'') \leq \|\omega' - \omega''\| , \ \forall \omega', \omega'' \in \Omega ,
$$

by the assumption of f. Thus, f and g constitute a feasible solution for the optimization problem (6.6b). Moreover, the two objective functions coincide when $g = -f$. Therefore, the optimal value of (6.6c) is the objective value of a feasible solution of (6.6b), which results in (6.6b) $\geq$ (6.6c).

2. Prove that (6.6b) $\leq$ (6.6c): We start from a feasible solution $f, g : \Omega \to \mathbb{R}$ of problem (6.6b), then $f(\omega') + g(\omega'') \leq \|\omega' - \omega''\|, \forall \omega', \omega'' \in \Omega$. It follows that:

$$
\begin{aligned}
f(\omega) &\leq \inf_{\omega' \in \Omega} \{\|\omega - \omega'\| - g(\omega')\} \ , \ \forall \omega \in \Omega \ , \\
g(\omega) &\leq \inf_{\omega' \in \Omega} \{\|\omega - \omega'\| - f(\omega')\} \ , \ \forall \omega \in \Omega \ .
\end{aligned}
$$

According to the monotonicity of expectation in Proposition 2.4, a larger f leads to a larger objective function value in (6.6c). Thus, we assume $f(\omega) = \inf_{\omega' \in \Omega}\{\|\omega - \omega'\| - g(\omega')\}, \forall \omega \in \Omega$ without loss of generality.

Let $\omega' = \omega''$, then $g(\omega'') \leq -f(\omega''), \forall \omega'' \in \Omega$. Substituting it into the objective function, we have:

$$
\int_\Omega f(\omega')\mathrm{d}\mathbb{P}_1(\omega') + \int_\Omega g(\omega'')\mathrm{d}\mathbb{P}_2(\omega'') \leq \int_\Omega f(\omega')\mathrm{d}\mathbb{P}_1(\omega') - \int_\Omega f(\omega'')\mathrm{d}\mathbb{P}_2(\omega'') \ ,
$$

where the right side equals the objective function of problem (6.6c). It remains to show that the function f is feasible for problem (6.6c).

Fix ω' and ω'' in Ω. According to $f(\omega'') = \inf_{\omega \in \Omega}\{\|\omega'' - \omega\| - g(\omega)\}$, for any $\varepsilon > 0$, there is $\omega_\varepsilon \in \Omega$ such that $\|\omega'' - \omega_\varepsilon\| - g(\omega_\varepsilon) < f(\omega'') + \varepsilon$. Thus,

$$
\begin{aligned}
f(\omega') - f(\omega'') &< \inf_{\omega \in \Omega}\{\|\omega' - \omega\| - g(\omega)\} + (\varepsilon - \|\omega'' - \omega_\varepsilon\| + g(\omega_\varepsilon)) \\
&\leq \|\omega' - \omega_\varepsilon\| - g(\omega_\varepsilon) + \varepsilon - \|\omega'' - \omega_\varepsilon\| + g(\omega_\varepsilon) \\
&= \|\omega' - \omega_\varepsilon\| - \|\omega'' - \omega_\varepsilon\| + \varepsilon \\
&\leq \|\omega' - \omega''\| + \varepsilon \ ,
\end{aligned}
$$

where the last inequality follows from the triangle inequality of norm. Let $\varepsilon \to 0$, then we have $f(\omega') - f(\omega'') \leq \|\omega' - \omega''\|$. Because $\omega', \omega'' \in \Omega$ are arbitrary, f is a feasible solution of problem (6.6c) and hence (6.6b) $\leq$ (6.6c). □

Remark 6.3 From the above proof, we can see that the conclusion of Theorem 6.1 still holds if the norm $\|\omega' - \omega''\|$ is replaced by any distance function $d(\omega', \omega'')$ on Ω.

Exercise 6.2 Compare the statement of Theorem 6.1 with the case of discrete probability distributions and show that the Kantorovich duality in (6.5b) for discrete probability distributions can be derived from Theorem 6.1.

6.1.4 Wasserstein Distance is a Distance Function

Theorem 6.2 says that the Wasserstein distance defined in (6.3) truly satisfies the requirements to be a distance function. Therefore, Wasserstein distance can measure the difference between any two probability distributions in $\mathcal{M}_1(\Omega)$.

Theorem 6.2 *Wasserstein distance in* (6.3) *is a distance function on the space* $\mathcal{M}_1(\Omega)$ *of probability distributions.*

Proof 1. (Nonnegativity) By Lemma 6.1, $d_W(\mathbb{P}_1, \mathbb{P}_2)$ always have a nonnegative and finite value.

2. (Point-separating) We need to prove $d_W(\mathbb{P}_1, \mathbb{P}_2) = 0$ if and only if $\mathbb{P}_1 = \mathbb{P}_2$ as probability distributions.

(a) When $\mathbb{P}_1 = \mathbb{P}_2$, we have $\mathbb{P}_1(S) = \mathbb{P}_2(S)$ for any measurable $S \subset \Omega$. Define a measure $\mathbb{P}_{1,2}$ on $\Omega \times \Omega$ as follows:

$$\mathbb{P}_{1,2}(A) := \mathbb{P}_1(\{\omega \in \Omega \mid (\omega, \omega) \in A\}) , \ \forall \text{ measurable } A \subset \Omega \times \Omega . \tag{6.11}$$

Therefore, $\mathbb{P}_{1,2}(A) \geq 0$ for any measurable A in $\Omega \times \Omega$ and $\mathbb{P}_{1,2}(\Omega \times \Omega) = \mathbb{P}_1(\Omega) = 1$. Suppose $S_1, S_2, \ldots, S_n, \ldots$ are disjoint measurable sets in $\Omega \times \Omega$. For $n \neq n'$, because $S_n \cap S_{n'} = \varnothing$, we have:

$$\begin{aligned} &\{\omega \in \Omega \mid (\omega, \omega) \in S_n\} \cap \{\omega \in \Omega \mid (\omega, \omega) \in S_{n'}\} \\ =& \{\omega \in \Omega \mid (\omega, \omega) \in S_n \cap S_{n'}\} = \varnothing . \end{aligned}$$

Thus,

$$\begin{aligned} \mathbb{P}_{1,2}\left(\bigcup_{n=1}^{+\infty} S_n\right) &= \mathbb{P}_1\left(\left\{\omega \in \Omega \mid (\omega, \omega) \in \bigcup_{n=1}^{+\infty} S_n\right\}\right) \\ &= \mathbb{P}_1\left(\bigcup_{n=1}^{+\infty} \{\omega \in \Omega \mid (\omega, \omega) \in S_n\}\right) \\ &= \sum_{n=1}^{+\infty} \mathbb{P}_1\left(\{\omega \in \Omega \mid (\omega, \omega) \in S_n\}\right) = \sum_{n=1}^{+\infty} \mathbb{P}_{1,2}(S_n) . \end{aligned}$$

Therefore, $\mathbb{P}_{1,2}$ is a probability distribution on $\Omega \times \Omega$.

For any measurable $S \subset \Omega$, $\mathbb{P}_{1,2}(S \times \Omega) = \mathbb{P}_1(\{\omega \in \Omega \mid (\omega, \omega) \in S \times \Omega\}) = \mathbb{P}_1(S)$. Similarly, $\mathbb{P}_{1,2}(\Omega \times S) = \mathbb{P}_1(S)$. Hence, $\mathbb{P}_{1,2}$ is a joint probability distribution of $\mathbb{P}_1$ and $\mathbb{P}_2$. Therefore,

$$0 \leq d_W(\mathbb{P}_1, \mathbb{P}_2) \leq \int_{\Omega \times \Omega} \|\omega' - \omega''\| \cdot \mathrm{d}\mathbb{P}_{1,2}(\omega', \omega'') .$$

According to the definition of $\mathbb{P}_{1,2}$ in (6.11),

$$\begin{aligned} &\mathbb{P}_{1,2}(\{(\omega', \omega'') \in \Omega \times \Omega \mid \|\omega' - \omega''\| > 0\}) \\ =& \mathbb{P}_1(\{\omega \in \Omega \mid \|\omega - \omega\| > 0\}) = \mathbb{P}_1(\varnothing) = 0 , \end{aligned}$$

so $\int_{\Omega \times \Omega} \|\omega' - \omega''\| \cdot \mathrm{d}\mathbb{P}_{1,2}(\omega', \omega'') = 0$ by Lemma 2.3 and then $d_W(\mathbb{P}_1, \mathbb{P}_2) = 0$.

(b) Assume $d_W(\mathbb{P}_1, \mathbb{P}_2) = 0$. By Remark 6.1, there is a joint probability distribution $\mathbb{P}_{1,2}$ with marginal probability distributions $\mathbb{P}_1$ and $\mathbb{P}_2$ so that:

$$\int_{\Omega\times\Omega} \|\omega' - \omega''\| \cdot \mathrm{d}\mathbb{P}_{1,2}(\omega', \omega'') = d_W(\mathbb{P}_1, \mathbb{P}_2) = 0 \, .$$

Then according to Proposition 2.10, $\mathbb{P}_{1,2}(\{(\omega', \omega'') \in \Omega \times \Omega \mid \|\omega' - \omega''\| > 0\}) = 0$. Thus, $\mathbb{P}_{1,2}(\{(\omega, \omega) \mid \omega \in \Omega\}) = \mathbb{P}_{1,2}(\{(\omega', \omega'') \in \Omega \times \Omega \mid \omega' = \omega''\}) = 1$.

Since $\mathbb{P}_{1,2}$ is a joint probability distribution of $\mathbb{P}_1$ and $\mathbb{P}_2$, for any measurable $S \subset \Omega$,

$$\begin{aligned}\mathbb{P}_1(S) &= \mathbb{P}_{1,2}(S \times \Omega) = \mathbb{P}_{1,2}((S \times \Omega) \cap \{(\omega, \omega) \mid \omega \in \Omega\}) \\ &= \mathbb{P}_{1,2}(\{(\omega, \omega) \mid \omega \in S\}) = \mathbb{P}_2(S) \, ,\end{aligned}$$

by Lemma 2.2. Since S is arbitrary, $\mathbb{P}_1 = \mathbb{P}_2$ as probability distributions on Ω.

3. (Symmetry) $d_W(\mathbb{P}_1, \mathbb{P}_2)$ is clearly symmetric according to its definition.

4. (Triangle inequality) We need to show the triangle inequality: For any $\mathbb{P}_1, \mathbb{P}_2, \mathbb{P}_3 \in \mathcal{M}_1(\Omega)$, $d_W(\mathbb{P}_1, \mathbb{P}_3) \le d_W(\mathbb{P}_1, \mathbb{P}_2) + d_W(\mathbb{P}_2, \mathbb{P}_3)$.

Consider $\mathbb{P}_{1,2} \in \mathcal{J}(\mathbb{P}_1, \mathbb{P}_2)$ and $\mathbb{P}_{2,3} \in \mathcal{J}(\mathbb{P}_2, \mathbb{P}_3)$. According to Proposition 2.11, we have a joint probability distribution $\mathbb{P}_{1,3} \in \mathcal{J}(\mathbb{P}_1, \mathbb{P}_3)$ defined as follows:

$$\mathbb{P}_{1,3}(S) := \int_S \int_\Omega \mathbb{P}_{1|2}(\mathrm{d}\omega'|\omega'')\mathrm{d}\mathbb{P}_{2,3}(\omega'', \omega''') \, , \ \forall \text{ measurable } S \subset \Omega \times \Omega \, ,$$

where $\mathbb{P}_{1|2}$ is the conditional probability distribution in Definition 2.35. Then

$$\begin{aligned}
&\int_{\Omega\times\Omega} \|\omega' - \omega'''\|\mathrm{d}\mathbb{P}_{1,3}(\omega', \omega''') && (6.12\text{a})\\
&= \int_{\Omega\times\Omega}\int_\Omega \|\omega' - \omega'''\|\mathbb{P}_{1|2}(\mathrm{d}\omega'|\omega'')\mathrm{d}\mathbb{P}_{2,3}(\omega'', \omega''') && (6.12\text{b})\\
&\le \int_{\Omega\times\Omega}\int_\Omega \|\omega' - \omega''\|\mathbb{P}_{1|2}(\mathrm{d}\omega'|\omega'')\mathrm{d}\mathbb{P}_{2,3}(\omega'', \omega''') \\
&\quad + \int_{\Omega\times\Omega}\int_\Omega \|\omega'' - \omega'''\|\mathbb{P}_{1|2}(\mathrm{d}\omega'|\omega'')\mathrm{d}\mathbb{P}_{2,3}(\omega'', \omega''') && (6.12\text{c})\\
&= \int_{\Omega\times\Omega} \|\omega' - \omega''\|\mathbb{P}_{1|2}(\mathrm{d}\omega'|\omega'')\mathbb{P}_{2,3}(\mathrm{d}\omega'', \Omega) \\
&\quad + \int_{\Omega\times\Omega} \|\omega'' - \omega'''\|\mathbb{P}_{1|2}(\Omega|\omega'')\mathrm{d}\mathbb{P}_{2,3}(\omega'', \omega''') && (6.12\text{d})\\
&= \int_{\Omega\times\Omega} \|\omega' - \omega''\|\mathrm{d}\mathbb{P}_{1,2}(\omega', \omega'') + \int_{\Omega\times\Omega} \|\omega'' - \omega'''\|\mathrm{d}\mathbb{P}_{2,3}(\omega'', \omega''') \, . && (6.12\text{e})
\end{aligned}$$

Equation (6.12c) follows from the triangle inequality of norm; from (6.12c) to (6.12d), Fubini's theorem is used so that we can first integrate ω_3 in the first term and integrate ω_1 in the second term; (6.12e) follows from $\mathbb{P}_{1|2}(\mathrm{d}\omega'|\omega'')\mathbb{P}_{2,3}(\mathrm{d}\omega'', \Omega) = \mathbb{P}_{1|2}(\mathrm{d}\omega'|\omega'')\mathbb{P}_2(\mathrm{d}\omega'') = \mathrm{d}\mathbb{P}_{1,2}(\omega', \omega'')$ and the unit mass property of $\mathbb{P}_{1|2}$. Taking infimum over $\mathbb{P}_{1,2} \in \mathcal{J}(\mathbb{P}_1, \mathbb{P}_2)$ and $\mathbb{P}_{2,3} \in \mathcal{J}(\mathbb{P}_2, \mathbb{P}_3)$ in (6.12), we have:

$$\begin{aligned} d_{\mathrm{W}}(\mathbb{P}_1, \mathbb{P}_3) &\le \inf_{\mathbb{P}_{1,2}\in\mathcal{J}(\mathbb{P}_1,\mathbb{P}_2),\mathbb{P}_{2,3}\in\mathcal{J}(\mathbb{P}_2,\mathbb{P}_3)} \int_{\Omega\times\Omega} \|\omega' - \omega'''\| \mathrm{d}\mathbb{P}_{1,3}(\omega', \omega''') \\ &\le \inf_{\mathbb{P}_{1,2}\in\mathcal{J}(\mathbb{P}_1,\mathbb{P}_2)} \int_{\Omega\times\Omega} \|\omega' - \omega''\| \mathrm{d}\mathbb{P}_{1,2}(\omega', \omega'') \\ &\quad + \inf_{\mathbb{P}_{2,3}\in\mathcal{J}(\mathbb{P}_2,\mathbb{P}_3)} \int_{\Omega\times\Omega} \|\omega'' - \omega'''\| \mathrm{d}\mathbb{P}_{2,3}(\omega'', \omega''') \\ &= d_{\mathrm{W}}(\mathbb{P}_1, \mathbb{P}_2) + d_{\mathrm{W}}(\mathbb{P}_2, \mathbb{P}_3) . \end{aligned}$$

Combining the above four properties, we know that Wasserstein distance is a distance function. □

Remark 6.4 As the proof shows, the conclusion of Theorem 6.2 still holds even if the norm $\|\omega' - \omega''\|$ is replaced by a distance function $d(\omega', \omega'')$ in (6.3).

Remark 6.5 Recall that we can regard marginal probability distributions as mass distributions, the joint probability distributions as transport plans between the mass distributions, and Wasserstein distance as the minimum total transport cost. Then Wasserstein distance fulfills the requirements to be a distance quite naturally: (Nonnegativity) The total transport cost must be nonnegative. (Point-separating) If the mass distribution does not change, no transport is needed so the transport cost is zero. Whenever the mass distributions are different, some mass has to be transported, which results in a positive cost. (Symmetry) Assume that we use the optimal transport plan to change mass distribution A into mass distribution B. Then we can reverse the plan and recover the mass so that B is changed into A while the transport cost remains the same. This reveals that there is a symmetry between the original mass distribution and the goal mass distribution. (Triangle inequality) We can put together a transport plan from A to B and another transport plan from B to C. After that, we get a transport plan from A to C and the total transport cost is the summation. Therefore, the optimal plans from A to B and from B to C give a plan from A to C and the minimum transport cost from A to C cannot be larger than the summation.

Remark 6.6 Kantorovich duality in Definition 6.1 gives an alternative way to define Wasserstein distance:

$$\begin{aligned} d_{\mathrm{W}}(\mathbb{P}_1, \mathbb{P}_2) = \sup_f \ & \int_{\Omega} f(\omega')\mathrm{d}\mathbb{P}_1(\omega') - \int_{\Omega} f(\omega'')\mathrm{d}\mathbb{P}_2(\omega'') \\ \text{s.t. } & f(\omega') - f(\omega'') \le \|\omega' - \omega''\| , \ \forall \omega', \omega'' \in \Omega . \end{aligned} \tag{6.13}$$

Nonnegativity and symmetry directly follow from the above definition, which is left as an exercise. Now we prove the triangle inequality in this way. By definition, for any $\varepsilon > 0$, there is a function $f_\varepsilon : \Omega \to \mathbb{R}$ so that $f_\varepsilon(\omega') - f_\varepsilon(\omega'') \le \|\omega' - \omega''\|$, $\forall \omega', \omega'' \in \Omega$ and

$$\begin{aligned} d_W(\mathbb{P}_1, \mathbb{P}_3) - \varepsilon &\le \int_\Omega f_\varepsilon(\omega') \mathrm{d}\mathbb{P}_1(\omega') - \int_\Omega f_\varepsilon(\omega'') \mathrm{d}\mathbb{P}_3(\omega'') \\ &= \mathbb{E}^{\mathbb{P}_1}[f_\varepsilon(\omega)] - \mathbb{E}^{\mathbb{P}_3}[f_\varepsilon(\omega)] \\ &= (\mathbb{E}^{\mathbb{P}_1}[f_\varepsilon(\omega)] - \mathbb{E}^{\mathbb{P}_2}[f_\varepsilon(\omega)]) + (\mathbb{E}^{\mathbb{P}_2}[f_\varepsilon(\omega)] - \mathbb{E}^{\mathbb{P}_3}[f_\varepsilon(\omega)]) \\ &\le d_W(\mathbb{P}_1, \mathbb{P}_2) + d_W(\mathbb{P}_2, \mathbb{P}_3) \ . \end{aligned}$$

Let $\varepsilon \to 0$ then the triangle inequality is obtained.

Exercise 6.3 Prove nonnegativity and symmetry properties of Wasserstein distance using the definition in (6.13).

6.1.5 Relation to f-Divergence

We introduce the relationship between Wasserstein distance and f-divergences by investigating the connection of total variation distance to the optimal transport problem. Wasserstein distance and f-divergences are popular tools that measure the difference between probability distributions. Readers can refer to Gibbs and Su (2002); Tsybakov (2009) for comprehensive reviews of various probability metrics.

6.1.5.1 Total Variation Distance and Optimal Transport Problem

Like Wasserstein distance, total variation distance is also a widely used distance function of probability distributions, which has a straightforward definition and has been introduced in Sect. 5.1.4.1. We restate it in the following for convenience: Assume $\mathbb{P}_1$ and $\mathbb{P}_2$ are probability distributions defined on $(\Omega, \mathcal{F})$. Then the *total variation distance* between $\mathbb{P}_1$ and $\mathbb{P}_2$ is:

$$d_{TV}(\mathbb{P}_1, \mathbb{P}_2) := \sup_{S \in \mathcal{F}} \{\mathbb{P}_1(S) - \mathbb{P}_2(S)\} \ . \tag{6.14}$$

In addition, by Theorem 5.8,

$$d_{TV}(\mathbb{P}_1, \mathbb{P}_2) = \frac{1}{2} \sup_{|f| \le 1} \left\{ \int_\Omega f(\omega) \mathrm{d}\mathbb{P}_1(\omega) - \int_\Omega f(\omega) \mathrm{d}\mathbb{P}_2(\omega) \right\} \ . \tag{6.15}$$

The right side in (6.15) is similar to formula (6.6c) that appears in the Kantorovich duality. If we replace the norm $\|\omega' - \omega''\|$ with the discrete metric d_{D} in Sect. 2.1.1.2, Kantorovich duality still holds according to Remark 6.3, which is:

$$\begin{aligned} &\inf_{\mathbb{P}_{1,2}\in\mathcal{J}(\mathbb{P}_1,\mathbb{P}_2)} \int_{\Omega\times\Omega} d_{\mathrm{D}}(\omega',\omega'')\cdot \mathrm{d}\mathbb{P}_{1,2}(\omega',\omega'') \\ &= \begin{cases} \sup\limits_f \displaystyle\int_\Omega f(\omega')\mathrm{d}\mathbb{P}_1(\omega') - \int_\Omega f(\omega'')\mathrm{d}\mathbb{P}_2(\omega'') \\ \text{s.t. } f(\omega') - f(\omega'') \le d_{\mathrm{D}}(\omega',\omega'')\ ,\ \forall \omega',\omega'' \in \Omega\ . \end{cases} \end{aligned} \tag{6.16}$$

Recall that $d_{\mathrm{D}}(\omega',\omega'') = 0$ if $\omega' = \omega''$ and $d_{\mathrm{D}}(\omega',\omega'') = 1$ otherwise. Thus,

$$\begin{aligned} &\inf_{\mathbb{P}_{1,2}\in\mathcal{J}(\mathbb{P}_1,\mathbb{P}_2)} \int_{\Omega\times\Omega} d_{\mathrm{D}}(\omega',\omega'')\cdot \mathrm{d}\mathbb{P}_{1,2}(\omega',\omega'') \\ &= \inf_{\mathbb{P}_{1,2}\in\mathcal{J}(\mathbb{P}_1,\mathbb{P}_2)} \mathbb{P}_{1,2}(\{(\omega',\omega'')\in\Omega\times\Omega \mid \omega' \ne \omega''\})\ , \end{aligned} \tag{6.17}$$

and

$$\begin{aligned} &\begin{cases} \sup\limits_f \displaystyle\int_\Omega f(\omega')\mathrm{d}\mathbb{P}_1(\omega') - \int_\Omega f(\omega'')\mathrm{d}\mathbb{P}_2(\omega'') \\ \text{s.t. } f(\omega') - f(\omega'') \le d_{\mathrm{D}}(\omega',\omega'')\ ,\ \forall \omega',\omega'' \in \Omega\ , \end{cases} \\ &= \begin{cases} \sup\limits_f \displaystyle\int_\Omega f(\omega')\mathrm{d}\mathbb{P}_1(\omega') - \int_\Omega f(\omega'')\mathrm{d}\mathbb{P}_2(\omega'') \\ \text{s.t. } f(\omega') - f(\omega'') \le 1\ ,\ \forall \omega',\omega'' \in \Omega\ . \end{cases} \end{aligned} \tag{6.18}$$

For a measurable function $f : \Omega \to \mathbb{R}$ subject to $f(\omega') - f(\omega'') \le 1, \forall \omega', \omega'' \in \Omega$, we have $f(\omega') \ge f(\omega) - 1, \forall \omega' \in \Omega$, so $\inf_{\omega'\in\Omega} f(\omega')$ is finite. Let $g : \Omega \to \mathbb{R}$ be defined by $g(\omega) := 2f(\omega) - 2\inf_{\omega'\in\Omega} f(\omega') - 1$. Then $g(\omega) \ge -1, \forall \omega \in \Omega$. We claim that:

$$f(\omega') - f(\omega'') \le 1\ ,\ \forall \omega',\omega'' \in \Omega \iff g(\omega) \le 1\ ,\ \forall \omega \in \Omega\ .$$

The left side clearly implies the right side. Conversely, when $g(\omega) \le 1, \forall \omega \in \Omega$, we have $f(\omega') - f(\omega'') \le f(\omega') - \inf_{\omega\in\Omega} f(\omega) = (g(\omega') + 1)/2 \le 1, \forall \omega', \omega'' \in \Omega$. This completes the proof of the claim. In addition, given function $g : \Omega \to \mathbb{R}$ with $|g(\omega)| \le 1, \forall \omega \in \Omega$, we can set a function $f : \Omega \to \mathbb{R}$ by $f(\omega) := (g(\omega) + 1)/2$, then $f(\omega') - f(\omega'') = (g(\omega') - g(\omega''))/2 \le 1, \forall \omega', \omega'' \in \Omega$. Simple calculation leads to:

$$
\begin{aligned}
&\int_{\Omega} g(\omega')\mathrm{d}\mathbb{P}_1(\omega') - \int_{\Omega} g(\omega'')\mathrm{d}\mathbb{P}_2(\omega'') \\
&= \int_{\Omega} \left(2f(\omega') - 2\inf_{\omega\in\Omega} f(\omega) - 1\right) \mathrm{d}\mathbb{P}_1(\omega') \\
&\quad - \int_{\Omega} \left(2f(\omega'') - 2\inf_{\omega\in\Omega} f(\omega) - 1\right) \mathrm{d}\mathbb{P}_2(\omega'') \\
&= 2\left(\int_{\Omega} f(\omega')\mathrm{d}\mathbb{P}_1(\omega') - \int_{\Omega} f(\omega'')\mathrm{d}\mathbb{P}_2(\omega'')\right) .
\end{aligned} \tag{6.19}
$$

By combining (6.16)–(6.19) and Theorem 5.8, Kantorovich duality comes down to the following equation in this case:

$$
\begin{aligned}
&\inf_{\mathbb{P}_{1,2}\in\mathcal{J}(\mathbb{P}_1,\mathbb{P}_2)} \mathbb{P}_{1,2}(\{(\omega',\omega'')\in\Omega\times\Omega \mid \omega'\neq\omega''\}) \\
&= \begin{cases} \sup\limits_{g} \dfrac{1}{2}\left(\displaystyle\int_{\Omega} g(\omega')\mathrm{d}\mathbb{P}_1(\omega') - \int_{\Omega} g(\omega'')\mathrm{d}\mathbb{P}_2(\omega'')\right) \\ \text{s.t. } |g(\omega)| \le 1 \ , \ \forall\omega\in\Omega \end{cases} \\
&= d_{\mathrm{TV}}(\mathbb{P}_1,\mathbb{P}_2) \ .
\end{aligned}
$$

Therefore, just like Wasserstein distance, total variation distance is also the optimal value of an optimal transport problem, but the discrete metric is used to measure the unit transport cost. By Remark 6.4, the proof of Theorem 6.2 is able to show that total variation distance is a distance function. Note that we also have proved it is a distance function directly in Theorem 5.7. To sum up, we have the following conclusion:

Theorem 6.3 *The total variation distance $d_{\mathrm{TV}}(\mathbb{P}_1,\mathbb{P}_2)$ defined in Definition 5.5 is a distance function on $\mathcal{M}(\Omega)$ and satisfies that*

$$
\begin{aligned}
d_{\mathrm{TV}}(\mathbb{P}_1,\mathbb{P}_2) &= \inf_{\mathbb{P}_{1,2}\in\mathcal{J}(\mathbb{P}_1,\mathbb{P}_2)} \mathbb{P}_{1,2}(\{(\omega',\omega'')\in\Omega\times\Omega \mid \omega'\neq\omega''\}) \\
&= \inf_{\mathbb{P}_{1,2}\in\mathcal{J}(\mathbb{P}_1,\mathbb{P}_2)} \int_{\Omega\times\Omega} d_{\mathrm{D}}(\omega',\omega'')\mathrm{d}\mathbb{P}_{1,2}(\omega',\omega'') \ .
\end{aligned} \tag{6.20}
$$

In (6.20), $d_{\mathrm{TV}}(\mathbb{P}_1,\mathbb{P}_2)$ equals the minimum probability of $\omega'\neq\omega''$ among all the joint probability distribution $\mathbb{P}_{1,2}$. This formula is another equivalent definition of total variation distance.

6.1.5.2 Relation to Total Variation Distance

From (6.3) and (6.20), we can see that d_{W} and d_{TV} differ from each other because $\|\omega'-\omega''\|$ and $d_{\mathrm{D}}(\omega',\omega'')$ are used to represent the unit transport costs in the optimal transport problems of Wasserstein distance and total variation distance, respectively. They are compared in Fig. 6.3 in the 1-dimensional case, which shows that $\|\omega'-\omega''\|$

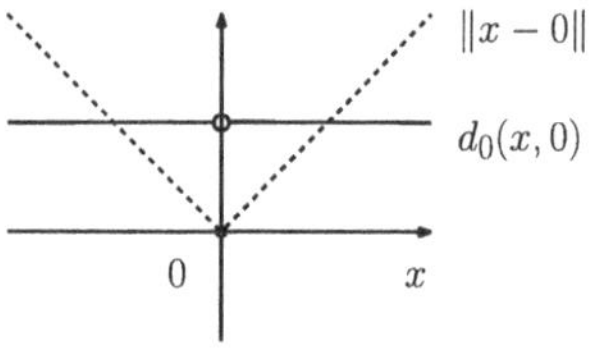

Fig. 6.3 Comparison of the norm $\|x - 0\|$ and the discrete metric $d_D(x, 0)$ for $x \in \mathbb{R}$

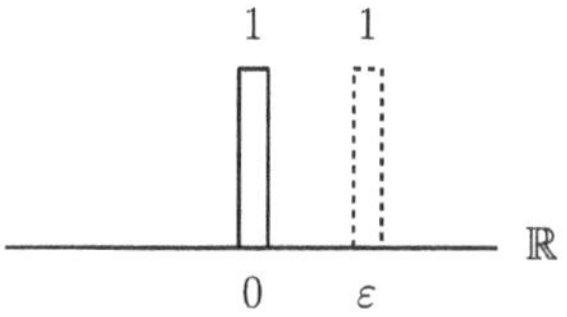

Fig. 6.4 For Dirac distributions δ_0 and δ_ε defined on $\mathbb{R}$ with $\varepsilon > 0$, we have $d_W(\delta_0, \delta_\varepsilon) = \varepsilon$ and $d_{TV}(\delta_0, \delta_\varepsilon) = 1$

can reflect how far the two points are to each other; in contrast, $d_D(\omega', \omega'')$ only detects whether the two points are different. This results in the different performances of Wasserstein distance and total variation distance, which will be elaborated by examples.

Example 6.3 (d_W *can be arbitrarily small while* d_{TV} *remains* 1) Let $\Omega = \mathbb{R}$ and $\varepsilon > 0$. Consider two Dirac distributions δ_0 and δ_ε with unit mass at 0 and ε, respectively, as shown in Fig. 6.4. Then we have $d_W(\delta_0, \delta_\varepsilon) = \varepsilon$ by (6.5b) and $d_{TV}(\delta_0, \delta_\varepsilon) = \delta_0(0) - \delta_\varepsilon(0) = 1$. Note that ε here can be arbitrarily small while $d_{TV}(\delta_0, \delta_\varepsilon)$ is always 1, the maximum total variation distance.

Example 6.4 (d_{TV} *can be arbitrarily small while* d_W *is arbitrarily large*) Let $\Omega = \mathbb{R}$, $\varepsilon > 0$, and $N > 0$. Set $\mathbb{P}_1 := \delta_0$, the Dirac distribution with unit mass at 0. Define a discrete probability distribution $\mathbb{P}_2$ as follows:

$$\mathbb{P}_2(0) = 1 - \varepsilon \ , \ \mathbb{P}_2(N/\varepsilon) = \varepsilon \ .$$

The two probability distributions are depicted in Fig. 6.5. Then $d_W(\mathbb{P}_1, \mathbb{P}_2) = N$ and $d_{TV}(\mathbb{P}_1, \mathbb{P}_2) = \mathbb{P}_1(0) - \mathbb{P}_2(0) = \varepsilon$. Because $\varepsilon > 0$ and $N > 0$ are arbitrary, d_{TV} can be arbitrarily close to 0 while d_W is arbitrarily large.

Exercise 6.4 Calculate and verify the distance values in Examples 6.3 and 6.4.

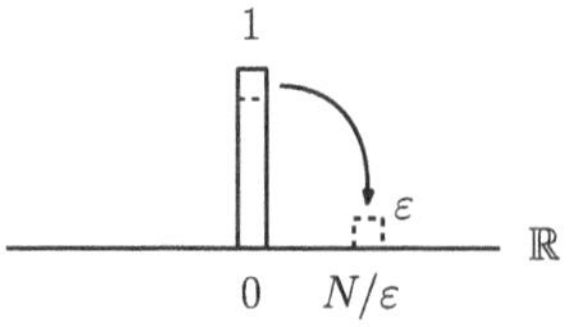

Fig. 6.5 For Dirac distribution $\mathbb{P}_1 = \delta_0$ and discrete distribution $\mathbb{P}_2$ with $\mathbb{P}_2(0) = 1 - \varepsilon$ and $\mathbb{P}_2(N/\varepsilon) = \varepsilon$, $d_W(\mathbb{P}_1, \mathbb{P}_2) = N$ and $d_{TV}(\mathbb{P}_1, \mathbb{P}_2) = \varepsilon$

From Examples 6.3 and 6.4, we see that Wasserstein distance and total variation distance may behave very differently on general sample spaces, which stems from the difference between $\|\omega' - \omega''\|$ and $d_{\mathrm{D}}(\omega', \omega'')$. For specific sample spaces, there may be a relationship between Wasserstein distance and total variation distance. Recall that the diameter of set Ω with respect to norm $\|\cdot\|$ is:

$$\mathrm{diam}(\Omega) := \sup_{\omega', \omega'' \in \Omega} \|\omega' - \omega''\| \ .$$

In addition, a set in an Euclidean space is bounded if and only if its diameter is finite. Proposition 6.1 says that if the sample space is bounded, Wasserstein distance is no larger than $\mathrm{diam}(\Omega)$ times the total variation distance.

Proposition 6.1 (Bounding d_{W} *using* d_{TV} and the diameter) *Assume* $\varnothing \neq \Omega \subset \mathbb{R}^M$ *and* $\mathit{diam}(\Omega) < +\infty$*, then*

$$d_{\mathrm{W}}(\mathbb{P}_1, \mathbb{P}_2) \leq \mathit{diam}(\Omega) \cdot d_{\mathrm{TV}}(\mathbb{P}_1, \mathbb{P}_2) \ , \ \forall \mathbb{P}_1, \mathbb{P}_2 \in \mathcal{M}(\Omega) \ .$$

In addition,

$$\forall 0 < N < \mathit{diam}(\Omega) \ , \ \exists \mathbb{P}_1, \mathbb{P}_2 \in \mathcal{M}(\Omega) \ , \ \textit{s.t.} \ d_{\mathrm{W}}(\mathbb{P}_1, \mathbb{P}_2) > N d_{\mathrm{TV}}(\mathbb{P}_1, \mathbb{P}_2) \ .$$

Proof 1. We observe that

$$\|\omega' - \omega''\| \leq \mathrm{diam}(\Omega) \cdot d_{\mathrm{D}}(\omega', \omega'') \ , \ \forall \omega', \omega'' \in \Omega \ .$$

Hence,

$$\begin{aligned}
&\inf_{\mathbb{P}_{1,2} \in \mathcal{J}(\mathbb{P}_1, \mathbb{P}_2)} \int_{\Omega \times \Omega} \|\omega' - \omega''\| \cdot \mathrm{d}\mathbb{P}_{1,2}(\omega', \omega'') \\
\leq &\inf_{\mathbb{P}_{1,2} \in \mathcal{J}(\mathbb{P}_1, \mathbb{P}_2)} \int_{\Omega \times \Omega} \mathrm{diam}(\Omega) \cdot d_{\mathrm{D}}(\omega', \omega'') \cdot \mathrm{d}\mathbb{P}_{1,2}(\omega', \omega'') \\
= &\ \mathrm{diam}(\Omega) \inf_{\mathbb{P}_{1,2} \in \mathcal{J}(\mathbb{P}_1, \mathbb{P}_2)} \int_{\Omega \times \Omega} d_{\mathrm{D}}(\omega', \omega'') \cdot \mathrm{d}\mathbb{P}_{1,2}(\omega', \omega'') \ , \ \forall \mathbb{P}_1, \mathbb{P}_2 \in \mathcal{M}(\Omega) \ .
\end{aligned}$$

Then the conclusion follows from the definitions of Wasserstein distance and total variation distance in (6.3) and (6.20).

2. Since $\mathrm{diam}(\Omega) = \sup_{\omega', \omega'' \in \Omega} \|\omega' - \omega''\|$, for any $0 < N < \mathrm{diam}(\Omega)$, there are $\omega', \omega'' \in \Omega$ so that $\|\omega' - \omega''\| > N$. Let $\mathbb{P}_1 = \delta_{\omega'}$ and $\mathbb{P}_2 = \delta_{\omega''}$. Then $d_{\mathrm{W}}(\mathbb{P}_1, \mathbb{P}_2) = \|\omega' - \omega''\| > N = N d_{\mathrm{TV}}(\mathbb{P}_1, \mathbb{P}_2)$. □

Proposition 6.1 means that Wasserstein distance can be bounded by total variation distance when the sample space is bounded. However, for general bounded sample space, the converse does not hold, i.e., total variation distance cannot be bounded by Wasserstein distance. Figure 6.4 gives such an example. Proposition 6.2 tells us when total variation distance can be bounded by Wasserstein distance.

Proposition 6.2 *Assume $\Omega \subset \mathbb{R}^M$, there are at least 2 elements in Ω, and*

$$d_{\min} := \inf\{\|\omega' - \omega''\| \mid \omega', \omega'' \in \Omega, \omega' \neq \omega''\} > 0 .$$

Then

$$d_{\min} \cdot d_{\mathrm{TV}}(\mathbb{P}_1, \mathbb{P}_2) \leq d_{\mathrm{W}}(\mathbb{P}_1, \mathbb{P}_2) \ , \ \forall \mathbb{P}_1, \mathbb{P}_2 \in \mathcal{M}_1(\Omega) .$$

In addition,

$$\forall N > d_{\min} \ , \ \exists \mathbb{P}_1, \mathbb{P}_2 \in \mathcal{M}_1(\Omega) \ , \ s.t. \ N d_{\mathrm{TV}}(\mathbb{P}_1, \mathbb{P}_2) > d_{\mathrm{W}}(\mathbb{P}_1, \mathbb{P}_2) .$$

Proof 1. For any $\mathbb{P}_1, \mathbb{P}_2 \in \mathcal{M}_1(\Omega)$,

$$\begin{aligned} d_{\mathrm{W}}(\mathbb{P}_1, \mathbb{P}_2) &= \inf_{\mathbb{P}_{1,2} \in \mathcal{J}(\mathbb{P}_1, \mathbb{P}_2)} \int_{\Omega \times \Omega} \|\omega' - \omega''\| \cdot \mathrm{d}\mathbb{P}_{1,2}(\omega', \omega'') \\ &= \inf_{\mathbb{P}_{1,2} \in \mathcal{J}(\mathbb{P}_1, \mathbb{P}_2)} \int_{\{(\omega', \omega'') \in \Omega \times \Omega \mid \omega' \neq \omega''\}} \|\omega' - \omega''\| \cdot \mathrm{d}\mathbb{P}_{1,2}(\omega', \omega'') \\ &\geq \inf_{\mathbb{P}_{1,2} \in \mathcal{J}(\mathbb{P}_1, \mathbb{P}_2)} \int_{\{(\omega', \omega'') \in \Omega \times \Omega \mid \omega' \neq \omega''\}} d_{\min} \cdot \mathrm{d}\mathbb{P}_{1,2}(\omega', \omega'') \\ &= d_{\min} \inf_{\mathbb{P}_{1,2} \in \mathcal{J}(\mathbb{P}_1, \mathbb{P}_2)} \int_{\Omega \times \Omega} d_{\mathrm{D}}(\omega', \omega'') \cdot \mathrm{d}\mathbb{P}_{1,2}(\omega', \omega'') \\ &= d_{\min} \cdot d_{\mathrm{TV}}(\mathbb{P}_1, \mathbb{P}_2) . \end{aligned}$$

2. For any $N > d_{\min}$, there are $\omega', \omega'' \in \Omega$ so that $\|\omega' - \omega''\| < N$. Set $\mathbb{P}_1 = \delta_{\omega'}$ and $\mathbb{P}_2 = \delta_{\omega''}$. Then $d_{\mathrm{W}}(\mathbb{P}_1, \mathbb{P}_2) = \|\omega' - \omega''\| < N = N d_{\mathrm{TV}}(\mathbb{P}_1, \mathbb{P}_2)$. □

6.1.5.3 Relation to Other f-Divergences

The conclusions of Propositions 6.1 and 6.2 can be denoted by:

$$d_{\min} \cdot d_{\mathrm{TV}} \leq d_{\mathrm{W}} \leq \mathrm{diam}(\Omega) \cdot d_{\mathrm{TV}} .$$

Recall the relationships between common f-divergences from Sect. 5.1.5:

$$\begin{aligned} &d_{\mathrm{TV}} \leq \sqrt{D_{\mathrm{KL}}/2} \ , \ d_{\mathrm{TV}} \leq 1 - \exp(-D_{\mathrm{KL}})/2 \ , \\ &d_{\mathrm{H}}^2/2 \leq d_{\mathrm{TV}} \leq d_{\mathrm{H}} \ , \ d_{\mathrm{H}}^2 \leq D_{\mathrm{KL}} \ , \\ &D_{\mathrm{KL}} \leq \log(1 + D_{\chi^2}) \leq D_{\chi^2} . \end{aligned}$$

Combine them and we have:

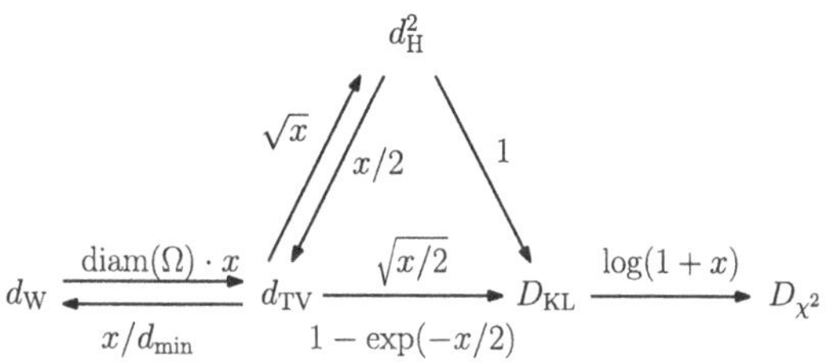

Fig. 6.6 The relationships between Wasserstein distance and some f-divergences. A directed arrow from A to B with a function $h(x)$ means that $A \le h(B)$

$$
\begin{aligned}
& d_{\mathrm{W}} \le \mathrm{diam}(\Omega)\cdot\sqrt{D_{\mathrm{KL}}/2}\ ,\ d_{\mathrm{W}} \le \mathrm{diam}(\Omega)\cdot(1-\exp(-D_{\mathrm{KL}})/2)\ ,\\
& d_{\min}\cdot d_{\mathrm{H}}^2/2 \le d_{\mathrm{W}} \le \mathrm{diam}(\Omega)\cdot d_{\mathrm{H}}\ ,\\
& d_{\mathrm{W}} \le \mathrm{diam}(\Omega)\cdot\sqrt{\log(1+D_{\chi^2})/2}\ .
\end{aligned}
$$

The relationships between Wasserstein distance and common f-divergences are summarized in Fig. 6.6 (Gibbs and Su 2002).

6.1.6 Convergence Rate of Empirical Probability Distribution Under Wasserstein Distance

When the real probability distribution $\mathbb{P}_*$ on $\mathbb{R}^M$ is unknown, we use the empirical probability distribution $\mathbb{P}_0 = \frac{1}{N}\sum_{n=1}^{N}\delta_{\omega_n}$ to approximate it, where $\omega_1, \omega_2, \ldots, \omega_N$ are N samples identically and independently drawn from $\mathbb{P}_*$. The reason is that by Glivenko-Cantelli theorem, $\mathbb{P}_0$ converges to $\mathbb{P}_*$. To be more specific, the probability of $\sup_{x\in\mathbb{R}^M}|F_0(x) - F_*(x)| \to 0$ as $N \to \infty$ is 1, where F_0 and F_* are the cumulative distribution functions (CDFs) of $\mathbb{P}_0$ and $\mathbb{P}_*$, respectively.

In order to construct a reasonable Wasserstein-distance ambiguity set, we are interested in the convergence rate of $\mathbb{P}_0$ to $\mathbb{P}_*$ under Wasserstein distance, i.e., how $d_{\mathrm{W}}(\mathbb{P}_0, \mathbb{P}_*)$ converges to 0 as $N \to \infty$. In the following, we introduce some theoretical results and the readers can refer to Fournier and Guillin (2015) for the proofs.

There are two common approaches to reflecting the convergence rate. The first one is called moment estimate, which finds a sequence converging to 0 that bounds the sequence of $\mathbb{E}[d_{\mathrm{W}}(\mathbb{P}_0, \mathbb{P}_*)]$. The second one is the concentration inequality, which estimates the probability $\mathbb{P}(d_{\mathrm{W}}(\mathbb{P}_0, \mathbb{P}_*) \ge d_0)$. The expectation and the probability are under the product distribution of $\mathbb{P}_*$, since $\omega_1, \omega_2, \ldots$ are identically and independently distributed with probability distribution $\mathbb{P}_*$.

The convergence rate results are related to moment conditions. We define:

$$
\begin{aligned}
M_q(\mathbb{P}_*) &:= \int_{\mathbb{R}^M} \|x\|^q \mathrm{d}\mathbb{P}_*(x)\ ,\\
\mathcal{E}_{\alpha,\gamma}(\mathbb{P}_*) &:= \int_{\mathbb{R}^M} \exp(\gamma\|x\|^\alpha)\mathrm{d}\mathbb{P}_*(x)\ ,
\end{aligned}
$$

where $q > 0$, $\alpha > 0$, and $\gamma > 0$.

Theorem 6.4 (Moment estimates of Wasserstein distance, Theorem 1 in Fournier and Guillin 2015) *Assume $M_q(\mathbb{P}_*) < \infty$ for some $q > 1$. Then there is a constant C depending on M and q such that for any $N \in \mathbb{N}_+$,*

$$\mathbb{E}[d_W(\mathbb{P}_0, \mathbb{P}_*)] \le C \cdot M_q^{1/q}(\mathbb{P}_*) \cdot a(N) \ ,$$

where

$$a(N) := \begin{cases} N^{-1/2} + N^{-(q-1)/q} \ , & \text{if } M = 1 \text{ and } q \neq 2 \ , \\ N^{-1/2}\log(1+N) + N^{-(q-1)/q} \ , & \text{if } M = 2 \text{ and } q \neq 2 \ , \\ N^{-1/M} + N^{-(q-1)/q} \ , & \text{if } M \ge 3 \text{ and } q \neq M/(M-1) \ . \end{cases}$$

Theorem 6.5 (Concentration inequalities of Wasserstein distance, Theorem 2 in Fournier and Guillin 2015) *Assume one of the following conditions hold:*

1. *There exist $\alpha > 1$ and $\gamma > 0$ such that $\mathcal{E}_{\alpha,\gamma}(\mathbb{P}_*)$ is finite.*
2. *There exist $\alpha \in (0, 1)$ and $\gamma > 0$ such that $\mathcal{E}_{\alpha,\gamma}(\mathbb{P}_*)$ is finite.*
3. *There exists $q > 2$ such that $M_q(\mathbb{P}_*)$ is finite.*

Then for any $N \in \mathbb{N}_+$ and $d_0 \in (0, +\infty)$,

$$\mathbb{P}(d_W(\mathbb{P}_0, \mathbb{P}_*) \ge d_0) \le a(N, d_0) \cdot 1_{(0,1]}(d_0) + b(N, d_0) \ ,$$

where

$$a(N, d_0) := C \cdot \begin{cases} \exp(-cNd_0^2) \ , & \text{if } M = 1 \ , \\ \exp(-cN(d_0/\log(2 + 1/d_0))^2) \ , & \text{if } M = 2 \ , \\ \exp(-cNd_0^M) \ , & \text{if } M \ge 3 \ , \end{cases}$$

and corresponding to the conditions above,

1. *$b(N, d_0) := C \cdot \exp(-cNd_0^\alpha) \cdot 1_{(1,+\infty)}(d_0)$ under condition 1.*
2. *$b(N, d_0) := C \cdot (\exp(-c(Nd_0)^{\alpha-\varepsilon}) \cdot 1_{(0,1]}(d_0) + \exp(-c(Nd_0)^\alpha) \cdot 1_{(1,+\infty)}(d_0))$ for any $\varepsilon \in (0, \alpha)$ under condition 2.*
3. *$b(N, d_0) := CN(Nd_0)^{-(q-\varepsilon)}$ for any $\varepsilon \in (0, q)$ under condition 3.*

C and c are constants possibly depending on d_0, α, γ, $\mathcal{E}_{\alpha,\gamma}(\mathbb{P}_*)$, ε, q and $M_q(\mathbb{P}_*)$.

In practice, the random variable is often bounded. In this case, the moment conditions naturally hold. Specifically, suppose the norm of the random variable is always no larger than some $R > 0$, then $\mathbb{P}_*(B(0, R)) = 1$, where $B(0, R) := \{x \in \mathbb{R}^M \mid \|x\| \le R\}$. Thus,

$$M_q(\mathbb{P}_*) = \int_{B(0,R)} \|x\|^q \mathrm{d}\mathbb{P}_*(x) \le R^q < +\infty \ , \ \forall q > 0 \ ,$$

$$\mathcal{E}_{\alpha,\gamma}(\mathbb{P}_*) = \int_{B(0,R)} \exp(\gamma\|x\|^\alpha)\mathrm{d}\mathbb{P}_*(x) \le \exp(\gamma R^\alpha) < +\infty \ , \ \forall \alpha > 0 \ , \ \gamma > 0 \ .$$

Therefore, in Theorem 6.4, q can be arbitrarily large. Then the term $N^{-(q-1)/q}$ converges to 0 faster than $N^{-1/2}$, $N^{-1/2}\log(1+N)$, and $N^{-1/M}$, so it can be removed, which leads to Corollary 6.1.

Corollary 6.1 *Assume $\mathbb{P}_*(B(0,R)) = 1$ for some $R > 0$, then there is a constant C depending on M such that for any $N \in \mathbb{N}_+$,*

$$\mathbb{E}[d_{\mathrm{W}}(\mathbb{P}_0, \mathbb{P}_*)] \le \begin{cases} CRN^{-1/2}\,, & \text{if } M = 1\,,\\ CRN^{-1/2}\log(1+N)\,, & \text{if } M = 2\,,\\ CRN^{-1/M}\,, & \text{if } M \ge 3\,. \end{cases}$$

Similarly, if $\mathbb{P}_*(B(0,R)) = 1$, we can let $\alpha = \max\{2, M\} > 1$ and then condition 1 holds in Theorem 6.5. Thus, we have Corollary 6.2.

Corollary 6.2 *Assume $\mathbb{P}_*(B(0,R)) = 1$ for some $R > 0$, then for any $N \in \mathbb{N}_+$ and $d_0 \in (0, +\infty)$,*

$$\mathbb{P}(d_{\mathrm{W}}(\mathbb{P}_0, \mathbb{P}_*) \ge d_0) \le C \times$$
$$\begin{cases} \exp(-cNd_0^2), & \text{if } M = 1,\\ \exp(-cN(d_0/\log(2+1/d_0))^2)\cdot 1_{(0,1]}(d_0) + \exp(-cNd_0^2)\cdot 1_{(1,+\infty)}(d_0), & \text{if } M = 2,\\ \exp(-cNd_0^M), & \text{if } M \ge 3, \end{cases}$$

where C and c are constants depending on d_0 and M.

6.1.7 Example: Wasserstein Distance Between 1-Dimensional Probability Distributions

In this subsection, we introduce a particular case where the solution of the optimal transport problem is known. Let $\mathbb{P}_1$ and $\mathbb{P}_2$ be two probability distributions on $\mathbb{R}$ and consider the optimal transport problem between them. An intuitive idea is the so-called monotone transport: Start from an endpoint of the original distribution and transport a small piece of mass to the same end of the destiny distribution. Then turn to the next small piece of mass and repeat. The procedure is illustrated in Fig. 6.7.

In the following, we will express the monotone transport mathematically and prove that it is optimal. To this end, recall the notion of CDF and generalized inverse distribution function from Sect. 2.2.1.2.

Proposition 6.3 *Let $\mathbb{U}$ be the uniform distribution on $[0, 1]$. Assume $\mathbb{P}_1$ and $\mathbb{P}_2$ are probability distributions on $\mathbb{R}$ with CDFs F_1 and F_2 and generalized inverse distribution functions F_1^{-1} and F_2^{-1}, respectively. Define a measure $\mathbb{P}_{\mathrm{mon}}$ on $\mathbb{R}^2$ by*

$$\mathbb{P}_{\mathrm{mon}}(S) := \mathbb{U}(\{p \in [0,1] \mid (F_1^{-1}(p), F_2^{-1}(p)) \in S\})\ ,\ \ \forall S \text{ measurable in } \mathbb{R}^2\,.$$

Then $\mathbb{P}_{\mathrm{mon}} \in \mathcal{J}(\mathbb{P}_1, \mathbb{P}_2)$ and for any $a, b \in \mathbb{R}$,

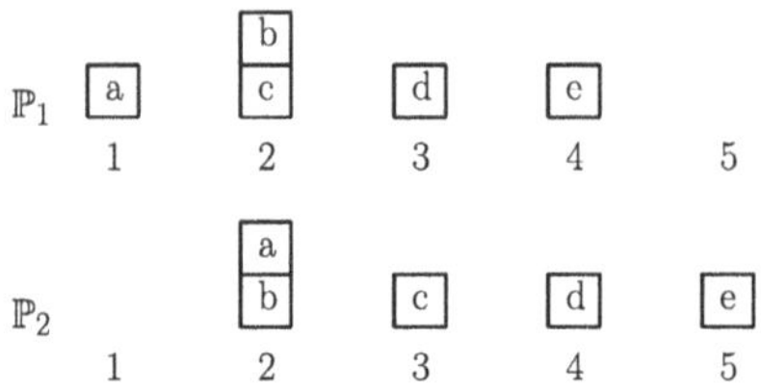

Fig. 6.7 An example of monotone transport: Start from the left side of distribution $\mathbb{P}_1$. Move the first box to the far-left position in $\mathbb{P}_2$. Then turn to the second box and repeat till all the boxes are moved to the positions in $\mathbb{P}_2$

$$\mathbb{P}_{\text{mon}}((-\infty, a] \times (-\infty, b]) = \min\{F_1(a), F_2(b)\} .$$

Proof Recall from Proposition 2.9 that for any $p \in [0, 1]$ and $a \in \mathbb{R}$, $F_1^{-1}(p) \leq a \iff p \leq F_1(a)$. Thus,

$$\begin{aligned}
&\mathbb{P}_{\text{mon}}((-\infty, a] \times (-\infty, b]) \\
&= \mathbb{U}(\{p \in [0, 1] \mid (F_1^{-1}(p), F_2^{-1}(p)) \in (-\infty, a] \times (-\infty, b]\}) \\
&= \mathbb{U}(\{p \in [0, 1] \mid F_1^{-1}(p) \leq a, F_2^{-1}(p) \leq b\}) \\
&= \mathbb{U}(\{p \in [0, 1] \mid p \leq F_1(a), p \leq F_2(b)\}) \\
&= \mathbb{U}([0, \min\{F_1(a), F_2(b)\}]) = \min\{F_1(a), F_2(b)\} .
\end{aligned}$$

Then $\mathbb{P}_{\text{mon}}(\mathbb{R} \times (-\infty, b]) = \min\{1, F_2(b)\} = F_2(b) = \mathbb{P}_2((-\infty, b])$. Similarly, $\mathbb{P}_{\text{mon}}((-\infty, a] \times \mathbb{R}) = \mathbb{P}_1((-\infty, a])$. Since a probability distribution on $\mathbb{R}$ can be determined by its CDF, we have $\mathbb{P}_{\text{mon}} \in \mathcal{J}(\mathbb{P}_1, \mathbb{P}_2)$. □

Remark 6.7 The joint probability distribution $\mathbb{P}_{\text{mon}}$ defined in Proposition 6.3 is the mathematical expression of the monotone transport plan. $\mathbb{P}_{\text{mon}}((-\infty, a] \times (-\infty, b])$ represents the volume of mass transported from $(-\infty, a]$ to $(-\infty, b]$. When $\mathbb{P}_1((-\infty, a]) < \mathbb{P}_2((-\infty, b])$, the mass in $(-\infty, a]$ are all transported into $(-\infty, b]$ in the monotone transport plan, so $\mathbb{P}_{\text{mon}}((-\infty, a] \times (-\infty, b]) = \mathbb{P}_1((-\infty, a])$. When $\mathbb{P}_1((-\infty, a]) > \mathbb{P}_2((-\infty, b])$, only the first $\mathbb{P}_2((-\infty, b])$ volume of mass can be transported into $(-\infty, b]$, so $\mathbb{P}_{\text{mon}}((-\infty, a] \times (-\infty, b]) = \mathbb{P}_2((-\infty, b])$.

Theorem 6.6 *For probability distributions $\mathbb{P}_1$ and $\mathbb{P}_2$ defined on $\mathbb{R}$, the joint probability distribution $\mathbb{P}_{\text{mon}}$ defined in Proposition 6.3 is optimal in the optimal transport problem:*

$$\inf_{\mathbb{P}_{1,2} \in \mathcal{J}(\mathbb{P}_1, \mathbb{P}_2)} \int_{\mathbb{R} \times \mathbb{R}} |x_1 - x_2| \cdot \mathrm{d}\mathbb{P}_{1,2}(x_1, x_2) .$$

Readers can refer to Santambrogio (2015) for the proof. Here we explain why Theorem 6.6 holds. If a transport plan is not monotone, there are $x_1 < x_1'$ transported to $x_2 > x_2'$, respectively. We change the plan so that x_1 is transported to x_2', x_1' is transported to x_2, and other parts remain the same, as illustrated in Fig. 6.8. Since

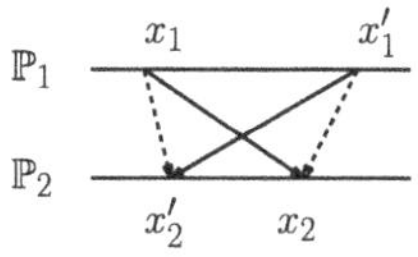

Fig. 6.8 The original transport plan is not monotone: x_1 is transported to x_2 and x_1' is transported to x_2'. Change it so that x_1 and x_1' are transported to x_2' and x_2, respectively

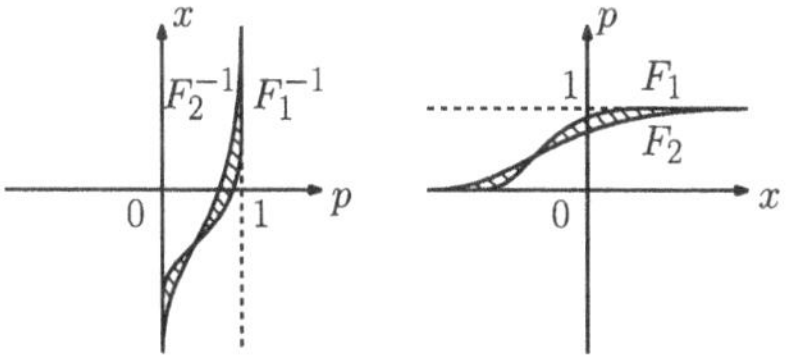

Fig. 6.9 On the left side, the *shaded area* is $\int_0^1 |F_1^{-1}(p) - F_2^{-1}(p)| \mathrm{d}p$. On the right side, the *shaded area* is $\int_{\mathbb{R}} |F_1(x) - F_2(x)| \mathrm{d}x$. By geometry relationship, the two areas are the same

$$|x_1 - x_2| + |x_1' - x_2'| \geq |x_1 - x_2'| + |x_1' - x_2| \,, \tag{6.21}$$

the total transport cost will not increase after the plan changes. Imagine doing this whenever the plan is not monotone. Then it leads to a monotone transport plan with a total cost no larger than the original transport plan. Therefore, the monotone transport plan is optimal.

Exercise 6.5 Assume $x_1 < x_1'$ and $x_2 > x_2'$ in $\mathbb{R}$. Prove (6.21).

Now we can calculate the Wasserstein distance between probability distributions $\mathbb{P}_1$ and $\mathbb{P}_2$ on $\mathbb{R}$:

$$d_{\mathrm{W}}(\mathbb{P}_1, \mathbb{P}_2) = \int_{\mathbb{R}\times\mathbb{R}} |x_1 - x_2| \cdot \mathrm{d}\mathbb{P}_{\mathrm{mon}}(x_1, x_2) \tag{6.22a}$$

$$= \int_{\mathbb{R}\times\mathbb{R}} |x_1 - x_2| \cdot \mathbb{U}(\{p \mid (F_1^{-1}(p), F_2^{-1}(p)) \in \mathrm{d}(x_1, x_2)\}) \tag{6.22b}$$

$$= \int_{[0,1]} |F_1^{-1}(p) - F_2^{-1}(p)| \mathrm{d}p \,, \tag{6.22c}$$

where from (6.22b) to (6.22c), we have changed variables. Formula (6.22c) can be further simplified to be:

$$d_{\mathrm{W}}(\mathbb{P}_1, \mathbb{P}_2) = \int_{\mathbb{R}} |F_1(x) - F_2(x)| \mathrm{d}x \,,$$

as illustrated in Fig. 6.9.

6.2 Wasserstein-Distance Ambiguity Set

In this section, we define the Wasserstein-distance ambiguity set and discuss the choices of radius and sample space.

6.2.1 Definition

Just like the case of f-divergences, the ambiguity sets based on Wasserstein distance consist of probability distributions near the empirical probability distribution, where the difference between probability distributions is measured by Wasserstein distance. The definition of the Wasserstein-distance ambiguity set is as follows:

Definition 6.2 Assume $\mathbb{P}_0$ is a probability distribution on the sample space $(\Omega, \mathcal{F})$ and d_0 is a nonnegative constant. Let $\mathcal{M}_1(\Omega)$ denote the set of all probability distributions on Ω that have finite first-order moments. Then the *Wasserstein-distance ambiguity set* is defined as:

$$\mathcal{P}_{\mathrm{W}}(\Omega, \mathbb{P}_0, d_0) := \{\mathbb{P} \in \mathcal{M}_1(\Omega) \mid d_{\mathrm{W}}(\mathbb{P}, \mathbb{P}_0) \leq d_0\} .$$

Therefore, the ambiguity set $\mathcal{P}_{\mathrm{W}}(\Omega, \mathbb{P}_0, d_0)$ contains all the probability distributions in $\mathcal{M}_1(\Omega)$ whose Wasserstein distance to $\mathbb{P}_0$ is no larger than d_0. The center of the ambiguity set, i.e., $\mathbb{P}_0$, is usually an empirical probability distribution. Let ω be a random variable defined on $\mathbb{R}^M$. There are N historical data $\omega_1, \omega_2, \ldots, \omega_N$ of ω. Then the empirical probability distribution of ω is:

$$\mathbb{P}_0 := \frac{1}{N} \sum_{n=1}^{N} \delta_{\omega_n} ,$$

where δ_{ω_n} is the Dirac distribution at point ω_n. Therefore, the empirical probability distribution $\mathbb{P}_0$ assigns probability $1/N$ to each historical data.

6.2.2 The Choice of Radius

The radius d_0 directly controls the size of the ambiguity set $\mathcal{P}_{\mathrm{W}}(\Omega, \mathbb{P}_0, d_0)$. When $d_0 = 0$, the ambiguity set only contains the center $\mathbb{P}_0$ because Wasserstein distance is a distance function and has the point-separating property. Larger d_0 leads to a larger ambiguity set, i.e., $\mathbb{P}_0 \in \mathcal{P}_{\mathrm{W}}(\Omega, \mathbb{P}_0, d_1) \subset \mathcal{P}_{\mathrm{W}}(\Omega, \mathbb{P}_0, d_2)$ whenever $0 \leq d_1 \leq d_2$. Since Wasserstein distance has a finite value for any probability distribution $\mathbb{P} \in \mathcal{M}_1(\Omega)$, there is a radius $d_0 \geq 0$ such that $\mathbb{P} \in \mathcal{P}_{\mathrm{W}}(\Omega, \mathbb{P}_0, d_0)$.

When choosing the radius, the number of historical data, the dimension and range of the random variable, and the risk appetite of the decision maker should be considered. In practice, the random variable is often bounded, whose range size can be reflected by the diameter $\mathrm{diam}(\Omega) := \sup_{\omega'\omega''\in\Omega} \|\omega' - \omega''\|$. By the definition in (6.3), Wasserstein distance has the same unit as the norm, which indicates that the radius should be proportional to $\mathrm{diam}(\Omega)$.

From Corollary 6.2 in Sect. 6.1.6, we can know that there exist constants C and c depending on d_0 and M such that

$$\mathbb{P}(d_W(\mathbb{P}_0, \mathbb{P}_*) \leq d_0) \geq 1 - C\exp(-cNd_0^{M'}) , \tag{6.23}$$

where $M' := M + 1$ when $M = 1, 2$ and $M' := M$ when $M \geq 3$. $\mathbb{P}_*$ denotes the real or actual probability distribution. Note that $\mathbb{P}(d_W(\mathbb{P}_0, \mathbb{P}_*) \leq d_0)$ in (6.23) is exactly the probability of the real probability distribution $\mathbb{P}_*$ being included in the ambiguity set $\mathcal{P}_W(\mathbb{R}^M, \mathbb{P}_0, d_0)$, so (6.23) can be regarded as a probability guarantee for the Wasserstein-distance ambiguity set. When (6.23) holds, with probability guarantee $1 - C\exp(-cNd_0^{M'})$, the performance of the optimal solution of the distributionally robust optimization (DRO) problem under the real probability distribution is no worse than the optimal value. To maintain a given probability threshold using (6.23), the radius d_0 should be proportional to $N^{-1/M'}$.

Instead of focusing on the probability $\mathbb{P}(d_W(\mathbb{P}_0, \mathbb{P}_*) \leq d_0)$, the difference between the worst-case expectation and the real expectation is estimated in Gao (2023) so that the impact of the loss function $g(x, \omega)$ is also considered, and it argues that the radius d_0 should be proportional to $1/\sqrt{N}$ under some appropriate properties such as bounded support and Lipschitz continuity of the loss function.

6.2.3 *The Choice of Sample Space*

The choice of the sample space also has a significant influence on the range of the ambiguity set $\mathcal{P}_W(\Omega, \mathbb{P}_0, d_0)$. First of all, the sample set should be no smaller than the set of historical data $\{\omega_n \mid n = 1, 2, \ldots, N\}$ in order to include $\mathbb{P}_0$ in the ambiguity set. In addition, $\mathcal{P}_W(\Omega_1, \mathbb{P}_0, d_0) \subset \mathcal{P}_W(\Omega_2, \mathbb{P}_0, d_0)$ if $\{\omega_n \mid n = 1, 2, \ldots, N\} \subset \Omega_1 \subset \Omega_2$.

For instance, suppose the 1-dimensional random variable ω has 1 historical datum: $\omega_1 = 1$, then the empirical probability distribution $\mathbb{P}_0 = \delta_1$. Consider two discrete probability distributions on $\mathbb{R}$: $\mathbb{P}_1 = \delta_0$ and $\mathbb{P}_2 = \delta_{-1}/2 + \delta_1/2$. It can be easily checked that $d_W(\mathbb{P}_1, \mathbb{P}_0) = d_W(\mathbb{P}_2, \mathbb{P}_0) = 1$. Therefore, $\mathbb{P}_1, \mathbb{P}_2 \in \mathcal{P}_W(\mathbb{R}, \mathbb{P}_0, 1)$. If we know that the random variable ω must take nonnegative values, then the sample space can be set as $\Omega = [0, +\infty)$. Any probability distribution $\mathbb{P}$ on Ω should satisfy $\mathbb{P}(\Omega) = 1$, so $\mathbb{P}_1 \in \mathcal{P}_W(\Omega, \mathbb{P}_0, 1)$ but $\mathbb{P}_2 \notin \mathcal{P}_W(\Omega, \mathbb{P}_0, 1)$. Therefore, compared with $\mathcal{P}_W(\mathbb{R}, \mathbb{P}_0, 1)$, the sample space $\mathcal{P}_W(\Omega, \mathbb{P}_0, 1)$ is smaller and hence less conservative. This shows the power of choosing sample space. In fact, strictly smaller sample space always leads to a strictly smaller ambiguity set as Proposition 6.4 shows.

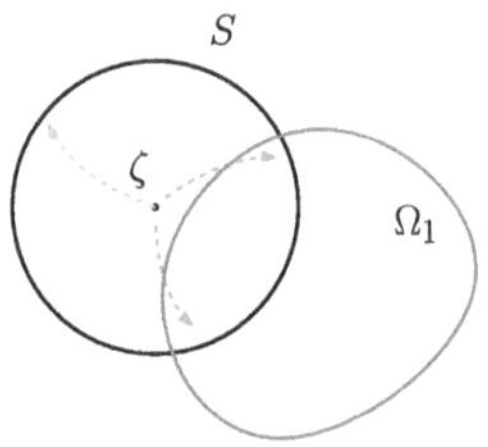

Fig. 6.10 Illustration of the transport plan in the proof of Proposition 6.4. ζ is outside of Ω_1. To transport the mass from $\mathbb{P}$ to $\mathbb{P}_0$, the mass outside of $\{\zeta\}$ is not moved and the mass at ζ is transported into $S \setminus \{\zeta\}$

Proposition 6.4 *Assume* $\varnothing \neq \Omega_1 \subsetneq \Omega_2 \subset \mathbb{R}^M$ *and* $\mathbb{P}_0$ *is a probability distribution on* Ω_1, *then for any* $d_0 > 0$,

$$\mathbb{P}_0 \in \mathcal{P}_W(\Omega_1, \mathbb{P}_0, d_0) \subsetneq \mathcal{P}_W(\Omega_2, \mathbb{P}_0, d_0) .$$

Proof Since $\mathbb{P}_0 \in \mathcal{M}_1(\Omega_1) \subset \mathcal{M}_1(\Omega_2)$, we have $\mathbb{P}_0 \in \mathcal{P}_W(\Omega_1, \mathbb{P}_0, d_0) \subset \mathcal{P}_W(\Omega_2, \mathbb{P}_0, d_0)$. Then we only need to prove that $\mathcal{P}_W(\Omega_1, \mathbb{P}_0, d_0) \neq \mathcal{P}_W(\Omega_2, \mathbb{P}_0, d_0)$, i.e., find $\mathbb{P} \in \mathcal{P}_W(\Omega_2, \mathbb{P}_0, d_0)$ such that $\mathbb{P} \notin \mathcal{P}_W(\Omega_1, \mathbb{P}_0, d_0)$.

Because $\Omega_1 \subsetneq \Omega_2$, there is a point $\zeta \in \Omega_2 \setminus \Omega_1$. By $\mathbb{P}_0(\mathbb{R}^M) = 1$ and $\bigcup_{N \in \mathbb{N}_+} \{x \in \mathbb{R}^M \mid \|x - \zeta\| \leq N\} = \mathbb{R}^M$, there is $N \in \mathbb{N}_+$ such that $\mathbb{P}_0(\{x \in \mathbb{R}^M \mid \|x - \zeta\| \leq N\}) > 0$. Let $S := \{x \in \mathbb{R}^M \mid \|x - \zeta\| \leq N\}$ and $p := \mathbb{P}_0(S)$, then $0 < p \leq 1$ and $\|x - \zeta\| \leq N, \forall x \in S$. Let $a := p \cdot \min\{d_0/N, 1\}$ and $b := \max\{1 - d_0/N, 0\}$, then $0 < a \leq 1, 0 \leq b < 1$, and $a + bp = p$. Define

$$\mathbb{P} := \mathbb{P}_0 1_{\mathbb{R}^M \setminus S} + a \cdot \delta_\zeta + b \cdot \mathbb{P}_0 1_S ,$$

where $1_{\mathbb{R}^M \setminus S}$ and 1_S are indicator functions of set $\mathbb{R}^M \setminus S$ and S, respectively. Since

$$\mathbb{P}(\mathbb{R}^M) = \mathbb{P}_0(\mathbb{R}^M \setminus S) + a + b\mathbb{P}_0(S) = (1 - p) + a + bp = 1 ,$$

and $a, b \geq 0$, $\mathbb{P} \in \mathcal{M}_1(\Omega_2)$. By $a > 0$ and $\zeta \notin \Omega_1$, we know $\mathbb{P} \notin \mathcal{M}_1(\Omega_1)$.

To estimate $d_W(\mathbb{P}, \mathbb{P}_0)$, consider the following transport plan from mass distribution $\mathbb{P}$ to $\mathbb{P}_0$: Do not move the mass outside of $\{\zeta\}$ and transport the mass at ζ into $S \setminus \{\zeta\}$ to get $\mathbb{P}_0$, which is illustrated in Fig. 6.10. Since $\|x - \zeta\| \leq N, \forall x \in S$, the cost of this plan is no larger than $aN \leq pd_0 \leq d_0$. Therefore, $d_W(\mathbb{P}, \mathbb{P}_0) \leq d_0$ and we have found $\mathbb{P} \in \mathcal{P}_W(\Omega_2, \mathbb{P}_0, d_0)$ with $\mathbb{P} \notin \mathcal{P}_W(\Omega_1, \mathbb{P}_0, d_0)$, which completes the proof. □

Therefore, choosing an appropriate sample space helps to rule out the impossible probability distributions from the ambiguity set and reduce the conservativeness of Wasserstein-distance DRO. In the following, we introduce some common choices for sample space.

Example 6.5 (*Euclidean space as the sample space*) The largest sample space is the Euclidean space $\mathbb{R}^M$. In this case, probability distributions of M-dimensional random variables that have finite first-order moments are all in $\mathcal{M}_1(\mathbb{R}^M)$, including discrete probability distributions, continuous probability distributions, and others. As long as $\mathbb{P}$'s Wasserstein distance to $\mathbb{P}_0$ is no larger than d_0, $\mathbb{P}$ is contained in the ambiguity set $\mathcal{P}_W(\mathbb{R}^M, \mathbb{P}_0, d_0)$.

Example 6.6 (*Sample space based on the bounds of random variable*) In practice, although the exact probability distribution of the random variable may be difficult to obtain, there are usually straightforward bounds for the range of the random variable. For example, if the random variable represents the available wind power, then it must be nonnegative and cannot exceed the capacity of the wind farm. This information can be utilized in the construction of the sample space to narrow the ambiguity set: Suppose the range of the random variable is in Ω, then the ambiguity set $\mathcal{P}_W(\Omega, \mathbb{P}_0, d_0)$ can be used instead of the larger one $\mathcal{P}_W(\mathbb{R}^M, \mathbb{P}_0, d_0)$.

The bounds usually can be expressed or approximated by linear constraints. In this circumstance, the sample space is polyhedral, which can be written as:

$$\Omega = \{x \in \mathbb{R}^M \mid Px \le Q\} ,$$

for some coefficient matrix P and vector Q. For example, if the random variable $\omega \in \mathbb{R}^M$ satisfies $\underline{\omega} \le \omega \le \overline{\omega}$ with bounds $\underline{\omega}$ and $\overline{\omega}$, then we can let:

$$P = \begin{pmatrix} I_M \\ -I_M \end{pmatrix} , \quad Q = \begin{pmatrix} \overline{\omega} \\ -\underline{\omega} \end{pmatrix} ,$$

where I_M is the identity matrix of size M.

Example 6.7 (*Sample space by the estimation based on the historical data*) The sample space can also be defined as the convex hull of historical data, i.e.,

$$\begin{aligned} \Omega &= \mathrm{conv}(\omega_1, \omega_2, \dots, \omega_N) \\ &= \left\{ \sum_{n=1}^{N} \lambda_n \omega_n \mid \sum_{n=1}^{N} \lambda_n = 1 \ , \ 0 \le \lambda_n \le 1 \ , \ n = 1, 2, \dots, N \right\} , \end{aligned} \tag{6.24}$$

so that any point in Ω is a convex combination of the historical data $\omega_1, \omega_2, \dots, \omega_N$.

There are some cases where the convex hull is suitable as the sample space: First, when we lack information on the random variable range and do not have bounds for it, estimating the range by historical data is a reasonable way. Many random variable ranges are convex, which means that the weighted average of possible values is also possible. Therefore, the convex hull is a reasonable choice for the sample space. Second, sometimes we have bounds for the random variable but there are some extreme and rare situations in it that we want to rule out to reduce the conservativeness. In this circumstance, using the convex hull of the historical data tends to only consider the relatively likely values.

The convex hull of a finite number of points is a polyhedron. To see this, note that by (6.24), Ω is the projection of the following polyhedron to the space $\mathbb{R}^M$ of x:

$$P := \left\{(x, \lambda) \in \mathbb{R}^{M+N} \mid 0_{N\times 1} \leq \lambda \leq 1_{N\times 1}\ ,\ 1_{1\times N}^\top \lambda = 1\ ,\ x = (\omega_1\ \omega_2\ \ldots\ \omega_N)\lambda\right\}\ ,$$

where x and λ are $M \times 1$ and $N \times 1$ vectors, respectively. To express the polyhedron of the convex hull in terms of hypersurfaces, or equivalently, linear inequalities, the free toolbox named Multi-Parametric Toolbox (MPT) (Herceg et al. 2013) can be used.

The sample space can also be constructed as an ellipsoid or ball containing the historical data, i.e.,

$$\Omega = \{x \in \mathbb{R}^M \mid (x - \omega_0)^\top A(x - \omega_0) \leq 1\}\ , \tag{6.25}$$

where ω_0 is the center point and A is a real, symmetric, positive-definite matrix. For the case of a ball, $A = kI_M$ for some $k > 0$. After obtaining the center point, the radius can be enlarged a little bit, if a margin is needed.

Example 6.8 (*Finite set as the sample space*) Sometimes the sample space Ω is a finite set. For instance, suppose an electric vehicle (EV) charging station has 5 chargers and regard the number of EVs charging in the station as a random variable. Then it has 6 possible values in total: 0, 1, 2, 3, 4, 5. Thus, the sample space $\Omega = \{0, 1, 2, 3, 4, 5\}$ is a finite set. Another situation where a finite sample space arises is when the range of historical data is considered as the sample space, i.e.,

$$\Omega = \{x \in \mathbb{R}^M \mid x = \omega_n \text{ for some } n \in \mathbb{N}_+\ ,\ 1 \leq n \leq N\}\ .$$

In either case, the sample space is a finite set. Wasserstein-distance DRO problems with a finite sample space are relatively easy to reformulate and solve, which will be revealed in Sect. 6.3.1. The reason is basically that the probability distributions in the ambiguity set have to be discrete if the sample space is a finite set.

6.3 Reformulation of Worst-Case Expectation

We focus on the worst-case expectation under a Wasserstein-distance ambiguity set in this section, i.e.,

$$\sup_{\mathbb{P}\in\mathcal{P}_W(\Omega,\mathbb{P}_0,d_0)} \mathbb{E}^{\mathbb{P}}[g(\omega)]\ , \tag{6.26}$$

where $\mathbb{P}_0 = \sum_{n=1}^N \delta_{\omega_n}/N$ is the empirical probability distribution; $d_0 \geq 0$ is the radius of the ambiguity set; Ω is the sample space of the random variable ω; $\mathcal{P}_W(\Omega, \mathbb{P}_0, d_0) = \{\mathbb{P} \in \mathcal{M}_1(\Omega) \mid d_W(\mathbb{P}, \mathbb{P}_0) \leq d_0\}$ is the Wasserstein-distance ambiguity set; the measurable function $g : \Omega \to \mathbb{R}$ is a function of ω representing the loss.

Thus, (6.26) is the worst-case expectation of loss under the probability distributions on Ω whose Wasserstein distance to $\mathbb{P}_0$ is no larger than d_0.

In DRO, the loss function g is usually also influenced by a decision variable x, and $x \in \mathcal{X}$ is optimized to minimize the worst-case expectation, i.e.,

$$\min_{x \in \mathcal{X}} \sup_{\mathbb{P} \in \mathcal{P}_{\mathrm{W}}(\Omega, \mathbb{P}_0, d_0)} \mathbb{E}^{\mathbb{P}}[g(x, \omega)] , \tag{6.27}$$

or the worst-case expectation is required to be no larger than a threshold C, i.e.,

$$\sup_{\mathbb{P} \in \mathcal{P}_{\mathrm{W}}(\Omega, \mathbb{P}_0, d_0)} \mathbb{E}^{\mathbb{P}}[g(x, \omega)] \leq C . \tag{6.28}$$

The reformulation methods of (6.26) can be used to make (6.27) and (6.28) tractable. We first investigate the case where the sample space is a finite set and then move to the general case, because the finite sample space case is much simpler.

6.3.1 Ambiguity Set on a Finite Sample Space

Suppose sample space Ω is a finite set as follows:

$$\Omega = \{\omega_1, \omega_2, \ldots, \omega_K\} \subset \mathbb{R}^M ,$$

satisfying that $\omega_k \in \Omega$, $k = 1, 2, \ldots, K$. Then any probability distribution $\mathbb{P}$ on Ω must be a discrete probability distribution and $\mathbb{E}^{\mathbb{P}}[g(\omega)] = \sum_{k=1}^{K} g(\omega_k) \cdot p_k$, where $p_k := \mathbb{P}(\omega_k)$, $k = 1, 2, \ldots, K$. Recall that we have used LP duality to deduce the Kantorovich duality for discrete probability distributions in Sect. 6.1.3. Here the idea is similar. Note that:

$$d_{\mathrm{W}}(\mathbb{P}, \mathbb{P}_0) = \inf_{\mathbb{P}_{\mathrm{J}} \in \mathcal{J}(\mathbb{P}, \mathbb{P}_0)} \int_{\Omega \times \Omega} \|\omega' - \omega''\| \mathrm{d}\mathbb{P}_{\mathrm{J}}(\omega', \omega'')$$

$$= \begin{cases} \min\limits_{\pi} \sum\limits_{k=1}^{K} \sum\limits_{l=1}^{K} \|\omega_k - \omega_l\| \cdot \pi_{kl} \\ \text{s.t. } \sum\limits_{l=1}^{K} \pi_{kl} = p_k , \ k = 1, 2, \ldots, K , \\ \sum\limits_{k=1}^{K} \pi_{kl} = \mathbb{P}_0(\omega_l) , \ l = 1, 2, \ldots, K , \\ \pi_{kl} \geq 0 , \ k, l = 1, 2, \ldots, K , \end{cases}$$

where π_{kl} is the probability of (ω_k, ω_l) under the joint probability distribution $\mathbb{P}_J$ of $\mathbb{P}$ and $\mathbb{P}_0$. As long as a joint probability distribution can be found such that $\sum_{k=1}^{K}\sum_{l=1}^{K}\|\omega_k - \omega_l\| \cdot \pi_{kl} \le d_0$, the probability distribution $\mathbb{P}$ is in the ambiguity set $\mathcal{P}(\Omega, \mathbb{P}_0, d_0)$. The converse is also true since the infimum in the definition of Wasserstein distance can be achieved by some probability distributions according to Remark 6.1. Therefore, the worst-case expectation $\sup_{\mathbb{P}\in\mathcal{P}_W(\Omega,\mathbb{P}_0,d_0)} \mathbb{E}^{\mathbb{P}}[g(\omega)]$ equals the optimal value of the following optimization problem:

$$
\begin{aligned}
\max_{p,\pi} \ & \sum_{k=1}^{K} g(\omega_k) \cdot p_k \\
\text{s.t.} \ & \sum_{k=1}^{K}\sum_{l=1}^{K} \|\omega_k - \omega_l\| \cdot \pi_{kl} \le d_0 \ , \\
& \sum_{l=1}^{K} \pi_{kl} = p_k \ , \ k = 1, 2, \ldots, K \ , \\
& \sum_{k=1}^{K} \pi_{kl} = \mathbb{P}_0(\omega_l) \ , \ l = 1, 2, \ldots, K \ , \\
& \pi_{kl} \ge 0 \ , \ k, l = 1, 2, \ldots, K \ ,
\end{aligned}
$$

which can be further written as:

$$
\sup_{\mathbb{P}\in\mathcal{P}_W(\Omega,\mathbb{P}_0,d_0)} \mathbb{E}^{\mathbb{P}}[g(\omega)] = \begin{cases} \max\limits_{\pi_{kl}\ge 0,\forall k,l} & \sum\limits_{k=1}^{K}\sum\limits_{l=1}^{K} g(\omega_k) \cdot \pi_{kl} \\ \text{s.t.} & \sum\limits_{k=1}^{K}\sum\limits_{l=1}^{K} \|\omega_k - \omega_l\| \cdot \pi_{kl} \le d_0 : \lambda \ , \\ & \sum\limits_{k=1}^{K} \pi_{kl} = \mathbb{P}_0(\omega_l) : \mu_l \ , \ l = 1, 2, \ldots, K \ . \end{cases} \tag{6.29}
$$

In problem (6.29), the decision variables are π_{kl}, $k, l = 1, 2, \ldots, K$. Therefore, (6.29) is an LP problem and thus is tractable.

Problem (6.29) can be used to calculate the worst-case expectation when the loss function g is fixed. Now we turn to the reformulation of (6.27) and (6.28), which are the forms of the worst-case expectations appearing in the DRO problems. Since the worst-case expectation is minimized in (6.27), the direct usage of (6.29) leads to a min-max problem which should be further reformulated before being solved. By LP duality, the optimal value of problem (6.29) equals that of its dual problem if it has a finite optimal value. Thus,

$$\sup_{\mathbb{P}\in\mathcal{P}_{\mathrm{W}}(\Omega,\mathbb{P}_0,d_0)} \mathbb{E}^{\mathbb{P}}[g(\omega)]$$
$$= \begin{cases} \min\limits_{\lambda\geq 0,\mu} \ d_0\lambda + \sum\limits_{l=1}^{K} \mathbb{P}_0(\omega_l)\cdot\mu_l \\ \text{s.t. } \|\omega_k-\omega_l\|\lambda+\mu_l \geq g(\omega_k)\ ,\ k,l=1,2,\ldots,K\ , \end{cases} \tag{6.30}$$

which also can calculate the worst-case expectation. Using (6.30), (6.27) can be equivalently transformed into a single-layer optimization problem with linear constraints as follows:

$$\min_{x\in\mathcal{X}} \sup_{\mathbb{P}\in\mathcal{P}_{\mathrm{W}}(\Omega,\mathbb{P}_0,d_0)} \mathbb{E}^{\mathbb{P}}[g(x,\omega)]$$
$$= \begin{cases} \min\limits_{x\in\mathcal{X},\lambda\geq 0,\mu} \ d_0\lambda + \sum\limits_{l=1}^{K} \mathbb{P}_0(\omega_l)\cdot\mu_l \\ \text{s.t. } \|\omega_k-\omega_l\|\lambda+\mu_l \geq g(x,\omega_k)\ ,\ k,l=1,2,\ldots,K\ . \end{cases} \tag{6.31}$$

If $\mathcal{X}$ is convex and $g(x,\omega_k)$ is convex with respect to x for any ω_k, then problem (6.31) is convex and hence tractable. The number of constraints is K^2, which becomes very large when the sample space Ω contains many points. The upper bound constraint (6.28) of the worst-case expectation can be handled in a similar way, which comes down to:

$$\sup_{\mathbb{P}\in\mathcal{P}_{\mathrm{W}}(\Omega,\mathbb{P}_0,d_0)} \mathbb{E}^{\mathbb{P}}[g(x,\omega)] \leq C$$
$$\iff \begin{cases} \exists\lambda\in\mathbb{R}_+\ ,\ \mu\in\mathbb{R}_K\ ,\ \text{s.t.} \\ d_0\lambda + \sum\limits_{l=1}^{K} \mathbb{P}_0(\omega_l)\cdot\mu_l \leq C\ , \\ \|\omega_k-\omega_l\|\lambda+\mu_l \geq g(x,\omega_k)\ ,\ k,l=1,2,\ldots,K\ . \end{cases} \tag{6.32}$$

Similarly, the set of (x,λ,μ) defined by (6.32) is convex when $\mathcal{X}$ is convex and $g(x,\omega_k)$ is convex in x for any ω_k.

Exercise 6.6 Show that the dual problem of (6.29) is (6.30). In addition,

$$\sup_{\mathbb{P}\in\mathcal{P}_{\mathrm{W}}(\Omega,\mathbb{P}_0,d_0)} \mathbb{E}^{\mathbb{P}}[g(\omega)]$$
$$= \min_{\lambda\geq 0}\left\{ d_0\lambda + \sum_{l=1}^{K} \mathbb{P}_0(\omega_l)\cdot\max\{g(\omega_k)-\|\omega_k-\omega_l\|\lambda \mid k=1,2,\ldots,K\}\right\}.$$

Exercise 6.7 Assume $g(x,\omega)=\max\{(A_i(x))^\top\omega+B_i(x)\mid i=1,2,\ldots,I\},\forall\omega\in\Omega,\ x\in\mathcal{X}$ with vectors $A_i(x)$, $i=1,2,\ldots,I$ and numbers $B_i(x)$, $i=1,2,\ldots,I$ depending on x. Show

$$
\begin{aligned}
&\min_{x\in\mathcal{X}} \sup_{\mathbb{P}\in\mathcal{P}_{\mathrm{W}}(\Omega,\mathbb{P}_0,d_0)} \mathbb{E}^{\mathbb{P}}[g(x,\omega)] \\
&= \min_{x\in\mathcal{X},\lambda\geq 0} \left\{ d_0\lambda + \sum_{l=1}^{K} \mathbb{P}_0(\omega_l)\cdot \max\{(A_i(x))^\top \omega_k + B_i(x) - \|\omega_k-\omega_l\|\lambda \mid \forall k, \forall i\} \right\} \\
&= \begin{cases} \min\limits_{x\in\mathcal{X},\lambda\geq 0,s} & d_0\lambda + \sum\limits_{l=1}^{K} \mathbb{P}_0(\omega_l)\cdot s_l \\ \text{s.t.} & s_l \geq (A_i(x))^\top \omega_k + B_i(x) - \|\omega_k-\omega_l\|\lambda \ , \ \forall k \ , \ \forall i \ , \ \forall l \ . \end{cases}
\end{aligned}
$$

In addition, when $A_i(x)$ and $B_i(x)$ are affine in x, the above problem is an LP problem.

Exercise 6.8 Assume that for any $x \in \mathcal{X}$ and $\omega \in \Omega$,

$$
g(x,\omega) = \begin{cases} \min\limits_{y} C^\top y \\ \text{s.t. } Ay \geq B(x)\omega + H(x) \ , \end{cases}
$$

with coefficient matrices and vectors A, $B(x)$, C, and $H(x)$, where $B(x)$ and $H(x)$ depend on x. Show that:

$$
\begin{aligned}
&\min_{x\in\mathcal{X}} \sup_{\mathbb{P}\in\mathcal{P}_{\mathrm{W}}(\Omega,\mathbb{P}_0,d_0)} \mathbb{E}^{\mathbb{P}}[g(x,\omega)] \\
&= \begin{cases} \min\limits_{x\in\mathcal{X},\lambda\geq 0,s,y} & d_0\lambda + \sum\limits_{l=1}^{K} \mathbb{P}_0(\omega_l)\cdot s_l \\ \text{s.t.} & s_l \geq C^\top y_k - \|\omega_k-\omega_l\|\lambda \ , \ k,l = 1,2,\ldots,K \ , \\ & Ay_k \geq B(x)\omega_k + H(x) \ , \ k = 1,2,\ldots,K \ . \end{cases}
\end{aligned}
$$

Moreover, if $B(x)$ and $H(x)$ are affine in x, the above problem is an LP problem.

6.3.2 Ambiguity Set on a General Sample Space

Now we consider the general sample space case. According to the reformulation technologies, we introduce the strong duality, Lipschitz constant, and discrete probability distribution reformulations.

6.3.2.1 Strong Duality Reformulation

We first reformulate the Wasserstein-distance worst-case expectation in Theorem 6.7 using the duality technique.

Theorem 6.7 (Duality reformulation) *Suppose* $\mathbb{P}_0 = \sum_{n=1}^{N} \delta_{\omega_n}/N$ *is a probability distribution on* Ω *and* $g : \Omega \to \mathbb{R}$ *is measurable. Then for any* $d_0 > 0$*, the worst-case expectation*

$$\sup_{\mathbb{P} \in \mathcal{P}_{\mathrm{W}}(\Omega, \mathbb{P}_0, d_0)} \mathbb{E}^{\mathbb{P}}[g(\omega)] = \inf_{\lambda \geq 0} \left\{ d_0 \lambda + \frac{1}{N} \sum_{n=1}^{N} \sup_{\omega \in \Omega} \{ g(\omega) - \|\omega - \omega_n\| \} \lambda \right\} .$$

Remark 6.8 Theorem 6.7 is analogous to the reformulation in Exercise 6.6 from a finite sample space to a general sample space. In fact, when $\{\omega_1, \omega_2, \ldots, \omega_N\} \subset \Omega = \{\omega_1, \omega_2, \ldots, \omega_K\}$, we have:

$$\begin{aligned} &\mathbb{P}_0(\omega_l) \cdot \max\{ g(\omega_k) - \|\omega_k - \omega_l\| \lambda \mid k = 1, 2, \ldots, K \} \\ &= \begin{cases} 0 , & \text{if } \omega_l \neq \omega_n \ , \ \forall 1 \leq n \leq N \ , \\ \frac{1}{N} \sup_{\omega \in \Omega} \{ g(\omega) - \|\omega - \omega_n\| \lambda \} \ , & \text{if } \omega_l = \omega_n \ . \end{cases} \end{aligned}$$

Proof By the definition of Wasserstein distance in (6.3),

$$\begin{aligned} &\sup_{\mathbb{P} \in \mathcal{P}_{\mathrm{W}}(\Omega, \mathbb{P}_0, d_0)} \mathbb{E}^{\mathbb{P}}[g(\omega)] \\ &= \begin{cases} \sup\limits_{\mathbb{P} \in \mathcal{M}_1(\Omega)} \displaystyle\int_\Omega g(\omega) \mathrm{d}\mathbb{P}(\omega) \\ \quad \text{s.t. } d_{\mathrm{W}}(\mathbb{P}, \mathbb{P}_0) \leq d_0 \end{cases} \\ &= \begin{cases} \sup\limits_{\mathbb{P} \in \mathcal{M}_1(\Omega), \mathbb{P}_{\mathrm{J}} \in \mathcal{J}(\mathbb{P}, \mathbb{P}_0)} \displaystyle\int_\Omega g(\omega) \mathrm{d}\mathbb{P}(\omega) \\ \quad \text{s.t. } \displaystyle\int_{\Omega \times \Omega} \|\omega' - \omega''\| \mathrm{d}\mathbb{P}_{\mathrm{J}}(\omega', \omega'') \leq d_0 \ . \end{cases} \end{aligned}$$

By the total probability formula and $\mathbb{P}_0 = \sum_{n=1}^{N} \delta_{\omega_n}/N$, the joint probability distribution $\mathbb{P}_{\mathrm{J}}$ of $\mathbb{P}$ and $\mathbb{P}_0$ can be written as the average of N product distributions, i.e., $\mathbb{P}_{\mathrm{J}} = \sum_{n=1}^{N} \mathbb{P}_n \times \delta_{\omega_n}/N$, where $\mathbb{P}_n$ is the conditional probability distribution of $\mathbb{P}$ under the condition $\omega'' = \omega_n$. Then $\mathbb{P} = \sum_{n=1}^{N} \mathbb{P}_n/N$. Noting that

$$\int_{\Omega \times \Omega} \|\omega' - \omega''\| \mathrm{d}(\mathbb{P}_n \times \delta_{\omega_n})(\omega', \omega'') = \int_\Omega \|\omega' - \omega_n\| \mathrm{d}\mathbb{P}_n(\omega') \ ,$$

we have:

$$\sup_{\mathbb{P} \in \mathcal{P}_{\mathrm{W}}(\Omega, \mathbb{P}_0, d_0)} \mathbb{E}^{\mathbb{P}}[g(\omega)] \tag{6.34a}$$

$$= \begin{cases} \sup_{\mathbb{P}_n \in \mathcal{M}_1(\Omega), \forall n} \frac{1}{N} \sum_{n=1}^{N} \int_{\Omega} g(\omega) \mathrm{d}\mathbb{P}_n(\omega) \\ \text{s.t.} \ \frac{1}{N} \sum_{n=1}^{N} \int_{\Omega} \|\omega - \omega_n\| \mathrm{d}\mathbb{P}_n(\omega) \le d_0 \end{cases} \tag{6.34b}$$

$$= \begin{cases} \sup_{\mathbb{P}_n \in \mathcal{M}_1(\Omega), \forall n} \inf_{\lambda \ge 0} \frac{1}{N} \sum_{n=1}^{N} \int_{\Omega} g(\omega) \mathrm{d}\mathbb{P}_n(\omega) \\ \quad + \lambda \left(d_0 - \frac{1}{N} \sum_{n=1}^{N} \int_{\Omega} \|\omega - \omega_n\| \mathrm{d}\mathbb{P}_n(\omega) \right) \end{cases} \tag{6.34c}$$

$$= \sup_{\mathbb{P}_n \in \mathcal{M}_1(\Omega), \forall n} \inf_{\lambda \ge 0} \left\{ d_0\lambda + \frac{1}{N} \sum_{n=1}^{N} \int_{\Omega} (g(\omega) - \lambda \|\omega - \omega_n\|) \mathrm{d}\mathbb{P}_n(\omega) \right\} \tag{6.34d}$$

$$\le \inf_{\lambda \ge 0} \sup_{\mathbb{P}_n \in \mathcal{M}_1(\Omega), \forall n} \left\{ d_0\lambda + \frac{1}{N} \sum_{n=1}^{N} \int_{\Omega} (g(\omega) - \lambda \|\omega - \omega_n\|) \mathrm{d}\mathbb{P}_n(\omega) \right\} \tag{6.34e}$$

$$= \inf_{\lambda \ge 0} \left\{ d_0\lambda + \frac{1}{N} \sum_{n=1}^{N} \sup_{\mathbb{P}_n \in \mathcal{M}_1(\Omega)} \int_{\Omega} (g(\omega) - \lambda \|\omega - \omega_n\|) \mathrm{d}\mathbb{P}_n(\omega) \right\} \tag{6.34f}$$

$$= \inf_{\lambda \ge 0} \left\{ d_0\lambda + \frac{1}{N} \sum_{n=1}^{N} \sup_{\omega \in \Omega} \{ g(\omega) - \lambda \|\omega - \omega_n\| \} \right\} . \tag{6.34g}$$

The Lagrangian dual is used from (6.34b) to (6.34c). $\sup_{\mathbb{P}_n \in \mathcal{M}_1(\Omega), \forall n}$ and $\inf_{\lambda \ge 0}$ are swapped from (6.34d) to (6.34e). The "$\le$" follows from:

$$\inf_{\lambda \ge 0} (\cdot) \le \inf_{\lambda \ge 0} \sup_{\mathbb{P}_n \in \mathcal{M}_1(\Omega), \forall n} (\cdot) .$$

From (6.34f) to (6.34g), we use the fact that $\mathcal{M}_1(\Omega)$ contains all the Dirac distributions on Ω.

The "$\le$" in (6.34e) is actually "$=$", which follows from the strong conic duality result by observing that the primal problem satisfies the Slater's condition because $d_W(\mathbb{P}_0, \mathbb{P}_0) = 0 < d_0$. □

Readers can refer to Gao and Kleywegt (2023) for a constructive proof and it also shows that the strong duality of Wasserstein-distance DRO holds under more generalized settings, such as Wasserstein p-distance and other forms of empirical probability distributions.

Theorem 6.7 successfully eliminates the supremum over the ambiguity set, although the resulting formulation is not in a tractable form. In fact, it has a minimax structure and can be regarded as a robust optimization (RO) problem and solved by methods such as column-and-constraint generation (C&CG) algorithm. However, different from ordinary RO, there are N worst-case scenarios to consider. In the following, we use Theorem 6.7 to deduce tractable reformulations under further assumptions.

Theorem 6.8 (Loss function being the maximum of concave functions, Theorem 4.2 in Esfahani and Kuhn 2018) *Assume $\Omega \subset \mathbb{R}^M$ is convex and closed. $\mathbb{P}_0 = \sum_{n=1}^{N} \delta_{\omega_n}/N$ is a probability distribution on Ω. Function $g(\omega) = \max_{1 \le k \le K} g_k(\omega)$, where $-g_k$, $1 \le k \le K$ are proper, convex, and lower semicontinuous functions on Ω. Then for any $d_0 \ge 0$, the worst-case expectation $\sup_{\mathbb{P} \in \mathcal{P}_W(\Omega,\mathbb{P}_0,d_0)} \mathbb{E}^{\mathbb{P}}[g(\omega)]$ equals the optimal value of the following convex program:*

$$
\begin{aligned}
&\min_{\lambda,s,z,v} \ d_0\lambda + \frac{1}{N}\sum_{n=1}^{N} s_n \\
&\quad s.t. \ [-g_k]^*(z_{nk} - v_{nk}) + h_\Omega(v_{nk}) - \omega_n^\top z_{nk} \le s_n \ , \ n = 1, \dots, N \ , \ k = 1, \dots, K \ , \\
&\qquad \|z_{nk}\|_* \le \lambda \ , \ n = 1, \dots, N \ , \ k = 1, \dots, K \ ,
\end{aligned} \tag{6.35}
$$

where $[-g_k]^*$ is the conjugate function of $-g_k$, h_Ω is the support function of Ω, and $\|\cdot\|_*$ is the dual norm of $\|\cdot\|$.

Proof By Theorem 6.7, we have:

$$
\begin{aligned}
&\sup_{\mathbb{P} \in \mathcal{P}_W(\Omega,\mathbb{P}_0,d_0)} \mathbb{E}^{\mathbb{P}}[g(\omega)] \\
&= \inf_{\lambda \ge 0} \left\{ d_0\lambda + \frac{1}{N}\sum_{n=1}^{N} \sup_{\omega\in\Omega}\{g(\omega) - \lambda\|\omega - \omega_n\|\} \right\} \\
&= \inf_{\lambda \ge 0} \left\{ d_0\lambda + \frac{1}{N}\sum_{n=1}^{N} \sup_{\omega\in\Omega}\left\{ \max_{1\le k\le K} g_k(\omega) - \lambda\|\omega - \omega_n\| \right\} \right\} \\
&= \inf_{\lambda \ge 0} \left\{ d_0\lambda + \frac{1}{N}\sum_{n=1}^{N} \max_{1\le k\le K} \sup_{\omega\in\Omega}\{g_k(\omega) - \lambda\|\omega - \omega_n\|\} \right\} \\
&= \left\{ \begin{aligned} &\min_{\lambda\ge 0,s} \ d_0\lambda + \frac{1}{N}\sum_{n=1}^{N} s_n \\ &\ \text{s.t.} \ \sup_{\omega\in\Omega}\{g_k(\omega) - \lambda\|\omega - \omega_n\|\} \le s_n \ , \ \forall n \ , \ \forall k \ . \end{aligned} \right.
\end{aligned}
$$

Using the property of dual norm, we know that:

$$
-\lambda\|\omega - \omega_n\| = \inf_{z_{nk}\in\mathbb{R}^M : \|z_{nk}\|_* \le \lambda} z_{nk}^\top(\omega - \omega_n) \ .
$$

Then

$$\begin{aligned}\sup_{\omega\in\Omega}\{g_k(\omega)-\lambda\|\omega-\omega_n\|\} &= \sup_{\omega\in\Omega}\left\{g_k(\omega)+\inf_{z_{nk}:\|z_{nk}\|_*\le\lambda} z_{nk}^\top(\omega-\omega_n)\right\}\\ &= \inf_{z_{nk}:\|z_{nk}\|_*\le\lambda}\ \sup_{\omega\in\Omega}\{g_k(\omega)+z_{nk}^\top(\omega-\omega_n)\}\ ,\end{aligned}$$

where the last equation follows from Theorem 2.15, noting that the set $\{z_{nk}\in\mathbb{R}^M \mid \|z_{nk}\|_*\le\lambda\}$ is compact. Therefore,

$$\begin{aligned}&\sup_{\mathbb{P}\in\mathcal{P}_W(\Omega,\mathbb{P}_0,d_0)} \mathbb{E}^{\mathbb{P}}[g(\omega)]\\ &=\begin{cases}\min\limits_{\lambda\ge 0,s}\ d_0\lambda+\dfrac{1}{N}\sum\limits_{n=1}^{N}s_n\\ \text{s.t.}\quad \inf\limits_{z_{nk}:\|z_{nk}\|_*\le\lambda}\ \sup\limits_{\omega\in\Omega}\{g_k(\omega)+z_{nk}^\top(\omega-\omega_n)\}\le s_n\ ,\ \forall n\ ,\ \forall k\end{cases}\\ &=\begin{cases}\min\limits_{\lambda,s}\ d_0\lambda+\dfrac{1}{N}\sum\limits_{n=1}^{N}s_n\\ \text{s.t.}\ \sup\limits_{\omega\in\Omega}\{g_k(\omega)+z_{nk}^\top\omega\}-z_{nk}^\top\omega_n\le s_n\ ,\ \forall n\ ,\ \forall k\ ,\\ \|z_{nk}\|_*\le\lambda\ ,\ \forall n\ ,\ \forall k\ .\end{cases}\end{aligned}\tag{6.36}$$

Let χ_Ω be the characteristic function of Ω, then

$$\sup_{\omega\in\Omega}\{g_k(\omega)+z_{nk}^\top\omega\}=\sup_{\omega\in\mathbb{R}^M}\{-\chi_\Omega(\omega)+g_k(\omega)+z_{nk}^\top\omega\}=[\chi_\Omega-g_k]^*(z_{nk})\ .\tag{6.37}$$

The function $\chi_\Omega - g_k$ is proper, convex, and lower semicontinuous. According to a result of the conjugate function of the summation of two functions (Theorem 11.23 in Rockafellar and Wets 1998), we have:

$$\begin{aligned}[\chi_\Omega-g_k]^*(z_{nk}) &= \inf_{v_{nk}}\{[-g_k]^*(z_{nk}-v_{nk})+[\chi_\Omega]^*(v_{nk})\}\\ &= \inf_{v_{nk}}\{[-g_k]^*(z_{nk}-v_{nk})+h_\Omega(v_{nk})\}\ .\end{aligned}\tag{6.38}$$

Combining (6.36)–(6.38), we have (6.35). □

If the sample space is the Euclidean space, the conclusion in Theorem 6.8 can be simplified using

$$h_{\mathbb{R}^M}(\theta)=\sup_{\omega\in\mathbb{R}^M}\{\theta^\top\omega-\chi_{\mathbb{R}^M}(\omega)\}=\sup_{\omega\in\mathbb{R}^M}\theta^\top\omega=\begin{cases}0\ , & \text{if }\theta=0\ ,\\ +\infty\ , & \text{otherwise}\ .\end{cases}$$

Corollary 6.3 *The assumptions are the same as those in Theorem 6.8 except that* $\Omega = \mathbb{R}^M$. *Then the worst-case expectation*

$$\sup_{\mathbb{P}\in\mathcal{P}_W(\mathbb{R}^M,\mathbb{P}_0,d_0)} \mathbb{E}^{\mathbb{P}}[g(\omega)]$$

$$= \begin{cases} \min\limits_{\lambda,s,z} d_0\lambda + \dfrac{1}{N}\sum\limits_{n=1}^{N} s_n \\ s.t.\ [-g_k]^*(z_{nk}) - \omega_n^\top z_{nk} \le s_n\ ,\ n = 1,\ldots,N\ ,\ k = 1,\ldots,K\ , \\ \qquad \|z_{nk}\|_* \le \lambda\ ,\ n = 1,\ldots,N\ ,\ k = 1,\ldots,K\ , \end{cases}$$

In Theorem 6.8 and Corollary 6.3, the loss function g is required to be the maximum of a finite number of concave functions. In fact, this form can cover a wide variety of loss functions and we can obtain a series of tractable reformulations of Wasserstein-distance worst-case expectation using Theorem 6.8. We first consider the case of piecewise affine loss functions.

Proposition 6.5 (Convex piecewise linear loss function and polyhedral sample space) *Assume that the sample space* $\Omega = \{\omega \in \mathbb{R}^M \mid P\omega \le Q\}$ *is a nonempty polyhedron, where* P *and* Q *are coefficients.* $\mathbb{P}_0 = \sum_{n=1}^{N} \delta_{\omega_n}/N$ *is a probability distribution on* Ω. *The loss function* $g(\omega) = \max_{1\le k\le K}\{A_k^\top\omega + B_k\}$, $\forall\omega \in \mathbb{R}^M$ *is the maximum of affine functions, where* A_k *and* B_k *are coefficients for any* $1 \le k \le K$. *Then the worst-case expectation*

$$\sup_{\mathbb{P}\in\mathcal{P}_W(\Omega,\mathbb{P}_0,d_0)} \mathbb{E}^{\mathbb{P}}[g(\omega)]$$

$$= \begin{cases} \min\limits_{\lambda,s,\gamma} d_0\lambda + \dfrac{1}{N}\sum\limits_{n=1}^{N} s_n \\ s.t.\ B_k + A_k^\top\omega_n + (Q - P\omega_n)^\top\gamma_{nk} \le s_n\ ,\ n = 1,\ldots,N\ ,\ k = 1,\ldots,K\ , \\ \qquad \|P^\top\gamma_{nk} - A_k\|_* \le \lambda\ ,\ n = 1,\ldots,N\ ,\ k = 1,\ldots,K\ , \\ \qquad \gamma_{nk} \ge 0\ ,\ n = 1,\ldots,N\ ,\ k = 1,\ldots,K\ . \end{cases} \tag{6.39}$$

Proof Affine functions are concave. Therefore, by Theorem 6.8, we have the formulation in (6.35). In addition,

$$[-A_k^\top\omega - B_k]^*(\theta) = \sup_{\omega\in\mathbb{R}^M}\{\theta^\top\omega - (-A_k^\top\omega - B_k)\} = \begin{cases} B_k\ , & \text{if } \theta = -A_k\ , \\ +\infty\ , & \text{otherwise}\ . \end{cases}$$

The support function of Ω is $h_\Omega(\theta) = \sup_{\omega\in\mathbb{R}^M}\{\theta^\top\omega - \chi_\Omega(\omega)\}$, where χ_Ω is the characteristic function, i.e., $\chi_\Omega(\omega) = 0$, if $P\omega \le Q$; $\chi_\Omega(\omega) = +\infty$, otherwise. Then

$$h_\Omega(\theta) = \sup_{\omega \in \mathbb{R}^M} \{\theta^\top \omega - \chi_\Omega(\omega)\} = \left\{ \begin{array}{l} \max\limits_{\omega} \theta^\top \omega \\ \text{s.t. } P\omega \le Q \end{array} \right. = \left\{ \begin{array}{l} \min\limits_{\gamma} Q^\top \gamma \\ \text{s.t. } P^\top \gamma = \theta \ , \ \gamma \ge 0 \ , \end{array} \right. \tag{6.40}$$

by LP duality and the assumption that Ω is nonempty. Therefore, (6.35) can be equivalently transformed into:

$$\begin{aligned} \min_{\lambda,s,z,v,\gamma} \ & d_0\lambda + \frac{1}{N}\sum_{n=1}^{N} s_n \\ \text{s.t. } & B_k + Q^\top \gamma_{nk} - \omega_n^\top z_{nk} \le s_n \ , \ z_{nk} - v_{nk} = -A_k \ , \ \forall n \ , \ \forall k \ , \\ & P^\top \gamma_{nk} = v_{nk} \ , \ \gamma_{nk} \ge 0 \ , \ \forall n \ , \ \forall k \ , \\ & \|z_{nk}\|_* \le \lambda \ , \ \forall n \ , \ \forall k \ . \end{aligned}$$

Eliminating z and v, then we have (6.39). □

Proposition 6.5 gives a tractable exact reformulation for the case where the sample space is a polyhedron and the loss function is the maximum of a finite number of affine functions. Such a loss function is convex. Moreover, any convex function can be approximated by a piecewise affine convex function, which can be written in the form of the maximum of some affine functions. Therefore, Proposition 6.5 can be used to approximate the convex loss function case and will give a good reformulation if the number of pieces is not too large.

Exercise 6.9 Under the assumptions of Proposition 6.5, further assume $A_k(x)$ and $B_k(x)$ are affine in x, $\mathcal{X}$ is a polyhedron, and the norm is either 1-norm or ∞-norm. Reformulate the Wasserstein-distance DRO problem (6.27) as an equivalent LP problem using Proposition 6.5.

Proposition 6.6 considers the minimum of some affine functions:

Proposition 6.6 (Concave piecewise linear loss function and polyhedral sample space) *The assumptions are the same as those in Proposition 6.5 except that* $g(\omega) = \min_{1 \le k \le K}\{A_k^\top \omega + B_k\}$, $\forall \omega \in \mathbb{R}^M$, *then the worst-case expectation*

$$\sup_{\mathbb{P} \in \mathcal{P}_W(\Omega, \mathbb{P}_0, d_0)} \mathbb{E}^{\mathbb{P}}[g(\omega)]$$

$$= \left\{ \begin{aligned} \min_{\lambda,s,\gamma,\rho} \ & d_0\lambda + \frac{1}{N}\sum_{n=1}^{N} s_n \\ s.t. \ & (B + A^\top \omega_n)^\top \rho_n + (Q - P\omega_n)^\top \gamma_n \le s_n \ , \ n = 1, \dots, N \ , \\ & \|P^\top \gamma_n - A\rho_n\|_* \le \lambda \ , \ n = 1, \dots, N \ , \\ & 1_{K\times 1}^\top \rho_n = 1 \ , \ \gamma_n \ge 0 \ , \ \rho_n \ge 0 \ , \ n = 1, \dots, N \ , \end{aligned} \right.$$

where $A = (A_1 \ \dots \ A_K)$ *and* B *is the column vector with the kth element being* B_k.

Proof Note that $g(\omega)$ is a concave function. Therefore, by Theorem 6.8, we have:

$$
\sup_{\mathbb{P}\in\mathcal{P}_{\mathrm{W}}(\Omega,\mathbb{P}_0,d_0)} \mathbb{E}^{\mathbb{P}}[g(\omega)]
= \begin{cases}
\min\limits_{\lambda,s,z,v} \ d_0\lambda + \dfrac{1}{N}\sum\limits_{n=1}^{N} s_n \\
\text{s.t. } [-g]^*(z_n - v_n) + h_\Omega(v_n) - \omega_n^\top z_n \le s_n \ , \ n = 1, \dots, N \ , \\
\quad \|z_n\|_* \le \lambda \ , \ n = 1, \dots, N \ .
\end{cases} \tag{6.41}
$$

Because

$$
\begin{aligned}
[-g]^*(\theta) &= \left[-\min_{1\le k\le K} \{A_k^\top \omega + B_k\} \right]^* (\theta) \\
&= \sup_{\omega\in\mathbb{R}^M} \left\{ \theta^\top \omega + \min_{1\le k\le K} \{A_k^\top \omega + B_k\} \right\} \\
&= \begin{cases}
\max\limits_{\omega,t} \ \theta^\top \omega + t \\
\text{s.t. } t \le A_k^\top \omega + B_k \ , \ k = 1, \dots, K
\end{cases} \\
&= \begin{cases}
\max\limits_{\omega,t} \ \theta^\top \omega + t \\
\text{s.t. } 1_{K\times 1} \cdot t - A^\top \omega \le B : \rho
\end{cases} \\
&= \begin{cases}
\min\limits_{\rho} \ B^\top \rho \\
\text{s.t. } 1_{K\times 1}^\top \rho = 1 \ , \ -A\rho = \theta \ , \ \rho \ge 0 \ ,
\end{cases}
\end{aligned} \tag{6.42}
$$

where the last inequality follows from the LP's strong duality and the feasibility of the primal problem. Using (6.40), (6.41), and (6.42), we have:

$$
\sup_{\mathbb{P}\in\mathcal{P}_{\mathrm{W}}(\Omega,\mathbb{P}_0,d_0)} \mathbb{E}^{\mathbb{P}}[g(\omega)]
= \begin{cases}
\min\limits_{\lambda,s,z,v,\gamma,\rho} \ d_0\lambda + \dfrac{1}{N}\sum\limits_{n=1}^{N} s_n \\
\text{s.t. } B^\top \rho_n + Q^\top \gamma_n - \omega_n^\top z_n \le s_n \ , \ \forall n \ , \\
\quad 1_{K\times 1}^\top \rho_n = 1 \ , \ -A\rho_n = z_n - v_n \ , \ \rho_n \ge 0 \ , \ \forall n \ , \\
\quad P^\top \gamma_n = v_n \ , \ \gamma_n \ge 0 \ , \ \forall n \ , \\
\quad \|z_n\|_* \le \lambda \ , \ \forall n \ .
\end{cases}
$$

The proof is completed after eliminating z and v. □

Now we study two-stage problems using Theorem 6.8.

Proposition 6.7 (Loss function defined by an LP problem with uncertainty in the objective) *Suppose that the sample space $\Omega = \{\omega \in \mathbb{R}^M \mid P\omega \leq Q\}$ is a nonempty polyhedron and $\mathbb{P}_0 = \sum_{n=1}^{N} \delta_{\omega_n}/N$ is a probability distribution on Ω. The loss function $g : \Omega \to \mathbb{R}$ is defined by:*

$$g(\omega) := \begin{cases} \min\limits_{y} \ (D\omega + E)^\top y \\ s.t. \ Ay \geq B \ , \end{cases}$$

where A, B, D, E, P, and Q are coefficients and the set $\{y \mid Ay \geq B\}$ is nonempty and compact. Then the worst-case expectation

$$\sup_{\mathbb{P}\in\mathcal{P}_W(\Omega,\mathbb{P}_0,d_0)} \mathbb{E}^{\mathbb{P}}[g(\omega)]$$

$$= \begin{cases} \min\limits_{\lambda,s,\gamma,y} \ d_0\lambda + \dfrac{1}{N}\sum\limits_{n=1}^{N} s_n \\ s.t. \ (D\omega_n + E)^\top y_n + (Q - P\omega_n)^\top \gamma_n \leq s_n \ , \ n = 1, \ldots, N \ , \\ \qquad Ay_n \geq B \ , \ \|D^\top y_n - P^\top \gamma_n\|_* \leq \lambda \ , \ \gamma_n \geq 0 \ , \ n = 1, \ldots, N \ . \end{cases}$$

Proof We observe that $g(\omega)$ is concave, and the proof is left as an exercise. In addition,

$$\begin{aligned} [-g]^*(\theta) &= \sup_{\omega\in\mathbb{R}^M} \left\{\theta^\top \omega + \inf_{y:Ay\geq B} (D\omega + E)^\top y\right\} \\ &= \inf_{y:Ay\geq B} \left\{ \sup_{\omega\in\mathbb{R}^M} \{\theta^\top \omega + (D\omega)^\top y\} + E^\top y\right\} \\ &= \begin{cases} \min\limits_{y} \ E^\top y \\ \text{s.t. } Ay \geq B \ , \ D^\top y + \theta = 0 \end{cases} \end{aligned}$$

where the interchange of $\inf_{y:Ay\geq B}$ and $\sup_{\omega\in\mathbb{R}^M}$ is due to Theorem 2.15 and the compactness of $\{y \mid Ay \geq B\}$. Combining it with the conclusion of Theorem 6.8 and (6.40), we have:

$$\sup_{\mathbb{P}\in\mathcal{P}_W(\Omega,\mathbb{P}_0,d_0)} \mathbb{E}^{\mathbb{P}}[g(\omega)]$$

$$= \begin{cases} \min\limits_{\lambda,s,z,v,\gamma,y} \; d_0\lambda + \dfrac{1}{N}\sum\limits_{n=1}^{N} s_n \\ \text{s.t. } E^\top y_n + Q^\top \gamma_n - \omega_n^\top z_n \le s_n \;,\; \forall n \;, \\ \quad Ay_n \ge B \;,\; D^\top y_n + z_n - v_n = 0 \;,\; \forall n \;, \\ \quad P^\top \gamma_n = v_n \;,\; \gamma_n \ge 0 \;,\; \forall n \;, \\ \quad \|z_n\|_* \le \lambda \;,\; \forall n \;. \end{cases}$$

The proof is completed after eliminating z and v. □

Exercise 6.10 Assume $\{y \mid Ay \ge B\}$ is compact. Prove that the function $g(\omega) = \inf_{y:Ay\ge B}(D\omega + E)^\top y$ is concave on $\mathbb{R}^M$.

Exercise 6.11 Under the assumptions of Proposition 6.7, further assume $B(x)$ is affine in x, $\mathcal{X}$ is a polyhedron, and the norm is either 1-norm or ∞-norm. Reformulate the Wasserstein-distance DRO problem (6.27) as an equivalent LP problem using Proposition 6.7.

Proposition 6.8 (Loss function defined by an LP problem with right-hand-side uncertainty) *Assume that* $\mathbb{P}_0 = \sum_{n=1}^{N} \delta_{\omega_n}/N$ *is a probability distribution on the nonempty sample space* $\Omega = \{\omega \in \mathbb{R}^M \mid P\omega \le Q\}$. *The loss function* $g : \Omega \to \mathbb{R}$ *is:*

$$g(\omega) := \begin{cases} \min\limits_{y} \; D^\top y \\ \text{s.t. } Ay \ge B\omega + H \;, \end{cases} \tag{6.43}$$

where A, B, D, H, P, *and* Q *are coefficients. Assume the dual-feasible set* $\{\nu \mid A^\top \nu = D, \nu \ge 0\}$ *is nonempty and compact with vertices* ν_k, $k = 1, \ldots, K$. *Then the worst-case expectation*

$$\sup_{\mathbb{P}\in\mathcal{P}_W(\Omega,\mathbb{P}_0,d_0)} \mathbb{E}^{\mathbb{P}}[g(\omega)]$$

$$= \begin{cases} \min\limits_{\lambda,s,\gamma} \; d_0\lambda + \dfrac{1}{N}\sum\limits_{n=1}^{N} s_n \\ \text{s.t. } H^\top \nu_k + \nu_k^\top B\omega_n + (Q - P\omega_n)^\top \gamma_{nk} \le s_n \;,\; n = 1, \ldots, N \;,\; k = 1, \ldots, K \;, \\ \quad \|P^\top \gamma_{nk} - B^\top \nu_k\|_* \le \lambda \;,\; \gamma_{nk} \ge 0 \;,\; n = 1, \ldots, N \;,\; k = 1, \ldots, K \;. \end{cases}$$

Proof According to the LP duality theory,

$K = 2$ $K = 4$ $K = 8$

Fig. 6.11 For the rectangle $\{x \in \mathbb{R}^M \mid 0 \leq x \leq 1\}$, the number of vertices is $K = 2^M$, which grows at an exponential rate as M increases

$$
\begin{aligned}
g(\omega) &= \begin{cases} \min\limits_{y} \; D^\top y \\ \text{s.t. } Ay \geq B\omega + H \end{cases} \\
&= \begin{cases} \max\limits_{\nu} \; (B\omega + H)^\top \nu \\ \text{s.t. } A^\top \nu = D \,, \; \nu \geq 0 \end{cases} \\
&= \max_{1 \leq k \leq K} (B\omega + H)^\top \nu_k = \max_{1 \leq k \leq K} \{(B^\top \nu_k)^\top \omega + H^\top \nu_k\} \,, \; \forall \omega \in \Omega \,.
\end{aligned}
$$

Then the conclusion follows from Proposition 6.5, where A_k and B_k in Proposition 6.5 are replaced by $B^\top \nu_k$ and $H^\top \nu_k$, respectively. □

Exercise 6.12 Under the assumptions of Proposition 6.8, further assume $B(x)$ and $H(x)$ are affine in x, $\mathcal{X}$ is a polyhedron, and the norm is either 1-norm or ∞-norm. Reformulate the Wasserstein-distance DRO problem (6.27) as an equivalent LP problem using Proposition 6.8.

Although Proposition 6.8 gives a tractable formulation for two-stage Wasserstein-distance DRO problems, it requires information on the vertices of the dual-feasible set. However, finding all the vertices of a polyhedron represented by some linear constraints is not trivial in general. Moreover, the number of vertices grows at an exponential rate as the dimension of the dual-feasible set increases due to the curse of dimensionality, as illustrated in Fig. 6.11. Therefore, it becomes impracticable to directly use Proposition 6.8 under high-dimensional situations.

Nonetheless, the loss function can be approximated by considering a part of the vertices, i.e.,

$$
g(\omega) = \max_{1 \leq k \leq K} \{(B^\top \nu_k)^\top \omega + H^\top \nu_k\} \geq \max_{k \in \mathcal{K}} \{(B^\top \nu_k)^\top \omega + H^\top \nu_k\} \,,
$$

where $\mathcal{K}$ is a subset of $\{1, \ldots, K\}$. Some critical vertices play an important role in $g(\omega)$, whereas others do not. Suppose $\Omega = \{\omega \in \mathbb{R}^M \mid P\omega \leq Q\}$ is compact and we have a set $\mathcal{K}$, the most important vertex in the remaining part can be found by solving

$$\max_{\omega\in\Omega}\left\{g(\omega)-\max_{k\in\mathcal{K}}\{(B^\top\nu_k)^\top\omega+H^\top\nu_k\}\right\}$$

$$=\begin{cases}\max\limits_{\omega,s}\ g(\omega)-s\\ \text{s.t. } s\geq(B^\top\nu_k)^\top\omega+H^\top\nu_k\ ,\ \forall k\in\mathcal{K}\end{cases}$$

$$=\begin{cases}\max\limits_{\omega,s,\tilde{\nu}}\ (B\omega+H)^\top\tilde{\nu}-s\\ \text{s.t. } s\geq(B^\top\nu_k)^\top\omega+H^\top\nu_k\ ,\ \forall k\in\mathcal{K}\ ,\\ \qquad A^\top\tilde{\nu}=D\ ,\ \tilde{\nu}\geq 0\ .\end{cases}$$

Although this problem has bilinear terms and thus is difficult to find the global optimum, a suboptimal vertex can also improve the accuracy of the approximation. By adding enough vertices into $\mathcal{K}$, we can control the approximation error to a desired degree and then obtain an accurate enough reformulation by Proposition 6.8.

When sample space Ω is the Euclidean space $\mathbb{R}^M$, the reformulation becomes easier.

Theorem 6.9 (Loss function defined by an LP problem with right-hand-side uncertainty and Euclidean sample space, Theorem 6 in Hanasusanto and Kuhn 2018) *Assume that $\mathbb{P}_0=\sum_{n=1}^N\delta_{\omega_n}/N$ is a probability distribution on $\mathbb{R}^M$. The loss function $g:\Omega\to\mathbb{R}$ is defined by* (6.43)*, i.e., $g(\omega)=\min_{y:Ay\geq B\omega+H}D^\top y$, which is assumed to be finite for any $\omega\in\mathbb{R}^M$. Then for any $d_0>0$, the worst-case expectation*

$$\sup_{\mathbb{P}\in\mathcal{P}_W(\mathbb{R}^M,\mathbb{P}_0,d_0)}\mathbb{E}^{\mathbb{P}}[g(\omega)]=\begin{cases}\min\limits_{y}\ d_0\left(\sup\limits_{\nu:A^\top\nu=D,\nu\geq0}\|B^\top\nu\|_*\right)+\dfrac{1}{N}\sum\limits_{n=1}^N D^\top y_n\\ \text{s.t. } Ay_n\geq B\omega_n+H\ ,\ n=1,\ldots,N\ .\end{cases}$$

Proof By Theorem 6.7,

$$\sup_{\mathbb{P}\in\mathcal{P}_W(\mathbb{R}^M,\mathbb{P}_0,d_0)}\mathbb{E}^{\mathbb{P}}[g(\omega)]=\inf_{\lambda\geq0}\left\{d_0\lambda+\frac{1}{N}\sum_{n=1}^N\sup_{\omega\in\mathbb{R}^M}\{g(\omega)-\lambda\|\omega-\omega_n\|\}\right\}.$$

We focus on $\sup_{\omega\in\mathbb{R}^M}\{g(\omega)-\lambda\|\omega-\omega_n\|\}$:

$$\sup_{\omega\in\mathbb{R}^M}\{g(\omega)-\lambda\|\omega-\omega_n\|\} \tag{6.44a}$$

$$=\sup_{\omega\in\mathbb{R}^M}\left\{\sup_{\nu:A^\top\nu=D,\nu\geq0}(B\omega+H)^\top\nu-\lambda\|\omega-\omega_n\|\right\} \tag{6.44b}$$

$$=\sup_{\nu:A^\top\nu=D,\nu\geq0}\ \sup_{\omega\in\mathbb{R}^M}\ \inf_{\zeta:\|\zeta\|_*\leq\lambda}\{(B\omega+H)^\top\nu-\zeta^\top\omega+\zeta^\top\omega_n\} \tag{6.44c}$$

$$=\sup_{\nu:A^\top\nu=D,\nu\geq0}\ \inf_{\zeta:\|\zeta\|_*\leq\lambda}\ \sup_{\omega\in\mathbb{R}^M}\{(B\omega+H)^\top\nu-\zeta^\top\omega+\zeta^\top\omega_n\} \tag{6.44d}$$

$$= \begin{cases} \sup_{\nu : A^\top \nu = D, \nu \geq 0} \{H^\top \nu + \omega_n^\top B^\top \nu\} , & \text{if } \|B^\top \nu\|_* \leq \lambda \ , \ \forall \nu \ , \ \text{s.t. } A^\top \nu = D, \nu \geq 0 \ , \\ +\infty \ , & \text{otherwise} \end{cases} \tag{6.44e}$$

$$= \begin{cases} \min_{y : Ay \geq B\omega_n + H} D^\top y \ , & \text{if } \lambda \geq \sup_{\nu : A^\top \nu = D, \nu \geq 0} \|B^\top \nu\|_* \ , \\ +\infty \ , & \text{otherwise} \ . \end{cases} \tag{6.44f}$$

From (6.44b) to (6.44c), we use the property of dual norm. (6.44d) follows from the compactness of $\{\zeta \mid \|\zeta\|_* \leq \lambda\}$ and the classical minimax theorem in Theorem 2.15. In (6.44b) and (6.44f), the LP strong duality is used. Combining (6.44f) and the duality result in Theorem 6.7, we complete the proof. □

Corollary 6.4 *Under the assumption of Theorem 6.9, further assume the norm is 1-norm, then*

$$\sup_{\mathbb{P} \in \mathcal{P}_W(\mathbb{R}^M, \mathbb{P}_0, d_0)} \mathbb{E}^{\mathbb{P}}[g(\omega)] = \begin{cases} \min_{\lambda, y, \phi, \psi} \ d_0 \lambda + \frac{1}{N} \sum_{n=1}^{N} D^\top y_n \\ s.t. \ Ay_n \geq B\omega_n + H \ , \ n = 1, \ldots, N \ , \\ \quad D^\top \phi_m \leq \lambda \ , \ A\phi_m \geq B_m \ , \ m = 1, \ldots, M \ , \\ \quad D^\top \psi_m \leq \lambda \ , \ A\psi_m \geq -B_m \ , \ m = 1, \ldots, M \ , \end{cases}$$

where $B = (B_1 \ \ldots \ B_M)$ *with column vectors* $B_1, \ldots, B_M$.

Proof By the strong duality of LP,

$$\sup_{\nu : A^\top \nu = D, \nu \geq 0} B_m^\top \nu = \min_{\phi_m : A\phi_m \geq B_m} D^\top \phi_m \ ,$$

$$\sup_{\nu : A^\top \nu = D, \nu \geq 0} -B_m^\top \nu = \min_{\psi_m : A\psi_m \geq -B_m} D^\top \psi_m \ .$$

Thus,

$$\sup_{\nu : A^\top \nu = D, \nu \geq 0} \|B^\top \nu\|_\infty \leq \lambda$$

$$\Longleftrightarrow \begin{cases} \sup_{\nu : A^\top \nu = D, \nu \geq 0} B_m^\top \nu \leq \lambda \ , \ m = 1, \ldots, M \ , \\ \sup_{\nu : A^\top \nu = D, \nu \geq 0} -B_m^\top \nu \leq \lambda \ , \ m = 1, \ldots, M \end{cases}$$

$$\Longleftrightarrow \begin{cases} \exists \phi_m \ , \ \text{s.t. } D^\top \phi_m \leq \lambda \ , \ A\phi_m \geq B_m \ , \ m = 1, \ldots, M \ , \\ \exists \psi_m \ , \ \text{s.t. } D^\top \psi_m \leq \lambda \ , \ A\psi_m \leq -B_m \ , \ m = 1, \ldots, M \ . \end{cases}$$

Then the conclusion follows from Theorem 6.9. □

Exercise 6.13 Under the assumptions of Corollary 6.4, further assume $B(x)$ and $H(x)$ are affine in x and $\mathcal{X}$ is a polyhedron. Reformulate the Wasserstein-distance DRO problem (6.27) as an equivalent LP problem using Corollary 6.4.

6.3.2.2 Lipschitz Constant Reformulation

From the Kantorovich duality in Theorem 6.1 and Remark 6.2, we know that Wasserstein distance has a close relationship to Lipschitz continuity. In fact, the worst-case expectation has an immediate estimation using the Lipschitz constant of the loss function as follows:

Theorem 6.10 (Lipschitz constant reformulation) *Suppose $\mathbb{P}_0$ is a probability distribution on $\Omega \subset \mathbb{R}^M$. The measurable function $g : \Omega \to \mathbb{R}$ is Lipschitz continuous with Lipschitz constant $L > 0$, i.e.,*

$$|g(\omega') - g(\omega'')| \leq L\|\omega' - \omega''\| \ , \ \forall \omega', \omega'' \in \Omega \ .$$

Then for any $d_0 \geq 0$, the worst-case expectation

$$\mathbb{E}^{\mathbb{P}_0}[g(\omega)] \leq \sup_{\mathbb{P} \in \mathcal{P}_W(\Omega, \mathbb{P}_0, d_0)} \mathbb{E}^{\mathbb{P}}[g(\omega)] \leq d_0 L + \mathbb{E}^{\mathbb{P}_0}[g(\omega)] \ . \tag{6.45}$$

Proof Because $\mathbb{P}_0 \in \mathcal{P}_W(\Omega, \mathbb{P}_0, d_0)$, we have the first inequality in (6.45). To prove the second one, we use the Kantorovich duality in Theorem 6.1 and Remark 6.2:

$$d_W(\mathbb{P}, \mathbb{P}_0) = \begin{cases} \sup\limits_{f:\Omega \to \mathbb{R}} \ \mathbb{E}^{\mathbb{P}}[f(\omega)] - \mathbb{E}^{\mathbb{P}_0}[f(\omega)] \\ \quad \text{s.t. } |f(\omega') - f(\omega'')| \leq \|\omega' - \omega''\| \ , \ \forall \omega', \omega'' \in \Omega \ . \end{cases}$$

Note that $|g(\omega')/L - g(\omega'')/L| \leq \|\omega' - \omega''\|, \forall \omega', \omega'' \in \Omega$. Therefore,

$$\mathbb{E}^{\mathbb{P}}[g(\omega)/L] - \mathbb{E}^{\mathbb{P}_0}[g(\omega)/L] \leq d_W(\mathbb{P}, \mathbb{P}_0) \leq d_0 \ , \ \forall \mathbb{P} \in \mathcal{P}_W(\Omega, \mathbb{P}_0, d_0) \ ,$$

which implies the second inequality in (6.45). □

The conclusion of Theorem 6.10 is rather general in that there is no restriction on the form of $\mathbb{P}_0$, Ω, and g. The result says that the worst-case expectation is between the expectation under the empirical probability distribution and a constant plus it, where the constant is proportional to the Wasserstein-distance threshold d_0 and the Lipschitz constant L. Therefore, the smaller L, the tighter the upper bound. The best Lipschitz constant is the smallest one, i.e., Lipschitz modulus $\text{Lip}(g)$ defined in Definition 2.14:

$$\text{Lip}(g) = \sup_{\omega', \omega'' \in \Omega, \omega' \neq \omega''} \frac{|g(\omega') - g(\omega'')|}{\|\omega' - \omega''\|} \ ,$$

There is a simple way to comprehend the form of the upper bound in Theorem 6.10. Suppose $\mathbb{P}_0(\omega') = 1$ and $g(\omega'') - g(\omega') = L\|\omega' - \omega''\| > 0$ for some $\omega'' \in \Omega$. Let $\mathbb{P}$ be the probability distribution generated from $\mathbb{P}_0$ by transforming probability ε at ω' to ω'', i.e., $\mathbb{P}(\omega') = 1 - \varepsilon$ and $\mathbb{P}(\omega'') = \varepsilon$. Then $d_W(\mathbb{P}, \mathbb{P}_0) = \varepsilon\|\omega' - \omega''\|$ and $\mathbb{E}^{\mathbb{P}}[g(\omega)] - \mathbb{E}^{\mathbb{P}_0}[g(\omega)] = -\varepsilon g(\omega') + \varepsilon g(\omega'') = \varepsilon L\|\omega' - \omega''\| = L \cdot d_W(\mathbb{P}, \mathbb{P}_0)$. Therefore, if the Lipschitz modulus can be obtained at some point with a large enough probability, then the upper bound based on the Lipschitz modulus in Theorem 6.10 is exact, i.e., $\sup_{\mathbb{P}\in\mathcal{P}_W(\Omega,\mathbb{P}_0,d_0)} \mathbb{E}^{\mathbb{P}}[g(\omega)] = d_0 \cdot \mathrm{Lip}(g) + \mathbb{E}^{\mathbb{P}_0}[g(\omega)]$.

For a convex function on the Euclidean space, the Lipschitz modulus can be calculated using the conjugate function:

Proposition 6.9 (Lipschitz modulus of a function on the Euclidean space) *Denote the conjugate function of function $g : \mathbb{R}^M \to \mathbb{R}$ by g^*. The dual norm of norm $\|\cdot\|$ is denoted by $\|\cdot\|_*$. Then the Lipschitz modulus under norm $\|\cdot\|$ satisfies:*

1. *Lip$(g) \geq \sup\{\|\theta\|_* \mid g^*(\theta) < +\infty\}$.*
2. *If furthermore g is convex, then Lip$(g) = \sup\{\|\theta\|_* \mid g^*(\theta) < +\infty\}$.*

Proof 1. Consider some θ such that $g^*(\theta) < +\infty$. Because $g^*(\theta) = \sup_{\omega\in\mathbb{R}^M} \{\theta^\top\omega - g(\omega)\}$, for any $\varepsilon > 0$, there exists $\omega_\varepsilon \in \mathbb{R}^M$ such that $g^*(\theta) \leq \theta^\top\omega_\varepsilon - g(\omega_\varepsilon) + \varepsilon$. By Fenchel's inequality in Theorem 2.19, $g^*(\theta) \geq \theta^\top\zeta - g(\zeta), \forall\zeta \in \mathbb{R}^M$, so

$$0 \leq g^*(\theta) + g(\zeta) - \theta^\top\zeta \leq \theta^\top\omega_\varepsilon - g(\omega_\varepsilon) + \varepsilon + g(\zeta) - \theta^\top\zeta\ ,\ \forall\zeta \in \mathbb{R}^M\ ,$$

which implies:

$$g(\zeta) - g(\omega_\varepsilon) \geq \theta^\top(\zeta - \omega_\varepsilon) - \varepsilon\ ,\ \forall\zeta \in \mathbb{R}^M\ .$$

There exists $\theta' \in \mathbb{R}^M$ such that $\|\theta'\| = 1$ and $\theta^\top\theta' = \|\theta\|_*$. Let $\zeta_\varepsilon = \omega_\varepsilon + \theta'$, then

$$g(\zeta_\varepsilon) - g(\omega_\varepsilon) \geq \theta^\top\theta' - \varepsilon = \|\theta\|_*\|\zeta_\varepsilon - \omega_\varepsilon\| - \varepsilon\ .$$

Let $\varepsilon \to 0$, then $\mathrm{Lip}(g) \geq \|\theta\|_*$. Taking supremum over $\{\theta \mid g^*(\theta) < +\infty\}$, we have $\mathrm{Lip}(g) \geq \sup\{\|\theta\|_* \mid g^*(\theta) < +\infty\}$.

2. Consider some θ such that $g^*(\theta) < +\infty$. By Fenchel's inequality, $\theta^\top\omega - g(\omega) \leq g^*(\theta), \forall\omega \in \mathbb{R}^M$. Thus, for any $\omega, \zeta \in \mathbb{R}^M$,

$$g(\zeta) - g(\omega) \leq g(\zeta) - \theta^\top\omega + g^*(\theta) = \theta^\top(\zeta - \omega) + g(\zeta) + g^*(\theta) - \theta^\top\zeta\ ,$$

which implies:

$$\begin{aligned}
g(\zeta) - g(\omega) &\leq \inf_{\theta: g^*(\theta) < +\infty} \{\theta^\top(\zeta - \omega) + g(\zeta) + g^*(\theta) - \theta^\top \zeta\} \\
&\leq \inf_{\theta: g^*(\theta) < +\infty} \left\{ g(\zeta) + g^*(\theta) - \theta^\top \zeta + \sup_{\tilde{\theta}: g^*(\tilde{\theta}) < +\infty} \tilde{\theta}^\top(\zeta - \omega) \right\} \\
&= g(\zeta) - \sup_{\theta: g^*(\theta) < +\infty} \{\theta^\top \zeta - g^*(\theta)\} + \sup_{\theta: g^*(\theta) < +\infty} \theta^\top(\zeta - \omega) \\
&= g(\zeta) - g^{**}(\zeta) + \sup_{\theta: g^*(\theta) < +\infty} \theta^\top(\zeta - \omega) \\
&= \sup_{\theta: g^*(\theta) < +\infty} \theta^\top(\zeta - \omega) ,
\end{aligned}$$

where g^{**} is the biconjugate function of g and the last equality follows from the Fenchel–Moreau theorem for convex functions in Theorem 2.20. Since $\theta^\top(\zeta - \omega) \leq \|\theta\|_* \|\zeta - \omega\|$ and ζ and ω are arbitrary, we have $\mathrm{Lip}(g) \leq \sup\{\|\theta\|_* \mid g^*(\theta) < +\infty\}$. Therefore, they are equal. □

Theorem 6.11 gives a special case when the upper bound in Theorem 6.10 is exact:

Theorem 6.11 (Convex loss function and Euclidean sample space, Theorem 6.3 in Esfahani and Kuhn 2018) *Suppose $\mathbb{P}_0$ is a probability distribution on $\mathbb{R}^M$. The function $g : \mathbb{R}^M \to \mathbb{R}$ is convex, its conjugate function is g^*, and its Lipschitz modulus under norm $\|\cdot\|$ is $Lip(g)$. The dual norm of $\|\cdot\|$ is $\|\cdot\|_*$. Then for any $d_0 \geq 0$, the worst-case expectation*

$$\begin{aligned}
\sup_{\mathbb{P} \in \mathcal{P}_W(\mathbb{R}^M, \mathbb{P}_0, d_0)} \mathbb{E}^{\mathbb{P}}[g(\omega)] &= d_0 \cdot Lip(g) + \mathbb{E}^{\mathbb{P}_0}[g(\omega)] \\
&= d_0 \cdot \sup\{\|\theta\|_* \mid g^*(\theta) < +\infty\} + \mathbb{E}^{\mathbb{P}_0}[g(\omega)] .
\end{aligned}$$

Proof We only need to prove the case $d_0 > 0$. According to Theorem 6.7,

$$\sup_{\mathbb{P} \in \mathcal{P}_W(\mathbb{R}^M, \mathbb{P}_0, d_0)} \mathbb{E}^{\mathbb{P}}[g(\omega)] = \inf_{\lambda \geq 0} \left\{ d_0 \lambda + \frac{1}{N} \sum_{n=1}^{N} \sup_{\omega \in \mathbb{R}^M} \{g(\omega) - \lambda \|\omega - \omega_n\|\} \right\} .$$

By Fenchel–Moreau theorem in Theorem 2.20, we have $g^{**} = g$. Let $\Theta := \{\theta \in \mathbb{R}^M \mid g^*(\theta) < +\infty\}$. Then

$$\begin{aligned}
&\sup_{\omega \in \mathbb{R}^M} \{g(\omega) - \lambda \|\omega - \omega_n\|\} \\
&= \sup_{\omega \in \mathbb{R}^M} \left\{ \sup_{\theta \in \mathbb{R}^M} \{\omega^\top \theta - g^*(\theta)\} - \lambda \|\omega - \omega_n\| \right\} \\
&= \sup_{\omega \in \mathbb{R}^M} \sup_{\theta \in \Theta} \left\{ \omega^\top \theta - g^*(\theta) - \sup_{\zeta: \|\zeta\|_* \leq \lambda} \zeta^\top(\omega - \omega_n) \right\}
\end{aligned}$$

$$
\begin{aligned}
&= \sup_{\omega\in\mathbb{R}^M} \sup_{\theta\in\Theta} \inf_{\zeta:\|\zeta\|_*\leq\lambda} \{\omega^\top\theta - g^*(\theta) - \zeta^\top(\omega-\omega_n)\} \\
&= \sup_{\theta\in\Theta} \sup_{\omega\in\mathbb{R}^M} \inf_{\zeta:\|\zeta\|_*\leq\lambda} \{(\theta-\zeta)^\top\omega - g^*(\theta) + \zeta^\top\omega_n\} .
\end{aligned}
$$

By Theorem 2.15, we can interchange $\sup_{\omega\in\mathbb{R}^M}$ and $\inf_{\zeta:\|\zeta\|_*\leq\lambda}$ and obtain:

$$
\begin{aligned}
&\sup_{\omega\in\mathbb{R}^M} \{g(\omega) - \lambda\|\omega-\omega_n\|\} \\
&= \sup_{\theta\in\Theta} \inf_{\zeta:\|\zeta\|_*\leq\lambda} \sup_{\omega\in\mathbb{R}^M} \{(\theta-\zeta)^\top\omega - g^*(\theta) + \zeta^\top\omega_n\} \\
&= \sup_{\theta\in\Theta} \inf_{\zeta:\|\zeta\|_*\leq\lambda} \begin{cases} -g^*(\theta) + \theta^\top\omega_n \ , & \text{if } \zeta = \theta \ , \\ +\infty \ , & \text{otherwise} \end{cases} \\
&= \sup_{\theta\in\Theta} \begin{cases} -g^*(\theta) + \theta^\top\omega_n \ , & \text{if } \|\theta\|_* \leq \lambda \ , \\ +\infty \ , & \text{otherwise} \end{cases} \\
&= \begin{cases} g^{**}(\omega_n) \ , & \text{if } \|\theta\|_* \leq \lambda \ , \ \forall\theta\in\Theta \ , \\ +\infty \ , & \text{otherwise} \end{cases} \\
&= \begin{cases} g(\omega_n) \ , & \text{if } \sup_{\theta\in\Theta}\|\theta\|_* \leq \lambda \ , \\ +\infty \ , & \text{otherwise} . \end{cases}
\end{aligned}
$$

Therefore,

$$
\begin{aligned}
&\sup_{\mathbb{P}\in\mathcal{P}_{\mathrm{W}}(\mathbb{R}^M,\mathbb{P}_0,d_0)} \mathbb{E}^{\mathbb{P}}[g(\omega)] \\
&= \inf_{\lambda\geq 0} \left\{ d_0\lambda + \frac{1}{N}\sum_{n=1}^{N} \sup_{\omega\in\mathbb{R}^M} \{g(\omega) - \lambda\|\omega-\omega_n\|\} \right\} \\
&= \inf_{\lambda\geq\sup_{\theta\in\Theta}\|\theta\|_*} \left\{ d_0\lambda + \frac{1}{N}\sum_{n=1}^{N} g(\omega_n) \right\} \\
&= d_0 \cdot \sup\{\|\theta\|_* \mid g^*(\theta) < +\infty\} + \mathbb{E}^{\mathbb{P}_0}[g(\omega)] .
\end{aligned}
$$

Combine it with the conclusion of Proposition 6.9, then the proof is complete. □

Exercise 6.14 The assumptions are the same as those in Proposition 6.5 except that $\Omega = \mathbb{R}^M$.

1. Use Corollary 6.3 to prove that the worst-case expectation

$$
\sup_{\mathbb{P}\in\mathcal{P}_{\mathrm{W}}(\mathbb{R}^M,\mathbb{P}_0,d_0)} \mathbb{E}^{\mathbb{P}}[g(\omega)] = \begin{cases} \min_{s} d_0\left(\max_{1\leq k\leq K}\|A_k\|_*\right) + \frac{1}{N}\sum_{n=1}^{N} s_n \\ \text{s.t. } s_n \geq B_k + \omega_n^\top A_k \ , \ \forall n \ , \ \forall k \ . \end{cases} \tag{6.46}
$$

2. Use Theorem 6.11 to prove (6.46).

In some applications, the Lipschitz constant can be obtained via analysis of the loss function, whereas sometimes it can only be estimated. In the following, we present such an example that estimates the Lipschitz constant and then conservatively approximates two-stage Wasserstein-distance DRO using Theorem 6.10.

Consider the problem:

$$\min_{x \in \mathcal{X}} \left\{ C^\top x + \sup_{\mathbb{P} \in \mathcal{P}_{\mathrm{W}}(\Omega, \mathbb{P}_0, d_0)} \mathbb{E}^{\mathbb{P}}[g(x, \omega)] \right\} , \tag{6.47}$$

where $\Omega \subset \mathbb{R}^M$, the norm is 1-norm, and

$$g(x, \omega) = \begin{cases} \min\limits_y D^\top y \\ \text{ s.t. } Ay \geq B(x)\omega + H(x) , \end{cases}$$

is finite for any $(x, \omega) \in \mathcal{X} \times \Omega$ and $B(x) = (B_1(x) \dots B_M(x))$. If we can find a function $L(x)$ such that for any $x \in \mathcal{X}$,

$$|g(x, \omega') - g(x, \omega'')| \leq L(x) \|\omega' - \omega''\|_1 , \quad \forall \omega', \omega'' \in \Omega , \tag{6.48}$$

then according to Theorem 6.10,

$$\sup_{\mathbb{P} \in \mathcal{P}_{\mathrm{W}}(\Omega, \mathbb{P}_0, d_0)} \mathbb{E}^{\mathbb{P}}[g(x, \omega)] \leq d_0 L(x) + \mathbb{E}^{\mathbb{P}_0}[g(x, \omega)] ,$$

and (6.47) can be conservatively approximated by:

$$\min_{x \in \mathcal{X}} \left\{ C^\top x + d_0 L(x) + \mathbb{E}^{\mathbb{P}_0}[g(x, \omega)] \right\} , \tag{6.49}$$

which has the form of two-stage stochastic programming (SP) and can be solved. The problem left is how to find the function $L(x)$ satisfying (6.48).

We will derive a function $L(x)$ using the dual variable of the LP problem, which can reflect the sensitivity of the optimal value with respect to the right-hand-side coefficients in the constraints. Since $g(x, \omega)$ is finite for any $(x, \omega) \in \mathcal{X} \times \Omega$, by LP duality we have:

$$g(x, \omega) = \begin{cases} \min\limits_y D^\top y \\ \text{ s.t. } Ay \geq B(x)\omega + H(x) \end{cases} = \begin{cases} \max\limits_\nu (B(x)\omega + H(x))^\top \nu \\ \text{ s.t. } A^\top \nu = D , \ \nu \geq 0 , \end{cases} \tag{6.50}$$

is continuous and piecewise affine in ω under fixed $x \in \mathcal{X}$. Moreover, in each piece,

$$\frac{\partial g}{\partial \omega}(x, \omega) = \frac{\partial((B(x)\omega + H(x))^\top \nu^*)}{\partial \omega}(x, \omega) = B(x)^\top \nu^* ,$$

where ν^* is a dual-optimal solution depending on x and ω.

Assume $\omega' = (\omega'_1, \ldots, \omega'_M)$ and $\omega'' = (\omega''_1, \ldots, \omega''_M)$, then

$$
\begin{aligned}
&|g(x, \omega') - g(x, \omega'')| \\
&\leq \sum_{m=1}^{M} |g(x, (\omega'_1, \ldots, \omega'_m, \omega''_{m+1}, \ldots, \omega''_M)) - g(x, (\omega'_1, \ldots, \omega'_{m-1}, \omega''_m, \ldots, \omega''_M))| \\
&\leq \sum_{m=1}^{M} |\omega'_m - \omega''_m| \cdot \sup_{\omega\in\Omega} \left|\frac{\partial g}{\partial \omega_m}(x, \omega)\right| \leq \left(\sum_{m=1}^{M} |\omega'_m - \omega''_m|\right) \cdot \max_m \sup_{\omega\in\Omega} \left|\frac{\partial g}{\partial \omega_m}(x, \omega)\right| \\
&\leq \|\omega' - \omega''\|_1 \sup_{\omega\in\Omega} \left\|\frac{\partial g}{\partial \omega}(x, \omega)\right\|_\infty .
\end{aligned}
$$

Therefore, we can choose the function $L(x)$ as follows to guarantee (6.48):

$$L(x) = \sup_{\omega\in\Omega} \|B(x)^\top \nu^*(x, \omega)\|_\infty .$$

To solve problem (6.49), we need a tractable formulation of $L(x)$ or its upper bound. A direct upper bound is:

$$L(x) \leq L_1(x) := \sup_{\nu: A^\top \nu = D, \nu \geq 0} \|B(x)^\top \nu\|_\infty ,$$

which is the same as the reformulation in Theorem 6.9 and Corollary 6.4. Because

$$\sup_{\nu: A^\top \nu = D, \nu\geq 0} B_m(x)^\top \nu = \min_{\phi: A\phi \geq B_m(x)} D^\top \phi ,$$

by LP duality, we have:

$$
\begin{aligned}
&\min_{x\in\mathcal{X}} \left\{C^\top x + d_0 L_1(x) + \mathbb{E}^{\mathbb{P}_0}[g(x, \omega)]\right\} \\
\iff &\left\{
\begin{aligned}
&\min_{x\in\mathcal{X}, l, \phi, \psi} \quad C^\top x + d_0 l + \mathbb{E}^{\mathbb{P}_0}[g(x, \omega)] \\
&\text{s.t. } l \geq D^\top \phi_m , \ A\phi_m \geq B_m(x) , \ m = 1, \ldots, M , \\
&\qquad l \geq D^\top \psi_m , \ A\psi_m \geq -B_m(x) , \ m = 1, \ldots, M .
\end{aligned}
\right.
\end{aligned}
$$

Another upper bound for $L(x)$ is:

$$L(x) \leq L_2(x) := \left\| |B(x)|^\top \sup_{x\in\mathcal{X}, \omega\in\Omega} |\nu^*(x, \omega)| \right\|_\infty ,$$

where the supremum operation is element-wise. The supremum of the kth element of the dual variable is:

$$\max_{x \in \mathcal{X}, \omega \in \Omega, y, \nu} \quad \nu_k$$
$$\text{s.t. } Ay \geq B(x)\omega + H(x) \, , \; A^\top \nu = D \, , \; \nu \geq 0 \, ,$$
$$\nu^\top (Ay - B(x)\omega - H(x)) = 0 \, ,$$

which can be approximated by an upper bound

$$\nu_k^u := \begin{cases} \max\limits_{\gamma \in \Gamma, y, \nu, z} \quad \nu_k \\ \quad \text{s.t. } A^\top \nu = D, 0 \leq \nu \leq M_L z, 0 \leq Ay - \gamma \leq M_L(1 - z), z \text{ is binary} \, , \end{cases}$$

where $\Gamma \supset \{B(x)\omega + H(x) \mid x \in \mathcal{X}, \omega \in \Omega\}$ and M_L is a large enough positive constant. Similarly, we can calculate a lower bound ν_k^l and then $L(x) \leq L_3(x) := \| |B(x)|^\top \bar{\nu} \|_\infty$, where $\bar{\nu}_k := \max\{|\nu_k^u|, |\nu_k^l|\}$. Moreover,

$$\min_{x \in \mathcal{X}} \left\{ C^\top x + d_0 L_3(x) + \mathbb{E}^{\mathbb{P}_0}[g(x, \omega)] \right\}$$
$$\Longleftrightarrow \begin{cases} \min\limits_{x \in \mathcal{X}, l, s} \; C^\top x + d_0 l + \mathbb{E}^{\mathbb{P}_0}[g(x, \omega)] \\ \quad \text{s.t. } l \geq s_m^\top \bar{\nu} \, , \; -s_m \leq B_m(x) \leq s_m \, , \; m = 1, \ldots, M \, . \end{cases}$$

6.3.2.3 Discrete Probability Distribution Reformulation

The probability distributions in the ambiguity set $\mathcal{P}_W(\Omega, \mathbb{P}_0, d_0)$ can be of any form. If we restrict them to discrete probability distributions, then we can obtain a better lower bound than that in Theorem 6.10 for the worst-case expectation:

Theorem 6.12 (Discrete probability distribution reformulation) *Suppose* $\mathbb{P}_0 = \sum_{n=1}^N \delta_{\omega_n}/N$ *is a probability distribution on* Ω. *Then*

$$\sup_{\mathbb{P} \in \mathcal{P}_W(\Omega, \mathbb{P}_0, d_0)} \mathbb{E}^{\mathbb{P}}[g(\omega)] \geq \begin{cases} \sup\limits_{\zeta} \dfrac{1}{N} \sum\limits_{n=1}^N g(\zeta_n) \\ \text{s.t. } \zeta_n \in \Omega \, , \; n = 1, \ldots, N \, , \\ \quad \dfrac{1}{N} \sum\limits_{n=1}^N \|\zeta_n - \omega_n\| \leq d_0 \, , \end{cases}$$
$$\geq \mathbb{E}^{\mathbb{P}_0}[g(\omega)] \, . \tag{6.51}$$

Proof The probability distribution $\mathbb{P} := \sum_{n=1}^N \delta_{\zeta_n}/N$ is in the ambiguity set $\mathcal{P}_W(\Omega, \mathbb{P}_0, d_0)$ if $\zeta_n \in \Omega, \forall n$ and $\sum_{n=1}^N \|\zeta_n - \omega_n\|/N \leq d_0$. Therefore, the worst-case expectation is no smaller than $\mathbb{E}^{\mathbb{P}}[g(\omega)]$. Take supremum with respect to $\zeta_n, \forall n$, then we have (6.51). □

In fact, the idea of the above proof can be extended to $\mathbb{P} := \sum_{n=1}^{N}\sum_{k=1}^{K} \delta_{\zeta_{nk}}/NK$, and then we have:

$$\sup_{\mathbb{P}\in\mathcal{P}_W(\Omega,\mathbb{P}_0,d_0)} \mathbb{E}^{\mathbb{P}}[g(\omega)] \geq \begin{cases} \sup\limits_{\zeta} \dfrac{1}{NK}\sum\limits_{n=1}^{N}\sum\limits_{k=1}^{K} g(\zeta_{nk}) \\ \text{s.t. } \zeta_{nk} \in \Omega \ , \\ \qquad \dfrac{1}{NK}\sum\limits_{n=1}^{N}\sum\limits_{k=1}^{K} \|\zeta_{nk} - \omega_n\| \leq d_0 \ , \end{cases} \tag{6.52}$$

which is no worse than the approximation in (6.51) when $K > 1$. Moreover, according to Corollary 2 in Gao and Kleywegt (2023), the right side of (6.52) converges to the left side as $K \to \infty$ if g is Lipschitz continuous.

Recall that any norm is convex. Therefore, when Ω is convex and $g : \Omega \to \mathbb{R}$ is concave, the optimization problems in (6.51) and (6.52) are convex. However, when g is the optimal value of a second-stage problem, it may be convex but not concave. In this case, (6.51) and (6.52) are not convex.

The reformulations in (6.51) and (6.52) are suitable for worst-case expectation estimation. If the loss function is concave and (6.51) and (6.52) are used to approximate DRO problems, the duality may be further employed to change the maximization into minimization.

Exercise 6.15 Suppose $g(x, \omega) = A(x)^\top\omega + B(x)$ with $A(x) \in \mathbb{R}^M$ and $B(x) \in \mathbb{R}$ affine in x, $\Omega = \{\omega \in \mathbb{R}^M \mid P\omega \leq Q\} \neq \varnothing$, the norm is either 1-norm or ∞-norm. Use (6.52) to derive an optimistic approximation for the Wasserstein-distance DRO problem (6.27) and find an equivalent LP reformulation.

6.4 Reformulation of Robust Chance Constraint

The reformulation methods introduced in Sect. 3.4 can be used to transform robust chance constraints and distributionally robust risk measures into worst-case expectations and then the methods in Sect. 6.3 can be further applied. In the following, we introduce some direct methods to reformulate the robust chance constraint, which leads to tractable forms in special cases. Theorem 6.13 considers a random event defined by a polyhedron.

Theorem 6.13 (Polyhedral random event and polyhedral sample space) *Assume that $\mathbb{P}_0 = \sum_{n=1}^{N} \delta_{\omega_n}/N$ is a probability distribution on sample space $\Omega = \{\omega \in \mathbb{R}^M \mid P\omega \leq Q\}$. Suppose the set $\{\omega \in \mathbb{R}^M \mid A^\top\omega < B\}$ is an open polyhedron, where $A = (A_1\ A_2\ \ldots\ A_K)$ with column vectors $A_1, \ldots, A_K$ and B is a column vector with kth entry being B_k. Assume for any $1 \leq k \leq K$, $\Omega \cap \{\omega \in \mathbb{R}^M \mid A_k^\top\omega \geq B_k\} \neq \varnothing$. Then the robust chance constraint*

$$\sup_{\mathbb{P}\in\mathcal{P}_W(\Omega,\mathbb{P}_0,d_0)} \mathbb{P}(\Omega \setminus \{\omega \in \mathbb{R}^M \mid A^\top \omega < B\}) \leq 1-\beta \iff$$

$$\begin{cases} \exists \lambda, s, \gamma, \nu \ , \ s.t. \ d_0\lambda + \dfrac{1}{N}\displaystyle\sum_{n=1}^{N} s_n \leq 1-\beta \ , \\ s_n \geq 0 \ , \ n = 1, \ldots, N \ , \\ 1-(B_k - A_k^\top \omega_n)^\top \nu_{nk} + (Q - P\omega_n)^\top \gamma_{nk} \leq s_n \ , \ n = 1, \ldots, N \ , \ k = 1, \ldots, K \ , \\ \|A_k \nu_{nk} - P^\top \gamma_{nk}\|_* \leq \lambda \ , \ \gamma_{nk} \geq 0 \ , \ \nu_{nk} \geq 0 \ , \ n = 1, \ldots, N \ , \ k = 1, \ldots, K \ . \end{cases}$$

Proof Let $S_k := \{\omega \in \mathbb{R}^M \mid A_k^\top \omega \geq B_k\}$, $k = 1, \ldots, K$ and $\tilde{S} := S_1 \cup \cdots \cup S_K$. Then

$$\begin{aligned} &\mathbb{P}(\Omega \setminus \{\omega \in \mathbb{R}^M \mid A^\top \omega < B\}) \\ &= \mathbb{P}(\exists 1 \leq k \leq K \ , \ \text{s.t. } A_k^\top \omega \geq B_k) \\ &= \mathbb{P}(\tilde{S}) = \mathbb{E}^{\mathbb{P}}[1_{\tilde{S}}(\omega)] \ , \end{aligned}$$

where $1_{\tilde{S}} : \mathbb{R}^M \to \mathbb{R}$ is the indicator function of $\tilde{S}$. We claim that

$$1_{\tilde{S}}(\omega) = \max\{1 - \chi_{S_1}(\omega), \ldots, 1 - \chi_{S_K}(\omega), 0\} \ .$$

In fact, if $\omega \in \tilde{S}$, then $\exists 1 \leq k \leq K$ such that $\omega \in S_k$. Thus, $1 - \chi_{S_k}(\omega) = 1 \geq 1 - \chi_{S_{k'}}(\omega)$ for $k' \neq k$ and $1_{\tilde{S}}(\omega) = 1$. If $\omega \notin \tilde{S}$, then $1 - \chi_{S_k}(\omega) = -\infty$, $1 \leq k \leq K$ and $1_{\tilde{S}}(\omega) = 0$.

Therefore, $1_{\tilde{S}}(\omega) = \max_{1\leq k\leq K+1} g_k(\omega)$, $\forall \omega \in \mathbb{R}^M$, where $g_k(\omega) = 1 - \chi_{S_k}(\omega)$, $1 \leq k \leq K$ and $g_{K+1}(\omega) = 0$ are concave functions. By LP duality,

$$\begin{aligned} [-g_k]^*(\theta) &= \sup_{\omega\in\mathbb{R}^M} \{\theta^\top \omega + 1 - \chi_{S_k}(\omega)\} \\ &= \sup_{\omega: A_k^\top \omega \geq B_k} \{\theta^\top \omega + 1\} \\ &= \inf_{\nu:\nu\geq 0, -A_k\nu=\theta} \{1 - B_k^\top \nu\} \ , \ k = 1, \ldots, K \ . \end{aligned}$$

Additionally,

$$[-g_{K+1}]^*(\theta) = \sup_{\omega\in\mathbb{R}^M} \theta^\top \omega = \begin{cases} 0 \ , & \text{if } \theta = 0 \ , \\ +\infty \ , & \text{otherwise} \ . \end{cases}$$

Then using Theorem 6.8 and (6.40), we have:

$$\sup_{\mathbb{P}\in\mathcal{P}_{\mathrm{W}}(\Omega,\mathbb{P}_0,d_0)} \mathbb{P}(\exists 1 \le k \le K\ ,\ \text{s.t. } A_k^\top \omega \ge B_k)$$

$$= \begin{cases} \min\limits_{\lambda,s,z,\upsilon,\gamma,\nu} & d_0\lambda + \dfrac{1}{N}\displaystyle\sum_{n=1}^{N} s_n \\ \text{s.t.} & 1 - B_k^\top \nu_{nk} + Q^\top \gamma_{nk} - \omega_n^\top z_{nk} \le s_n\ ,\ n = 1,\dots,N\ ,\ k = 1,\dots,K\ , \\ & \nu_{nk} \ge 0\ ,\ -A_k \nu_{nk} = z_{nk} - \upsilon_{nk}\ ,\ n = 1,\dots,N\ ,\ k = 1,\dots,K\ , \\ & Q^\top \gamma_{n(K+1)} - \omega_n^\top z_{n(K+1)} \le s_n\ ,\ n = 1,\dots,N\ , \\ & z_{n(K+1)} - \upsilon_{n(K+1)} = 0\ ,\ n = 1,\dots,N\ , \\ & P^\top \gamma_{nk} = \upsilon_{nk}\ ,\ \gamma_{nk} \ge 0\ ,\ n = 1,\dots,N\ ,\ k = 1,\dots,K+1\ , \\ & \|z_{nk}\|_* \le \lambda\ ,\ n = 1,\dots,N\ ,\ k = 1,\dots,K+1\ . \end{cases}$$

The proof is completed after substituting the worst-case expectation into the robust chance constraint, eliminating z and υ, and observing that $\gamma_{n(K+1)} = 0$ is optimal. □

Although the reformulation in Theorem 6.13 is a group of linear constraints and conic constraints, it becomes nonlinear and nonconvex if the parameter A and B of the random event depend on the decision variable x, due to the bilinear term $B_k^\top \nu_{nk}$ and $A_k \nu_{nk}$, which may cause problems in solving Wasserstein-distance DRO problem

$$\begin{aligned} &\min_{x\in\mathcal{X}}\ C^\top x \\ &\text{s.t.} \sup_{\mathbb{P}\in\mathcal{P}_{\mathrm{W}}(\Omega,\mathbb{P}_0,d_0)} \mathbb{P}(\Omega \setminus S(x)) \le 1-\beta\ , \end{aligned} \tag{6.53}$$

where $S(x)$ is the random event that we would like to ensure with a high probability.

Theorem 6.14 considers the case of Euclidean sample space and gives an exact reformulation of the robust chance constraint using conditional value-at-risk (CVaR) under the empirical probability distribution.

Theorem 6.14 (Euclidean sample space, Theorem 3 in Chen et al. 2024 and Corollary 1 in Xie 2021) *Suppose that* $\mathbb{P}_0 = \sum_{n=1}^{N} \delta_{\omega_n}/N$ *is a probability distribution on* $\mathbb{R}^M$, *S is a measurable set in* $\mathbb{R}^M$, $S \ne \varnothing$, $S \ne \mathbb{R}^M$, $d_0 > 0$, *and* $\beta \in (0, 1)$. *Then the robust chance constraint*

$$\sup_{\mathbb{P}\in\mathcal{P}_{\mathrm{W}}(\mathbb{R}^M,\mathbb{P}_0,d_0)} \mathbb{P}(\mathbb{R}^M \setminus S) \le 1-\beta \tag{6.54a}$$

$$\Longleftrightarrow \begin{cases} \exists \theta, t\ ,\ s.t.\ d_0 + \dfrac{1}{N}\displaystyle\sum_{n=1}^{N} t_n \le (1-\beta)\theta\ , \\ t_n \ge 0\ ,\ t_n \ge \theta - d(\omega_n, \mathbb{R}^M \setminus S)\ ,\ n = 1,\dots,N \end{cases} \tag{6.54b}$$

$$\Longleftrightarrow \frac{d_0}{1-\beta} + \beta\text{-}CVaR^{\mathbb{P}_0}(-d(\omega, \mathbb{R}^M \setminus S)) \le 0\ , \tag{6.54c}$$

where

$$d(\omega, A) := \inf_{\omega' \in A} \|\omega - \omega'\| \ , \ \forall \omega \in \mathbb{R}^M \ , \ A \subset \mathbb{R}^M \ .$$

Proof Using Theorem 6.7, we have:

$$\begin{aligned} & \sup_{\mathbb{P} \in \mathcal{P}_W(\mathbb{R}^M, \mathbb{P}_0, d_0)} \mathbb{P}(\mathbb{R}^M \setminus S) \\ = & \sup_{\mathbb{P} \in \mathcal{P}_W(\mathbb{R}^M, \mathbb{P}_0, d_0)} \mathbb{E}^{\mathbb{P}}[1_{\mathbb{R}^M \setminus S}(\omega)] \\ = & \inf_{\lambda \geq 0} \left\{ d_0 \lambda + \frac{1}{N} \sum_{n=1}^{N} \sup_{\omega \in \mathbb{R}^M} \{1_{\mathbb{R}^M \setminus S}(\omega) - \lambda \|\omega - \omega_n\|\} \right\} \ . \end{aligned}$$

Let $h_n(\lambda) := \sup_{\omega \in \mathbb{R}^M} \{1_{\mathbb{R}^M \setminus S}(\omega) - \lambda \|\omega - \omega_n\|\}$ for $\lambda \geq 0$ and $n = 1, \ldots, N$. Then $h_n(\lambda) \leq \sup_{\omega \in \mathbb{R}^M} 1_{\mathbb{R}^M \setminus S}(\omega) = 1$ and $h_n(\lambda) \geq 1_{\mathbb{R}^M \setminus S}(\omega_n) - 0 \geq 0$ for any $\lambda \geq 0$ and $n = 1, \ldots, N$. $h_n(0) = 1$. If $\lambda > 0$ and $\omega_n \notin S$, then $h_n(\lambda) \geq 1_{\mathbb{R}^M \setminus S}(\omega_n) - 0 = 1$, which implies $h_n(\lambda) = 1$. If $\lambda > 0$ and $\omega_n \in S$, we have $1_{\mathbb{R}^M \setminus S}(\omega) - \lambda \|\omega - \omega_n\| \leq 0$ for $\omega \in S$, then $h_n(\lambda) = \sup_{\omega \in \mathbb{R}^M \setminus S} \{1 - \lambda \|\omega - \omega_n\|\} = 1 - \lambda \inf_{\omega \in \mathbb{R}^M \setminus S} \|\omega - \omega_n\| = 1 - \lambda d(\omega_n, \mathbb{R}^M \setminus S)$. To sum up,

$$\begin{aligned} & \sup_{\omega \in \mathbb{R}^M} \{1_{\mathbb{R}^M \setminus S}(\omega) - \lambda \|\omega - \omega_n\|\} \\ = & \max\{0, 1 - \lambda d(\omega_n, \mathbb{R}^M \setminus S)\} \ , \ \forall \lambda \geq 0 \ , \ n = 1, \ldots, N \ . \end{aligned}$$

Therefore,

$$\sup_{\mathbb{P} \in \mathcal{P}_W(\mathbb{R}^M, \mathbb{P}_0, d_0)} \mathbb{P}(\mathbb{R}^M \setminus S) = \inf_{\lambda \geq 0} \left\{ d_0 \lambda + \frac{1}{N} \sum_{n=1}^{N} \max\{0, 1 - \lambda d(\omega_n, \mathbb{R}^M \setminus S)\} \right\} \ ,$$

and the robust chance constraint (6.54a) is equivalent to:

$$\exists \lambda \geq 0 \ , \ \text{s.t.} \ d_0 \lambda + \frac{1}{N} \sum_{n=1}^{N} \max\{0, 1 - \lambda d(\omega_n, \mathbb{R}^M \setminus S)\} \leq 1 - \beta \ . \tag{6.55}$$

We observe that the left side is continuous in λ, equals $1 > 1 - \beta$ when $\lambda = 0$, and is larger than 1 if $\lambda > 1/d_0$. Therefore, (6.55) is equivalent to:

$$\begin{aligned} & \exists \lambda > 0 \ , \ \text{s.t.} \ d_0 \lambda + \frac{1}{N} \sum_{n=1}^{N} \max\{0, 1 - \lambda d(\omega_n, \mathbb{R}^M \setminus S)\} \leq 1 - \beta \\ \iff & \begin{cases} \exists \lambda, s \ , \ \text{s.t.} \ \lambda > 0 \ , \ d_0 \lambda + \dfrac{1}{N} \displaystyle\sum_{n=1}^{N} s_n \leq 1 - \beta \ , \\ s_n \geq 0 \ , \ s_n \geq 1 - \lambda d(\omega_n, \mathbb{R}^M \setminus S) \ , \ n = 1, \ldots, N \end{cases} \iff (6.54b) \ , \end{aligned}$$

where $\theta = 1/\lambda$ and $t_n = s_n/\lambda$. $\theta \geq 0$ follows from $\theta \geq (d_0 + \sum_{n=1}^{N} t_n/N)/(1-\beta)$ and $t_n \geq 0, n = 1, \ldots, N$.

Moreover, (6.54b) can be written as:

$$\frac{d_0}{1-\beta} + (-\theta) + \frac{1}{1-\beta}\frac{1}{N}\sum_{n=1}^{N} \max\{0, -d(\omega_n, \mathbb{R}^M \setminus S) - (-\theta)\} \leq 0 ,$$

which is equivalent to (6.54c) according to the property of CVaR in Theorem 2.6. □

Unfortunately, the formulation in (6.54b) is generally nonconvex, because the function $d(\omega, \mathbb{R}^M \setminus S)$ is not concave. Nonetheless, it can be reformulated as mixed-integer programming in some cases, as shown in Propositions 6.10 and 6.11.

Proposition 6.10 (Halfspace random event) *Suppose that $\mathbb{P}_0 = \sum_{n=1}^{N} \delta_{\omega_n}/N$ is a probability distribution on $\mathbb{R}^M$, $0 \neq A \in \mathbb{R}^M$, $B \in \mathbb{R}$, $d_0 > 0$, and $\beta \in (0, 1)$. Then the robust chance constraint*

$$\sup_{\mathbb{P}\in\mathcal{P}_W(\mathbb{R}^M,\mathbb{P}_0,d_0)} \mathbb{P}(A^\top\omega > B) \leq 1-\beta$$

$$\Longleftrightarrow \begin{cases} \exists\theta, t, z\ ,\ \text{s.t. } d_0\|A\|_* + \dfrac{1}{N}\displaystyle\sum_{n=1}^{N} t_n \leq (1-\beta)\theta\ , \\ B - A^\top\omega_n + M_L z_n \geq \theta - t_n\ ,\ M_L(1-z_n) \geq \theta - t_n\ ,\ n = 1, \ldots, N\ , \\ t_n \geq 0\ ,\ z_n \in \{0, 1\}\ ,\ n = 1, \ldots, N\ , \end{cases}$$

where M_L is a large enough positive constant.

Proof Let $S = \{\omega \in \mathbb{R}^M \mid A^\top\omega \leq B\}$, then we have the formulation (6.54b). Now we need to calculate $d(\omega, \mathbb{R}^M \setminus S)$. If $\omega \in \mathbb{R}^M \setminus S$, i.e., $A^\top\omega > B$, then $d(\omega, \mathbb{R}^M \setminus S) = 0$. If $\omega \in S$, let p be the projection of ω to the halfspace S, then $p - \omega = kA$ for some $k > 0$ and $A^\top p = B$, which implies $k = (B - A^\top\omega)/(A^\top A)$ and $d(\omega, \mathbb{R}^M \setminus S) = \|\omega - p\| = (B - A^\top\omega)\|A\|/(A^\top A) = (B - A^\top\omega)/\|A\|_*$. To sum up,

$$d(\omega, \mathbb{R}^M \setminus S) = \max\left\{0, \frac{B - A^\top\omega}{\|A\|_*}\right\} = \frac{1}{\|A\|_*}\max\{0, B - A^\top\omega\}\ . \qquad (6.56)$$

Then (6.54b) can be written as:

$$\begin{cases} \exists\theta, t\ ,\ \text{s.t. } d_0 + \dfrac{1}{N}\displaystyle\sum_{n=1}^{N} t_n \leq (1-\beta)\theta\ , \\ t_n \geq 0\ ,\ \max\{0, B - A^\top\omega_n\} \geq \theta\|A\|_* - t_n\|A\|_*\ ,\ n = 1, \ldots, N\ , \end{cases}$$

which is as follows after multiplying θ and t by $\|A\|_*$:

$$\begin{cases} \exists \theta, t \ , \ \text{s.t. } d_0 \|A\|_* + \frac{1}{N} \sum_{n=1}^{N} t_n \le (1-\beta)\theta \ , \\ t_n \ge 0 \ , \ \max\{0, B - A^\top \omega_n\} \ge \theta - t_n \ , \ n = 1, \ldots, N \ , \end{cases}$$

It can be verified that for a large enough constant $M_L > 0$,

$$\begin{aligned} &\max\{0, B - A^\top \omega_n\} \ge \theta - t_n \\ \Longleftrightarrow\ &\exists z_n \in \{0,1\} \ , \ \text{s.t. } B - A^\top \omega_n + M_L z_n \ge \theta - t_n \ , \ M_L(1 - z_n) \ge \theta - t_n \ , \end{aligned}$$

by checking the cases $z_n = 0$ and $z_n = 1$. Then the conclusion follows. □

Exercise 6.16 Under the assumption of Proposition 6.10, suppose that $A(x)$ and $B(x)$ are affine in x, $\mathcal{X}$ is a polyhedron, and the norm is either 1-norm or ∞-norm. Use Proposition 6.10 to prove that (6.53) can be formulated as an equivalent mixed-integer linear programming (MILP) problem.

Proposition 6.11 (Polyhedral random event) *Assume that* $\mathbb{P}_0 = \sum_{n=1}^{N} \delta_{\omega_n}/N$ *is a probability distribution on* $\mathbb{R}^M$, $A = (A_1\ A_2\ \ldots\ A_K)$ *and* B *are coefficients with* $0 \ne A_k \in \mathbb{R}^M$, $k = 1, \ldots, K$, $d_0 > 0$, *and* $\beta \in (0,1)$. *Then the robust chance constraint*

$$\sup_{\mathbb{P} \in \mathcal{P}_W(\mathbb{R}^M, \mathbb{P}_0, d_0)} \mathbb{P}(\exists 1 \le k \le K \ , \ s.t.\ A_k^\top \omega > B_k) \le 1 - \beta$$

$$\Longleftrightarrow \begin{cases} \exists \theta, t, z \ , \ s.t.\ d_0 + \frac{1}{N} \sum_{n=1}^{N} t_n \le (1-\beta)\theta \ , \\ \frac{B_k - A_k^\top \omega_n}{\|A_k\|_*} + M_L z_n \ge \theta - t_n \ , \ k = 1, \ldots, K \ , \ n = 1, \ldots, N \ , \\ M_L(1 - z_n) \ge \theta - t_n \ , \ t_n \ge 0 \ , \ z_n \in \{0,1\} \ , \ n = 1, \ldots, N \ , \end{cases}$$

where M_L *is a large enough positive constant.*

Proof Let $S_k := \{\omega \in \mathbb{R}^M \mid A_k^\top \omega \le B_k\}$ and $S := \bigcap_{k=1}^{K} S_k = \{\omega \in \mathbb{R}^M \mid A^\top \omega \le B\}$. Then

$$\begin{aligned} &\{\omega \in \mathbb{R}^M \mid \exists 1 \le k \le K \ , \ \text{s.t. } A_k^\top \omega > B_k\} \\ = &\ \mathbb{R}^M \setminus S = \mathbb{R}^M \setminus \bigcap_{k=1}^{K} S_k = \bigcup_{k=1}^{K} (\mathbb{R}^M \setminus S_k) \ . \end{aligned}$$

Thus, using (6.56), we have:

$$
\begin{aligned}
d(\omega, \mathbb{R}^M \setminus S) &= \inf_{\omega' \in \mathbb{R}^M \setminus S} \|\omega - \omega'\| = \min_{1 \le k \le K} \inf_{\omega' \in \mathbb{R}^M \setminus S_k} \|\omega - \omega'\| \\
&= \min_{1 \le k \le K} d(\omega, \mathbb{R}^M \setminus S_k) = \min_{1 \le k \le K} \max \left\{ 0, \frac{B_k - A_k^\top \omega}{\|A_k\|_*} \right\} \\
&= \max \left\{ 0, \min_{1 \le k \le K} \frac{B_k - A_k^\top \omega}{\|A_k\|_*} \right\} .
\end{aligned}
$$

By

$$
\begin{aligned}
& \max \left\{ 0, \min_{1 \le k \le K} \frac{B_k - A_k^\top \omega_n}{\|A_k\|_*} \right\} \ge \theta - t_n \\
\iff & \begin{cases} \exists z_n \in \{0, 1\}, \text{ s.t. } M_L(1 - z_n) \ge \theta - t_n , \\ M_L z_n + \min\limits_{1 \le k \le K} \dfrac{B_k - A_k^\top \omega_n}{\|A_k\|_*} \ge \theta - t_n \end{cases} \\
\iff & \begin{cases} \exists z_n \in \{0, 1\}, \text{ s.t. } M_L(1 - z_n) \ge \theta - t_n , \\ M_L z_n + \dfrac{B_k - A_k^\top \omega_n}{\|A_k\|_*} \ge \theta - t_n , \ k = 1, \ldots, K , \end{cases}
\end{aligned}
$$

the proof is completed. □

Exercise 6.17 Under the assumption of Proposition 6.11, suppose that $B_k(x)$ is affine in x for $k = 1, \ldots, K$ and $\mathcal{X}$ is a polyhedron. Use Proposition 6.11 to prove that (6.53) can be formulated as an equivalent MILP problem.

Readers can refer to Ho-Nguyen et al. (2022) and Ho-Nguyen et al. (2023) for more formulations of Wasserstein-distance DRO problems with robust chance constraints.

References

Chen Z, Kuhn D, Wiesemann W (2024) Data-driven chance constrained programs over Wasserstein balls. Oper Res 72(1):410–424

Esfahani PM, Kuhn D (2018) Data-driven distributionally robust optimization using the Wasserstein metric: performance guarantees and tractable reformulations. Math Program 171:115–166

Fournier N, Guillin A (2015) On the rate of convergence in Wasserstein distance of the empirical measure. Probab Theory Related Fields 162:707–738

Gao R (2023) Finite-sample guarantees for Wasserstein distributionally robust optimization: breaking the curse of dimensionality. Oper Res 71(6):2291–2306

Gao R, Kleywegt A (2023) Distributionally robust stochastic optimization with Wasserstein distance. Math Oper Res 48(2):603–655

Gibbs AL, Su FE (2002) On choosing and bounding probability metrics. Int Stat Rev 70(3):419–435

Hanasusanto GA, Kuhn D (2018) Conic programming reformulations of two-stage distributionally robust linear programs over Wasserstein balls. Oper Res 66(3):849–869

Herceg M, Kvasnica M, Jones C, et al (2013) Multi-parametric toolbox 3.0. In: Proceedings of the European control conference, Zürich, Switzerland, pp 502–510

Ho-Nguyen N, Kılınç-Karzan F, Küçükyavuz S et al (2022) Distributionally robust chance-constrained programs with right-hand side uncertainty under Wasserstein ambiguity. Math Program 196:641–672
Ho-Nguyen N, Kılınç-Karzan F, Küçükyavuz S et al (2023) Strong formulations for distributionally robust chance-constrained programs with left-hand side uncertainty under Wasserstein ambiguity. INFORMS J Opt 5(2):211–232
Kantorovich LV, Rubinshtein GS (1958) On a space of totally additive functions. Vestnik Leningrad Univ 13:52–59
Kantorovich LV (1960) Mathematical methods of organizing and planning production. Manag Sci 6(4):366–422
Rockafellar RT, Wets RJB (1998) Variational analysis. Grundlehren der mathematischen Wissenschaften, Springer, Berlin, Heidelberg
Santambrogio F (2015) Optimal transport for applied mathematicians: calculus of variations, PDEs, and modeling. Progress in nonlinear differential equations and their applications. Birkhäuser Cham, Switzerland
Tsybakov AB (2009) Introduction to nonparametric estimation. Springer series in statistics. Springer, New York, NY
Villani C (2003) Topics in optimal transportation. Graduate studies in mathematics. American Mathematical Society, Rhode Island
Xie W (2021) On distributionally robust chance constrained programs with Wasserstein distance. Math Program 186:115–155

Part II
Application

Chapter 7
Sizing Renewable and Energy Storage in Fully Renewable Stand-Alone Microgrids

Abstract This chapter studies the optimal sizing of renewable and storage capacities in a stand-alone microgrid powered by renewable energy. It proposes a bi-objective optimization model to minimize load shedding Shortfall risk and total investment cost, taking into account the uncertainty of renewable energy generation and load. To account for the inaccuracy of empirical probability distributions, an ambiguity set of probability distributions is constructed based on the Wasserstein distance, and a distributionally robust Shortfall risk model of load shedding is constructed. It is equivalently transformed into a worst-case expectation form and further conservatively approximated using a method based on the Lipschitz constant, transforming the problem into bi-objective linear programming (LP). According to the theory of parametric LP, it is explained that the Pareto frontier is piecewise affine, and an algorithm for solving the analytical expression of the Pareto frontier is proposed. This chapter introduces the work in Xie et al. (2022).

7.1 Introduction

In order to reduce fossil fuel consumption and carbon emissions, renewable energy generation is being increasingly adopted worldwide. Stand-alone microgrids powered entirely by renewable energy are a competitive choice for remote areas with access to renewable energy resources and far from existing power grids. Such microgrids offer the benefit of zero carbon emissions during power generation. However, due to the strong fluctuations and randomness in wind and solar power generation, it is necessary to equip energy storage (ES) systems to mitigate the impact of uncertainty and to compensate for the difference between renewable generation and load demands.

In traditional power systems, conventional generators such as coal-fired or gas-fired units are controllable. It is typically required that the installed capacity meet load demand under all conditions. However, for stand-alone microgrids, renewable energy generation is highly random, and the unit cost of ES is relatively high. If it is still required to avoid all possible load shedding, the investment cost will be very high.

R. Xie and W. Wei, *Distributionally Robust Optimization and its Applications in Power System Energy Storage Sizing*,
https://doi.org/10.1007/978-981-97-2566-3_7

Therefore, it is necessary to strike a suitable balance between cost and risk, enabling decision-makers to effectively adjust the system margin by setting a risk threshold. If a sufficiently low risk threshold is set, the stand-alone microgrid will be able to meet load demand in almost all scenarios, although it still cannot guarantee all situations. In addition, the load shedding only occurs in extreme operational scenarios, not normal ones. Typical extreme operational scenarios would be, for example, low power outputs from renewable energy sources due to rare and sustained bad weather conditions.

This chapter aims to propose a coordinated optimal planning method for renewable generation and ES capacities in an entirely renewable energy-powered stand-alone microgrid. This method takes into consideration the inexactness of the empirical probability distribution and its impact on the load-shedding Shortfall risk. The innovations of this chapter lie in the following two aspects.

Firstly, a distributionally robust model for load shedding Shortfall risk is proposed. In this model, the probability distribution ambiguity sets are composed of all probability distributions that are close to the empirical probability distribution, including discrete and continuous probability distributions, with the distance being measured via the Wasserstein distance. Uncertainties come from renewable energy generation and load, and the Shortfall risk is defined by the optimum of the operational optimization model. It is shown that the distributionally robust Shortfall risk can be expressed by the worst-case expectation and that it can be conservatively approximated by the optimum of a linear programming (LP) problem, allowing for effective solutions.

Secondly, a bi-objective optimization model is established for the planning of renewable energy generation and ES capacity in stand-alone microgrids, balancing the distributionally robust Shortfall risk and investment costs. Based on the ϵ-constraint method, it is demonstrated through parametric LP theory that the Pareto frontier of the bi-objective optimization model is a piecewise affine function, and an algorithm for solving the analytical expression of the Pareto frontier is proposed.

This chapter first establishes the mathematical model for optimal sizing in Sect. 7.2, then presents the solution method in Sect. 7.3, verifies the proposed model and method through case studies in Sect. 7.4, and finally concludes in Sect. 7.5.

7.2 Mathematical Formulation

This section establishes a specific mathematical model for the optimal sizing problem, which consists of three components: the operational optimization model that minimizes load shedding, the probability distribution ambiguity set of uncertainties, and the distributionally robust Shortfall risk. Finally, the three components are combined into a capacity optimization model.

7.2.1 Operation Model and Load Shedding Function

The considered stand-alone microgrid is fully powered by renewable energy. It includes renewable energy generation equipment, ES, and loads. The energy flow relationships are illustrated in Fig. 7.1. The operation process is divided into T time periods, denoted as $S_T := \{1, 2, \dots, T\}$. The length of the time period is denoted as Δ_t. The subscript t represents the t-th time period. The load is denoted as ξ_t^c and is a random variable. Suppose that the power output of the unit installed capacity of renewable energy generation is $\xi_t^g \in [0, 1]$, which is also a random variable. Let x_g represent the installed capacity of the renewable energy generation equipment, so the power of renewable energy generation is $\xi_t^g x_g$. Let p_t^{gl}, p_t^{gs}, and p_t^{sl} represent the power from renewable energy generation to load, from renewable energy generation to ES, and from ES to load, respectively, as shown in Fig. 7.1.

Let p_t^c and p_t^l represent the curtailed renewable power and load shedding power, respectively. Let x_p and x_e represent the power and energy capacities of the ES, respectively, and use e_t to denote the energy stored in the ES at the end of the t-th period. The operational optimization model is established as follows:

$$g(x, \xi) := \min \sum_{t \in S_T} p_t^l \Delta_t \tag{7.1a}$$

$$\text{s.t. } p_t^{gl}, p_t^{gs}, p_t^{sl}, p_t^c, p_t^l \geq 0 \ , \ \forall t \in S_T \ , \tag{7.1b}$$

$$\xi_t^g x_g = p_t^{gl} + p_t^{gs} + p_t^c \ , \ \forall t \in S_T \ , \tag{7.1c}$$

$$\xi_t^c - p_t^l = p_t^{gl} + p_t^{sl} \ , \ \forall t \in S_T \ , \tag{7.1d}$$

$$e_t = e_0 + \sum_{s=1}^{t} \left(\eta_c p_s^{gs} - \frac{p_s^{sl}}{\eta_d} \right) \Delta_t \ , \ \forall t \in S_T \ , \tag{7.1e}$$

$$\omega_l x_e \leq e_t \leq \omega_h x_e \ , \ \forall t \in S_T \ , \tag{7.1f}$$

$$p_t^{gs} \leq x_p \ , \ p_t^{sl} \leq x_p \ , \ \forall t \in S_T \ , \ e_T = e_0 \ , \tag{7.1g}$$

$$p_t^{gs} p_t^{sl} = 0 \ , \ \forall t \in S_T \ , \tag{7.1h}$$

where Eq. (7.1a) presents the objective function to minimize total load shedding; Eq. (7.1b) stipulates the nonnegativity of the variables; Eqs. (7.1c) and (7.1d) describe the energy flow relationships in Fig. 7.1; Eq. (7.1e) is the ES charging and discharging

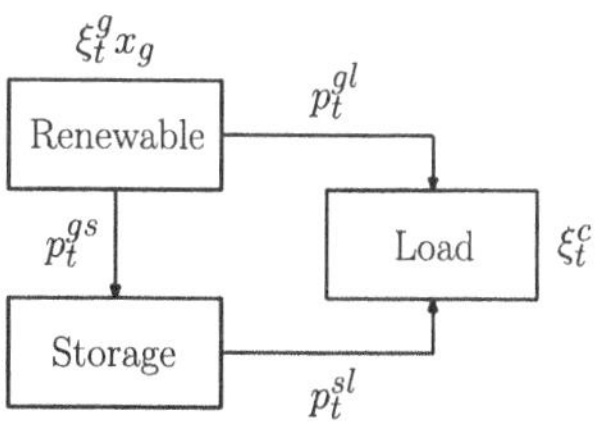

Fig. 7.1 Energy flow relationships in a fully-renewable stand-alone microgrid (Xie et al. 2022)

dynamic equation, where η_c and η_d represent the ES charging and discharging efficiency parameters, respectively; Eq. (7.1f) is the upper and lower bound constraint on the storage energy, where ω_l and ω_h are the lower and upper bounds limiting the state of charge (SoC); Eq. (7.1g) restricts the maximum charging and discharging power and stipulates that the energy stored in the ES at the beginning and end is the same, allowing for continuous operation of the system; Eq. (7.1h) is the complementarity constraint to prevent simultaneous charging and discharging of the ES.

Equation (7.1) forms the operational optimization model, where the decision variables include $p_t^{pl}, p_t^{gs}, p_t^{sl}, p_t^c, p_t^l, e_t$, whereas the capacity variables $x := (x_g, x_e, x_p)$ are determined during the planning stage, and the random variables $\xi := (\xi^c, \xi^g)$ are considered as fixed parameters. When (x, ξ) and other parameters are given, the minimum load shedding can be obtained by solving the optimum of the operational optimization model in (7.1). Thus, the load shedding is considered as a function of x and ξ, and is denoted as $g(x, \xi)$. The nonlinearity in the operational optimization model of (7.1) only exists in the ES non-simultaneous charging and discharging constraint in (7.1h).

7.2.2 Wasserstein-Distance Ambiguity Set

The load shedding is affected by $\xi = (\xi^c, \xi^g)$, which includes the uncertainty of renewable energy generation and load demand. This uncertainty is described by defining an ambiguity set.

Based on load forecasting data and renewable energy resource data, a set of sample data $\xi_n = (\xi_n^c, \xi_n^g), n \in S_N := \{1, 2, \ldots, N\}$ can be obtained for the random variable ξ. The range of values of the random variable ξ is established as follows:

$$\Xi = \left\{ (\xi^c, \xi^g) \in \mathbb{R}^{2T} \,\middle|\, 0 \le \xi_t^c \le \bar{\xi^c} \;,\; 0 \le \xi_t^g \le 1 \;,\; \forall t \in S_T \right\} , \tag{7.2}$$

where $\bar{\xi^c}$ represents the maximum possible load power. Following the method in Sect. 6.2, a discrete empirical probability distribution $\mathbb{P}_0$ can be defined by:

$$\mathbb{P}_0 := \frac{1}{N} \sum_n \delta_{\xi_n} , \tag{7.3}$$

where δ_{ξ_n} is the probability distribution with unit mass at ξ_n. We use Wasserstein 1-distance d_W in (6.3) under 1-norm to measure the difference between probability distributions. Then the ambiguity set $\mathcal{B}(\mathbb{P}_0, \varepsilon)$ is defined by:

$$\mathcal{B}(\mathbb{P}_0, \varepsilon) := \{ \mathbb{P} \in \mathcal{M}_1(\Xi) \mid d_W(\mathbb{P}, \mathbb{P}_0) \le \varepsilon \} . \tag{7.4}$$

7.2.3 Distributionally Robust Shortfall Risk of Load Shedding

The operation optimization model in (7.1) gives the load shedding $g(x, \xi)$. The Shortfall risk measure is used to assess the degree of load shedding under uncertainty. The Shortfall risk is given in Definition 2.41, and on this basis, the distributionally robust Shortfall risk of load shedding is defined as follows:

$$r_0(x) := \sup_{\mathbb{P} \in \mathcal{B}(\mathbb{P}_0, \varepsilon)} \mathrm{SR}^{\mathbb{P}}_{l,\lambda}(g(x, \xi)) , \tag{7.5}$$

where $l(r) := \max\{r, 0\}$ and λ is the acceptable load shedding. The distributionally robust Shortfall risk of load shedding $r_0(x)$ is a function of capacity variable x, and is the upper-bound of the Shortfall risk when the probability distribution varies in the ambiguity set $\mathcal{B}(\mathbb{P}_0, \varepsilon)$, i.e., the worst-case Shortfall risk. Therefore, as long as the actual probability distribution is in $\mathcal{B}(\mathbb{P}_0, \varepsilon)$, the actual Shortfall risk will not exceed $r_0(x)$.

There are three reasons to use $l(r) := \max\{r, 0\}$ to measure the degree of loss: First, this function meets the requirements for the function l in Definition 2.41 of Shortfall risk, i.e., it is convex and increasing. Second, this function conforms to the physical meaning of loss. Specifically, when the random variable representing the loss does not exceed 0, the loss degree is 0, and the loss degree grows linearly when the random variable is greater than 0. Third, this function is piecewise linear, so the method in Sect. 3.4.5 can be used to transform $r_0(x)$ into a linear model expression. The function l can also take other forms, and as long as it is approximated by a piecewise linear function, the subsequent methods in this chapter will still be effective. The specific method can be seen in Sect. 3.4.5.

7.2.4 Sizing Model

The bi-objective sizing model for renewable and ES is:

$$\begin{aligned} &\min_{x,r} \{r_0(x), C_g x_g + C_e x_e + C_p x_p\} \\ &\text{s.t. } x = (x_g, x_e, x_p) \geq 0 . \end{aligned} \tag{7.6}$$

The first objective function minimizes the distributionally robust Shortfall risk of load shedding, and the second objective function minimizes the total investment cost, where C_g, C_e, and C_p are the unit capacity costs of renewable energy generation, ES energy capacity, and ES power capacity, respectively. This objective function applies to cases where the cost of ES capacity and power capacity are relatively independent. For the ES types where the energy capacity and power capacity are closely linked, the value of x_e/x_p can be fixed, making the energy capacity proportional to the power capacity. This is equivalent to adding a linear constraint $x_e = C_{ep}x_p$, where C_{ep} is the ES duration parameter. Adding this linear constraint does not change the modeling of other parts and the solution method of the optimization model.

This chapter is dedicated to the problem of optimal capacity planning in a stand-alone microgrid fully powered by renewable energy sources. Since renewable energy generation does not consume fuel, its operation cost can be ignored compared to the investment cost (Liu et al. 2019). Load shedding is taken into account in risk considerations, and the problem of capacity sizing is modeled as a bi-objective optimization to simultaneously minimize investment cost and load shedding risk. In some existing studies, load shedding is modeled as a penalty term and is minimized together with cost (Diab et al. 2019). However, the penalty coefficient has a crucial impact on results and the decision-maker's attitude towards risk may be quite vague, making it difficult to determine the appropriate penalty coefficient. To this end, this chapter uses a bi-objective optimization modeling approach and proposes a solution method for calculating the analytical expression of the Pareto frontier to assist the decision-making process.

7.3 Solution Method

This section will first transform the distributionally robust Shortfall risk of load shedding so that the sizing model in (7.6) becomes a traditional bi-objective LP. Then, the method for calculating the analytical expression of the Pareto frontier is given according to the parametric LP.

7.3.1 *Relaxation of Charging and Discharging Complementary Constraint*

The constraint of ES not charging and discharging at the same time in (7.1h) is the only nonlinear part in the operation optimization model in (7.1). Remove (7.1h), leave the objective function and other constraints, and define

$$\begin{aligned} g_l(x,\xi) := \min \; & \sum_{t\in S_T} p_t^l \Delta_t \\ \text{s.t.} \; & (7.1\text{b})-(7.1\text{g})\,. \end{aligned} \tag{7.7}$$

Equation (7.7) is an LP problem. In fact, the following proposition asserts that Eq. (7.1h) is redundant when calculating the load shedding amount.

Proposition 7.1 *For any $x \geq 0$ and $\xi \in \Xi$, we have $g(x,\xi) = g_l(x,\xi)$.*

Proof For fixed $x \geq 0$ and $\xi \in \Xi$, because $g_l(x, \xi)$ is obtained by relaxing (7.1h) in $g(x, \xi)$, so $g_l(x, \xi) \leq g(x, \xi)$. We only need to prove $g_l(x, \xi) \geq g(x, \xi)$. According to (7.1b), $p_t^l \geq 0$, so $0 \leq g_l(x, \xi) \leq g(x, \xi)$. If $g_l(x, \xi) = +\infty$, then $g(x, \xi) = g_l(x, \xi)$. Therefore, we only need to consider the case where $g_l(x, \xi)$ is finite. From an optimal solution $(\hat{p}_t^{gl}, \hat{p}_t^{gs}, \hat{p}_t^{sl}, \hat{p}_t^c, \hat{p}_t^l, \hat{e}_t; t \in S_T)$ of problem (7.7), we will construct a feasible solution for (7.1) and make its objective value equal $g_l(x, \xi)$.

If $(\hat{p}_t^{gl}, \hat{p}_t^{gs}, \hat{p}_t^{sl}, \hat{p}_t^c, \hat{p}_t^l, \hat{e}_t; t \in S_T)$ is infeasible in (7.1), then only Eq. (7.1h) does not hold. Assume that for some t_0, $\hat{p}_{t_0}^{gs}\hat{p}_{t_0}^{sl} \neq 0$, then $\hat{p}_{t_0}^{gs} > 0$ and $\hat{p}_{t_0}^{sl} > 0$. Consider the following solution:

$$
\begin{aligned}
&(\tilde{p}_t^{gl}, \tilde{p}_t^{gs}, \tilde{p}_t^{sl}, \tilde{p}_t^c, \tilde{p}_t^l, \tilde{e}_t) = (\hat{p}_t^{gl}, \hat{p}_t^{gs}, \hat{p}_t^{sl}, \hat{p}_t^c, \hat{p}_t^l, \hat{e}_t)\ , \ \forall t \neq t_0\ , \\
&\tilde{p}_{t_0}^{gl} = \hat{p}_{t_0}^{gl} + \eta_c\eta_d\varepsilon\ , \ \tilde{p}_{t_0}^{gs} = \hat{p}_{t_0}^{gs} - \varepsilon\ , \ \tilde{p}_{t_0}^{sl} = \hat{p}_{t_0}^{sl} - \eta_c\eta_d\varepsilon\ , \\
&\tilde{p}_{t_0}^{c} = \hat{p}_{t_0}^{c} + (1 - \eta_c\eta_d)\varepsilon\ , \ \tilde{p}_{t_0}^{l} = \hat{p}_{t_0}^{l}\ , \ \tilde{e}_{t_0} = \hat{e}_{t_0}\ ,
\end{aligned}
$$

where $\varepsilon > 0$. Because the objective function only depends on p_t^l, so the objective function at $(\tilde{p}_t^{gl}, \tilde{p}_t^{gs}, \tilde{p}_t^{sl}, \tilde{p}_t^c, \tilde{p}_t^l, \tilde{e}_t; t \in S_T)$ equals $g_l(x, \xi)$.

Check whether $(\tilde{p}_t^{gl}, \tilde{p}_t^{gs}, \tilde{p}_t^{sl}, \tilde{p}_t^c, \tilde{p}_t^l, \tilde{e}_t; t \in S_T)$ satisfies (7.1b)–(7.1h). If $\varepsilon \leq \min\{\hat{p}_{t_0}^{gs}, \hat{p}_{t_0}^{sl}/(\eta_c\eta_d)\}$, then Eq. (7.1b) holds. According to the construction process, Eqs. (7.1c) and (7.1d) also hold. Because $\eta_c\tilde{p}_t^{gs} - \tilde{p}_t^{sl}/\eta_d = \eta_c\hat{p}_t^{gs} - \hat{p}_t^{sl}/\eta_d, \forall t \in S_T$, Eq. (7.1e) is satisfied. Note that $\tilde{e}_t = \hat{e}_t$, $\tilde{p}_t^{gs} \leq \hat{p}^{gs}$, $\tilde{p}_t^{sl} \leq \hat{p}^{sl}, \forall t \in S_T$, so Eqs. (7.1f) and (7.1g) hold. Therefore, so long as $\varepsilon = \min\{\hat{p}_{t_0}^{gs}, \hat{p}_{t_0}^{sl}/(\eta_c\eta_d)\}$, the constructed solution still satisfies (7.1b)–(7.1g), and $\tilde{p}_{t_0}^{gs}\tilde{p}_{t_0}^{sl} = 0$. Thus, the ES does not charge and discharge simultaneously in period t_0.

Repeat the above steps for each time period until all time periods satisfy (7.1h). Then, a feasible solution for (7.1) is obtained, and the objective function value at this point is equal to $g_l(x, \xi)$. $g(x, \xi)$ is not greater than the objective function value at any feasible solution, so $g_l(x, \xi) \geq g(x, \xi)$.

In conclusion, $g(x, \xi) = g_l(x, \xi)$ holds. □

The main idea of the proof is to start from an optimal solution of the relaxed problem that ignores the complementary constraints and then seek a feasible solution for the unrelaxed problem while keeping the objective function value unchanged. The method is to reduce the charging and discharging power at the same time. In fact, the reason for the validity of Proposition 7.1 is that allowing ES to charge and discharge at the same time cannot help reduce the load shedding, because simultaneous charging and discharging can only lead to more energy consumption due to efficiency reasons, further reducing the electricity that the load may obtain. This allows the removal of the complementary constraint that ES cannot charge and discharge at the same time.

Therefore, we will use $g_l(x, \xi)$, which has removed the complementary constraint of ES charging and discharging power, as the load shedding function used later. In the application process, if the result does not satisfy (7.1h), a solution that satisfies (7.1h) can be found through the steps in the proof of Proposition 7.1.

7.3.2 Reformulation of the Distributionally Robust Sizing Model

After $g(x,\xi)$ is replaced by $g_l(x,\xi)$, in order to use the method in Chap. 6 to transform the distributionally robust Shortfall risk $r_0(x)$, we first investigate the properties of the loss function $g_l(x,\xi)$.

Proposition 7.2 *For any $x \geq 0$ and $\xi \in \Xi$, $g_l(x,\xi)$ is finite.*

Proof For the given $x \geq 0$ and $\xi \in \Xi$, first construct a feasible solution for (7.7). Let

$$\tilde{p}_t^{gl} = \tilde{p}_t^{gs} = \tilde{p}_t^{sl} = 0\ ,\ \ \tilde{p}_t^{c} = \xi_t^{g} x_g\ ,\ \ \tilde{p}_t^{l} = \xi_t^{c}\ ,\ \ \tilde{e}_t = e_0\ ,\ \ \forall t \in S_T\ . \tag{7.8}$$

It means that ES is not charged or discharged. Renewable energy generation is completely abandoned and does not supply any load. It is easy to verify that the solution $(\tilde{p}_t^{gl}, \tilde{p}_t^{gs}, \tilde{p}_t^{sl}, \tilde{p}_t^{c}, \tilde{p}_t^{l}, \tilde{e}_t; t \in S_T)$ satisfies (7.1b)–(7.1h), and therefore it is a feasible solution. Thereby $g_l(x,\xi) \leq \sum_{t\in S_T} \tilde{p}_t^l \Delta_t = \sum_{t\in S_T} \xi_t^c \Delta_t$. In addition, for any feasible solution, Eq. (7.1b) gives the objective function $\sum_{t\in S_T} p_t^l \Delta_t \geq 0$, so that $g_l(x,\xi) \geq 0$. Based on the above discussion, it can be concluded that $g_l(x,\xi)$ is finite. □

The physical meaning of Proposition 7.2 is that the load shedding is finite. The conclusion is proved by showing that the load-shedding amount is nonnegative and does not exceed the total load, so the conclusion of Proposition 7.2 is natural.

By the reformulation method in Sect. 3.4.5, the distributionally robust Shortfall risk of load shedding is transformed into:

$$r_0(x) = \sup_{\mathbb{P}\in\mathcal{B}(\mathbb{P}_0,\varepsilon)} \mathrm{SR}_{l,\lambda}^{\mathbb{P}}(g_l(x,\xi)) \tag{7.9a}$$

$$= \begin{cases} \min\limits_{r\in\mathbb{R}} r \\ \text{s.t.} \quad \sup\limits_{\mathbb{P}\in\mathcal{B}(\mathbb{P}_0,\varepsilon)} \mathbb{E}^{\mathbb{P}}[l(g_l(x,\xi)-r)] \leq \lambda\ . \end{cases} \tag{7.9b}$$

We are about to use the Lipschitz constant-based method in Sect. 6.3 to further reformulate the worst-case expectation in (7.9b). To this end, we need to find the Lipschitz constant of $l(g_l(x,\xi)-r)$ with respect to ξ. The following proposition gives this Lipschitz constant.

Proposition 7.3 *For any $x \geq 0$,*

$$|g_l(x,\xi) - g_l(x,\xi')| \leq L(x)\|\xi - \xi'\|_1\ ,\ \ \forall \xi, \xi' \in \Xi\ , \tag{7.10}$$

where $L(x) = \max\{x_g, 1\}\Delta_t$.

Proof Define

$$
\begin{aligned}
\xi^{(2T)} &:= \xi = (\xi_1^c, \xi_2^c \dots, \xi_T^c, \xi_1^g, \dots, \xi_T^g)\,,\\
\xi^{(0)} &:= \xi' = (\xi_1^c + \nu_1^c, \xi_2^c + \nu_2^c, \dots, \xi_T^c + \nu_T^c, \xi_1^g + \nu_1^g, \dots, \xi_T^g + \nu_T^g)\,,\\
\xi^{(i)} &:= (\xi_1^c, \dots, \xi_{i-1}^c, \xi_i^c + \nu_i^c, \dots, \xi_T^c + \nu_T^c, \xi_1^g + \nu_1^g, \dots, \xi_T^g + \nu_T^g)\,,\ 1 \le i \le T\,,\\
\xi^{(i)} &:= (\xi_1^c, \dots, \xi_T^c, \xi_1^g, \dots, \xi_{i-T-1}^g, \xi_{i-T}^g + \nu_{i-T}^g, \dots, \xi_T^g + \nu_T^g)\,,\ T+1 \le i \le 2T\,.
\end{aligned}
$$

So long as $|g_l(x, \xi^{(i)}) - g_l(x, \xi^{(i+1)})| \le L(x)\|\xi^{(i)} - \xi^{(i+1)}\|_1$ hold for $i = 0, 1, \dots, 2T-1$, the desired conclusion is obtained, because

$$
\begin{aligned}
|g_l(x,\xi) - g_l(x,\xi')| &= \left|\sum_{i=0}^{2T-1} (g_l(x,\xi^{(i)}) - g_l(x,\xi^{(i+1)}))\right|\\
&\le \sum_{i=0}^{2T-1} \left|g_l(x,\xi^{(i)}) - g_l(x,\xi^{(i+1)})\right|\\
&\le \sum_{i=0}^{2T-1} L(x)\|\xi^{(i)} - \xi^{(i+1)}\|_1\\
&= L(x)\|\xi - \xi'\|_1\,.
\end{aligned}
$$

In fact, $|g_l(x, \xi^{(i)}) - g_l(x, \xi^{(i+1)})| \le L(x)\|\xi^{(i)} - \xi^{(i+1)}\|_1$ is a direct corollary of the following two conclusions: For any $t_0 \in S_T$ and $\varepsilon > 0$,

$$
\begin{aligned}
&|g_l(x,\xi) - g_l(x,\xi')| \le \Delta_t \varepsilon\,,\ \text{ where } \xi, \xi' \in \Xi\,,\\
&\xi = (\xi_1^c, \xi_2^c, \dots, \xi_T^c, \xi_1^g, \xi_2^g, \dots, \xi_T^g)\,,\\
&\xi' = (\xi_1^c, \xi_2^c, \dots, \xi_{t_0-1}^c, \xi_{t_0}^c - \varepsilon, \xi_{t_0+1}^c, \dots, \xi_T^c, \xi_1^g, \xi_2^g, \dots, \xi_T^g)\,,
\end{aligned}
\tag{7.11}
$$

$$
\begin{aligned}
&|g_l(x,\xi) - g_l(x,\xi')| \le x_g \Delta_t \varepsilon\,,\ \text{ where } \xi, \xi' \in \Xi\,,\\
&\xi = (\xi_1^c, \xi_2^c, \dots, \xi_T^c, \xi_1^g, \xi_2^g, \dots, \xi_T^g)\,,\\
&\xi' = (\xi_1^c, \xi_2^c, \dots, \xi_T^c, \xi_1^g, \xi_2^g, \dots, \xi_{t_0-1}^g, \xi_{t_0}^g - \varepsilon, \xi_{t_0+1}^g, \dots, \xi_T^g)\,.
\end{aligned}
\tag{7.12}
$$

1. Prove (7.11).

Assume that $(\hat{p}_t^{gl}, \hat{p}_t^{gs}, \hat{p}_t^{sl}, \hat{p}_t^c, \hat{p}_t^l, \hat{e}_t; t \in S_T)$ is an optimal solution of (7.7) with respect to (x, ξ'). Let

$$
\begin{aligned}
&\tilde{p}_t^{gl} = \hat{p}_t^{gl}\,,\ \tilde{p}_t^{gs} = \hat{p}_t^{gs}\,,\ \tilde{p}_t^{sl} = \hat{p}_t^{sl}\,,\ \tilde{p}_t^c = \hat{p}_t^c\,,\ \tilde{e}_t = \hat{e}_t\,,\ \forall t \in S_T\,,\\
&\tilde{p}_{t_0}^l = \hat{p}_{t_0}^l + \varepsilon\,,\ \tilde{p}_t^l = \hat{p}_t^l\,,\ \forall t \ne t_0\,,
\end{aligned}
$$

then $(\tilde{p}_t^{gl}, \tilde{p}_t^{gs}, \tilde{p}_t^{sl}, \tilde{p}_t^c, \tilde{p}_t^l, \tilde{e}_t; t \in S_T)$ is a feasible solution with respect to (x, ξ), and its objective function satisfies:

$$g_l(x,\xi') + \Delta_t\varepsilon = \sum_{t\in S_T} \hat{p}_t^l \Delta_t + \Delta_t\varepsilon = \sum_{t\in S_T} \tilde{p}_t^l \Delta_t \geq g_l(x,\xi) \ .$$

If $g_l(x,\xi') \leq g_l(x,\xi)$ is proven, then we get (7.11). Thus, we just need to consider the case where $\varepsilon > 0$ is small enough. Assume that $(\check{p}_t^{gl}, \check{p}_t^{gs}, \check{p}_t^{sl}, \check{p}_t^{c}, \check{p}_t^{l}, \check{e}_t; t \in S_T)$ is an optimal solution for (7.7) with respect to (x,ξ). According to Proposition 7.1, it can be assumed that Eq. (7.1h) is satisfied. Considering that $\check{p}_{t_0}^{gl} + \check{p}_{t_0}^{sl} + \check{p}_{t_0}^{l} = \xi_{t_0}^{c} > 0$, there are three possible cases:

Case 1: If $\check{p}_{t_0}^{l} > 0$, let

$$\begin{aligned} &\tilde{p}_t^{gl} = \check{p}_t^{gl} \ , \ \tilde{p}_t^{gs} = \check{p}_t^{gs} \ , \ \tilde{p}_t^{sl} = \check{p}_t^{sl} \ , \ \tilde{p}_t^{c} = \check{p}_t^{c} \ , \ \tilde{e}_t = \check{e}_t \ , \ \forall t \in S_T \ , \\ &\tilde{p}_{t_0}^{l} = \check{p}_{t_0}^{l} - \varepsilon \ , \ \tilde{p}_t^{l} = \check{p}_t^{l} \ , \ \forall t \neq t_0 \ . \end{aligned}$$

Because ε is small enough, $(\tilde{p}_t^{gl}, \tilde{p}_t^{gs}, \tilde{p}_t^{sl}, \tilde{p}_t^{c}, \tilde{p}_t^{l}, \tilde{e}_t; t \in S_T)$ is a feasible solution with respect to (x,ξ'), and its objective function satisfies

$$g_l(x,\xi') \leq \sum_{t\in S_T} \tilde{p}_t^l \Delta_t = \sum_{t\in S_T} \check{p}_t^l \Delta_t - \Delta_t\varepsilon = g_l(x,\xi) - \Delta_t\varepsilon \leq g_l(x,\xi) \ . \quad (7.13)$$

Case 2: If $\check{p}_{t_0}^{l} = 0$ and $\check{p}_{t_0}^{gl} > 0$, let

$$\begin{aligned} &\tilde{p}_t^{gs} = \check{p}_t^{gs} \ , \ \tilde{p}_t^{sl} = \check{p}_t^{sl} \ , \ \tilde{p}_t^{l} = \check{p}_t^{l} \ , \ \tilde{e}_t = \check{e}_t \ , \ \forall t \in S_T \ , \\ &\tilde{p}_{t_0}^{gl} = \check{p}_{t_0}^{gl} - \varepsilon \ , \ \tilde{p}_{t_0}^{c} = \check{p}_{t_0}^{c} + \varepsilon \ , \ \tilde{p}_t^{gl} = \check{p}_t^{gl} \ , \ \tilde{p}_t^{c} = \check{p}_t^{c} \ , \ \forall t \neq t_0 \ . \end{aligned}$$

Then $(\tilde{p}_t^{gl}, \tilde{p}_t^{gs}, \tilde{p}_t^{sl}, \tilde{p}_t^{c}, \tilde{p}_t^{l}, \tilde{e}_t; t \in S_T)$ is a feasible solution with respect to (x,ξ'), and satisfies

$$g_l(x,\xi') \leq \sum_{t\in S_T} \tilde{p}_t^l \Delta_t = \sum_{t\in S_T} \check{p}_t^l \Delta_t = g_l(x,\xi) \ . \quad (7.14)$$

Case 3: If $\check{p}_{t_0}^{l} = \check{p}_{t_0}^{gl} = 0$ and $\check{p}_{t_0}^{sl} > 0$, let

$$\begin{aligned} &\tilde{p}_t^{gl} = \check{p}_t^{gl} \ , \ \tilde{p}_t^{l} = \check{p}_t^{l} \ , \ \forall t \in S_T \ , \\ &\tilde{p}_{t_0}^{sl} = \check{p}_{t_0}^{sl} - \varepsilon \ , \ \tilde{p}_t^{sl} = \check{p}_t^{sl} \ , \ \forall t \neq t_0 \ . \end{aligned}$$

There are two possible scenarios below:

The first scenario: If there exists $t_1 > t_0$ such that $\check{p}_{t_1}^{gs} > 0$, choose t_1 so that $\check{e}_{t_1} \geq \check{e}_{t_2}, \forall t_2 > t_0$. The approach is to first find the time period $\text{argmax}\{\check{e}_t \mid t > t_0\}$ during which the SoC reaches its maximum, then find t_1 in it that makes $\check{p}_{t_1}^{gs} > 0$. Such t_1 must exist, because according to (7.1e) and (7.1h), $\check{e}_{t-1} < \check{e}_t$ is equavalent to $\check{p}_t^{gs} > 0$. Let

$$\tilde{p}_{t_1}^{gs} = \check{p}_{t_1}^{gs} - \varepsilon/(\eta_c\eta_d) \ , \ \ \tilde{p}_{t_1}^{c} = \check{p}_{t_1}^{c} + \varepsilon/(\eta_c\eta_d) \ ,$$
$$\tilde{p}_{t}^{gs} = \check{p}_{t}^{gs} \ , \ \ \tilde{p}_{t}^{c} = \check{p}_{t}^{c} \ , \ \ \forall t \neq t_1 \ .$$

Then we can determine $\tilde{e}_t$ by (7.1e), and $\tilde{e}_0 = \check{e}_0$. Thus, $\tilde{e}_T = \tilde{e}_0$ and $\omega_l x_e \leq \min_s\{\check{e}_s\} \leq \tilde{e}_t \leq \max_s\{\check{e}_s\} \leq \omega_h x_e$. Hence, $(\tilde{p}_t^{gl}, \tilde{p}_t^{gs}, \tilde{p}_t^{sl}, \tilde{p}_t^{c}, \tilde{p}_t^{l}, \tilde{e}_t; t \in S_T)$ is a feasible solution with respect to (x, ξ'), then $g_l(x, \xi') \leq g_l(x, \xi)$.

The second scenario: If for any $t_1 > t_0$, $\check{p}_{t_1}^{gs} = 0$, then $\check{e}_{t_0} \geq \check{e}_T = \check{e}_0$. Because $\check{p}_{t_0}^{sl} > 0$, $\check{e}_{t_0-1} > \check{e}_{t_0} \geq \check{e}_0$, there exists $1 \leq t < t_0$ such that $\check{p}_t^{gs} > 0$. Choose $t_2 = \max\{t \mid 1 \leq t < t_0, \check{p}_t^{gs} > 0\}$. Let

$$\tilde{p}_{t_2}^{gs} = \check{p}_{t_2}^{gs} - \varepsilon/(\eta_c\eta_d) \ , \ \ \tilde{p}_{t_2}^{c} = \check{p}_{t_2}^{c} + \varepsilon/(\eta_c\eta_d) \ ,$$
$$\tilde{p}_{t}^{gs} = \check{p}_{t}^{gs} \ , \ \ \tilde{p}_{t}^{c} = \check{p}_{t}^{c} \ , \ \ \forall t \neq t_2 \ ,$$

and we can determine $\tilde{e}_t$ by (7.1e) and $\tilde{e}_0 = \check{e}_0$. Then $(\tilde{p}_t^{gl}, \tilde{p}_t^{gs}, \tilde{p}_t^{sl}, \tilde{p}_t^{c}, \tilde{p}_t^{l}, \tilde{e}_t; t \in S_T)$ is a feasible solution with respect to (x, ξ'), and thus $g_l(x, \xi') \leq g_l(x, \xi)$.
2. The proof of (7.12) is similar to that of (7.11). □

In fact, the best or the smallest Lipschitz constant is exactly the value $L(x) = \max\{x_g, 1\}\Delta_t$ in Proposition 7.3. To show it, we only need to prove that $|g_l(x, \xi) - g_l(x, \xi')| \leq L(x)\|\xi - \xi'\|_1, \forall \xi, \xi' \in \Xi$ and there exist $\xi_0, \xi_0' \in \Xi$ such that $g_l(x, \xi_0) - g_l(x, \xi_0') = L(x)\|\xi_0 - \xi_0'\|_1$. The former has been proven in Proposition 7.3. The reason for the latter is as follows: $L(x)$ is the maximum of Δ_t and $x_g\Delta_t$. In the proof of Proposition 7.3, case 1 gives an example with $g_l(x, \xi) - g_l(x, \xi') = \Delta_t\|\xi - \xi'\|_1$. Similarly, for $x_g\Delta_t$ we can find $\xi, \xi' \in \Xi$ such that $g_l(x, \xi) - g_l(x, \xi') = x_g\Delta_t\|\xi - \xi'\|_1$. At least one of the above two cases can indicate that $L(x)$ is the smallest Lipschitz constant.

In terms of physical meaning, the Lipschitz constant $L(x)$ reflects how much the load shedding will change when the random variable changes. The random variable $\xi = (\xi^c, \xi^g)$ includes load demand and renewable energy generation, and because this paper uses the 1-norm, the uncertainties of load and renewable energy generation can be considered separately. If the load power increases by 1 unit at a certain time period, the load shedding will not decrease, and at most, it will increase by Δ_t. If the renewable energy generation per unit capacity increases by 1 at a certain time period, the load shedding will not increase, and it will decrease by $x_g\Delta_t$ at most. The value of $L(x)$ is the combination of Δ_t and $x_g\Delta_t$.

Because $l(r) = \max\{r, 0\}$, we have $|l(r) - l(r')| \leq |r - r'|, \forall r, r' \in \mathbb{R}$, and thus

$$|l(g_l(x, \xi) - r) - l(g_l(x, \xi') - r)| \leq L(x)\|\xi - \xi'\|_1 \ , \ \ \forall \xi, \xi' \in \Xi \ . \qquad (7.15)$$

Therefore, $l(g_l(x, \xi) - r)$ satisfies the conditions of Theorem 6.10 as a function of ξ, and then

$$\sup_{\mathbb{P} \in \mathcal{B}(\mathbb{P}_0, \varepsilon)} \mathbb{E}^{\mathbb{P}}[l(g_l(x, \xi) - r)] \leq \varepsilon L(x) + \mathbb{E}^{\mathbb{P}_0}[l(g_l(x, \xi) - r)] \ , \qquad (7.16)$$

where the right side is a conservative estimation of the worst-case expectation, based on which we obtain a conservative approximation for the distributionally robust Shortfall risk as follows:

$$\begin{aligned} r_c(x) := & \min_{r \in \mathbb{R}} r \\ & \text{s.t. } \varepsilon L(x) + \mathbb{E}^{\mathbb{P}_0}[l(g_l(x,\xi) - r)] \le \lambda \,. \end{aligned} \tag{7.17}$$

By the sample average approximation (SAA) and $L(x) = \max\{x_g, 1\}\Delta_t$, $r_c(x)$ is expressed as the optimal value of the following LP problem:

$$r_c(x) = \min r \tag{7.18a}$$

$$\text{s.t. } \varepsilon L + \frac{1}{N}\sum_{n \in S_N} s_n \le \lambda \,, \tag{7.18b}$$

$$L \ge \Delta_t \,,\; L \ge x_g \Delta_t \,, \tag{7.18c}$$

$$s_n \ge 0 \,,\; s_n \ge \sum_{t \in S_T} p_{t,n}^l \Delta_t - r \,,\; \forall n \in S_N \,, \tag{7.18d}$$

$$(7.1\text{b})\text{–}(7.1\text{g}) \,,\; \forall n \in S_N \,, \tag{7.18e}$$

where L and $s_n, n \in S_N$ are auxiliary variables. L represents $L(x) = \max\{x_g, 1\}\Delta_t$ and s_n denotes $l(g_l(x,\xi) - r)$.

Using (7.18), the sizing problem in (7.6) is approximately transformed into the following traditional bi-objective LP problem:

$$\begin{aligned} & \min \; \{r, C_g x_g + C_e x_e + C_p x_p\} \\ & \text{s.t. } x_g, x_e, x_p \ge 0 \,,\; (7.18\text{b})\text{–}(7.18\text{e}) \,. \end{aligned} \tag{7.19}$$

7.3.3 Calculation of the Pareto Frontier

We are about to propose an algorithm to calculate the Pareto frontier of problem (7.19) based on parametric LP. Denote the feasible objective value set of problem (7.19) by $\mathcal{P}$, then

$$\mathcal{P} := \left\{ (r, f) \in \mathbb{R}^2 \;\middle|\; \begin{array}{l} f \ge C_g x_g + C_e x_e + C_p x_p \,, \\ x \ge 0 \,,\; (7.18\text{b})\text{–}(7.18\text{e}) \end{array} \right\} . \tag{7.20}$$

Because the constraints are all linear, $\mathcal{P}$ is the projection of a polyhedron into the (r, f) surface, so $\mathcal{P}$ is also a polyhedron. The Pareto frontier consists of all feasible points that are not worse than any other feasible point, i.e.,

$$\mathcal{F} := \left\{ (r, f) \in \mathcal{P} \;\middle|\; \text{if } (r, f) \neq (r', f') \in \mathcal{P} \text{ then } r < r' \text{ or } f < f' \right\} . \tag{7.21}$$

$\mathcal{F}$ is a subset of the boundary of $\mathcal{P}$. According to the properties of Pareto frontier and ϵ-constraint method introduced in Sect. 2.6.2, $\mathcal{F}$ has the following equivalent model:

$$\mathcal{F} = \mathcal{F}_1 \cap \mathcal{F}_2 , \tag{7.22a}$$

$$\mathcal{F}_1 := \left\{ (r_*, f_*) \in \mathbb{R}^2 \;\middle|\; \begin{aligned} f_* = \min\ & f \\ \text{s.t. } & r \le r_* ,\ (r, f) \in \mathcal{P} \end{aligned} \right\} , \tag{7.22b}$$

$$\mathcal{F}_2 := \left\{ (r_*, f_*) \in \mathbb{R}^2 \;\middle|\; \begin{aligned} r_* = \min\ & r \\ \text{s.t. } & f \le f_* ,\ (r, f) \in \mathcal{P} \end{aligned} \right\} . \tag{7.22c}$$

$\mathcal{F}_1$ and $\mathcal{F}_2$ have similar forms, where the LP problems can be regarded as the following parametric LP problem:

$$\begin{aligned} v(\theta) := \min_y\ & c^\top y \\ \text{s.t. } & Ay \le b + b'\theta . \end{aligned} \tag{7.23}$$

For example, in (7.22b), $r_* = \theta$ is a parameter, $f_* = v(\theta)$ is the optimal value dependent on the parameter, and Eq. (7.23) contains the constraints $r \le r_*$ and $(r, f) \in \mathcal{P}$. Therefore, to calculate the Pareto frontier $\mathcal{F}$, we only need to obtain the analytic expression of $v(\theta)$. According to the theory of parametric LP in Sect. 2.5.4, $v(\theta)$ is a piecewise affine function.

For a fixed $\theta_0 \in \mathbb{R}$, assume that Eq. (7.23) has an optimal solution y_0. The constraints can be divided into two parts according to whether an equality is taken at y_0, i.e.,

$$A_1 y_0 = b_1 + b_1'\theta_0 , \tag{7.24a}$$

$$A_2 y_0 < b_2 + b_2'\theta_0 , \tag{7.24b}$$

$$A = \begin{pmatrix} A_1 \\ A_2 \end{pmatrix} ,\ b = \begin{pmatrix} b_1 \\ b_2 \end{pmatrix} ,\ b' = \begin{pmatrix} b_1' \\ b_2' \end{pmatrix} . \tag{7.24c}$$

If this segmentation remains unchanged in a neighborhood of θ_0, then there exist d_1 and d_1' such that

$$A_1 d_1 = b_1 ,\ A_1 d_1' = b_1' , \tag{7.25a}$$

$$y_0 = d_1 + d_1'\theta_0 . \tag{7.25b}$$

According to parametric LP, in a critical region of the parameter θ, the segmentation in (7.24) does not change. In the critical region near θ_0, $y_* := d_1 + d_1'\theta$ is the optimal solution. Substituting it into (7.24b), we get the following critical region:

$$\{\theta \mid (A_2 d_1' - b_2') \cdot \theta < b_2 - A_2 d_1\} . \quad (7.26)$$

Note that $\theta \in \mathbb{R}$, so there exist $\theta^l < \theta_0 < \theta^h$ such that Eq. (7.26) is equivalent to $\theta \in (\theta^l, \theta^h)$. When θ is in the critical region (θ^l, θ^h), $v(\theta) = c^\top y_* = c^\top d_1 + c^\top d_1' \theta$ takes the form of an affine function.

Let the range of θ be denoted as $[\theta_{\min}, \theta_{\max}]$. For instance, in (7.22c), $f_* = \theta$ and the range is $[0, F]$, where F is the upper bound of the total investment cost. In (7.22b), $r_* = \theta$ and the endpoints of $[r_{\min}, r_{\max}]$ can be obtained by solving the optimization problem in (7.22c) under $f_* = F$ and $f_* = 0$, respectively. In order to compute the analytical expression of $v(\theta)$, as shown in Algorithm 7.1, find the critical regions and coefficients of affine functions along $[\theta_{\min}, \theta_{\max}]$.

Algorithm 7.1 Calculate the analytical expression of Pareto frontier $v(\theta)$

Input: $A, b, b', c, [\theta_{\min}, \theta_{\max}]$, and searching step length $\delta > 0$.
Output: $v(\theta) = w_k + w_k' \theta, \theta \in (\theta_k^l, \theta_k^h), k = 1, 2, \ldots, M$.
1: Initiation: Let $M \leftarrow 1, \theta_0 \leftarrow \theta_{\min} + \delta$.
2: For $\theta = \theta_0$, solve (7.23) and obtain the coefficient matrices A_1, A_2 and coefficient vertices b_1, b_2, b_1', b_2' in (7.24a) and (7.24b).
3: Solve (7.25) and obtain d_1 and d_1'.
4: Solve (7.26) and obtain the interval (θ_M^l, θ_M^h). Let $w_M \leftarrow c^\top d_1$, $w_M' \leftarrow c^\top d_1'$.
5: Let $\theta_0 \leftarrow \theta_M^h + \delta$. If $\theta_0 \geq \theta_{\max}$, output $v(\theta) = w_k + w_k' \theta, \theta \in (\theta_k^l, \theta_k^h), k = 1, 2, \ldots, M$ and terminate; otherwise, let $M \leftarrow M + 1$ and turn to Step 2.

As θ_0 increases by at least δ in each iteration, Algorithm 7.1 terminates after at most $(\theta_{\max} - \theta_{\min})/\delta$ iterations, i.e., $M \leq (\theta_{\max} - \theta_{\min})/\delta$. When δ is chosen small enough, all critical regions will be searched, M will equal the number of critical regions, and the output of the algorithm will be independent of δ.

7.4 Case Study

7.4.1 Settings

Consider a stand-alone microgrid equipped with photovoltaic (PV) power generation. The sunlight data comes from the National Solar Radiation Database (NREL) (Sengupta et al. 2018). Figure 7.2 shows the degree of dispersion of PV power generation and Fig. 7.3 depicts the average load per hour. The load data is generated according to the method in Alharbi and Bhattacharya (2018), where the standard deviation is set to 0.1 per unit (p.u.), and the lower and upper bounds are 0 and 2 p.u., respectively, and the power base value is 2 MW.

To reflect the seasonal changes in PV power generation and load, each set of data consists of 4 typical days, and these 4 typical days come from 4 seasons. Based on 7 years of PV power generation and load data, 637 sets of data were obtained,

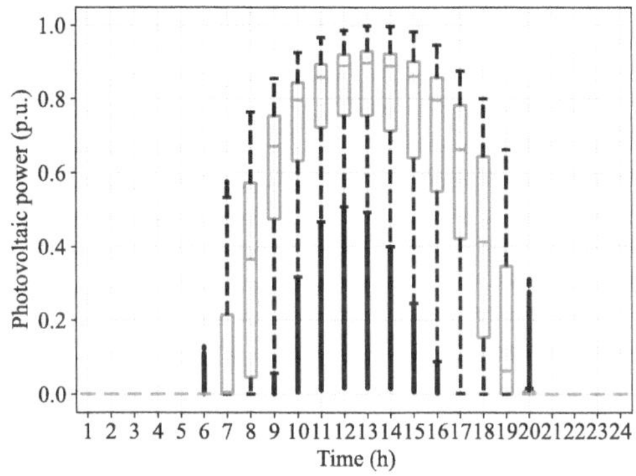

Fig. 7.2 Boxplot of photovoltaic power generation data (the *long box* represents the 50% confidence interval, the *line inside the box* represents the median, and the *points outside the box* and the *dashed line* represent outliers) (Xie et al. 2022)

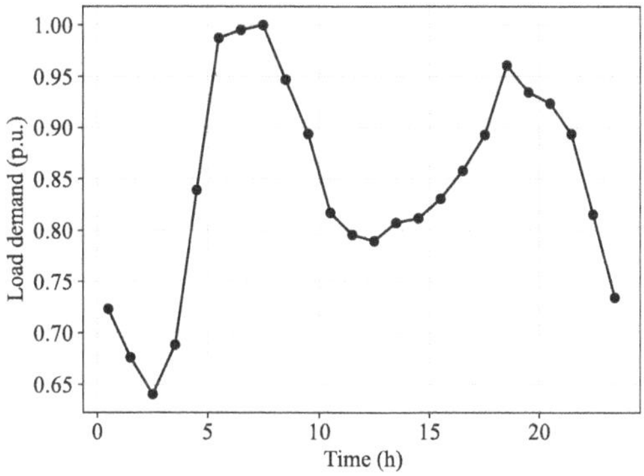

Fig. 7.3 Average load demand curve (Xie et al. 2022)

among which 455 sets were used to establish the empirical probability distribution $\mathbb{P}_0$, and the remaining 182 sets of data were used for testing. In the basic case, $T = 96$, $N = 182$, $\Delta_t = 1$ h, and the data obtained are $\xi_n = (\xi^c_{t,n}, \xi^g_{t,n}; t \in S_T), n \in S_N$. Other parameters include $\eta_c = 95\%$, $\eta_d = 95\%$, $\omega_l = 10\%$, $\omega_h = 90\%$, $C_g = 2.4 \times 10^6$ CNY/MW, $C_p = 10^6$ CNY/MW, $C_e = 1.2 \times 10^6$ CNY/MWh (Chen et al. 2022), $\varepsilon = 0.02$, and λ is set to 1% of the total load. The solving process is carried out on a laptop with an Intel Core i5-5200U processor and 8 GB of memory. The LP problem is programmed in the MATLAB environment using the YALMIP toolbox (Lofberg 2004), and solved by Gurobi 9.1 (Gurobi Optimization 2023).

7.4.2 Other Models for Comparison

Six methods are considered and compared in the case study of this chapter:

Base case study: The proposed method, denoted as DRO-C;

The first comparison method: The stochastic programming (SP) method, denoted as SP;

The second comparison method: The robust optimization (RO) method, denoted as RO;

The third comparison method: The distributionally robust optimization (DRO) method based on Wasserstein distance but only considering discrete probability distributions, denoted as DRO-D;

The fourth comparison method: The method based on time series simulation and search, denoted as CS;

The fifth comparison method: The DRO method based on Kullback-Leibler (KL) divergence, denoted as DRO-KL.

In the following, we will describe each comparison method in detail.

7.4.2.1 Stochastic Programming Method

Treating the empirical probability distribution $\mathbb{P}_0$ as the real probability distribution, the Shortfall risk under $\mathbb{P}_0$ is calculated, i.e., $r_s(x) := \mathrm{SR}_{l,\lambda}^{\mathbb{P}_0}(g(x,\xi))$. According to the analysis in Sect. 7.3, $r_s(x)$ can be calculated as follows:

$$r_s(x) = \min_{r,s_n} r \tag{7.27a}$$

$$\text{s.t.}\ \frac{1}{N}\sum_{n\in S_N} s_n \le \lambda\ ,\ s_n \ge 0\ ,\ \forall n \in S_N\ , \tag{7.27b}$$

$$s_n \ge g_l(x,\xi_n) - r\ ,\ \forall n \in S_N\ . \tag{7.27c}$$

The corresponding sizing model boils down to the following traditional bi-objective LP problem:

$$\begin{aligned}&\min\ \{r, C_g x_g + C_e x_e + C_p x_p\}\\&\text{s.t.}\ x_g, x_e, x_p \ge 0\ ,\ (7.18e)\ ,\ (7.27b)\ .\end{aligned}$$

The Pareto frontier can then be solved using Algorithm 7.1.

7.4.2.2 Robust Optimization Method

The sampled data set is regarded as the range of random variable values, and load shedding is not allowed. In this circumstance, the sizing model is as follows:

$$\begin{aligned}&\min_x\ C_g x_g + C_e x_e + C_p x_p\\&\text{s.t.}\ x_g, x_e, x_p \ge 0\ ,\\&\qquad g(x,\xi_n) = 0\ ,\ \forall n \in S_N\ ,\end{aligned}$$

which is equivalent to the following LP problem:

$$\begin{aligned} &\min\ C_g x_g + C_e x_e + C_p x_p \\ &\text{s.t. } x_g, x_e, x_p \geq 0\ , \\ &\quad (7.1\text{b})-(7.1\text{g})\ ,\ \forall n \in S_N\ , \\ &\quad p^l_{t,n} = 0\ ,\ \forall t \in S_T\ ,\ n \in S_N\ . \end{aligned}$$

7.4.2.3 Distributionally Robust Optimization Method Based on Wasserstein Distance Considering Only Discrete Probability Distributions

The probability distribution ambiguity set $\mathcal{B}_d(\mathbb{P}_0, \varepsilon)$ is defined according to the Wasserstein distance, but it only contains discrete probability distributions with the same support set $\Xi_0 := \{\xi_n \mid n \in S_N\}$ as the empirical probability distribution, defined as follows:

$$\mathcal{B}_d(\mathbb{P}_0, \varepsilon) := \left\{\mathbb{P} \in \mathcal{M}_1(\Xi_0) \;\middle|\; d_W(\mathbb{P}, \mathbb{P}_0) \leq \varepsilon\right\}\ , \tag{7.28}$$

so the corresponding distributionally robust Shortfall risk is:

$$r_d(x) := \sup_{\mathbb{P} \in \mathcal{B}_d(\mathbb{P}_0,\varepsilon)} \mathrm{SR}^{\mathbb{P}}_{l,\lambda}(g(x,\xi)) \tag{7.29a}$$

$$= \begin{cases} \min\limits_{r \in \mathbb{R}} r \\ \text{s.t.} \quad \sup\limits_{\mathbb{P} \in \mathcal{B}_d(\mathbb{P}_0,\varepsilon)} \mathbb{E}^{\mathbb{P}}[l(g_l(x,\xi) - r)] \leq \lambda\ . \end{cases} \tag{7.29b}$$

Using the reformulation method in Sect. 6.3.1, the above worst-case expectation can be equivalently reformulated. The final form of the sizing model to be solved is:

$$\min\ \{r, C_g x_g + C_e x_e + C_p x_p\} \tag{7.30a}$$

$$\text{s.t. } x_g, x_e, x_p \geq 0\ , \tag{7.30b}$$

$$\varepsilon\rho + \sum_{n \in S_N} \frac{\gamma_n}{N} + \tau \leq \lambda\ ,\ \rho \geq 0\ , \tag{7.30c}$$

$$\rho \|\xi_m - \xi_n\|_1 + \mu_m + \gamma_n \geq 0\ ,\ \forall m, n \in S_N\ , \tag{7.30d}$$

$$-\mu_n + \tau \geq \sum_{t \in S_T} p^l_{t,n} \Delta_t - y\ ,\ -\mu_n + \tau \geq 0\ ,\ \forall n \in S_N\ , \tag{7.30e}$$

$$(7.1\text{b})-(7.1\text{g})\ ,\ \forall n \in S_N\ . \tag{7.30f}$$

Because the probability distribution ambiguity set $\mathcal{B}_d(\mathbb{P}_0, \varepsilon) \subsetneq \mathcal{B}(\mathbb{P}_0, \varepsilon)$, Eq. (7.30) is an optimistic approximation of the sizing model in (7.6), so the distributionally robust Shortfall risk of load shedding $r_0(x)$ defined in this chapter satisfies:

$$r_d(x) \le r_0(x) \le r_c(x) \; . \tag{7.31}$$

Because Eq. (7.30) is a traditional bi-objective LP problem, the Pareto frontier can also be solved using the method in Algorithm 7.1.

7.4.2.4 Time Series Simulation and Search-Based Method

In the planning sample data, load shedding is not allowed, and the total investment cost $C_g x_g + C_e x_e + C_p x_p$ is minimized using time series simulation (Castro and Ferreira 2001) and the search method in MATLAB's global optimization toolbox (The MathWorks Inc 2004). For each capacity variable value $x = (x_g, x_e, x_p)$, it checks through time series simulation whether there would be no load shedding for all planning sample data.

In the time series simulation, the operation strategy of ES is: When the power of renewable energy generation is greater than load power, ES charges until the SoC reaches the upper limit or the charging power reaches the upper limit; when the power of renewable energy generation is not enough to meet the load, ES discharges until the SoC reaches the lower limit or the discharging power reaches the upper limit. If the sum of the power of renewable energy generation and the discharging power of ES is less than the load power, load shedding occurs, and the capacity variable value in this case is marked as infeasible. To allow for continuous operation of ES, the constraint $e_T \ge e_0$ is also considered.

7.4.2.5 Distributionally Robust Optimization Method Based on Kullback-Leibler Divergence

The KL divergence is used to measure the difference from the probability distribution to the empirical probability distribution $\mathbb{P}_0$ in order to establish a probability distribution ambiguity set. Noting that $\mathbb{P}_0$ is a discrete probability distribution, the probability distribution ambiguity set established only contains discrete probability distributions with the same support set $\Xi_0 = \{\xi_n \mid n \in S_N\}$ as $\mathbb{P}_0$. The definition of KL divergence D_{KL} is in Definition 5.4. The probability distribution ambiguity set based on KL divergence consists of all probability distributions whose KL divergence from $\mathbb{P}_0$ does not exceed the upper limit, i.e.,

$$\mathcal{B}_{KL}(\mathbb{P}_0, \varepsilon_{KL}) := \{\mathbb{P} \in \mathcal{M}_1(\Xi_0) \mid D_{\mathrm{KL}}(\mathbb{P}, \mathbb{P}_0) \le \varepsilon_{KL}\}$$
$$= \left\{ \mathbb{P} = \sum_{n \in S_N} q_n \mathbf{1}_{\xi_n} \;\middle|\; \begin{array}{l} q_n \ge 0 \ , \ \forall n \in S_N \ , \\ \sum_{n \in S_N} q_n = 1 \ , \ D_{\mathrm{KL}}(\mathbb{P}, \mathbb{P}_0) \le \varepsilon_{KL} \end{array} \right\} ,$$

where ε_{KL} is the upper limit of KL divergence.

The sizing model is established using a robust chance constraint as follows:

$$\min C_g x_g + C_e x_e + C_p x_p \tag{7.32a}$$
$$\text{s.t.} \quad \inf_{\mathbb{P} \in \mathcal{B}_{KL}(\mathbb{P}_0, \varepsilon)} \mathbb{P}[g(x, \xi) \le \lambda] \ge 1 - \alpha \ , \tag{7.32b}$$

where Eq. (7.32b) is the robust chance constraint for load shedding not exceeding the upper limit λ. For all probability distributions in the probability distribution ambiguity set, the probability of load shedding not exceeding the upper limit is required to be no less than $1 - \alpha$. The total investment cost is minimized under the robust chance constraint for load shedding.

According to Theorem 5.14, Eq. (7.32b) is equivalent to:

$$\mathbb{P}_0[g(x, \xi) \le \lambda] \ge 1 - \alpha_{1+} \ , \tag{7.33}$$

where α_{1+} satisfies:

$$\alpha_{1+} := \max\{\alpha_1, 0\} \ ,$$
$$\alpha_1 := 1 - \inf_{y \in (0,1)} \left\{ \frac{\exp(-\varepsilon_{KL}) y^{1-\alpha} - 1}{y - 1} \right\} .$$

Equation (7.33) is a traditional chance constraint under the empirical probability distribution $\mathbb{P}_0$. Using the SAA, we obtain the equivalent constraint of (7.33) as follows:

$$\begin{cases} \dfrac{1}{N} \sum_{n \in S_N} z_n \le \alpha_{1+} \ , \\ z_n \in \{0, 1\} \ , \ g_l(x, \xi_n) - \lambda \le M_{KL} \cdot z_n \ , \ \forall n \in S_N \ , \end{cases} \tag{7.34}$$

where M_{KL} is a sufficiently large positive number; $z_n, n \in S_N$ are auxiliary Boolean variables, 0 indicates that the load shedding under scenario ξ_n does not exceed the limit, and 1 indicates it exceeds the limit. It is required that the frequency of exceeding the limit does not exceed α_{1+}. Therefore, the sizing model in (7.32) is equivalent to the following mixed-integer linear programming (MILP) problem:

$$
\begin{aligned}
&\min\ C_g x_g + C_e x_e + C_p x_p \\
&\text{s.t.}\ \sum_{n \in S_N} z_n \leq \alpha_{1+}\ , \\
&\quad z_n \in \{0, 1\}\ ,\ \sum_{t \in S_T} p^l_{t,n} \Delta_t - \lambda \leq M_{KL} \cdot z_n\ ,\ \forall n \in S_N\ , \\
&\quad (7.1b)\text{–}(7.1g)\ ,\ \forall n \in S_N\ .
\end{aligned}
$$

7.4.3 Results and Comparison

Figure 7.4 shows the Pareto frontier of load shedding Shortfall risk and total investment cost under several methods, where two types of curves, namely those within a larger range and load shedding risks near 0 are plotted. In most scenarios, the differences among the three methods in the former kind are not evident, whereas more prominent differences are seen in the latter kind. According to the definition of Shortfall risk, it represents the additional input required to keep the expected loss within limits, so it can reflect the gap from the current state to the acceptable state of expected loss. Thus, in the decision-making process where an acceptable expected loss is hoped for, the focus is mainly near the load-shedding risk of 0, where the three methods differ. The curves from top to bottom correspond to DRO-C, DRO-D, and SP, respectively. This suggests that if the budget is given, then the load shedding risk meets the relationship in (7.31), validating that the proposed method is more conservative than the DRO considering only discrete probability distributions, while traditional SP is the most optimistic among these three methods. For the RO models and methods based on time series simulation, since no load shedding is allowed, there is no balance between load shedding and investment cost.

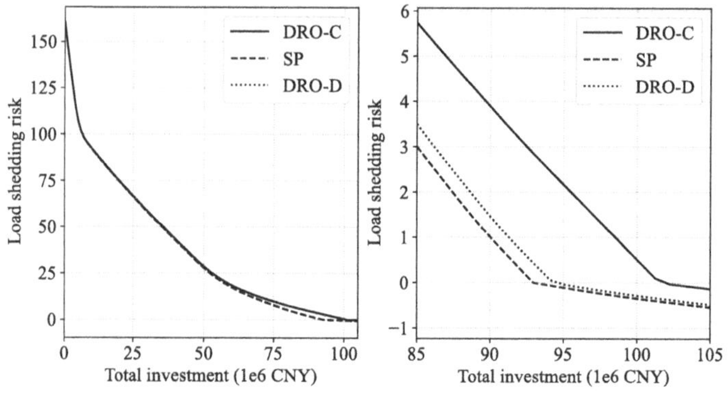

Fig. 7.4 The Pareto frontier of load shedding Shortfall risk and total investment cost in the stand-alone microgrid (Xie et al. 2022)

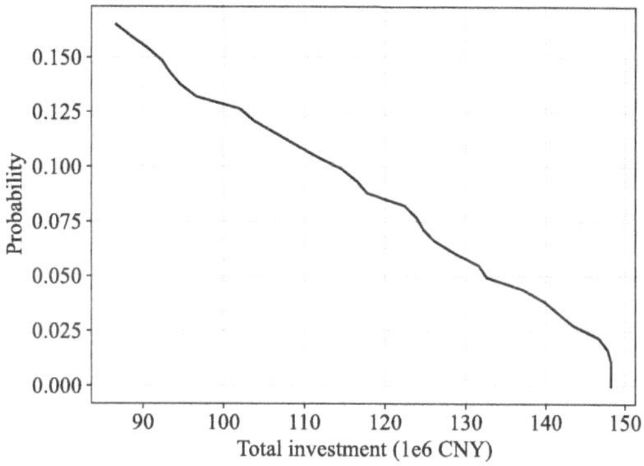

Fig. 7.5 The Pareto frontier of the probability of the load shedding exceeding the upper limit and the total investment cost in the stand-alone microgrid (Xie et al. 2022)

The KL divergence and Wasserstein distance are somewhat related according to Chaps. 5 and 6, and accordingly, we set $\varepsilon_{KL} = 0.005$. By employing the ϵ-constraint technique, the Pareto frontier between the probability α of load shedding exceeding the upper limit and the total investment cost is solved for the DRO model based on KL divergence, as shown in Fig. 7.5. Because the model based on KL divergence is in the form of a MILP problem, the Pareto frontier is not a convex function. The probability of load shedding exceeding the limit and the Shortfall risk of load shedding reflect the load shedding from different perspectives. Compared with the probability of load shedding exceeding the limit, the Shortfall risk of load shedding is more concerned with the load shedding degree.

The average total load for four typical days is 160.6 MWh, setting the allowable load shedding $\lambda = 1.606$ MWh, i.e., 1% of the total load. In addition, the upper limit of Shortfall risk is set to 0, meaning the expected shortfall must be within an acceptable range. Table 7.1 lists the sizing results under various methods. Among them, the DRO-C, SP, and RO methods all boil down to solving LP with about 210,000 constraints and 110,000 variables, the DRO-D method boils down to solving LP with about 240,000 constraints and 110,000 variables, and the DRO-KL method needs to solve MILP with about 210,000 constraints and 110,000 variables (including 182 Boolean variables). The Shortfall risk was tested with the 182 sets of sample data used for testing, and the results are displayed in the last row of Table 7.1.

As expected, the results of the RO model are more conservative than those of SP and the 3 DRO methods, meaning that no load shedding will occur even if the load increases slightly. The SP model has the lowest cost, but the test risk reaches 2.672, which is significantly higher than the upper limit of 0. This shows that the SP method can achieve the desired effect in the planning data, but not in the test data. This is because the amount of data is not large enough, and the empirical probability distribution represented by the planning data is different from the true probability distribution. Although both the planning data and the test data come from the sampling of the real probability distribution, SP only optimizes for the planning data, and its results do not reach the robustness required for the test data. Therefore, it is

Table 7.1 Sizing results of the stand-alone microgrid under different methods (Xie et al. 2022)

Method	DRO-C	SP	RO	DRO-D	CS	DRO-KL
Renewable capacity x_g (MW)	19.78	18.52	19.62	18.73	29.57	17.98
ES power capacity x_p (MW)	6.459	5.843	6.178	6.032	8.934	5.540
ES energy capacity x_e (MWh)	39.69	35.56	83.39	36.09	72.32	45.38
Total investment f (10^6 CNY)	101.6	92.96	153.3	94.30	166.7	103.1
Test Shortfall risk	−0.075	2.672	−1.473	2.153	−1.545	−0.079

necessary to use DRO methods to consider the inexactness of empirical probability distributions. For the DRO-D method that only considers discrete probability distributions, the test risk is still higher than 0, indicating the need to consider general forms of probability distribution. The proposed method offers a compromise: the total investment cost is only about 66% of the RO and the test risk is –0.075, which is still less than the upper limit. This again validates (7.31).

Although the method based on time series simulation, like the RO method, stipulates no load shedding in the test data and achieves the same risk value, its result is sub-optimal compared to RO, so it is not Pareto optimal, reflecting the limitations of the time series simulation method. Moreover, although these two methods do not have load shedding in the planning data, load shedding can still occur in the test data. The reason is the inaccuracy of the empirical probability distribution, which once again demonstrates the need to use DRO methods.

In Table 7.1, the DRO-KL result is at $\alpha = 0.126$, where $\alpha_{1+} \approx 0.096$. The test load-shedding risk of DRO-KL is very close to DRO-C, but the total investment cost is higher. The reason is that the proposed method optimizes the load shedding risk, while the method based on KL divergence and chance constraints cares about the probability of the load shedding exceeding the upper limit. Therefore, to achieve a similar load-shedding risk performance, the total investment cost paid by the method based on KL divergence is slightly higher. However, in reality, the probability distribution represented by the test data does not have the same support as the empirical probability distribution based on the planning data, so their KL divergence is undefined. This leads to the probability distribution ambiguity set constructed based on the KL divergence not including the probability distribution represented by the test data.

7.4.4 Parameter Sensitivity Analysis

The influence of certain parameters on the results of the proposed method is examined, including the radius ε of the probability distribution ambiguity set, the amount N of data used for planning, the acceptable amount of load shedding λ, and the ES unit capacity costs C_p and C_e. In the base scenario, $\varepsilon = 0.02$, $\lambda = 1.606$ MWh, $N = 182$, $C_p = 10^6$ CNY/MW, and $C_e = 1.2 \times 10^6$ CNY/MWh.

Table 7.2 shows the sizing results of the proposed method under different radii ε of the probability distribution ambiguity set. When ε increases, the size of the ambiguity set increases, thus the conservatism of the results increases, leading to higher total investment cost and lower load shedding risk. It should be noted that when $\varepsilon = 0$, 0.002, 0.005, and 0.01, the test risk is greater than 0, indicating that these radii are not sufficient to guarantee that the ambiguity set includes the probability distribution represented by the test data. When $\varepsilon = 0$, the proposed method degenerates into SP, and the results are exactly the same as the SP model.

Table 7.3 shows the impact of the volume N of planning data. Generally speaking, the more data there are, the closer the empirical probability distribution built upon it comes to the real probability distribution. Therefore, when selecting the radius ε of the ambiguity set, one should take into account both the volume N of data and the risk preference of the decision maker. In fact, Sect. 6.2.2 provides theoretical results about how to choose the radius ε according to N.

Table 7.4 shows the sizing results under different acceptable load shedding λ. Because the definition of Shortfall risk depends on λ, robustness is reflected through the average load shedding in the test data. The results show that the larger λ is, the higher the tolerance for load shedding, thus the total investment cost decreases, and the average load shedding in the test increases, usually approaching λ.

With the development and maturity of ES-related technologies, the cost of ES may decrease in the future. In order to study the impact of the unit capacity cost of ES on the sizing results, C_p and C_e are multiplied by a coefficient. The results when the coefficient changes are shown in Table 7.5. The base case corresponds to the

Table 7.2 Stand-alone microgrid sizing results under different ambiguity set radii (Xie et al. 2022)

Ambiguity set radius ε	0	0.002	0.005	0.01	0.02	0.04
Renewable capacity x_g (MW)	18.52	18.29	18.67	19.42	19.78	17.79
ES power capacity x_p (MW)	5.843	5.851	6.008	6.098	6.459	6.832
ES energy capacity x_e (MWh)	35.56	36.61	36.60	36.65	39.69	51.44
Total investment f (10^6 CNY)	92.96	93.67	94.74	96.69	101.6	111.3
Test Shortfall risk	2.672	2.496	2.024	1.198	−0.075	−0.418

Table 7.3 Stand-alone microgrid sizing results under different numbers of data (Xie et al. 2022)

Data number N	91	182	273	364	455
Renewable capacity x_g (MW)	19.60	19.78	18.54	19.22	18.91
ES power capacity x_p (MW)	6.538	6.459	5.898	5.823	6.040
ES energy capacity x_e (MWh)	45.11	39.69	47.23	46.97	48.39
Total investment f (10^6 CNY)	107.7	101.6	107.1	108.3	109.5
Test Shortfall risk	−0.346	−0.075	−0.284	−0.360	−0.397

Table 7.4 Stand-alone microgrid sizing results under different acceptable load shedding values (Xie et al. 2022)

Acceptable load shedding λ (MWh)	0.9636	1.606	2.409	3.212	4.015
Renewable capacity x_g (MW)	15.47	19.78	15.64	13.40	11.96
ES power capacity x_p (MW)	4.896	6.459	5.620	4.932	4.567
ES energy capacity x_e (MWh)	70.16	39.69	35.70	34.37	33.29
Total investment f (10^6 CNY)	126.2	101.6	86.00	78.34	73.21
Test average load shedding (MWh)	0.729	1.546	2.755	3.750	4.780

Table 7.5 Stand-alone microgrid sizing results under different unit ES capacity costs (Xie et al. 2022)

Coefficient multiplied by the unit capacity cost of ES	0.4	0.6	0.8	1.0	1.2
Renewable capacity x_g (MW)	9.072	11.68	15.64	19.78	20.38
ES power capacity x_p (MW)	4.744	4.120	5.620	6.459	6.200
ES energy capacity x_e (MWh)	72.08	62.55	35.70	39.69	38.76
Total investment f (10^6 CNY)	58.27	75.54	86.00	101.6	112.2

situation where the unit capacity cost of ES is multiplied by the coefficient of 1.0. As the unit capacity cost of ES decreases, the installed capacity of renewable energy generation decreases, the power and energy capacities of ES increase, and the total investment cost also decreases.

7.4.5 Discussion on the Conservatism

The conservatism potentially introduced by the proposed solution method in this chapter originates from the Lipschitz constant-based solution method. The conservatism may arise from two aspects: First, applying Theorem 6.10 gives a conservative approximation of the worst-case expectation, namely (7.16) in this chapter; second, additional conservatism might be introduced if the chosen Lipschitz constant is not the smallest.

In fact, the solution method proposed in this chapter does not have the second kind of extra conservatism. Proposition 7.3 and subsequent analysis have proved that the adopted Lipschitz constant is the smallest. Regarding the first type of additional conservatism, since the range of random variable values $\Xi \neq \mathbb{R}^{2T}$, extra conservatism may exist. Although it is difficult to calculate the exact value of the worst-case expectation, the degree of conservatism can be estimated by checking the upper and lower bounds. According to (7.31), $r_d(x) \leq r_0(x) \leq r_c(x)$, that is, the accurate risk is between the risk of the DRO-D model and the risk of the proposed solution method. If the risk upper limit is set to 0, then the order of sizing results from the most optimistic to the most conservative should be the result of the proposed solution method, the accurate result, and the DRO-D result. The result of the proposed solution method is 101.6×10^6 CNY, which is the upper bound of the accurate result.

In order to obtain a better lower bound using the DRO-D model, the ambiguity set of the discrete probability distribution is expanded to include the range of all 637 sets of data, and the result is 95.76×10^6 CNY. If the ambiguity set is further expanded, perhaps a better lower bound can be obtained. But it can be concluded now that the accurate result is between 95.76×10^6 and 101.6×10^6, so the conservatism of the proposed solution method in this example does not exceed 6.1%.

7.5 Conclusion

This chapter proposes a bi-objective model for the optimal planning of renewable energy generation and ES capacities in a stand-alone microgrid that considers the distributionally robust Shortfall risk of load shedding. The distributionally robust Shortfall risk defined under a probability distribution ambiguity set based on Wasserstein distance is approximated by the optimal value of an LP problem. The ϵ-constraint method is used to transform the problem of solving the Pareto frontier into a parametric LP problem, based on which an algorithm for computing the analytical expression of the Pareto frontier is proposed.

In the case study, the proposed method is compared with the traditional SP, RO, DRO based on Wasserstein distance but only considering discrete probability distributions, methods based on time-series simulation, and DRO based on KL divergence. Generally speaking, the proposed method can balance optimality and robustness well: It is more cost-effective compared to RO and methods based on time-series simulation, and it is more robust compared to SP and DRO methods that only consider discrete probability distributions. Increasing the radius of the probability distribution ambiguity set, increasing the amount of data, decreasing the acceptable load shedding amount, etc., will make the results more conservative. When the unit cost of ES decreases, the optimal sizing model results in a decrease in the capacity of renewable energy generation and an increase in ES capacity.

References

Alharbi H, Bhattacharya K (2018) Stochastic optimal planning of battery energy storage systems for isolated microgrids. IEEE Trans Sustain Energy 9(1):211–227

Castro R, Ferreira L (2001) A comparison between chronological and probabilistic methods to estimate wind power capacity credit. IEEE Trans Power Syst 16(4):904–909

Chen Y, Wei W, Wang C et al (2022) Storage and transmission capacity requirements of a remote solar power generation system. IEEE Syst J 16(2):3175–3178

Diab AAZ, Sultan HM, Mohamed IS et al (2019) Application of different optimization algorithms for optimal sizing of PV/wind/diesel/battery storage stand-alone hybrid microgrid. IEEE Access 7:119223–119245

Gurobi Optimization, LLC (2023) Gurobi optimizer reference manual. https://www.gurobi.com/documentation/current/refman/index.html[2023-12-20]

Liu P, Cai Z, Xie P et al (2019) A decomposition-coordination planning method for flexible generation resources in isolated microgrids. IEEE Access 7:76720–76730

Lofberg J (2004) YALMIP: a toolbox for modeling and optimization in MATLAB. In: 2004 IEEE international conference on robotics and automation, pp 284–289

Sengupta M, Xie Y, Lopez A et al (2018) The national solar radiation data base (NSRDB). Renew Sustain Energy Rev 89:51–60

The MathWorks Inc (2004) Direct search - Matlab and Simulink - Mathworks. https://www.mathworks.com/help/gads/directsearch.html[2023-12-20]

Xie R, Wei W, Shahidehpour M et al (2022) Sizing renewable generation and energy storage in stand-alone microgrids considering distributionally robust shortfall risk. IEEE Trans Power Syst 37(5):4054–4066

Chapter 8
Sizing Transmission and Energy Storage for Remote Large-Scale Renewable Power Plants

Abstract This chapter studies the optimal sizing of transmission and energy storage capacities for remote renewable power plants, minimizing total investment costs while considering distributionally robust renewable energy curtailment conditional value-at-risk constraints. The amount of renewable energy curtailment is given by the optimal value of an operating optimization problem. The uncertainty of renewable energy generation is modeled by a data-driven ambiguity set based on Wasserstein distance. Through piecewise-affine function expressions, the proposed sizing model is transformed into a linear programming problem. This chapter introduces the work in Xie et al. (2022).

8.1 Introduction

Renewable energy resources are sometimes far from the load center. For example, the northwest region of China is rich in wind and photovoltaic (PV) resources, but it has a lower population density and lower electric load. Therefore, the electric energy is transmitted to the load center in the east, resulting in the long-distance transmission of renewable energy. Affected by real-time weather, the output of PV and wind power generation is highly random and volatile (Gonzalez-Gonzalez et al. 2020), and the average power generation is usually much smaller than the installed capacity. Thus, if the transmission line capacity is consistent with the installed capacity, the utilization rate of the transmission line is very low.

Energy storage (ES) can be equipped at the renewable power plant (Tang et al. 2019) to smooth the renewable energy output. The advantages of this approach lie in two aspects: First, the renewable power plant becomes partially controllable, thereby reducing the uncertainty that the power grid needs to deal with (Sheibani et al. 2018); second, after smoothing the renewable energy output curve, the required capacity of the transmission line is reduced. Given the long distance of the transmission corridor and the high cost per unit capacity of the transmission lines, deploying ES at renewable power plants may greatly help to reduce investment costs. Therefore, the optimal coordination of the transmission line and ES capacities is needed.

R. Xie and W. Wei, *Distributionally Robust Optimization and its Applications in Power System Energy Storage Sizing*,
https://doi.org/10.1007/978-981-97-2566-3_8

This chapter investigates the planning problem of ES and transmission line capacities for remote renewable power plants, with a focus on optimizing the economy, including limiting renewable energy curtailment and minimizing total investment costs. To handle the uncertainty of renewable energy generation and accommodate the inexactness of empirical probability distributions, the distributionally robust optimization (DRO) method based on the Wasserstein distance is used to model the sizing problem, and it is equivalently transformed into linear programming (LP) for solving.

There are two main innovations in this chapter.

Firstly, a DRO model is proposed for coordinating and optimally sizing ES and transmission line capacities in remote renewable power plants, reducing the required transmission line capacity, lowering investment costs, and alleviating the problem of renewable energy fluctuations faced by the power grid. Renewable energy curtailment is limited by the distributionally robust conditional value-at-risk (CVaR) constraint, and the ambiguity of empirical probability distribution is considered. The proposed model can achieve high utilization of renewable energy, reduce total investment costs, and mitigate uncertainties simultaneously.

Secondly, a systematic method is proposed to solve the distributionally robust sizing model, with the key being the transformation of the distributionally robust CVaR constraint of renewable energy curtailment. For this purpose, the operation optimization problem of minimizing the renewable energy curtailment is transformed into parametric LP with the renewable energy generation random variable and the transmission-ES equipment capacities as parameters. A method based on LP is constructed to express the renewable energy curtailment as a piecewise affine function of the parameters. Then, using the properties of CVaR, the distributionally robust CVaR constraint of renewable energy curtailment is transformed into a linear model. Finally, the proposed distributionally robust sizing problem is reduced to solving an LP problem.

In this chapter, the mathematical model of the distributionally robust sizing problem is first established in Sect. 8.2, then the solution method is proposed in Sect. 8.3, followed by verification and analysis through case studies in Sect. 8.4, and finally, conclusions are drawn in Sect. 8.5.

8.2 Mathematical Formulation

In this section, we first establish the operation optimization model, then define the probability distribution ambiguity set, and based on this, construct the renewable energy curtailment constraint. Finally, we propose the mathematical model for the distributionally robust sizing problem.

8.2.1 Operation Model and Curtailment Function

The remote renewable power plant under consideration is configured with ES and connected to the power grid through transmission lines, as shown in Fig. 8.1. In the operation optimization problem, consider T time periods each of length Δ_t. The time period set is denoted as $S_T := \{1, 2, \ldots, T\}$. The maximum renewable energy generating power in the t-th time period is denoted as ξ_t, which is a random variable, and the power of renewable energy curtailed is denoted as p_t^c. The power from generation to ES, from generation to transmission, and from ES to transmission in the t-th time period are denoted as p_t^{rs}, p_t^{rg}, and p_t^{sg}, respectively.

Let x_p, x_e, and x_l represent the ES power capacity, ES energy capacity, and transmission line capacity, respectively. The energy stored in the ES at the end of the t-th time period is denoted as e_t, and e_0 represents the energy stored in the initial state. Therefore, the operation optimization problem is as follows:

$$f(x, \xi) := \min \sum_{t \in S_T} p_t^c \Delta_t \tag{8.1a}$$

$$\text{s.t. } p_t^{rs}, p_t^{rg}, p_t^{sg}, p_t^c \geq 0 , \ \forall t \in S_T , \tag{8.1b}$$

$$\xi_t^g = p_t^{rs} + p_t^{rg} + p_t^c , \ \forall t \in S_T , \tag{8.1c}$$

$$p_t^{rg} + p_t^{sg} \leq x_l , \ \forall t \in S_T , \tag{8.1d}$$

$$e_t = e_0 + \sum_{s=1}^{t} \left(\eta_c p_s^{rs} - \frac{p_s^{sg}}{\eta_d} \right) \Delta_t , \ \forall t \in S_T , \tag{8.1e}$$

$$\omega_l x_e \leq e_t \leq \omega_h x_e , \ \forall t \in S_T , \ e_T = e_0 , \tag{8.1f}$$

$$p_t^{rs} \leq x_p , \ p_t^{sg} \leq x_p , \ \forall t \in S_T , \tag{8.1g}$$

$$p_t^{rs} + p_t^{sg} \leq x_p , \ \forall t \in S_T , \tag{8.1h}$$

where Eq. (8.1a) expresses the optimization objective to minimize the curtailment of renewable energy; Eq. (8.1b) specifies the power flow direction and nonnegativity of the variables; Eq. (8.1c) represents the power balance at the generation point; Eq. (8.1d) imposes transmission line capacity constraints; Eqs. (8.1e)–(8.1h) constitute the ES operation model, which has been introduced in (1.14) in Sect. 1.3.2. Here, η_c and η_d are the charging and discharging efficiencies of the ES respectively; Eq. (8.1e) describes the dynamics of state-of-charge (SoC) of the ES; ω_l and ω_h are the lower

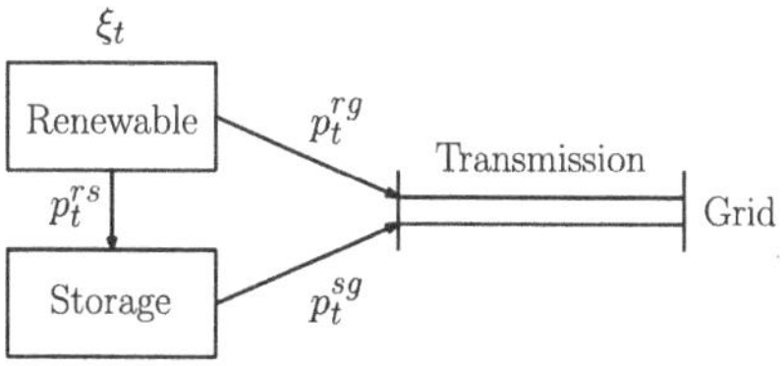

Fig. 8.1 Energy flow relationships of a remote renewable power plant (Xie et al. 2022)

and upper bounds of the SoC; Eq. (8.1f) limits the range of the SoC, and requires the initial and final stored energy to be the same for continuous operation; Eq. (8.1g) expresses charging and discharging power range constraints; Eq. (8.1h) permits each time period to be divided into charging sub-intervals and discharging sub-intervals, thereby making physically feasible charging and discharging powers.

In the operation optimization problem presented in (8.1), the capacity variables $x := (x_p, x_e, x_l)$ and renewable energy generation $\xi := (\xi_t; t \in S_T)$ are treated as parameters, while the decision variables are p_t^{rs}, p_t^{rg}, p_t^{sg}, p_t^{c}, and e_t with $t \in S_T$. Therefore, the optimal value of (8.1), which is the renewable energy curtailment, can be regarded as a function of x and ξ and is denoted as $f(x, \xi)$.

8.2.2 Distributionally Robust Conditional Value-at-Risk of Curtailment

The uncertainty comes from the renewable energy generation ξ, which consists of T random numbers ranging from 0 to $\bar{P}$, where $\bar{P}$ is the installed capacity of the renewable power plant. Therefore, the range Ξ of the random variable ξ is defined as follows:

$$\Xi := \left\{ \xi \in \mathbb{R}^T \,\middle|\, 0 \le \xi_t \le \bar{P} \ , \ \forall t \in S_T \right\} . \tag{8.2}$$

Based on the historical weather data and parameters of the renewable power plant, N groups of sampled data for ξ are generated and denoted as $\xi_n := (\xi_{t,n}; t \in S_T), n \in S_N$, where $S_N := \{1, 2, \ldots, N\}$. Using the method described in Sect. 6.2, an empirical probability distribution $\mathbb{P}_0$ is established based on the sampled data, and then the probability distribution ambiguity set $\mathcal{B}(\mathbb{P}_0, \varepsilon)$ is defined by Wasserstein 1-distance under 1-norm in (6.3) as follows:

$$\mathcal{B}(\mathbb{P}_0, \varepsilon) := \{\mathbb{P} \in \mathcal{M}_1(\Xi) \mid d_W(\mathbb{P}, \mathbb{P}_0) \le \varepsilon\} . \tag{8.3}$$

The acceptable rate of renewable energy curtailment is denoted as λ, where $\lambda \in [0, 1]$. The acceptable event for renewable energy curtailment can be expressed as $f(x, \xi) \le \lambda \sum_{t \in S_T} \xi_t \Delta_t$. Because of the uncertainty of renewable energy generation ξ, this is a random event. A common practice is to consider the probability of the random event, thereby obtaining the following robust chance constraint:

$$\inf_{\mathbb{P} \in \mathcal{B}(\mathbb{P}_0, \varepsilon)} \mathbb{P} \left[f(x, \xi) \le \lambda \sum_{t \in S_T} \xi_t \Delta_t \right] \ge \beta , \tag{8.4}$$

where β is the acceptable minimum probability. Therefore, Eq. (8.4) expresses that for all the probability distributions in the ambiguity set $\mathcal{B}(\mathbb{P}_0, \varepsilon)$, the probability that the renewable energy curtailment is acceptable is not less than β.

Equation (8.4) can be regarded as a distributionally robust risk constraint. In fact, by defining a function representing loss and using value-at-risk (VaR), Eq. (8.4) can be equivalently transformed into a distributionally robust VaR constraint. The function representing loss is defined as follows:

$$g(x,\xi) := f(x,\xi) - \lambda \sum_{t\in S_T} \xi_t \Delta_t \ , \tag{8.5}$$

which is the renewable energy curtailment obtained from operation optimization minus the acceptable renewable energy curtailment. Because $g(x,\xi) \le 0$ is equivalent to $f(x,\xi) \le \lambda \sum_{t\in S_T} \xi_t \Delta_t$, Eq. (8.4) is equivalent to:

$$\inf_{\mathbb{P}\in\mathcal{B}(\mathbb{P}_0,\varepsilon)} \mathbb{P}\left[g(x,\xi) \le 0\right] \ge \beta \ . \tag{8.6}$$

By the reformulation methods in Sect. 3.4, (8.6) is equivalent to the following distributionally robust VaR constraint

$$\sup_{\mathbb{P}\in\mathcal{B}(\mathbb{P}_0,\varepsilon)} \beta\text{-VaR}^{\mathbb{P}}(g(x,\xi)) \le 0 \ , \tag{8.7}$$

and the following distributionally robust CVaR constraint is a conservative approximation:

$$\sup_{\mathbb{P}\in\mathcal{B}(\mathbb{P}_0,\varepsilon)} \beta\text{-CVaR}^{\mathbb{P}}(g(x,\xi)) \le 0 \ . \tag{8.8}$$

8.2.3 Sizing Model

The coordinated optimal sizing problem for ES and transmission line capacities in remote renewable power plants is established as follows:

$$\min_{x} \ C_p x_p + C_e x_e + C_l x_l \tag{8.9a}$$

$$\text{s.t.} \ x = (x_p, x_e, x_l) \ge 0 \ , \tag{8.9b}$$

$$\sup_{\mathbb{P}\in\mathcal{B}(\mathbb{P}_0,\varepsilon)} \beta\text{-CVaR}^{\mathbb{P}}(g(x,\xi)) \le 0 \ , \tag{8.9c}$$

where Eq. (8.9a) indicates that the optimization goal is to minimize the total investment cost of ES and transmission lines; Eq. (8.9b) is the nonnegativity constraint for capacities; Eq. (8.9c) is the constraint for renewable energy curtailment. The objective function in (8.9a) applies to the ES types for which energy capacity and power capacity can be planned separately. For the ES where energy and power capacity are closely linked, the ES duration parameter can be fixed, i.e., let x_e be a fixed multiple of x_p. This is equivalent to adding a linear constraint, which does not change the main body of the optimal sizing model, and the subsequent solution method is still applicable.

Although generally speaking, the service life of ES and transmission lines is different, it can be converted into operational costs within the same time scale, and the total cost can be calculated by reasonable weighting. This process only requires adjusting the weight coefficients and does not need to change the proposed model and solution method.

The objective function in (8.9a) assumes that the cost of transmission lines is proportional to capacity (Chen et al. 2022). This assumption is rather rough, as the cost of transmission lines in actual scenarios is jointly determined by various factors such as voltage level, line type, line length, and geographical conditions. However, at the planning stage, before determining the specific details, the sizing model based on the objective function of (8.9a) can be used to estimate the approximate capacity of the transmission line, thus determining the range of candidate schemes. Afterward, the form of the objective function can be modified, and by solving mixed-integer linear programming (MILP), the optimal scheme and the corresponding ES capacity can be found among the candidate transmission line schemes. The specific method is discussed in Sect. 8.4.

Equation (8.9) contains a distributionally robust CVaR constraint in the optimization problem and the probability distribution ambiguity set is based on the definition of Wasserstein distance. Another challenge in solving comes from the function $g(x, \xi)$ representing loss, which contains the function $f(x, \xi)$ defined by the optimal value of LP problem in (8.1), and its analytical expression is currently unknown. These two challenges make it difficult to directly solve the sizing problem in (8.9).

8.3 Solution Method

In this section, we first analyze the analytical expression of the renewable energy curtailment function $f(x, \xi)$ according to parametric LP, then propose an algorithm for calculating the approximate expression, and based on this, transform the sizing problem in (8.9) into an LP problem form that can be directly solved by a commercial solver.

8.3.1 Curtailment Function Approximation

In this subsection, we first introduce the accurate analytical expression of the renewable energy curtailment function, then analyze the approximation error, propose an approximation method that gradually reduces the error through iteration, and finally summarize the algorithm for approximating the expression of the renewable energy curtailment function.

8.3.1.1 Analytical Expression of the Renewable Energy Curtailment Function

For simplicity, the LP problem that defines the renewable energy curtailment function $f(x, \xi)$ in (8.1) is written in the following compact form:

$$\begin{aligned} f(\theta) := \min_{y}\ & c^\top y \\ \text{s.t. } & A_1 y \le b_1 + B_1\theta \ , \\ & A_2 y = b_2 + B_2\theta \ , \end{aligned} \tag{8.10}$$

where $\theta := (x, \xi)$ is a column vector of parameter; y is the decision variable including p_t^{rs}, p_t^{rg}, p_t^{sg}, p_t^{c}, and $e_t, t \in S_T$; A_1, A_2, b_1, b_2, B_1, B_2, and c are all coefficient matrices or vectors. Equation (8.10) represents a parametric LP problem with parameters in the constraint.

Let $\bar{x}_p$, $\bar{x}_e$, and $\bar{x}_l$ represent the upper bounds of x_p, x_e, and x_l, respectively. The range X of the capacity variable x is defined as follows:

$$X := \left\{ x = (x_p, x_e, x_l) \in \mathbb{R}^3 \;\middle|\; 0 \le x_p \le \bar{x}_p, 0 \le x_e \le \bar{x}_e, 0 \le x_l \le \bar{x}_l \right\} \ . \tag{8.11}$$

By integrating the range X of x and the range Ξ of ξ, we obtain the definition of the range Θ of θ as follows:

$$\Theta := \left\{ (x, \xi) \;\middle|\; x \in X, \xi \in \Xi \right\} \ . \tag{8.12}$$

In the following, we derive the analytical expression of the function $f(\theta), \theta \in \Theta$.

Proposition 8.1 *For any $\theta \in \Theta$, $f(\theta)$ is finite.*

Proof The LP problem defining $f(x, \xi)$ is in (8.1). On the one hand, we can construct a feasible solution: Let

$$\tilde{p}_t^{rs} = \tilde{p}_t^{rg} = \tilde{p}_t^{sg} = 0 \ , \ \tilde{p}_t^{c} = \xi_t^{g} \ , \ \tilde{e}_t = e_0 \ , \ \forall t \in S_T \ . \tag{8.13}$$

Then it is easy to verify that $(\tilde{p}_t^{rs}, \tilde{p}_t^{rg}, \tilde{p}_t^{sg}, \tilde{p}_t^{c}, \tilde{e}_t; t \in S_T)$ is a feasible solution of (8.1). In addition, the objective function value is $\sum_{t \in S_T} \xi_t^g \Delta_t$, so we have obtained an upper bound of $f(x, \xi)$. On the other hand, for any feasible solution of (8.1), the objective function value must be greater than or equal to 0 according to (8.1b). In conclusion, $f(x, \xi)$ is finite. □

The physical meaning of Proposition 8.1 is that the amount of renewable energy curtailment is nonnegative and less than or equal to the maximum power generation of renewable energy. The feasible solution constructed in the proof is equivalent to abandoning all renewable energy power generation, so the conclusion of Proposition 8.1 is intuitive.

The dual problem of the LP problem in (8.10) is as follows:

$$\begin{aligned} &\max_{\gamma,\mu}\ (b_1 + B_1\theta)^\top \gamma + (b_2 + B_2\theta)^\top \mu \ , \\ &\text{s.t. } A_1^\top \gamma + A_2^\top \mu = c\ ,\ \gamma \le 0\ . \end{aligned} \tag{8.14}$$

Proposition 8.1 indicates that the LP problem in (8.10) has a finite optimal value for any $\theta \in \Theta$. Therefore, according to the duality theory of LP, Eq. (8.14) has the same optimal value as (8.10), so

$$f(\theta) = \max_{(\gamma,\mu)\in\Gamma} \left\{(b_1 + B_1\theta)^\top \gamma + (b_2 + B_2\theta)^\top \mu\right\}\ , \tag{8.15}$$

where the dual feasibility region is:

$$\Gamma := \left\{(\gamma,\mu)\ \middle|\ A_1^\top \gamma + A_2^\top \mu = c\ ,\ \gamma \le 0\right\}\ . \tag{8.16}$$

Therefore, Γ is a polyhedron. According to Proposition 8.1, $f(\theta)$ is finite, therefore the maximum value in (8.15) can be obtained at the vertices of Γ. Denote the set of vertices of Γ by $V(\Gamma) := \{(\gamma_k, \mu_k) \mid k = 1, 2, \ldots, K\}$, then

$$\begin{aligned} f(\theta) &= \max_{(\gamma_k,\mu_k)\in V(\Gamma)} \left\{(b_1 + B_1\theta)^\top \gamma_k + (b_2 + B_2\theta)^\top \mu_k\right\} \\ &= \max_{(\gamma_k,\mu_k)\in V(\Gamma)} \left\{(\gamma_k^\top B_1 + \mu_k^\top B_2)\cdot\theta + (\gamma_k^\top b_1 + \mu_k^\top b_2)\right\} \\ &= \max_{(\gamma_k,\mu_k)\in V(\Gamma)} f_k(\theta)\ , \end{aligned} \tag{8.17}$$

where

$$f_k(\theta) := (\gamma_k^\top B_1 + \mu_k^\top B_2)\cdot\theta + (\gamma_k^\top b_1 + \mu_k^\top b_2)\ , \tag{8.18}$$

is a piecewise affine function with respect to θ. Equation (8.17) shows that $f(\theta)$ is the pointwise maximum of a finite number of affine functions. Therefore, $f(\theta)$ is a piecewise affine convex function according to Proposition 2.21.

The number of constraints in (8.1) determines the dimensions of the dual variables (γ, μ). As the operation optimization problem with ES involves multiple time periods, the dimensions of (γ, μ) are inevitably high, leading to a large number of vertices in Γ. Therefore, it is impractical to list all the vertices in $V(\Gamma)$. In fact, we only need to include a subset of the vertices in (8.17).

8.3.1.2 Approximation Error Analysis

For any subset $W \subseteq V(\Gamma)$, we have:

$$f(\theta) \geq f_W(\theta) := \max_{(\gamma_k, \mu_k) \in W} f_k(\theta) , \tag{8.19}$$

where $f_W(\theta)$ is an approximation to $f(\theta)$, and the larger W, the more precise the approximation. Because $f_W(\theta)$ is piecewise affine, it is more convenient to study it in its affine regions. In such regions, some affine function attains the maximum among all affine functions, i.e.,

$$\Theta_{W,k} := \left\{ \theta \in \Theta \;\middle|\; f_k(\theta) \geq f_l(\theta) \ , \ \forall (\gamma_l, \mu_l) \in W \ , \ l \neq k \right\} . \tag{8.20}$$

When $\theta \in \Theta_{W,k}$, the k-th affine function attains the maximum, and $f_W(\theta) = f_k(\theta)$. Because both $f_k(\theta)$ and $f_l(\theta)$ are affine functions, $\Theta_{W,k}$ is a polyhedron. If $f_W(\theta)$ is used to approximate $f(\theta)$, then the maximum error in $\Theta_{W,k}$ can be calculated as follows:

$$\begin{aligned} &\max_{\theta \in \Theta_{W,k}} \{f(\theta) - f_k(\theta)\} \\ &= \begin{cases} \max\limits_{\theta, \gamma, \mu} v(\theta, \gamma, \mu) \\ \text{s.t. } \theta \in \Theta_{W,k} \ , \ (\gamma, \mu) \in \Gamma \ . \end{cases} \end{aligned} \tag{8.21}$$

The $v(\theta, \gamma, \mu)$ is given below according to the dual problem form in (8.14):

$$\begin{aligned} v(\theta, \gamma, \mu) :=& (b_1 + B_1\theta)^\top \gamma + (b_2 + B_2\theta)^\top \mu \\ &- (\gamma_k^\top B_1 + \mu_k^\top B_2) \cdot \theta - (\gamma_k^\top b_1 + \mu_k^\top b_2) \ . \end{aligned} \tag{8.22}$$

Although the feasible region in (8.21) is a polyhedron, the objective function $v(\theta, \gamma, \mu)$ contains product terms $\theta^\top B_1^\top \gamma$ and $\theta^\top B_2^\top \mu$. The Mountain Climbing method (Konno 1976) is used to solve the optimization problem in (8.21). As shown in Algorithm 8.1, this method alternates between optimizing θ and (γ, μ), and solves (8.21) by continuously solving LP problems.

Algorithm 8.1 Mountain climbing (Konno 1976)

Input: Parameters of the optimization problem in (8.21), initial feasible point $\theta^{(0)} \in \Theta_{W,k}$, and convergence criterion parameter $\delta > 0$.

Output: A solution $(\theta^{(i)}, \gamma^{(i)}, \mu^{(i)})$ of the optimization problem in (8.21).

1: Initiation: Let $i \leftarrow 1$.

2: Solve the LP $\max_{(\gamma,\mu)\in\Gamma} v(\theta^{(i-1)}, \gamma, \mu)$ and obtain the optimal solution $(\gamma^{(i)}, \mu^{(i)})$.

3: Solve the LP $\max_{\theta\in\Theta_{W,k}} v(\theta, \gamma^{(i)}, \mu^{(i)})$ and obtain the optimal solution $\theta^{(i)}$.

4: If $v(\theta^{(i)}, \gamma^{(i)}, \mu^{(i)}) - v(\theta^{(i-1)}, \gamma^{(i)}, \mu^{(i)}) \leq \delta$, then terminate and output the solution; otherwise, update $i \leftarrow i + 1$ and turn to Step 2.

To provide an initial feasible point $\theta^{(0)} \in \Theta_{W,k}$ for the optimization in (8.21), calculate the Chebyshev center of $\Theta_{W,k}$, which is the center of the largest ball in the polyhedron $\Theta_{W,k}$ and has been introduced in Sect. 2.5.4. If Eq. (8.20) is written in the form $\Theta_{W,k} = \{\theta \mid u_i^\top \theta \leq w_i, \forall i\}$, where u_i and w_i are constant vectors or constants, then the Chebyshev center can be found by solving the following LP problem:

$$\begin{aligned} &\max_{r,\theta} r \\ &\text{s.t. } u_i^\top \theta + r\|u_i\|_2 \leq w_i \ , \ \forall i \ , \end{aligned} \tag{8.23}$$

where $\|\cdot\|_2$ represents the 2-norm, and the Chebyshev center is the θ value in the optimal solution.

8.3.1.3 Algorithm for Calculating the Approximate Expression of the Renewable Energy Curtailment Function

Combining the above steps, we get the algorithm for calculating the approximate expression of the renewable energy curtailment function, as shown in Algorithm 8.2.

Algorithm 8.2 Calculating the approximate expression of the renewable energy curtailment

---

Input: Parameters of the optimization problem in (8.10), the initial point sets Θ_0 consisting of typical values of θ, and convergence criterion parameter $\delta > 0$.
Output: An approximate function $f_W(\theta)$ of $f(\theta)$.
1: Initiation: Let $W \leftarrow \emptyset$. For every point $\theta_k \in \Theta_0$, solve the LP problem in (8.14) for $\theta = \theta_k$ and obtain an optimal solution (γ_k, μ_k). If $(\gamma_k, \mu_k) \notin W$, update $W \leftarrow W \cup \{(\gamma_k, \mu_k)\}$.
2: Let $W' \leftarrow W$. For every $(\gamma_k, \mu_k) \in W$, find $\Theta_{W,k}$ by (8.20). Solve the LP problem in (8.23) and obtain $\theta^{(0)} \in \Theta_{W,k}$. Regarding $\theta^{(0)}$ as the initial feasible point, use Algorithm 8.1 to find a solution $(\theta^*, \gamma^*, \mu^*)$ of (8.21). If $v(\theta^*, \gamma^*, \mu^*) > \delta$, then solve the LP problem in (8.14) at $\theta = \theta^*$ and obtain an optimal solution (γ', μ'). If $(\gamma', \mu') \notin W'$, update $W' \leftarrow W' \cup \{(\gamma', \mu')\}$.
3: If $W' = W$, then output $f_W(\theta)$ and terminate; otherwise, update $W \leftarrow W'$ and turn to Step 2.

The idea of Algorithm 8.2 is to find the necessary dual feasible domain vertices by solving a series of LP problems. The constructions in Step 1 and Step 2 ensure that $f_k(\theta_k) = f(\theta_k)$, so for any $(\gamma_k, \mu_k) \in W$, $\Theta_{W,k}$ is non-empty. In addition, since the number of elements in $V(\Gamma)$ is finite, Algorithm 8.2 is guaranteed to converge.

8.3.2 *Reformulation of the Distributionally Robust Sizing Model*

Due to the distributionally robust CVaR constraint in (8.9c), the proposed sizing model cannot be directly solved. According to Sect. 3.4, the distributionally robust CVaR constraint in (8.9c) is equivalent to the following constraint of the worst expectation:

$$\alpha + \frac{1}{1-\beta} \sup_{\mathbb{P} \in \mathcal{B}(\mathbb{P}_0, \varepsilon)} \mathbb{E}^{\mathbb{P}}[\max\{g(x, \xi) - \alpha, 0\}] \leq 0 \,. \tag{8.24}$$

Algorithm 8.2 provides an approximate piecewise affine expression for $f(x, \xi)$, so $g(x, \xi) = f(x, \xi) - \lambda \sum_{t \in S_T} \xi_t \Delta_t$ has the same form, that is,

$$g(x, \xi) = \max_{k \in S_K} \left\{ d_{1,k}^{\top} x + d_{2,k}^{\top} \xi + d_{3,k} \right\} , \tag{8.25}$$

where $d_{1,k}$, $d_{2,k}$, and $d_{3,k}$ are constant vectors or constants, and S_K is the index set. Therefore, we have

$$\begin{aligned} &\max \{g(x, \xi) - \alpha, 0\} \\ &= \max \left\{ \max_{k \in S_K} \left\{ d_{2,k}^{\top} \xi + (d_{1,k}^{\top} x + d_{3,k} - \alpha) \right\} , 0 \right\} , \end{aligned} \tag{8.26}$$

is the pointwise maximum of some affine functions of (x, ξ).

Write the range Ξ of random variable values in the form $\Xi = \left\{ \xi \in \mathbb{R}^T \;\middle|\; H\xi \leq h \right\}$, where H and h are constant matrices or vectors. According to Proposition 6.8, the worst-case expectation $\sup_{\mathbb{P} \in \mathcal{B}(\mathbb{P}_0, \varepsilon)} \mathbb{E}^{\mathbb{P}}[\max\{g(x, \xi) - \alpha, 0\}]$ is equal to:

$$\inf_{\rho, s, \sigma} \; \varepsilon \rho + \frac{1}{N} \sum_{n \in S_N} s_n \tag{8.27a}$$

$$\text{s.t.}\; \sigma_{nk} \geq 0 \,, \; \forall n \in S_N \,, \; k \in S_K \,, \tag{8.27b}$$

$$d_{1,k}^{\top} x + d_{3,k} - \alpha + d_{2,k}^{\top} \xi_n + (h - H\xi_n)^{\top} \sigma_{nk} \leq s_n \,, \; \forall n \in S_N \,, \; k \in S_K \,, \tag{8.27c}$$

$$\sigma_{n0} \geq 0 \,, \; (h - H\xi_n)^{\top} \sigma_{n0} \leq s_n \,, \; \forall n \in S_N \,, \tag{8.27d}$$

$$H^{\top} \sigma_{nk} - d_{2,k} \leq \rho 1_T \,, \; -H^{\top} \sigma_{nk} + d_{2,k} \leq \rho 1_T \,, \; \forall n \in S_N \,, \; k \in S_K \,, \tag{8.27e}$$

$$H^{\top} \sigma_{n0} \leq \rho 1_T \,, \; -H^{\top} \sigma_{n0} \leq \rho 1_T \,, \; \forall n \in S_N \,, \tag{8.27f}$$

where 1_T represents the T-dimensional all-one vector.

Substituting (8.24) and (8.27a)–(8.27f), we obtain the approximate LP problem for the sizing problem in (8.9) as follows:

$$\min C_p x_p + C_e x_e + C_l x_l \tag{8.28a}$$

$$\text{s.t. } x_p, x_e, x_l \geq 0 , \tag{8.28b}$$

$$\alpha + \frac{1}{1-\beta}\left(\varepsilon\rho + \frac{1}{N}\sum_{n\in S_N} s_n\right) \leq 0 , \tag{8.28c}$$

$$(8.27b)\text{–}(8.27f) . \tag{8.28d}$$

Therefore, after calculating the approximate expression of the renewable energy curtailment function according to Algorithm 8.2, the optimal sizing result can be obtained by solving the above LP problem.

8.4 Case Study

8.4.1 Settings

To verify the models and methods proposed in this chapter, consider a remote PV power plant with an installed capacity of 1 GW. Hourly solar radiation data is obtained from the U.S. National Solar Radiation Database (Sengupta et al. 2018), based on which hourly renewable energy generation data is generated. Ten typical days are selected from each month, making up a total of 120 days of planning data, and an additional 245 days of data are selected for testing. In the base case, $T = 24$, $N = 120$, $\Delta_t = 1$ h, and $\bar{P} = 1$ GW. $\xi_{t,n}, t \in S_T, n \in S_N$ is derived from the 120 days of planning data. The parameters of the ES are $\eta_c = \eta_d = 0.95$, $\omega_l = 0.1$, $\omega_h = 0.9$, $C_p = 10^6$ CNY/MW, and $C_e = 1.2 \times 10^6$ CNY/MWh (Chen et al. 2022). The transmission line parameter is $C_l = 2 \times 10^7$ CNY/MW. Other parameters are $\varepsilon = 0.005$, $\lambda = 5\%$, and $\beta = 0.9$. The proposed method is implemented in the MATLAB environment on a laptop computer equipped with an Intel Core i5-5200U processor and 8 GB of memory, using the YALMIP toolbox (Lofberg 2004), and the optimization problem is solved with Gurobi 9.1 (Gurobi Optimization 2023).

8.4.2 Approximate Curtailment Function

The first step of the proposed method is to find an approximate expression for the renewable energy curtailment function, where the independent variables are renewable energy generation $\xi = (\xi_t; t \in S_T)$ and capacity variables $x = (x_p, x_e, x_l)$. Based on the information that the installed capacity of PV power generation is 1 GW, the domain Θ of the renewable energy curtailment function is set as:

$$\Theta := \left\{ \theta = (x, \xi) \;\middle|\; \begin{array}{l} 0 \le \xi_t \le 1\,\mathrm{GW}\ ,\ \forall t \in S_T\ , \\ 0 \le x_p \le 1\,\mathrm{GW}\ ,\ 0 \le x_e \le 12\,\mathrm{GWh}\ ,\ 0 \le x_l \le 1\,\mathrm{GW} \end{array} \right\} . \tag{8.29}$$

For the initial point set of Algorithm 8.2, ξ uses 120 days of renewable energy generation data, while x is initialized with random numbers. Specifically, the expectations are set to $x_p = 0.85$ GW, $x_e = 5.4$ GWh, and $x_l = 0.39$ GW, and then multiplied by independently and uniformly randomly generated numbers between 0.85 and 1.15. The expectation values are selected based on the results of the traditional optimization method, which will be detailed later.

Through 120 initial points of the parameter θ, 19 different vertices of $V(\Gamma)$ are obtained. Then, Algorithm 8.2 is used, repeatedly solving LP problems in the process, the scale of which does not exceed 1,000 constraints and 1,000 variables. After 6 rounds of iterations, which takes a total of 248 s of computation time, a subset $W \subseteq V(\Gamma)$ containing 46 vertices is obtained, as well as the approximate expression of the renewable energy curtailment function $f_W(x, \xi)$, which is the pointwise maximum of 46 affine functions.

To test the accuracy of the approximate expression $f_W(x, \xi)$, 245 days of test renewable energy generation data and 245 sets of typical capacity values are regenerated. The approximate renewable energy curtailment provided by $f_W(x, \xi)$ is compared with the exact value of renewable energy curtailment obtained by solving the operation optimization problem of (8.1), and the results are shown in Fig. 8.2. The approximate renewable energy curtailment is very close to the exact value in most test data.

The independent variable of the renewable energy curtailment function is 27-dimensional, including 24-dimensional renewable energy generation random variables and 3-dimensional capacity variables. Fix the value of ξ with the renewable energy generation data of a typical day, and then visualize the obtained approximate expression of the renewable energy curtailment function. Figure 8.3 shows the

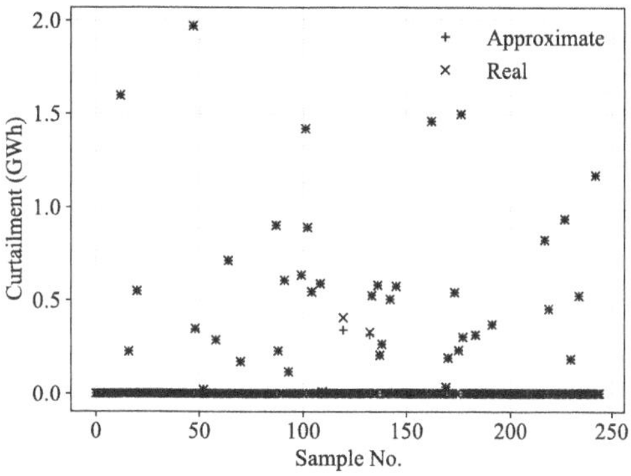

Fig. 8.2 Approximate accuracy of photovoltaic (PV) power plant curtailment function (Xie et al. 2022)

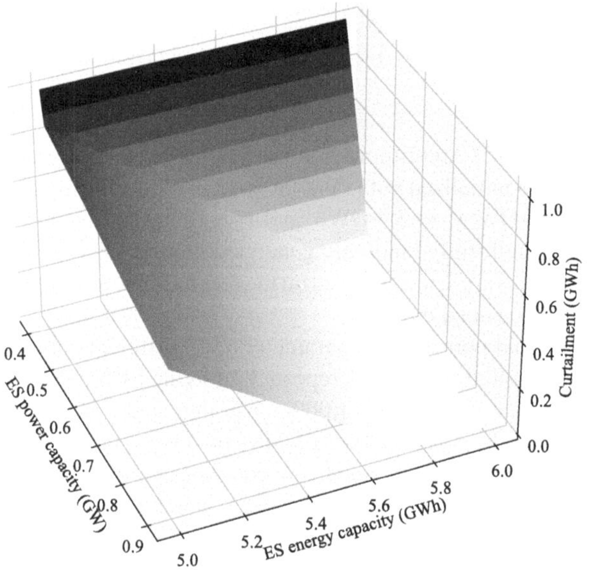

Fig. 8.3 Renewable curtailment function visualization under fixed transmission line capacity 0.39 GW (Xie et al. 2022)

Fig. 8.4 Renewable curtailment function visualization under fixed $x_e/x_p = 6\,\text{h}$ (Xie et al. 2022)

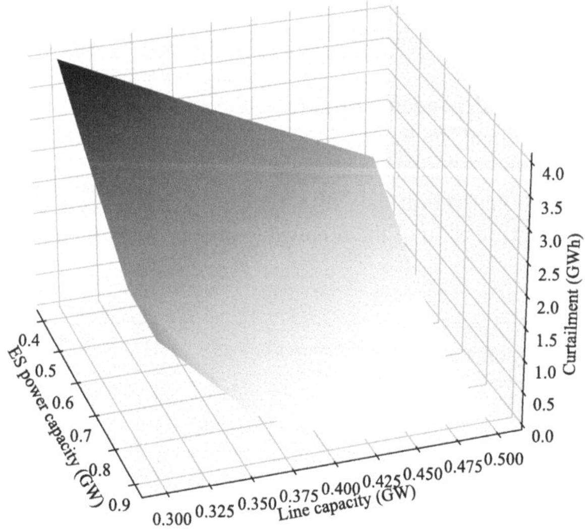

change of curtailment with the ES energy capacity x_e and power capacity x_p when the transmission line capacity x_l is fixed at 0.39 GW. Figure 8.4 shows the change of curtailment with x_p and x_l when $x_e/x_p = 6$ h is fixed. As x_p, x_e, or x_l increases, the curtailment of renewable energy decreases.

8.4.3 *Other Methods for Comparison*

The proposed method is denoted as DRO-C. In addition, three methods are considered for comparison.

The first comparison method: The method based on CVaR and stochastic programming (SP), denoted as SP;

The second comparison method: The robust optimization (RO) method, denoted as RO;

The third comparison method: The DRO method based on Wasserstein distance considering only discrete probability distribution, denoted as DRO-D.

In the following, we describe these three comparison methods individually.

8.4.3.1 Method Based on Conditional Value-at-Risk and Stochastic Programming

Ignore the possible inaccuracy of the empirical probability distribution $\mathbb{P}_0$ and regard $\mathbb{P}_0$ as the true probability distribution. Therefore, the distributionally robust CVaR constraint in the sizing model in (8.9) is replaced by the traditional CVaR constraint under $\mathbb{P}_0$, that is,

$$\begin{aligned} \min_{x} \; & C_p x_p + C_e x_e + C_l x_l \\ \text{s.t. } & x = (x_p, x_e, x_l) \geq 0 \, , \\ & \beta\text{-CVaR}^{\mathbb{P}_0}(g(x, \xi)) \leq 0 \, . \end{aligned}$$

Using Theorem 2.6, the above sizing model is equivalent to the following LP problem:

$$\min C_p x_p + C_e x_e + C_l x_l \tag{8.30a}$$

$$\text{s.t. } x_p, x_e, x_l \geq 0 \, , \tag{8.30b}$$

$$\alpha + \frac{1}{N(1-\beta)} \sum_{n \in S_N} s_n \leq 0 \, , \tag{8.30c}$$

$$s_n \geq \sum_{t \in S_T} p^c_{t,n} \Delta_t - \lambda \sum_{t \in S_T} \xi_{t,n} \Delta_t - \alpha \, , \; s_n \geq 0 \, , \; \forall n \in S_N \, , \tag{8.30d}$$

$$(8.1b)\text{–}(8.1h) \, , \; \forall n \in S_N \, , \tag{8.30e}$$

where $s_n, n \in S_N$ is auxiliary variable.

8.4.3.2 Robust Optimization

The range of sampling data is regarded as the range of random variable values, that is, $\Xi_0 := \left\{\xi_n \,\middle|\, n \in S_N\right\}$. It is required that for all $\xi \in \Xi_0$, the renewable energy curtailment rate does not exceed λ, that is:

$$
\begin{aligned}
\min\ & C_p x_p + C_e x_e + C_l x_l \\
\text{s.t. } & x_p, x_e, x_l \geq 0 , \\
& g(x, \xi) \leq 0 , \ \forall \xi \in \Xi_0 ,
\end{aligned}
$$

where the renewable energy curtailment constraint is equivalent to $g(x, \xi_n) \leq 0, \forall n \in S_N$, so in this case, the equivalent LP problem of the sizing problem is as follows:

$$
\begin{aligned}
\min\ & C_p x_p + C_e x_e + C_l x_l \\
\text{s.t. } & x_p, x_e, x_l \geq 0 , \\
& \sum_{t \in S_T} p^c_{t,n} \Delta_t - \lambda \sum_{t \in S_T} \xi_{t,n} \Delta_t \leq 0 , \ \forall n \in S_N , \\
& (8.1b)\text{–}(8.1h) , \ \forall n \in S_N .
\end{aligned}
\tag{8.31}
$$

8.4.3.3 Distributionally Robust Optimization Method Based on Wasserstein Distance Considering only Discrete Probability Distributions

The probability distribution ambiguity set $\mathcal{B}_d(\mathbb{P}_0, \varepsilon)$ only includes discrete probability distributions with support set on Ξ_0, and requires that the Wasserstein distance between these probability distributions and the empirical probability distribution $\mathbb{P}_0$ does not exceed ε:

$$
\mathcal{B}_d(\mathbb{P}_0, \varepsilon) := \left\{\mathbb{P} \in \mathcal{M}_1(\Xi_0) \,\middle|\, d_W(\mathbb{P}, \mathbb{P}_0) \leq \varepsilon\right\} . \tag{8.32}
$$

According to the method in Sect. 6.3, the sizing model then comes down to the following LP problem:

$$
\begin{aligned}
&\min \; C_p x_p + C_e x_e + C_l x_l \\
&\text{s.t. } x_p, x_e, x_l \geq 0 \,, \\
&\quad \alpha + \frac{1}{1-\beta}\left(\varepsilon\rho + \sum_{n\in S_N}\frac{\sigma_n}{N} + \tau\right) \leq 0 \,, \; \rho \geq 0 \,, \\
&\quad \rho\|\xi_m - \xi_n\|_1 + \nu_m + \sigma_n \geq 0 \,, \; \forall m, n \in S_N \,, \\
&\quad -\nu_n + \tau \geq \sum_{t\in S_T} p_{t,n}^{c}\Delta_t - \lambda \sum_{t\in S_T} \xi_{t,n}\Delta_t - \alpha \,, \; \forall n \in S_N \,, \\
&\quad -\nu_n + \tau \geq 0 \,, \; \forall n \in S_N \,, \\
&\quad (8.1b)\text{–}(8.1h) \,, \; \forall n \in S_N \,.
\end{aligned}
$$

8.4.4 Result and Comparison

The sizing results under different methods are compared in Table 8.1. The DRO-C method needs to solve an LP problem with 440,000 constraints and 220,000 variables, the SP and RO methods solve LP problems with no more than 30,000 constraints and 20,000 variables, and the DRO-D method solves an LP problem with no more than 50,000 constraints and 20,000 variables. To test the degree of conservatism, the β-CVaR of the loss function g is calculated in the 245 days of test data using Theorem 2.6.

The results in Table 8.1 show that the test CVaR is less than 0 in the base case, which verifies the effectiveness of the proposed method. In the SP method, the inaccuracy of the empirical probability distribution is not considered, so the results are optimistic, reflected as positive test CVaR. This indicates that the results of SP are only valid for the empirical probability distribution. Although both the planning data and test data are sampled data from the true probability distribution, SP has low robustness and cannot achieve the required low risk in the probability distribution

Table 8.1 Transmission-ES sizing results of PV power plant under different methods (Xie et al. 2022)

Method	DRO-C	SP	RO	DRO-D
ES power capacity x_p (GW)	0.8508	0.8487	0.8878	0.8834
ES energy capacity x_e (GWh)	5.4617	5.3783	5.5408	5.4336
Transmission line capacity x_l (GW)	0.3956	0.3935	0.3982	0.3937
Total investment (10^9 CNY)	15.3175	15.1722	15.5002	15.2777
Test CVaR	−0.0074	0.0327	−0.0493	0.0050
Computation time (s)	25	3	3	5

represented by the test data. RO has the highest total investment cost and the lowest test CVaR among all the methods compared, indicating that RO has the strongest conservatism. However, when using the sizing model of RO, there will still be cases where the curtailment of renewable energy exceeds the acceptable range in the test data because the support set of the empirical probability distribution does not cover all possible extreme cases. In fact, if the RO method is applied to all 365 days of data, the total investment cost will reach 15.9769×10^9 CNY, which is significantly higher than the result of using only 120 days of data in the table. The result of the DRO-D method is more conservative than SP, but more optimistic than RO, because according to the definition of the ambiguity set, $\mathbb{P}_0 \in \mathcal{B}_d(\mathbb{P}_0, \varepsilon) \subseteq \mathcal{B}(\mathbb{P}_0, \varepsilon)$. The result of the proposed method is different from the DRO-D method, which further illustrates $\mathcal{B}_d(\mathbb{P}_0, \varepsilon) \subsetneq \mathcal{B}(\mathbb{P}_0, \varepsilon)$, and the worst-case probability distribution in $\mathcal{B}(\mathbb{P}_0, \varepsilon)$ is not included in $\mathcal{B}_d(\mathbb{P}_0, \varepsilon)$. Therefore, the DRO-D method, which only considers discrete probability distributions, has its limitations, and it is necessary to adopt the proposed method considering general probability distributions. The last line in Table 8.1 is the calculation time of different methods. The calculation time of the proposed method is much longer than that of the SP method, and the DRO-D method also takes significantly longer than the SP method. Nevertheless, for planning problems, the computational efficiency of all methods is already high enough.

8.4.5 Sensitivity Analysis

Study the impact of some parameters on the results of the sizing problem, including the radius ε of the probability distribution ambiguity set, the amount of planning data N, the acceptable renewable energy curtailment rate λ, the acceptable probability β in CVaR, and the unit capacity cost C_l of the transmission line.

Table 8.2 shows the sizing results under different radii ε of the ambiguity set. When ε increases, the results become more conservative, because as a Wasserstein-distance ball centered on $\mathbb{P}_0$, the ambiguity set becomes larger, making it include worse probability distributions. In the case of $\varepsilon = 0$, the ambiguity set degenerates into a single-point set $\{\mathbb{P}_0\}$ that only contains the empirical probability distribution, so the proposed method also degenerates into the SP case. The result of the proposed method is the same as that of SP when $\varepsilon = 0$, which once again verifies that the accuracy of the renewable energy curtailment function is high enough.

Table 8.3 shows the sizing results under different planning data volumes N. The calculation time increases with the increase of N. Generally speaking, the larger the data volume, the closer the empirical probability distribution $\mathbb{P}_0$ is to the true probability distribution. Therefore, when selecting the radius ε, we should consider the conservative preference and data volume N comprehensively.

Table 8.4 shows the impact of the acceptable renewable energy curtailment rate λ on the sizing results. Since the definitions of the loss function g and CVaR depend on λ, the average renewable energy curtailment rate in the test data is used to reflect the degree of conservatism of the results. The larger λ is, the lower the requirement

Table 8.2 Transmission-ES sizing results of PV power plant under different ambiguity set radii (Xie et al. 2022)

Ambiguity set radius ε	0	0.001	0.002	0.005	0.010
ES power capacity x_p (GW)	0.8487	0.8493	0.8494	0.8508	0.8901
ES energy capacity x_e (GWh)	5.3783	5.3876	5.4092	5.4617	5.5309
Transmission line capacity x_l (GW)	0.3935	0.3941	0.3942	0.3956	0.4004
Total investment (10^9 CNY)	15.1722	15.1970	15.2249	15.3175	15.5355
Test CVaR	0.0327	0.0241	0.0169	−0.0074	−0.0622
Computation time (s)	24	24	24	25	24

Table 8.3 Transmission-ES sizing results of PV power plant under different planning data numbers (Xie et al. 2022)

Planning data number N	48	72	96	120
ES power capacity x_p (GW)	0.8351	0.8348	0.8506	0.8508
ES energy capacity x_e (GWh)	5.4058	5.3906	5.4580	5.4617
Transmission line capacity x_l (GW)	0.3946	0.3943	0.3954	0.3956
Total investment (10^9 CNY)	15.2143	15.1901	15.3077	15.3175
Test CVaR	0.0192	0.0264	−0.0048	−0.0074
Computation time (s)	5	7	15	25

Table 8.4 Transmission-ES sizing results of PV power plant under different acceptable curtailment rates (Xie et al. 2022)

Acceptable curtailment rate λ	3%	5%	7%	10%	15%
ES power capacity x_p (GW)	0.8959	0.8508	0.8427	0.8290	0.7868
ES energy capacity x_e (GWh)	5.6158	5.4617	5.2923	5.0934	4.7760
Transmission line capacity x_l (GW)	0.4063	0.3956	0.3875	0.3738	0.3519
Total investment (10^9 CNY)	15.7606	15.3175	14.9444	14.4169	13.5554
Test average curtailment rate	0.0023	0.0064	0.0125	0.0235	0.0480
Computation time (s)	25	25	25	25	25

for renewable energy curtailment, so the total investment cost decreases, and the test average renewable energy curtailment rate increases. Because the proposed method maintains the acceptable renewable energy curtailment rate λ with a probability of not less than $\beta = 0.90$ under the worst-case probability distribution, the test average renewable energy curtailment rate is significantly less than λ.

Table 8.5 shows the sizing results under different acceptable probabilities β of CVaR parameters. Since the definition of CVaR depends on β, the acceptable

Table 8.5 Transmission-ES sizing results of PV power plant under different acceptable probability values in CVaR (Xie et al. 2022)

Acceptable probability in CVaR β	0.80	0.84	0.87	0.90	0.93
ES power capacity x_p (GW)	0.4694	0.8265	0.8346	0.8508	0.8902
ES energy capacity x_e (GWh)	5.4360	5.3502	5.3872	5.4617	5.5333
Transmission line capacity x_l (GW)	0.3975	0.3915	0.3941	0.3956	0.4006
Total investment (10^9 CNY)	14.9423	15.0774	15.1813	15.3175	15.5420
Test acceptable curtailment rate	0.9224	0.9510	0.9592	0.9796	0.9878
Computation time (s)	25	24	25	25	26

Table 8.6 Transmission-ES sizing results of PV power plant under different unit cost values of the transmission line (Xie et al. 2022)

Unit cost of transmission line C_l (10^7 CNY/MW)	1.2	1.4	1.6	1.8	2.0
ES power capacity x_p (GW)	0	0.0961	0.1991	0.3741	0.8508
ES energy capacity x_e (GWh)	0	1.0273	2.1274	4.3855	5.4617
Transmission line capacity x_l (GW)	0.8481	0.7520	0.6579	0.4882	0.3956
Total investment (10^9 CNY)	10.1771	11.8564	13.2787	14.4236	15.3175
Computation time (s)	26	26	25	25	25

frequency of renewable energy curtailment in the test data is used to reflect the effectiveness of the proposed method. The total investment cost and the acceptable frequency of renewable energy curtailment in the test results decrease with the decrease of β. In addition, the acceptable frequency of renewable energy curtailment in the test is always greater than β, because the proposed method maintains an acceptable probability of β under the worst-case probability distribution.

The important influencing factor of the unit capacity cost C_l of the transmission line is the length of the transmission corridor. The impact of C_l on the sizing results is shown in Table 8.6. When the transmission line is relatively cheap, the optimal solution is to build a transmission line only without configuring storage. In this case, the required capacity of the transmission line is still less than 1 GW. When $C_l \geq 1.4 \times 10^7$ CNY/MW, sizing storage can help reduce the total investment cost. In this case, $C_l : C_p : C_e = 14 : 1 : 1.2$. The larger the C_l, the larger the storage capacity, the smaller the transmission line capacity, and the higher the total investment cost.

8.4.6 Lipschitz Constant-Based Method

Section 8.3 transforms the worst-case expectation into a linear model using the method based on piecewise affine function expressions and Proposition 6.8. In fact, the method based on Lipschitz constants and Theorem 6.10 can also be used. To do this, we need to first find the Lipschitz constant of the loss function $g(x, \xi)$.

Proposition 8.2 *For any $x \in X$ and $\xi, \xi' \in \Xi$, we have:*

$$|g(x, \xi) - g(x, \xi')| \leq L\|\xi - \xi'\|_1 , \tag{8.33}$$

where $L := \max\{\lambda, (1-\lambda)\}\Delta_t$.

Proof Assumed that

$$\begin{aligned}
\xi^{(T)} &:= \xi = (\xi_1, \xi_2, \ldots, \xi_T) ,\\
\xi^{(0)} &:= \xi' = (\xi_1 + \nu_1, \xi_2 + \nu_2, \ldots, \xi_T + \nu_T) ,\\
\xi^{(i)} &:= (\xi_1, \xi_2, \ldots, \xi_{i-1}, \xi_i + \nu_i, \ldots, \xi_T + \nu_T) , \ 1 \leq i \leq T .
\end{aligned}$$

So long as $|g(x, \xi^{(i)}) - g(x, \xi^{(i+1)})| \leq L\|\xi^{(i)} - \xi^{(i+1)}\|_1$ holds for $i = 0, 1, \ldots, T-1$, we have the required conclusion, because

$$\begin{aligned}
|g(x, \xi) - g(x, \xi')| &= \left|\sum_{i=0}^{T-1}(g(x, \xi^{(i)}) - g(x, \xi^{(i+1)}))\right|\\
&\leq \sum_{i=0}^{T-1} |g(x, \xi^{(i)}) - g(x, \xi^{(i+1)})|\\
&\leq \sum_{i=0}^{T-1} L\|\xi^{(i)} - \xi^{(i+1)}\|_1\\
&= L\|\xi - \xi'\|_1 .
\end{aligned}$$

Therefore, we only need to prove that for any $t_0 \in S_T$ and $\varepsilon > 0$, there is

$$\begin{aligned}
&|g(x, \xi) - g(x, \xi')| \leq L\varepsilon , \text{ where } \xi, \xi' \in \Xi ,\\
&\xi = (\xi_1, \xi_2, \ldots, \xi_T) , \ \xi' = (\xi_1, \xi_2, \ldots, \xi_{t_0-1}, \xi_{t_0} - \varepsilon, \xi_{t_0+1}, \ldots, \xi_T) .
\end{aligned} \tag{8.34}$$

In order to prove (8.34), first prove the following two inequalities separately:

$$f(x, \xi) \leq f(x, \xi') + \varepsilon\Delta_t , \tag{8.35}$$

$$f(x, \xi) \geq f(x, \xi') . \tag{8.36}$$

1. Prove (8.35): Assume that $(\hat{p}_t^{rs}, \hat{p}_t^{rg}, \hat{p}_t^{sg}, \hat{p}_t^c, \hat{e}_t; t \in S_T)$ is an optimal solution of (8.1) with respect to (x, ξ'). Let

$$\begin{aligned}&\tilde{p}_t^{rs} = \hat{p}_t^{rs}\ ,\ \tilde{p}_t^{rg} = \hat{p}_t^{rg}\ ,\ \tilde{p}_t^{sg} = \hat{p}_t^{sg}\ ,\ \tilde{e}_t = \hat{e}_t\ ,\ \forall t \in S_T\ ,\\ &\tilde{p}_t^c = \hat{p}_t^c\ ,\ \forall t \neq t_0\ ,\ \tilde{p}_{t_0}^c = \hat{p}_{t_0}^c + \varepsilon\ ,\end{aligned}$$

then $(\tilde{p}_t^{rs}, \tilde{p}_t^{rg}, \tilde{p}_t^{sg}, \tilde{p}_t^c, \tilde{e}_t; t \in S_T)$ is a feasible solution with respect to (x, ξ), and therefore we have:

$$f(x, \xi) \leq \sum_{t \in S_T} \tilde{p}_t^c \Delta_t = \sum_{t \in S_T} \hat{p}_t^c \Delta_t + \varepsilon \Delta_t = f(x, \xi') + \varepsilon \Delta_t\ .$$

2. Prove (8.36): We only need to consider the case where $\varepsilon > 0$ is small enough. Assume that $(\check{p}_t^{rs}, \check{p}_t^{rg}, \check{p}_t^{sg}, \check{p}_t^c, \check{e}_t; t \in S_T)$ is an optimal solution of (8.1) with respect to (x, ξ). Because $\check{p}_{t_0}^{rs} + \check{p}_{t_0}^{rg} + \check{p}_{t_0}^c = \xi_{t_0} > 0$, there are three possible cases:

Case 1: If $\check{p}_{t_0}^c > 0$, let

$$\begin{aligned}&\tilde{p}_t^{rs} = \check{p}_t^{rs}\ ,\ \tilde{p}_t^{rg} = \check{p}_t^{rg}\ ,\ \tilde{p}_t^{sg} = \check{p}_t^{sg}\ ,\ \tilde{e}_t = \check{e}_t\ ,\ \forall t \in S_T\ ,\\ &\tilde{p}_t^c = \check{p}_t^c\ ,\ \forall t \neq t_0\ ,\ \tilde{p}_{t_0}^c = \check{p}_{t_0}^c - \varepsilon\ ,\end{aligned}$$

then $(\tilde{p}_t^{rs}, \tilde{p}_t^{rg}, \tilde{p}_t^{sg}, \tilde{p}_t^c, \tilde{e}_t; t \in S_T)$ is a feasible solution with respect to (x, ξ'). Therefore,

$$f(x, \xi') \leq \sum_{t \in S_T} \tilde{p}_t^c \Delta_t = \sum_{t \in S_T} \check{p}_t^c \Delta_t - \varepsilon \Delta_t = f(x, \xi) - \varepsilon \Delta_t \leq f(x, \xi)\ .$$

Case 2: If $\check{p}_{t_0}^c = 0$ and $\check{p}_{t_0}^{rg} > 0$, let

$$\begin{aligned}&\tilde{p}_t^{rs} = \check{p}_t^{rs}\ ,\ \tilde{p}_t^{sg} = \check{p}_t^{sg}\ ,\ \tilde{p}_t^c = \check{p}_t^c\ ,\ \tilde{e}_t = \check{e}_t\ ,\ t \in S_T\ ,\\ &\tilde{p}_t^{rg} = \check{p}_t^{rg}\ ,\ \forall t \neq t_0\ ,\ \tilde{p}_{t_0}^{rg} = \check{p}_{t_0}^{rg} - \varepsilon\ ,\end{aligned}$$

then $(\tilde{p}_t^{rs}, \tilde{p}_t^{rg}, \tilde{p}_t^{sg}, \tilde{p}_t^c, \tilde{e}_t; t \in S_T)$ is a feasible solution with respect to (x, ξ'). Therefore,

$$f(x, \xi') \leq \sum_{t \in S_T} \tilde{p}_t^c \Delta_t = \sum_{t \in S_T} \check{p}_t^c \Delta_t = f(x, \xi)\ . \tag{8.37}$$

Case 3: If $\check{p}_{t_0}^c = \check{p}_{t_0}^{rg} = 0$ and $\check{p}_{t_0}^{rs} > 0$, let

$$\begin{aligned}&\tilde{p}_t^{rg} = \check{p}_t^{rg}\ ,\ \tilde{p}_t^c = \check{p}_t^c\ ,\ \forall t \in S_T\ ,\\ &\tilde{p}_{t_0}^{rs} = \check{p}_{t_0}^{rs} - \varepsilon\ ,\ \tilde{p}_t^{rs} = \check{p}_t^{rs}\ ,\ \forall t \neq t_0\ .\end{aligned}$$

There are two possible scenarios:

The first scenario: There exist $t_1 \geq t_0$ such that $\check{p}_{t_1}^{sg} > 0$. Then select t_1 as the minimum value to meet the requirements, i.e., $t_1 := \min\{t \mid t_0 \leq t \leq T, \check{p}_{t_1}^{sg} > 0\}$. Let

$$\tilde{p}_{t_1}^{sg} = \check{p}_{t_1}^{sg} - \eta_c \eta_d \varepsilon \ , \ \tilde{p}_t^{sg} = \check{p}_t^{sg} \ , \ \forall t \neq t_1 \ .$$

We can determine $\tilde{e}_t$ by (8.1e) and $\tilde{e}_0 = \check{e}_0$. Hence, $\tilde{e}_T = \tilde{e}_0$ and $\omega_l x_e \leq \min_s\{\check{e}_s\} \leq \tilde{e}_t \leq \max_s\{\check{e}_s\} \leq \omega_h x_e$. Thus, $(\tilde{p}_t^{rs}, \tilde{p}_t^{rg}, \tilde{p}_t^{sg}, \tilde{p}_t^{c}, \tilde{e}_t; t \in S_T)$ is a feasible solution with respect to (x, ξ'), and then similar to (8.37), we have $f(x, \xi') \leq f(x, \xi)$.

The second scenario: When $\check{p}_t^{sg} = 0, \forall t \geq t_0$, there exists $t_1 < t_0$ such that $\check{p}_{t_1}^{sg} > 0$. Then select t_1 as the maximum value to meet the requirements, i.e., $t_1 := \max\{t \mid 1 \leq t < t_0, \check{p}_{t_1}^{sg} > 0\}$. Let

$$\tilde{p}_{t_1}^{sg} = \check{p}_{t_1}^{sg} - \eta_c \eta_d \varepsilon \ , \ \tilde{p}_t^{sg} = \check{p}_t^{sg} \ , \ \forall t \neq t_1 \ .$$

Similar to the first scenario, we determine $\tilde{e}_t$ according to (8.1e), then we get $(\tilde{p}_t^{rs}, \tilde{p}_t^{rg}, \tilde{p}_t^{sg}, \tilde{p}_t^{c}, \tilde{e}_t; t \in S_T)$, which is a feasible solution with respect to (x, ξ'). Then we have $f(x, \xi') \leq f(x, \xi)$.

Therefore, Eq. (8.36) holds.

According to (8.35) and (8.36), we have:

$$-\lambda \varepsilon \Delta_t \leq f(x, \xi) - f(x, \xi') - \lambda \varepsilon \Delta_t \leq \varepsilon \Delta_t - \lambda \varepsilon \Delta_t \ .$$

Note that the acceptable renewable energy curtailment rate $\lambda \in [0, 1]$, then

$$\begin{aligned}
&|g(x, \xi) - g(x, \xi')| \\
&= \left| \left(f(x, \xi) - \lambda \sum_{t \in S_T} \xi_t \Delta_t \right) - \left(f(x, \xi') - \lambda \sum_{t \in S_T} \xi_t' \Delta_t \right) \right| \\
&= |f(x, \xi) - f(x, \xi') - \lambda \varepsilon \Delta_t| \\
&\leq \max\{|-\lambda \varepsilon \Delta_t|, |\varepsilon \Delta_t - \lambda \varepsilon \Delta_t|\} \\
&= \max\{\lambda, 1 - \lambda\} \Delta_t \cdot \varepsilon \\
&= L\varepsilon \ .
\end{aligned}$$

Therefore, (8.34) holds and the proof is completed. □

Note that λ represents the acceptable renewable energy curtailment rate, so in practical applications, $\lambda < 1/2$, for example, $\lambda = 0.05$ in the base case, thus $L = \max\{\lambda, 1 - \lambda\}\Delta_t = (1 - \lambda)\Delta_t$. In the proof of Proposition 8.2, case 1 in (8.36) provides an example of $|g(x, \xi) - g(x, \xi')| = (1 - \lambda)\Delta_t \cdot \|\xi - \xi'\|_1$, so $L = (1 - \lambda)\Delta_t$ is the best or minimal Lipschitz constant of the loss function $g(x, \xi)$.

In addition, the proof process of Proposition 8.2 can obtain the Lipschitz constant of the renewable energy curtailment function $f(x, \xi)$ as Δ_t. From the physical meaning point of view, when the renewable energy power generation increases by 1 unit in a certain period, the renewable energy curtailment will not decrease, and it will increase by up to Δ_t.

Using the Lipschitz constant $L = (1-\lambda)\Delta_t$ of the loss function $g(x, \xi)$, Eq. (8.24) is transformed into:

$$\alpha + \frac{1}{1-\beta}\left(\varepsilon L + \mathbb{E}^{\mathbb{P}_0}[\max\{g(x,\xi) - \alpha, 0\}]\right) \leq 0 \,.$$

Therefore, the sizing problem of (8.9) is transformed into the following LP problem:

$$\min C_p x_p + C_e x_e + C_l x_l \tag{8.38a}$$

$$\text{s.t. } x_p, x_e, x_l \geq 0 \,, \tag{8.38b}$$

$$\alpha + \frac{1}{1-\beta}\left(\varepsilon L + \frac{1}{N}\sum_{n \in S_N} s_n\right) \leq 0 \,, \tag{8.38c}$$

$$s_n \geq \sum_{t \in S_T} p_{t,n}^{C}\Delta_t - \lambda \sum_{t \in S_T} \xi_{t,n}\Delta_t - \alpha \,, \; s_n \geq 0 \,, \; \forall n \in S_N \,, \tag{8.38d}$$

$$(8.1b)\text{–}(8.1h) \,, \; \forall n \in S_N \,. \tag{8.38e}$$

8.4.7 Transmission–Storage Sizing Method in Remote Renewable Power Plants Based on Candidate Transmission Line Schemes

Different from the sizing model proposed in (8.9), the transmission-storage optimal configuration method based on transmission line candidate schemes for remote renewable power plants no longer assumes that the cost of transmission lines is proportional to capacity, but selects transmission line configuration schemes from some candidate schemes.

Let the set of transmission line candidate schemes be denoted by I. For any scheme $i \in I$, let the transmission line cost be $C_{l,i}$, and the transmission line capacity be $P_{l,i}$. Introduce the Boolean variable $z_{l,i}$ to indicate whether to choose the transmission line candidate scheme i, with $z_{l,i} = 1$ for the selection of scheme i. The mathematical model of the transmission-storage optimization configuration of remote renewable power plants is as follows:

$$\min_{x,z_{l,i}} C_p x_p + C_e x_e + \sum_{i \in I} C_{l,i} z_{l,i} \tag{8.39a}$$

$$\text{s.t. } x = (x_p, x_e, x_l) \geq 0 \ , \ x_l = \sum_{i \in I} P_{l,i} z_{l,i} \ , \ \sum_{i \in I} z_{l,i} = 1 \ , \tag{8.39b}$$

$$\sup_{\mathbb{P} \in \mathcal{B}(\mathbb{P}_0, \varepsilon)} \quad \beta\text{-CVaR}^{\mathbb{P}}(g(x, \xi)) \leq 0 \ , \tag{8.39c}$$

$$z_{l,i} \in \{0, 1\} \ , \ \forall i \in I \ . \tag{8.39d}$$

Since $z_{l,i}$ can only be 0 or 1 and the sum is stipulated to be 1 in (8.39b), only one transmission line scheme is selected. Therefore, $\sum_{i \in I} C_{l,i} z_{l,i}$ equals the transmission line cost of the selected scheme, $\sum_{i \in I} P_{l,i} z_{l,i}$ equals the transmission line capacity of the selected scheme. Thus, the optimization configuration model in (8.39) minimizes the total cost of storage and transmission lines under the distributionally robust CVaR constraint of renewable energy curtailment. According to the methods in Sect. 8.3, this optimization configuration model is transformed into solving the following MILP problem:

$$\begin{aligned} &\min C_p x_p + C_e x_e + \sum_{i \in I} C_{l,i} z_{l,i} \\ &\text{s.t. } (8.38c)\text{–}(8.38e) \ , \ (8.39b) \ , \ (8.39d) \ . \end{aligned}$$

To verify the proposed transmission-storage optimization configuration model considering transmission line candidate schemes, we test it based on the benchmark case. It is further assumed that there are 3 transmission line candidate schemes, which are the 5 billion CNY and 0.30 GW transmission line scheme, the 9 billion CNY and 0.45 GW transmission line scheme, and the 12 billion CNY and 0.60 GW transmission line scheme. The proposed method needs to solve a MILP with about 30,000 constraints and 15,000 variables (including 3 Boolean variables). The calculation time is 6 s. The optimization result is to choose the 0.45 GW transmission line scheme and configure the storage with a power capacity of 0.7748 GW and an energy capacity of 4.8211 GWh, with a total investment cost of 15.5591 billion CNY. Therefore, the proposed transmission-storage optimization configuration method for remote renewable power plants based on transmission line candidate schemes is effective.

8.5 Conclusion

This chapter is dedicated to solving the joint optimal sizing problem of transmission line and ES capacities for remote renewable power plants. An operation optimization problem is established to minimize the curtailment of renewable energy, and the optimal value is regarded as the renewable energy curtailment function of capacity variables and renewable energy generation random variables. It is demonstrated that the renewable energy curtailment function is piecewise affine, and a method is proposed to calculate the approximate expression of the renewable energy curtailment function by solving LP. A distributionally robust CVaR constraint for renewable energy curtailment is established using a probability distribution ambiguity set based on the Wasserstein distance. On this basis, a distributionally robust sizing model is proposed to minimize investment costs, and it is transformed into LP for an effective solution. The conclusions of the case analysis part include:

Firstly, in this test case, the worst-case probability distribution in the probability distribution ambiguity set based on the Wasserstein distance is not a discrete probability distribution with the same support as the empirical probability distribution, so it is necessary to consider the general form of probability distribution in the ambiguity set.

Secondly, the radius of the probability distribution ambiguity set should be selected according to the amount of data and the risk preference of the decision-maker.

Thirdly, the unit capacity cost of the transmission line has a significant impact on the joint optimal sizing results of the transmission-storage capacities of the remote renewable power plant.

Fourthly, in this test case, although the range of random variable values is restricted, the worst-case expectation transformation method based on the Lipschitz constant in Theorem 6.10 does not introduce additional conservatism.

References

Chen Y, Wei W, Wang C et al (2022) Storage and transmission capacity requirements of a remote solar power generation system. IEEE Syst. J. 16(2):3175–3178

Gonzalez-Gonzalez JM, Martin S, Lopez P et al (2020) Hybrid battery-ultracapacitor storage system sizing for renewable energy network integration. IET Renew Power Gener 14(13):2367–2375

Gurobi Optimization, LLC (2023) Gurobi optimizer reference manual. https://www.gurobi.com/documentation/current/refman/index.html. Accessed 20 Dec 2023

Konno H (1976) A cutting plane algorithm for solving bilinear programs. Math. Program. 11:14–27

Lofberg J (2004) YALMIP: a toolbox for modeling and optimization in MATLAB. In: 2004 IEEE international conference on robotics and automation, pp 284–289

Sengupta M, Xie Y, Lopez A et al (2018) The national solar radiation data base (NSRDB). Renew. Sustain. Energy Rev. 89:51–60

Sheibani MR, Yousefi GR, Latify MA et al (2018) Energy storage system expansion planning in power systems: a review. IET Renew. Power Gener. 12(11):1203–1221

Tang Z, Liu J, Liu Y et al (2019) Risk awareness enabled sizing approach for hybrid energy storage system in distribution network. IET Gener. Transm. & Distrib. 13(17):3814–3822

Xie R, Wei W, Ge MF et al (2022) Coordinate sizing of energy storage and transmission line for a remote renewable power plant. IET Renew Power Gener 16(12):2508–2520

Chapter 9
Storage Sizing in Power Networks to Reduce Renewable Generation Curtailment

Abstract This chapter tackles the issue of sizing energy storage in bulk power systems. To accurately represent the operational status of the power system, a specialized power flow model incorporating voltage and reactive power is utilized. The uncertainty of renewable energy production is expressed through inexact probability distributions, encapsulated in a data-driven ambiguity set based on Wasserstein distance. Using this data, the rate of renewable energy curtailment is formulated as a distributionally robust chance constraint. The aim is to minimize the overall investment cost, leading to a distributionally robust chance-constrained program for the optimal sizing problem. This is then reformulated into a manageable linear programming problem through conservative approximation. This chapter introduces the work in Guo et al. (2020).

9.1 Introduction

The incorporation of renewable energy into bulk power systems faces challenges due to its fluctuating nature and transmission congestion. To minimize renewable generation wastage, China's government implemented policies limiting the annual renewable power spillage rate in specific regions. In regions abundant in renewables, power spillage is a frequent issue. Complementing large wind farms or solar stations with energy storage (ES) has proven to be an effective strategy in reducing renewable power curtailment. This approach is typically more cost-effective and quicker than expanding transmission corridors, which can be both expensive and time-consuming, and can also lead to underutilization of the transmission system.

This work, in the context of distributionally robust optimization (DRO), addresses the issue of sizing storage to reduce renewable power curtailment while considering network power flow, with two significant contributions:

First, a DRO model is proposed for sizing ESs considering network power flow, which ensures the renewable power curtailment rate does not exceed a predefined limit. A detailed power flow model accounting for voltage and reactive power is employed, which is more precise than the conventional direct-current (DC) power

R. Xie and W. Wei, *Distributionally Robust Optimization and its Applications in Power System Energy Storage Sizing*,
https://doi.org/10.1007/978-981-97-2566-3_9

flow model. The uncertainty of available renewable power is illustrated by a Wasserstein distance-based ambiguity set, established in a data-driven manner. The maximum limit of renewable energy curtailment rate is imposed by a distributionally robust chance constraint (DRCC), ultimately giving rise to a distributionally robust chance-constrained program (DRCCP) for the storage sizing problem.

Second, a two-step reformulation procedure is employed on the DRCC for renewable energy curtailment rates. The initial step involves transforming the DRCC into a conditional value-at-risk (CVaR) expression. The CVaR is further substituted with a series of linear constraints, leading to a linear programming (LP) problem. These actions are dependent on two sufficient conditions that guarantee the model's robustness, albeit this might introduce a degree of conservatism. However, this reformulation effectively manages to address such a DRCCP where uncertain parameters are found not only in DRCC but also in other constraints. Notably, this final LP model can be efficiently solved, even with the introduction of a substantial number of scenario-dependent variables.

This chapter is structured as follows: Sect. 9.2 presents the network power flow model, the ambiguity set, and the DRCCP model for ES sizing. In Sect. 9.3, a solution method based on CVaR and LP reformulation is developed. Section 9.4 conducts case studies for practical demonstration of the presented models, followed by conclusions in Sect. 9.5.

9.2 Mathematical Formulation

Essentially, the assumption is that each renewable plant is equipped with an ES, whose capacity is to be determined. This section will provide mathematical models of the network power flow and the DRCCP for the sizing of ES.

9.2.1 Network Power Flow Model

The DC power flow model is commonly employed in various research contexts due to its linearity. Nonetheless, it overlooks bus voltage and reactive power, potentially resulting in a pessimistic solution when renewable energy penetration is high. Consequently, this chapter uses a dedicated power flow model proposed in Yang et al. (2018), which incorporates both reactive power and voltage while maintaining linearity. To illustrate the model, the active and reactive power flows in line $(i, j) \in \mathbb{L}$ can be expressed as follows:

$$P_{ij} = g_{ij}(v_i^2 - v_i v_j \cos\theta_{ij}) - b_{ij} v_i v_j \sin\theta_{ij} \ , \tag{9.1a}$$

$$Q_{ij} = -b_{ij}(v_i^2 - v_i v_j \cos\theta_{ij}) - g_{ij} v_i v_j \sin\theta_{ij} \ . \tag{9.1b}$$

In these expressions, $\theta_{ij} = \theta_i - \theta_j$. By assuming that $\theta_{ij} \ll 1$ and $v \approx 1.0$ per unit (p.u.), we can simplify these expressions to obtain the following approximations:

$$\sin\theta_{ij} \approx \theta_{ij} \ , \ \cos\theta_{ij} \approx 1 - \frac{\theta_{ij}^2}{2} \ , \tag{9.2a}$$

$$v_i v_j \theta_{ij} \approx \theta_{ij} \ , \ v_i v_j \theta_{ij}^2 \approx \theta_{ij}^2 \ . \tag{9.2b}$$

By substituting (9.2) into (9.1) and observing that $v_i^2 - v_i v_j = ((v_i^2 - v_j^2)/2) + (v_{ij}^2/2)$, where $v_{ij} = v_i - v_j$, we derive the following linearized power flow model for any $t \in \mathbb{T}$:

$$P_{ij,t} = g_{ij}\frac{v_{i,t}^2 - v_{j,t}^2}{2} - b_{ij}\theta_{ij,t} + P_{ij,t}^l \ , \ \forall (i,j) \in \mathbb{L} \ , \ t \in \mathbb{T} \ , \tag{9.3a}$$

$$Q_{ij,t} = -b_{ij}\frac{v_{i,t}^2 - v_{j,t}^2}{2} - g_{ij}\theta_{ij,t} + Q_{ij,t}^l \ , \ \forall (i,j) \in \mathbb{L} \ , \ t \in \mathbb{T} \ , \tag{9.3b}$$

$$P_{ij,t}^l = \frac{1}{2}g_{ij}\theta_{ij,t}^2 + \frac{1}{2}g_{ij}v_{ij,t}^2 \ , \ \forall (i,j) \in \mathbb{L} \ , \ t \in \mathbb{T} \ , \tag{9.3c}$$

$$Q_{ij,t}^l = \frac{1}{2}(-b_{ij})\theta_{ij,t}^2 + \frac{1}{2}(-b_{ij})v_{ij,t}^2 \ , \ \forall (i,j) \in \mathbb{L} \ , \ t \in \mathbb{T} \ . \tag{9.3d}$$

In these expressions, Eqs. (9.3a) and (9.3b) dictate the active and reactive power flows in the line (i, j), whereas (9.3c) and (9.3d) represent the loss terms, which share the same form. Taking $P_{ij,t}^l$ as an example, its linearization process is illustrated, and the same procedure applies to the reactive power loss term. The nodal voltage in the base case condition must be provided initially, which is accessible via the MATPOWER Toolbox (Zimmerman et al. 2011) through a standard AC power flow calculation. The magnitude or angle is denoted by $v_{i,0}$ and $\theta_{i,0}$ for every i node, resulting in $\theta_{ij,0} = \theta_{i,0} - \theta_{j,0}$. With $\theta_{ij,0}$ as the reference, the approximation $(\theta_{ij} - \theta_{ij,0})^2 \approx 0$ holds, leading to $\theta_{ij}^2 \approx 2\theta_{ij,0}\theta_{ij} - \theta_{ij,0}^2$, which can be used to express the first term of the active power loss $P_{ij,t}^l$ as follows:

$$P_{ij,t}^{l1} = \frac{1}{2}g_{ij}\theta_{ij,t}^2 \approx g_{ij}\theta_{ij,0}\theta_{ij,t} - \frac{1}{2}g_{ij}\theta_{ij,0}^2 \ .$$

Similarly, under the assumption that $V_i - V_j = v_i^2 - v_j^2 \approx v_{i,0}^2 - v_{j,0}^2$ and $(v_{i,t} - v_{j,t})^2 \approx (v_{i,0}^2 - v_{j,0}^2)^2$ around v_0, we have:

$$(v_{i,t} - v_{j,t})^2 \approx \frac{v_{i,0} - v_{j,0}}{v_{i,0} + v_{j,0}}(v_{i,0} - v_{j,0})^2 \approx \frac{v_{i,0} - v_{j,0}}{v_{i,0} + v_{j,0}}(V_i - V_j) \ .$$

Consequently, leveraging the squared voltage magnitude variable V, the second term of the active power loss $P_{ij,t}^l$ can be efficiently linearized.

$$
\begin{aligned}
P_{ij,t}^{l2} &= \frac{1}{2} g_{ij} v_{ij}^2 \\
&\approx g_{ij} \frac{v_{i,0} - v_{j,0}}{v_{i,0} + v_{j,0}} (V_i - V_j) - \frac{g_{ij}}{2} (v_{i,0} - v_{j,0})^2 \, .
\end{aligned}
$$

Regarding V as an independent variable, we can express the linearized power flow model as follows:

$$P_{ij,t} = g_{ij} \frac{V_{i,t} - V_{j,t}}{2} - b_{ij}\theta_{ij,t} + P_{ij,t}^l \, , \ \forall (i, j) \in \mathbb{L} \, , \ t \in \mathbb{T} \, , \tag{9.4a}$$

$$Q_{ij,t} = -b_{ij} \frac{V_{i,t} - V_{j,t}}{2} - g_{ij}\theta_{ij,t} + Q_{ij,t}^l \, , \ \forall (i, j) \in \mathbb{L} \, , \ t \in \mathbb{T} \, , \tag{9.4b}$$

$$P_{i,t}^n = \sum_{(i,j)\in\mathbb{L}} P_{ij,t} + V_{i,t} \sum_{j=1}^{N} G_{ij} \, , \ \forall i \in \mathbb{N} \, , \ t \in \mathbb{T} \, , \tag{9.4c}$$

$$Q_{i,t}^n = \sum_{(i,j)\in\mathbb{L}} Q_{ij,t} - V_{i,t} \sum_{j=1}^{N} B_{ij} \, , \ \forall i \in \mathbb{N} \, , \ t \in \mathbb{T} \, , \tag{9.4d}$$

$$\text{Bounds for power flow variables} \, . \tag{9.4e}$$

$P_{ij,t}^l$ and $Q_{ij,t}^l$ denote the aforementioned loss terms. The power balance at each node is governed by (9.4c) and (9.4d). The rightmost terms represent the power flow on shunt elements. $P_{i,t}^n/Q_{i,t}^n$ symbolize the net active/reactive power injection varying based on the generators' output $P_{i,t}^g/Q_{i,t}^g$, power demand $P_{i,t}^d/Q_{i,t}^d$, renewable power generation $P_{i,t}^{rg}$, and the output of ES $P_{i,t}^c/P_{i,t}^{dc}$ if exists. We consider the demand to be deterministic. However, it is important to note that uncertain demand can be dealt with in the same manner as renewable energy. Constraints on power flow variables fall within (9.4e), which contains all upper and lower bounds. Limits on generation capacity and ramping of traditional synchronized generators are encompassed within constraint (9.4e). Owing to their common knowledge, these constraints are omitted here for the purpose of conciseness.

Renewable energy generator (REG) constraints are:

$$P_{i,t}^r \le \xi_{i,t} \, , \ \forall i \in \Omega \, , \ t \in \mathbb{T} \, , \tag{9.5a}$$

$$P_{i,t}^r = P_{i,t}^{rs} + P_{i,t}^{rg} \, , \ \forall i \in \Omega \, , \ t \in \mathbb{T} \, . \tag{9.5b}$$

Equation (9.5a) suggests that the renewable power production denoted as $P_{i,t}^r$ must not exceed the dispatchable quantity $\xi_{i,t}$, that is subject to the randomness of the weather conditions. Equation (9.5b) denotes that the renewable power can either be used to charge the ES or supplied to the grid, represented by $P_{i,t}^{rs}$ and $P_{i,t}^{rg}$, respectively. To keep things concise, Eqs. (9.4) and (9.5) will be collectively referred to as Cons-PFM.

The model of ES at bus $i \in \Omega$ is:

$$E_{i,t+1} = E_{i,t}(1-\eta_{sd}) + P^{c}_{i,t}\eta_c\Delta - \frac{P^{dc}_{i,t}\Delta}{\eta_{dc}}\ , \ \forall i \in \Omega\ ,\ t \in \mathbb{T}\ , \tag{9.6a}$$

$$P^{c}_{i,t} = P^{rs}_{i,t} + P^{gs}_{i,t}\ , \ \forall i \in \Omega\ ,\ t \in \mathbb{T}\ , \tag{9.6b}$$

$$P^{dc}_{i,t} = P^{sg}_{i,t}\ , \ \forall i \in \Omega\ ,\ t \in \mathbb{T}\ , \tag{9.6c}$$

$$0 \le P^{c}_{i,t}\Delta \le C_i/h\ , \ \forall i \in \Omega\ ,\ t \in \mathbb{T}\ , \tag{9.6d}$$

$$0 \le P^{dc}_{i,t}\Delta \le C_i/h\ , \ \forall i \in \Omega\ ,\ t \in \mathbb{T}\ , \tag{9.6e}$$

$$C_i\eta_{lb} \le E_{i,t} \le C_i\ , \ \forall i \in \Omega\ ,\ t \in \mathbb{T}\ , \tag{9.6f}$$

$$E_{i,1} = E_{i,T+1}\ , \ \forall i \in \Omega\ , \tag{9.6g}$$

$$\text{Bounds for ES variables}\ . \tag{9.6h}$$

Equation (9.6a) illustrates the dynamics of the State of Charge (SoC), while Δ refers to the duration of each time slot, which is typically an hour. Equations (9.6b) and (9.6c) define the ES' charging and discharging power, with $P^{rs}_{i,t}/P^{gs}_{i,t}/P^{sg}_{i,t}$ representing power transition from renewable energy sources (RESs) to storage, storage to grid, and grid to storage, respectively. Furthermore, (9.6d) and (9.6e) impose limits on the maximum charging and discharging powers. The variable h is a predefined parameter reliant on the rated charging or discharging time. The ES' capacity, denoted by C_i, is subject to optimization. Equation (9.6f) sets the bounds for SoC, while (9.6g) introduces the cycling constraint, which stipulates that the final SoC should be equivalent to the initial one. Lastly, Eq. (9.6h) encapsulates all upper and lower limitations related to the ES variables. For simplicity, Eq. (9.6) will be referred to as Cons-ES.

9.2.2 Deterministic Sizing Model

First, it is necessary to define the renewable energy curtailment rate. For a given trajectory of available renewable power, denoted as $\{\xi_{i,t}\}$, $\forall i, t$, the curtailment rate associated with the renewable plant at node i can be computed by

$$D^{i}_{ct} = \frac{\sum_{t\in\mathbb{T}}(\xi_{i,t} - P^{r}_{i,t})}{\sum_{t\in\mathbb{T}}\xi_{i,t}}\ , \ \forall i \in \Omega\ . \tag{9.7}$$

Based on the aforementioned definition and a preset cap $D^{\max}_{ct}$, the deterministic ES sizing problem is outlined as follows:

$$\begin{aligned} &\min \sum_{i\in\Omega} I_{inv}C_i \\ &\text{s.t. Cons-PFM}\ ,\ \text{Cons-ES}\ , \\ &\qquad D^{i}_{ct} \le D^{\max}_{ct}\ ,\ \forall i \in \Omega\ . \end{aligned} \tag{9.8}$$

In the above formulation, the aim is to minimize the investment cost of the ES, which is represented by the objective function. The final constraint ensures that the renewable energy curtailment rate at each renewable plant does not exceed the specified cap. In the absence of uncertainty, problem (9.8) results in LP. However, renewable power, denoted by $\xi_{i,t}$, is challenging to predict with a high degree of accuracy, particularly for long-term planning. Applying the strategy provided by (9.8) might lead to the actual renewable energy curtailment rate exceeding the cap.

The following includes discussions regarding the storage sizing model.

Firstly, in (9.8), the operational cost of thermal generators is not considered. Instead, it is assumed that a government entity imposes a mandatory cap on the rate of renewable energy curtailment, and the planner's objective is to meet this curtailment cap with the least investment possible. The operational cost is omitted in the proposed model because it has a direct correlation to the renewable curtailment rate. In the context of energy balancing, assuming constant total demand, an increase in the renewable power curtailment rate implies a higher production from conventional generators, and thus, a higher production cost.

Secondly, in a scenario without a mandatory cap, it is possible to include a term in the objective function that penalizes renewable power spillage, as seen in Dui et al. (2018). However, implementing an appropriate penalty coefficient that considers the interests of all stakeholders is not an easy task. For instance, in Dui et al. (2018), the total wind curtailment rate reaches 13.85%, a level that would not be acceptable in China.

Thirdly, according to multi-objective optimization, the penalty approach corresponds to a weighted sum formulation, while problem (9.8) signifies a ϵ-constraint formulation for the identical multi-objective optimization problem. There exists a specific relationship between the penalty coefficient and the cap, in such a way that both formulations yield the same optimal solution.

9.2.3 Distributionally Robust Chance-Constrained Program Model

The ambiguity set, constructed on the basis of the Wasserstein distance, is used to model the uncertainty of renewable generation. This set takes into account a series of probability distributions that lie very close to the empirical probability distribution. Using historical data of available renewable power, we obtain the following empirical probability distribution:

$$\hat{P} = \frac{1}{K} \sum_{k=1}^{K} \delta_{\hat{\xi}^k} , \tag{9.9}$$

where $\delta_{\hat{\xi}^k}$ is the Dirac distribution on $\hat{\xi}^k$, and $\hat{\xi}^k = \{\hat{\xi}^k_{i,t}\}, \forall i \in \Omega, t \in \mathbb{T}$. K is the size of historical data. Let d_W be the Wasserstein 1-distance in (6.3) under 1-norm. The ambiguity set is:

$$\mathcal{M}^\varepsilon = \{P \in \mathcal{M}_1(\Xi) \mid d_W(P, \hat{P}) \leq \varepsilon\} . \tag{9.10}$$

During the planning stage, the time series of dispatchable renewable power, denoted by $\{\xi_{i,t}\}, \forall i, t$, is uncertain. So is the curtailment rate. Given the probability distribution of renewable power, the distribution of D_{ct} and the probability of $D_{ct} \leq D^{\max}$ can be acquired through Monte Carlo simulation. A cautious approach involves consideration of the worst-case probability distribution:

$$\inf_{P \in \mathcal{M}_i^\varepsilon} \Pr\{D^i_{ct} \leq D^{\max}_{ct}\} \geq 1 - \alpha \ , \ \forall i \in \Omega .$$

When the distribution of renewable power fluctuates within the ambiguity set $\mathcal{M}^\varepsilon$, the minimum probability for D_{ct} not exceeding $D^{\max}_{ct}$ should be at least $1 - \alpha$. This condition leads to a DRCCP formulation

$$\begin{aligned} \min \ & \sum_{i \in \Omega} I_{inv} C_i \\ \text{s.t.} \ & \text{Cons-PFM} \ , \ \text{Cons-ES} \ , \\ & \inf_{P \in \mathcal{M}_i^\varepsilon} \Pr\{D^i_{ct} \leq D^{\max}_{ct}\} \geq 1 - \alpha \ , \ \forall i \in \Omega . \end{aligned} \tag{9.11}$$

In the equation above, the DRCC guarantees that the curtailment limit is sustained with a minimum probability of $1 - \alpha$. Here, α denotes a predetermined risk tolerance. Since an ES is allocated for every renewable plant, the curtailment rate D^i_{ct} can be managed independently. Consequently, each plant aligns with one individual DRCC, thereby avoiding the complexities associated with joint chance constraints.

9.3 Solution Method

The DRCC in (9.11) makes direct solving unfeasible. We will reformulate (9.11) into a form that can be solved using LP.

Firstly, we rephrase the DRCC in (9.11):

$$\sup_{P \in \mathcal{M}_i^\varepsilon} \Pr\{d\xi_{\Sigma,i} - P^r_{\Sigma,i} \geq 0\} \leq \alpha \ , \ \forall i \in \Omega \ , \tag{9.12}$$

where

$$d = 1 - D_{ct}^{\max} , \ \xi_{\Sigma,i} = \sum_{t \in \mathbb{T}} \xi_{i,t} , \ P_{\Sigma,i}^{r} = \sum_{t \in \mathbb{T}} P_{i,t}^{r} .$$

For simplicity, we temporarily omit the subscript i and define a loss function as follows:

$$L(\xi_\Sigma, P_\Sigma^r) = d\xi_\Sigma - P_\Sigma^r . \tag{9.13}$$

Therefore, (9.12) is rewritten as:

$$\sup_{P \in \mathcal{M}_i^\varepsilon} \Pr\{L(\xi_\Sigma, P_\Sigma^r) \geq 0\} \leq \alpha . \tag{9.14}$$

According to Sect. 2.3.2, the following is a sufficient condition of (9.14).

$$\sup_{P \in \mathcal{M}_i^\varepsilon} \text{CVaR}_{1-\alpha}^{P}(L(\xi_\Sigma, P_\Sigma^r)) \leq 0 , \tag{9.15}$$

where the CVaR equals the optimal value of the following optimization problem:

$$\inf_{\gamma \in R} \alpha^{-1} \mathbb{E}_P[L(\xi_\Sigma, P_\Sigma^r) + \gamma]_+ - \gamma . \tag{9.16}$$

Here $(\cdot)_+ = \max\{\cdot, 0\}$. By Lemma V.8 in Hota et al. (2019), a sufficient condition of (9.15) is:

$$\varepsilon H_L + \inf_{\gamma \in R} \left\{ \frac{1}{K} \sum_{k=1}^{K} \left[L\left(\hat{\xi}_\Sigma^k, P_\Sigma^{r,k}\right) + \gamma \right]_+ - \gamma\alpha \right\} \leq 0 . \tag{9.17}$$

Here, H_L denotes the Lipschitz constant. Additionally, we affirm that the loss function $L(\xi_\Sigma, P_\Sigma^r)$ complies with the necessary assumptions stipulated by Lemma V.8, including boundedness, convexity, and Lipschitz continuity.

The boundedness of $L(\xi_\Sigma, P_\Sigma^r)$ is ensured by its physical interpretation; it is linear and hence convex in ξ_Σ. Furthermore, in instances when no renewable power is available, that is, $\xi_\Sigma = 0$, the $L(0, P_\Sigma^r)$ is guaranteed to be 0 because there isn't any energy to be curtailed. For the case where $L(\xi_\Sigma, P_\Sigma^r)$ and $\xi_\Sigma \geq 0$, the following applies:

$$-D_{ct}^{\max}\xi_\Sigma \leq L(\xi_\Sigma, P_\Sigma^r) - L(0, P_\Sigma^r) \leq d\xi_\Sigma . \tag{9.18}$$

In general, $D_{ct}^{\max} \ll d$ and (9.18) leads to

$$\|L(\xi_\Sigma, P_\Sigma^r) - L(0, P_\Sigma^r)\| \leq d\xi_\Sigma . \tag{9.19}$$

Therefore, $L(\xi_\Sigma, P^r_\Sigma)$ is Lipschitz continuous and we set H_L as d. Substituting DRCC (9.11) into (9.17), then

$$\begin{aligned}
&\min \sum_{i\in\Omega} I_{inv} C_i \\
&\text{s.t. Cons-PFM}_k\ ,\ \text{Cons-ES}_k(C_i)\ ,\ \forall k \in [K]\ , \\
&\varepsilon d + \inf_{\gamma_i \in R} \left\{ \frac{1}{K} \sum_{k=1}^{K} \left[L\left(\hat{\xi}^k_\Sigma, P^{r,k}_{\Sigma,i}\right) + \gamma_i \right]_+ - \gamma_i \alpha \right\} \le 0\ ,\ \forall i \in \Omega\ .
\end{aligned} \tag{9.20}$$

Here $[K] = \{1, 2, \ldots, K\}$. Problem (9.20) is equivalently reformulated as the following LP:

$$\begin{aligned}
&\min \sum_{i\in\Omega} I_{inv} C_i \\
&\text{s.t. Cons-PFM}_k\ ,\ \text{Cons-ES}_k(C_i)\ ,\ \forall k \in [K]\ , \\
&\varepsilon d + \frac{1}{K} \sum_{k=1}^{K} s_i^k - \gamma_i \alpha \le 0\ ,\ \forall i \in \Omega\ , \\
&s_i^k \ge d\xi^k_{\Sigma,i} - P^{r,k}_{\Sigma,i} + \gamma_i\ ,\ \forall k \in [K]\ ,\ i \in \Omega\ , \\
&s_i^k \ge 0\ ,\ \forall k \in [K]\ ,\ i \in \Omega\ .
\end{aligned} \tag{9.21}$$

For problem (9.21), we have the following remarks.

First, in the process of reshaping the DRCC, Eq. (9.15) acts as a sufficient stipulation for (9.14), paralleled by (9.17). In effect, the LP model represented by (9.21) becomes a conservative approximation of the original DRCCP illustrated by (9.11). As a result, the optimal solution will likely demonstrate a certain level of prudence. Despite the inherent conservatism, this reformulation proves effective in addressing complex DRCCP instances.

Secondly, it is important to note that despite the inclusion of CVaR in problem (9.21), the approach proposed herein significantly differs from conventional stochastic programming (SP) methods. This divergence stems from the fact that our principal aim lies in solving the DRCCP, as denoted by (9.11), with CVaR merely functioning as an approximation tool for the DRCC.

Thirdly, the reformulation pertains specifically to subscript k due to the dependence of operational constraints Cons-PFM and Cons-ES on the renewable power $\xi_{i,t}$, as dictated by (9.5a). As the number of taken samples $\xi^k_{i,t}$ increases, it inherently leads to an increase in the count of variables and constraints. However, it is crucial to note that Eq. (9.21) represents an LP problem. Therefore, even given this increase, its computational efficiency remains satisfactory, making it especially suitable for application in planning problems.

9.4 Case Studies

9.4.1 30-Bus System

To test the proposed model and methodology, a 30-bus testing system is employed. Leveraging the IEEE 30-bus system in the MATPOWER toolbox (Zimmerman et al. 2011), a wind farm and a solar station, rated at 75 and 100 MW respectively, have been placed at bus 10 and bus 28. Correspondingly, ES1 and ES2 are also allocated to these two buses. The hourly wind power metrics are procured from the actual historical data for 2018 from Qinghai Province, China. Meanwhile, photovoltaic (PV) power data are obtained from the Global Solar Atlas for the year 2014 (World Bank Group 2023). Four representative days, each corresponding to spring, summer, autumn, and winter, are chosen to approximate a yearly cycle, thereby presenting a total timeline of 96 h. The load curve along with a sample of wind and solar power data are depicted in Fig. 9.1, in per unit value. For wind and solar power, the base value is determined by the capacity of the respective wind farm and solar station. Meanwhile, for the load, the base value refers to the rated demand for each individual bus. The parameters are shown in Table 9.1. The LP problems have been resolved using CPLEX 12.6, implemented on a laptop equipped with an Intel i5-8250U CPU and 8 GB of memory. It is noteworthy that the solution efficiency of LP is exceptionally high. In our specific instance, the computation time fluctuates based on the scenario size, which is determined by the value of K. For a scenario with $K = 100$, the average solver time is about 10 s.

In the benchmark, we employ 100 trajectories of historical wind and solar power data, therefore, K is set as 100. We set $D_{ct}^{\max}$ as 5% and use $\varepsilon = 0.015$. The optimal sizes determined for ES1 and ES2 are 71.58 MWh and 95.44 MWh, respectively. The collective capacity amounts to 167.0 MWh, leading to an investment cost near 2.5×10^8 CNY. We will then examine the influence of various factors on the optimal size of the ES.

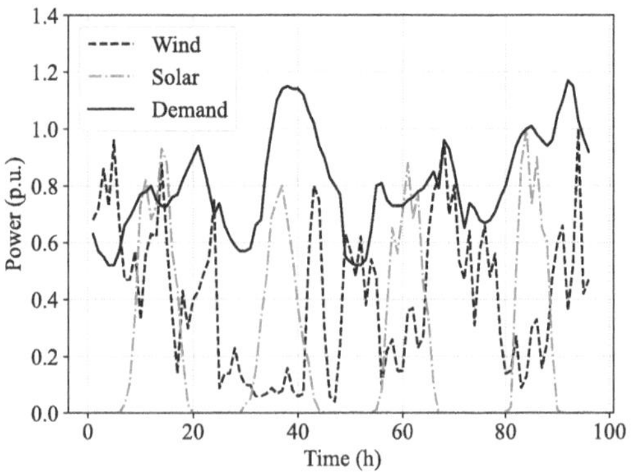

Fig. 9.1 Renewable power and demand power (Guo et al. 2020)

Table 9.1 Parameters (Guo et al. 2020)

Parameter	Value	Parameter	Value
η_{sd}	0.01	h	4 h
η_c	0.9	α	0.05
η_{dc}	0.9	β	0.05
η_{lb}	0.2	I_{inv}	1.5×10^6 CNY/MWh

To test the out-of-sample performance of the optimal sizing strategy, we calculate the curtailment rate across 500 new samples. The average wind and solar energy curtailment rates, represented by D^w_{av} and D^s_{av}, are subsequently put under analysis.

9.4.1.1 Impact of Ambiguity Set Radius

In DRO, the parameter ε is important as it dictates the size of the ambiguity set and manipulates the model's conservatism. We set $D^{\max}_{ct} = 0.05$, which translates to a maximum curtailment of 5% of renewable generation. Then we analyze the optimal size of the ES across varying values of ε and its corresponding out-of-sample performance.

The results are encapsulated in Table 9.2. It illustrates that the capacity of ESs and the total investment cost increase consistently with the expansion of ε. This phenomenon occurs because a larger ambiguity set includes more probability distributions, which can potentially increase the curtailment rate. Consequently, ESs with larger capacities are required to maintain the stipulated renewable energy curtailment rate. In instances where ε is small, the system's robustness is compromised, leading to the average solar energy curtailment rate approaching 5%. However, as ε expands, the system's robustness improves, and the observed average curtailment rates D^w_{av} and D^s_{av} maintain a certain degree of separation from the cap, providing a margin of safety.

Table 9.2 Impact of the ambiguity set radius (Guo et al. 2020)

ε	0.005	0.01	0.015	0.02	0.03
C_1 (MWh)	59.65	66.32	71.58	77.11	88.41
C_2 (MWh)	84.36	88.43	95.44	102.8	117.9
Total capacity (MWh)	144.0	154.6	167.0	179.9	206.3
Investment cost (10^8 CNY)	2.160	2.321	2.505	2.699	3.095
D^w_{av} (%)	3.16	2.98	2.92	2.39	1.85
D^s_{av} (%)	4.91	4.24	3.46	2.78	1.83

9.4.1.2 Impact of the Curtailment Rate Cap

In order to analyze the effect of $D_{ct}^{\max}$ on sizing strategies, we performed four distinct tests using varying values of $D_{ct}^{\max}$, while keeping both $\varepsilon = 0.015$ and $K = 100$ constant. The outcomes of these tests are presented in Table 9.3.

As anticipated, the capacity of the ES decreases rapidly with the increase of $D_{ct}^{\max}$. The last two rows of Table 9.3 reveal that the average wind/solar curtailment rate D_{av}^{w}/D_{av}^{s} in the sampled scenarios is significantly lower than the stipulated cap. This can be attributed to two main reasons. Firstly, the DRO method takes into account the worst-case distribution in the ambiguity set, thus necessitating a certain margin to be maintained in the chance constraint of the sampled scenario. Secondly, during the model reformulation, we rely on two sufficient conditions (9.15) and (9.17), which lead to increased robustness and conservatism.

One might conjecture that when the curtailment rate cap is sufficiently large, there is no need for an ES to facilitate the utilization of renewable power. The corresponding $D_{ct}^{\max}$ reflects the renewable accommodation capability of the power system. When considering a wider range of $D_{ct}^{\max}$, the patterns of the optimal capacity of the ES are depicted in Fig. 9.2. The exact values under which the wind farm/solar station does not require an ES are $D_{ct}^{\max} = 0.11/0.14$. These pivotal values suggest that under the current network configuration, the system can accommodate 89% of wind energy and 86% of solar energy without the aid of an ES even under the most adverse distribution.

In addition, we examine the investment cost in correlation with $D_{ct}^{\max}$, ranging from 10 to 3%. To reduce the curtailment rate by 3%, i.e., from 10 to 7%, an additional investment cost of 0.918×10^8 CNY is required. Further, two successive 2% decreases in $D_{ct}^{\max}$ incur costs of 0.998×10^8 and 1.199×10^8 CNY respectively. This trend suggests that the marginal cost progressively increases as $D_{ct}^{\max}$ decreases, indicating that a demanding requirement on curtailment rate is economically impractical and places a significant financial burden on the investor.

From the above analysis, it becomes apparent that an appropriate $D_{ct}^{\max}$ is crucial from a policy-making standpoint. Utilizing the Nash bargaining criterion (Nash 1950), the renewable energy curtailment rate and total capacity can be viewed as two conflicting objectives, each seeking to distance itself as far as possible from

Table 9.3 Impact of the curtailment cap (Guo et al. 2020)

$D_{ct}^{\max}$	3%	5%	7%	10%
C_1 (MWh)	105.8	71.58	23.76	3.550
C_2 (MWh)	141.1	95.44	76.72	35.74
Total capacity (MWh)	246.9	167.0	100.5	39.29
Investment cost (10^8 CNY)	3.704	2.505	1.507	0.589
D_{av}^{w} (%)	1.49	2.92	4.26	6.59
D_{av}^{s} (%)	0.87	3.46	5.49	8.01

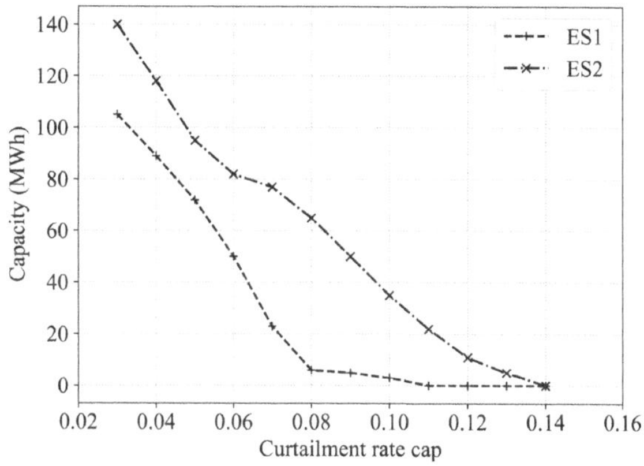

Fig. 9.2 ES capacity under different curtailment rate caps (Guo et al. 2020)

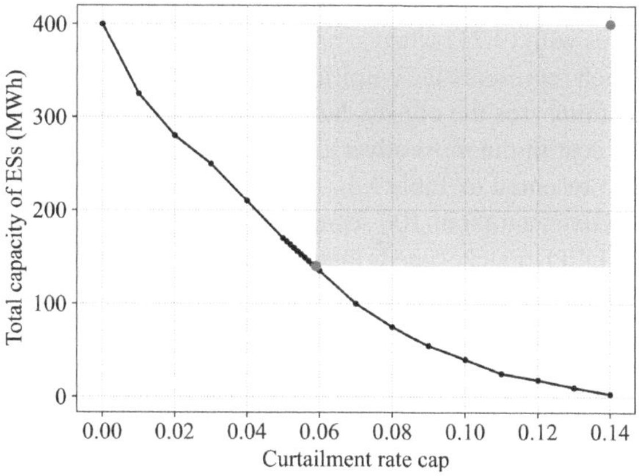

Fig. 9.3 Nash bargaining of curtailment rate and total capacity of ESs (Guo et al. 2020)

its worst value. The Nash bargaining scheme mediates between them, resulting in a fair tradeoff without the need for any manually supplied parameters that may be subjective. This is graphically illustrated in Fig. 9.3. From Fig. 9.2, it is evident that $D_{ct}^{\max} = 0.14$ corresponds to the point. Consequently, we plot the curve of total capacity, altering with $D_{ct}^{\max}$ as the Pareto frontier, adhering to the ϵ-constraint method in multi-objective optimization. The Nash Bargaining criterion aims to maximize the area of the shaded rectangle. Such a point can be discovered when $D_{ct}^{\max}$ falls within the range [0.05, 0.06]. A more detailed examination with fine discretization confirms that $D_{ct}^{\max} = 5.90\%$ at the bargaining point.

9.4.1.3 Impact of Optimization Method and Data Amount

We conducted a comparative analysis of the results derived from various optimization models and the availability of data. The models tested included the proposed DRCCP model, along with SP and robust optimization (RO). The SP model is synony-

Table 9.4 Impact of optimization method and data amount (Guo et al. 2020)

Method	Item	$K = 100$	$K = 200$	$K = 500$
DRO	$\sum_i C_i$ (MWh)	167.0	152.4	142.9
	D^w_{av}	2.92%	2.99%	3.18%
	D^s_{av}	3.46%	4.47%	4.92%
SP	$\sum_i C_i$ (MWh)	133.3	140.5	141.1
	D^w_{av}	3.86%	3.44%	3.36%
	D^s_{av}	5.39%	5.05%	5.02%
RO	$\sum_i C_i$ (MWh)	139.3	180.2	180.2
	D^w_{av}	3.56%	2.39%	2.39%
	D^s_{av}	5.15%	2.77%	2.77%

mous with (9.21) when $\varepsilon = 0$. In this case, the ambiguity set reduces to a singleton, which represents the empirical distribution. As for the RO model, the uncertainty set encompasses the convex hull of all sampled scenarios, with the aim of minimizing the cost in the most adverse scenario. The outcomes of these comparative analyses are presented in Table 9.4.

Given a total of 100 scenarios ($K = 100$), the optimal size of the ES provided by the DRO model exceeds that offered by the SP model. This exemplifies the robustness and conservatism of DRO, whose strategy effectively avoids constraint violations brought about by fluctuations in the probability distribution. On the other hand, the SP model tends to make more optimistic decisions. While it benefits from a lower investment cost, it sacrifices robustness, resulting in an average solar curtailment rate exceeding 5% in the out-of-sample test. Meanwhile, the RO model, in this instance, yields a solution with less robustness than DRO. This can be attributed to the inadequacy of the data size, which omits some extreme scenarios.

As the data size increases, the total capacity of ESs designed by DRO decreases and converges to the result of SP. Increased availability of historical data enhances confidence in the precision of the empirical distribution, leading to a reduction in the ambiguity set. Consequently, there is a decrease in conservatism. As the size of historical data tends toward infinity, the radius ε should decrease to 0, causing DRO to approach SP. This underlines the significance of data availability. For RO, pronounced robustness starts to surface as extreme scenarios are explored. In out-of-sample tests, DRO's performance is moderate. While it does not outperform all other methods, it balances cost and risk, improving robustness compared to SP and reducing conservatism in comparison with RO.

9.4.1.4 Impact of the Power Flow Model

In order to substantiate the worth of the utilized power flow model, we contrast the sizing strategies proposed by the DRCCP model with those of the linearized AC power flow model (9.4) and the conventional DC power flow model.

Firstly, to establish the validity of model (9.4), we compare it with the AC power flow model. We proceed by solving the proposed DRCCP model and recording the output curve for both active and reactive power. These outputs are then utilized to calculate the AC power flow via the Newton–Raphson method, followed by a check for any voltage violations. Throughout our tests, we did not encounter any voltage violations, thereby validating the effectiveness of model (9.4).

Next, we examine the results obtained when the DC power flow model is used in place of (9.4). The process is divided into three steps:

Step 1: Solve a DRCCP problem employing the DC power flow model.

Step 2: Implement a resolution for the initial DRCCP, utilizing the active power nodal injection acquired from Step 1 as parameter. Retain the voltage magnitude, voltage angle, and reactive power as decision variables. Meanwhile, introduce flexibility into the upper and lower bounds on voltage.

Step 3: Evaluate the voltage magnitude outcomes obtained from Step 2. Should any voltage magnitude exceed or fall short of the range outlined by the original constraints, it is identified as a voltage violation.

The quantity of hours featuring voltage violations at each bus in a given scenario is shown in Table 9.5. We observed voltage violations at three separate buses, most notably at bus 7 and bus 8. Within the IEEE 30-bus system, bus 8 is directly linked to bus 28, the connection point for the solar station. Bus 7 is near bus 28, and buses 7, 6, and 28 are interconnected via a transmission pathway. From an intuitive standpoint, it can be expected that during certain periods, a specific volume of renewable energy is not dispatchable due to constraints on reactive power and voltage. Such energy should ideally be stored in the ES to decrease the curtailment rate. As a result, the size of the ES is increased when the power flow model (9.4) is utilized, as demonstrated in Table 9.6. By and large, it is essential to consider the impacts of reactive power and voltage incorporated in the model (9.4), as discussed in this chapter.

Table 9.5 Voltage violation (Guo et al. 2020)

Bus	Bus 7	Bus 8	Bus 19	Other buses
Number of violations	11	15	2	0
Violation rate (%)	11.46	15.63	2.08	0

Table 9.6 Impact of power flow model (Guo et al. 2020)

Power flow model	Linearized AC power flow	DC power flow
C_1 (MWh)	71.58	54.23
C_2 (MWh)	95.44	89.10
Total capacity (MWh)	167.0	143.3

9.4.1.5 Analysis of Generation Cost

In Sect. 9.2, we delve into a qualitative discussion regarding the rationale behind the assumption of excluding the generation cost of thermal units in (9.8), and similarly in (9.11). To further substantiate the validity of this assumption, we conduct comparative analyses between model (9.11) and a subsequent model that does take into account the generation cost:

$$\begin{aligned} \min \; & \sum_{i \in \Omega} I_{inv} C_i + \delta \sum_{j \in G, t \in \mathbb{T}} f(P^g_{i,t}) \\ \text{s.t.} \; & \text{Cons-PFM} \; , \; \text{Cons-ES} \; , \\ & \inf_{P \in \mathcal{M}^\varepsilon_i} \Pr\{D^i_{ct} \leq D^{\max}_{ct}\} \geq 1 - \alpha \; , \; \forall i \in \Omega \; . \end{aligned} \tag{9.22}$$

In the given equation, the term $\delta = 3650/4$ signifies the ratio of the ES lifespan to the optimization horizon. The cost of generation is represented by the function $f(P^g_{i,t}) = a_i + b_i P^g_{i,t}, \forall i \in G$, wherein G symbolizes the set of nodes equipped with generators. The parameters a_i and b_i are established at 1,533 CNY and 117 CNY/MWh respectively (Wu et al. 2007).

Upon the resolution of (9.11) and (9.22), utilizing $D^{\max}_{ct} = 0.05$ and $\varepsilon = 0.015$ within a 30-bus system, the corresponding investment and generation costs (averaged across scenarios) are recorded. The derived result, as encapsulated in Table 9.7, indicates that the investment strategy for ESs remains unaltered despite considerations of generation cost. Moreover, the generation cost derived from (9.22) is merely 0.07% lower than the cost derived from (9.11). These findings verify the validity of disregarding generation cost as a reasonable assumption in this context.

Table 9.7 Comparison of (9.11) and (9.22) (Guo et al. 2020)

Model	Investment cost (CNY)	Generation cost (CNY)
Equation (9.11)	2.505×10^8	5.486×10^8
Equation (9.22)	2.505×10^8	5.482×10^8

Table 9.8 Result of 118-bus system (Guo et al. 2020)

D_{ct}^{max}	Capacity	Penetration level			
		80%	100%	120%	150%
0.03	C_1 (MWh)	165.6	216.3	250.6	316.9
	C_2 (MWh)	138.0	164.5	208.9	264.1
0.05	C_1 (MWh)	114.9	130.5	173.7	217.9
	C_2 (MWh)	95.72	134.0	144.8	181.6
0.07	C_1 (MWh)	64.08	80.88	98.10	132.6
	C_2 (MWh)	53.40	67.40	81.75	110.5

9.4.2 118-Bus System

A modified IEEE 118-bus system is employed to assess the scalability of this model. The system sees an average demand of 4,242 MW, with a 180-MW solar plant and a 150-MW wind farm situated at bus-7 and bus-82, respectively. The penetration level, in this context, is defined as the ratio of the REG's output versus that in the base case.

Taking into account the varying rate caps of curtailment, D_{ct}^{max}, and the penetration levels of REG, we can compute the optimal sizing strategy for ESs by solving the LP in (9.21). These results are concisely summarized in Table 9.8. When the penetration level is held constant, it is observed that the capacity of ESs tends to increase as D_{ct}^{max} decreases, which is a pattern that is consistent with observations from the 30-bus system test. Moreover, as the level of renewable power continues to rise, there is a clear indication for the requirement of additional capacity.

Owing to the expansion of the system size, the scale of (9.21) in the 118-bus system is significantly larger. For instance, with $K = 100$, the average solution time is approximately 100 s, which is an order of magnitude longer than that of the 30-bus system. However, such efficiency is still satisfactory for a planning problem, largely due to the linearity of (9.21).

9.5 Conclusion

This chapter tackles the challenge of ES sizing in bulk power systems employing a DRO approach. The principal discoveries can be summarized as follows:

Firstly, a distributionally robust model for optimally sizing ES has been established. This model strives to ensure a DRCC on the renewable energy curtailment rate, while simultaneously minimizing investment costs. Furthermore, a linear network model incorporating reactive power and voltage has been utilized, which more accurately captures the operating conditions of the power system.

Secondly, the uncertainty is characterized by an approximate empirical distribution, with an ambiguity set defined by the Wasserstein distance. The DRCC is then reinterpreted as a series of linear constraints, leveraging the relationship between chance constraints and risk measures. Consequently, the proposed model is transformed into a tractable LP.

Lastly, comprehensive case studies performed on the test system have explicitly showcased the efficacy of the suggested method, spanning a wide range of renewable energy curtailment rate caps and data volumes. It is worth noting that the method's conservatism may be attributed to the approximation scheme used in the reformulation of the DRCC. Through the lens of policy-making and in alignment with the Nash bargaining criterion, an appropriate cap falls between 5 and 6%.

References

Dui X, Zhu G, Yao L (2018) Two-stage optimization of battery energy storage capacity to decrease wind power curtailment in grid-connected wind farms. IEEE Trans Power Syst 33(3):3296–3305

Guo Z, Wei W, Chen L et al (2020) Sizing energy storage to reduce renewable power curtailment considering network power flows: a distributionally robust optimisation approach. IET Renew Power Gener 14(16):3273–3280

Hota AR, Cherukuri A, Lygeros J (2019) Data-driven chance constrained optimization under Wasserstein ambiguity sets. In: 2019 American control conference (ACC), pp 1501–1506

Nash JF (1950) The bargaining problem. Econometrica 18(2):155–162

World Bank Group (2023) Global solar atlas. https://globalsolaratlas.info/[2023-12-28]

Wu L, Shahidehpour M, Li T (2007) Stochastic security-constrained unit commitment. IEEE Trans Power Syst 22(2):800–811

Yang Z, Zhong H, Bose A et al (2018) A linearized OPF model with reactive power and voltage magnitude: a pathway to improve the MW-only DC OPF. IEEE Trans Power Syst 33(2):1734–1745

Zimmerman RD, Murillo-Sánchez CE, Thomas RJ (2011) MATPOWER: steady-state operations, planning, and analysis tools for power systems research and education. IEEE Trans Power Syst 26(1):12–19

Chapter 10
Sizing Renewable, Transmission, and Energy Storage in Low-Carbon Power Systems

Abstract This chapter studies the optimal sizing of renewable, transmission, and energy storage capacities in a low-carbon electricity grid. Data-driven ambiguity sets based on Wasserstein distance are used for modeling renewable energy generation and load demand uncertainties. A distributionally robust bi-objective sizing model is proposed that minimizes the total investment cost and the worst-case expectation of carbon emissions in normal situations while considering the distributionally robust risk constraint of load shedding in extreme situations. Using a method based on the Lipschitz constant, worst-case carbon emission expectation and distributionally robust risk of load shedding are transformed into linear models. This makes the proposed sizing model a matter of solving mixed-integer linear programming problems. This chapter introduces the work in Xie et al. (Energy 263:125653, 2023), (2023).

10.1 Introduction

The energy crisis and climate change have caused widespread concern around the world, and many countries and regions have developed clear plans to achieve carbon neutrality in the next few decades (Weng et al. 2021). Utilizing emerging renewable energy generation technologies, represented by photovoltaic (PV) and wind power generation, the power system can reduce carbon emissions during power generation by integrating more renewable energy sources (Chaube et al. 2021).

The real-time output of PV and wind power generation depends on weather conditions. Hence, it is fluctuating and random (Dong et al. 2022). When power generation and load are mismatched, sufficient reserves are needed to compensate for the unbalanced power. In a low-carbon power grid with a high penetration of renewable energy, the proportion of dispatchable units is relatively low, and the uncertainty is stronger, making it necessary to use energy storage (ES) to help peak shaving and maintain power balance. Pumped storage is the most technologically mature and widely used, but its use largely depends on geographical conditions. Electrochemical ES has attracted increasing attention in recent years, but the cost of electrochemical ES is still relatively high. To achieve a reasonable balance between investment cost,

R. Xie and W. Wei, *Distributionally Robust Optimization and its Applications in Power System Energy Storage Sizing*,
https://doi.org/10.1007/978-981-97-2566-3_10

carbon emissions, and system reliability, it is very important to optimize the capacity configuration of all parts of the system.

This chapter considers the sizing problem for a low-carbon power grid. At a certain future moment, a part of the conventional thermal power units will be retired, and in order to reduce carbon emissions, there will be no further investment in new fossil energy generation. At the same time, the growth of the load requires renewable energy generation to supply clean electricity. The changing generation and growing load trigger the need for transmission network expansion and also require the configuration of ES to maintain the power balance and compensate for the mismatch between renewable energy output and load curve. Based on the expected growth of the load, the capacity of renewable energy generation is determined under the premise of the given site selection, along with the quantity of transmission line extensions and ES capacity.

The innovations of this chapter are in the following two aspects.

First, a bi-objective distributionally robust optimization (DRO) model for a low-carbon power grid is proposed, which simultaneously optimizes the capacities of renewable energy generation, transmission lines, and ES. The uncertainty of renewable energy generation and load is modeled using probability distribution ambiguity sets based on the Wasserstein distance. Under the requirement of acceptable distributionally robust Shortfall risk in extreme situations, the total investment cost and the worst-case expectation of carbon emissions in conventional situations are balanced.

Second, a solving algorithm is proposed to calculate the Pareto frontier of the proposed bi-objective DRO model. The distributionally robust Shortfall risk is equivalently transformed into the worst-case expectation, and then by finding the Lipschitz constants of the load-shedding function and carbon emission function, the worst-case expectations in the load-shedding constraint and objective function are calculated, respectively. Finally, the optimal sizing problem is reduced to solving a bi-objective mixed-integer linear programming (MILP) problem. Using the ϵ-constraint method, the Pareto frontier is obtained.

In this chapter, we first establish the mathematical model for the joint optimal sizing problem in Sect. 10.2, then propose a solution algorithm in Sect. 10.3, perform case analysis in Sect. 10.4, and finally draw conclusions in Sect. 10.5.

10.2 Mathematical Model

10.2.1 Power System Model

Represent the set of buses in the power system as S_B and the set of time periods as $S_T = \{1, 2, \ldots, T\}$. For the conventional thermal power unit at bus i, the power generation $p_{i,t}^G$ during the t-th time period satisfies the following constraints:

$$P_i^{Gl} \le p_{i,t}^G \le P_i^{Gu} \ , \ \forall i \in S_B \ , \ t \in S_T \ , \tag{10.1a}$$

$$R_i^{Gl} \Delta_t \le p_{i,t}^G - p_{i,t-1}^G \le R_i^{Gu} \Delta_t \ , \ \forall i \in S_B \ , \ t = 2, 3, \dots T \ , \tag{10.1b}$$

$$c_{i,t}^G = F_i^G p_{i,t}^G \Delta_t \ , \ \forall i \in S_B \ , \ t \in S_T \ , \tag{10.1c}$$

where Eq. (10.1a) represents the power generation capacity constraint. P_i^{Gl} and P_i^{Gu} are the lower and upper bounds of power generation respectively. Δ_t is the length of time period, so Eq. (10.1b) restricts the change in generator output between two adjacent periods, which are ramp-up and ramp-down constraints. R_i^{Gl} and R_i^{Gu} are the limits of the ramp rate. F_i^G is the carbon emission coefficient. Therefore, Eq. (10.1c) computes the carbon emission $c_{i,t}^G$.

Let x_i^R represent the capacity of renewable power generation connected at bus i, and let $\xi_{i,t}^R$ represent the maximum output of renewable power generation per unit capacity in the t-th time period. $\xi_{i,t}^R$ is a random variable taking values within [0, 1]. Therefore, the renewable power generating power $p_{i,t}^R$ is:

$$p_{i,t}^R = x_i^R \xi_{i,t}^R \ , \ \forall i \in S_B \ , \ t \in S_T \ . \tag{10.2}$$

The operating constraint of ES includes:

$$0 \le p_{i,t}^{Sc} \le x_i^{Sp} \ , \ 0 \le p_{i,t}^{Sd} \le x_i^{Sp} \ , \ \forall i \in S_B \ , \ t \in S_T \ , \tag{10.3a}$$

$$e_{i,t}^S = e_{i,t-1}^S + \left(\eta^{Sc} p_{i,t}^{Sc} - \frac{p_{i,t}^{Sd}}{\eta^{Sd}} \right) \Delta_t \ , \ \forall i \in S_B \ , \ t \in S_T \ , \tag{10.3b}$$

$$\omega^{Sl} x_i^{Se} \le e_{i,t}^S \le \omega^{Su} x_i^{Se} \ , \ \forall i \in S_B \ , \ t \in S_T \ , \tag{10.3c}$$

$$e_{i,T}^S = e_{i,0}^S \ , \ \forall i \in S_B \ , \tag{10.3d}$$

$$p_{i,t}^{Sc} p_{i,t}^{Sd} = 0 \ , \ \forall i \in S_B \ , \ t \in S_T \ , \tag{10.3e}$$

where x_i^{Sp} represents the power capacity of ES at bus i, so Eq. (10.3a) restricts the charging power $p_{i,t}^{Sc}$ and discharging power $p_{i,t}^{Sd}$ of ES; $e_{i,t}^S$ represents the electric energy stored at the end of the t-th period, $e_{i,0}^S$ is the initial stored energy, η^{Sc} and η^{Sd} represent the charging and discharging efficiency of ES, respectively, so Eq. (10.3b) is the dynamic model of ES state-of-charge (SoC); x_i^{Se} represents the energy capacity of the ES, ω^{Sl} and ω^{Su} represent the lower and upper bounds of the SoC of ES, respectively, so Eq. (10.3c) is the constraint on the feasible range of SoC; Eq. (10.3d) requires that the initial and final states of ES be the same for continuous operation; Eq. (10.3e) is the complementary constraint of charging and discharging power prohibiting ES from charging and discharging at the same time, and since it contains the product of operation variables, it is nonlinear.

The direct current (DC) power flow model is used for power grid modeling. The set of branches is denoted as S_L. In the t-th time period, the active power flow in a transmission line from bus i to bus j is denoted as $p_{ij,t}^L$, and the voltage angle at bus i is denoted as $\theta_{i,t}^N$. Let X_{ij}^L be the line reactance. Then the DC power flow model is:

$$p_{ij,t}^L = \frac{\theta_{i,t}^N - \theta_{j,t}^N}{X_{ij}^L} \ , \ \forall (i,j) \in S_L \ , \ t \in S_T \ , \tag{10.4a}$$

$$- S_{ij}^L \le p_{ij,t}^L \le S_{ij}^L \ , \ \forall (i,j) \in S_L \ , \ t \in S_T \ , \tag{10.4b}$$

where Eq. (10.4a) represents the active power flow in this transmission line; S_{ij}^L is the capacity of a transmission line, so Eq. (10.4b) is the transmission line capacity.

In power grid expansion, it is assumed that the new transmission line has the same parameters as the existing transmission line. The number of existing and new transmission lines from bus i to j are denoted as N_{ij}^L and n_{ij}^L, respectively. Since every transmission line from i to j has the same parameters, they have the same power flow, so the active power from i to j is $(N_{ij}^L + n_{ij}^L) p_{ij,t}^L$.

Let $p_{i,t}^A$ represent the abandoned power at bus i during the t-th time period, then $p_{i,t}^A$ is nonnegative, i.e.,

$$p_{i,t}^A \ge 0 \ , \ \forall i \in S_B \ , \ t \in S_T \ . \tag{10.5}$$

Let $\xi_{i,t}^D$ represent the load power at bus i during the t-th time period, and $p_{i,t}^D$ represent the load shedding power, then $p_{i,t}^D$ is nonnegative and does not exceed the load power, i.e.,

$$0 \le p_{i,t}^D \le \xi_{i,t}^D \ , \ \forall i \in S_B \ , \ t \in S_T \ . \tag{10.6}$$

The nodal power balance equation is as follows:

$$\begin{aligned} &\sum_{(k,i) \in S_L} (N_{ki}^L + n_{ki}^L) p_{ki,t}^L + p_{i,t}^G + p_{i,t}^R + p_{i,t}^{Sd} - p_{i,t}^A \\ = &\sum_{(i,j) \in S_L} (N_{ij}^L + n_{ij}^L) p_{ij,t}^L + p_{i,t}^{Sc} + \xi_{i,t}^D - p_{i,t}^D \ , \ \forall i \in S_B \ , \ t \in S_T \ . \end{aligned} \tag{10.7}$$

Since n_{ki}^L, n_{ij}^L, $p_{ki,t}^L$, $p_{ij,t}^L$ are all variables, Eq. (10.7) is a nonlinear constraint.

10.2.2 Operation Optimization Model

Two types of operational optimization problems are considered: The first type minimizes the load shedding in extreme situations and will be used in the load shedding risk constraint. The second type minimizes carbon emissions in normal situations, does not allow load shedding, and will appear in the objective function of the optimal sizing model.

10.2.2.1 Operation Optimization Model to Minimize Load Shedding

One of the most fundamental requirements of the power system is to meet the load demand. Under extreme weather conditions, load shedding in a power system with a high proportion of renewable energy may be unavoidable. In this case, the purpose of operational optimization is to minimize load shedding. Since extreme weather conditions are rare and last for a short time, their impact on carbon emissions can be ignored. The operational optimization model for minimizing load shedding under extreme conditions is established as follows:

$$\begin{aligned} g^E(x,\xi) := \min & \sum_{t\in S_T}\sum_{i\in S_B} p_{i,t}^D \Delta_t \\ \text{s.t. } & (10.1)\text{–}(10.7)\,. \end{aligned} \tag{10.8}$$

In the optimization problem of (10.8), the capacity variable

$$x := (x_i^R, x_i^{Sp}, x_i^{Se}, n_{ij}^L; i \in S_B, (i,j) \in S_L) \tag{10.9}$$

and the random variable $\xi := (\xi_{i,t}^R, \xi_{i,t}^D; i \in S_B, t \in S_T)$ are fixed, while the decision variables include $p_{i,t}^G$, $c_{i,t}^G$, $p_{i,t}^A$, $p_{i,t}^{Sc}$, $p_{i,t}^{Sd}$, $e_{i,t}^S$, $p_{ij,t}^L$, $\theta_{i,t}^N$, and $p_{i,t}^D$. The optimization goal is to minimize the total load shedding, and its optimal value depends on x and ξ, so it is denoted as function $g^E(x,\xi)$, which represents the minimum load shedding under given x and ξ.

10.2.2.2 Operation Optimization Model to Minimize Carbon Emissions

Under regular conditions, load shedding is not allowed, and the goal is to minimize carbon emissions. In this case, the operation optimization model is as follows:

$$\begin{aligned} g^N(x,\xi) := \min & \sum_{t\in S_T}\sum_{i\in S_B} c_{i,t}^G \\ \text{s.t. } & p_{i,t}^D = 0\,,\ \forall i \in S_B\,,\ t \in S_T\,, \\ & (10.1)\text{–}(10.5)\,,\ (10.7)\,. \end{aligned} \tag{10.10}$$

Similar to the operation optimization model that minimizes load shedding, the minimum carbon emissions $g^N(x,\xi)$ is the optimal value of the optimization problem in (10.10), so it is a function of x and ξ. Because the operating cost of renewable energy generation is usually very low, it is not considered in a low-carbon electricity grid.

10.2.3 Ambiguity Set of Probability Distribution

Create ambiguity sets for extreme and regular scenarios, respectively.

10.2.3.1 Ambiguity Set in Extreme Cases

Let S_E be the index set of the sample data for extreme scenarios. For any $n \in S_E$, ξ_n^E is a sample of the random variable ξ, representing an extreme scenario, where ξ includes the power generation from renewable sources and the load power. Based on the sample data, the empirical probability distribution $\mathbb{P}^E$ for extreme scenarios is established as follows:

$$\mathbb{P}^E = \frac{1}{|S_E|} \sum_{n \in S_E} \mathbf{1}_{\xi_n^E} . \tag{10.11}$$

The extreme scenario ambiguity set $\mathcal{B}(\mathbb{P}^E, \varepsilon^E)$ contains the probability distributions whose Wasserstein 1-distance under 1-norm to the empirical probability distribution $\mathbb{P}^E$ does not exceed ε^E, that is:

$$\mathcal{B}(\mathbb{P}^E, \varepsilon^E) := \{\mathbb{P} \in \mathcal{M}_1(\Xi^E) \mid d_W(\mathbb{P}, \mathbb{P}^E) \leq \varepsilon^E\} , \tag{10.12}$$

where Ξ^E is

$$\Xi^E := \{\xi \mid 0 \leq \xi_{i,t}^R \leq 1, 0 \leq \xi_{i,t}^D \leq \xi_i^{Du}, \forall i \in S_B, t \in S_T\} . \tag{10.13}$$

ξ_i^{Du} is the possible maximum value of the load at bus i. The set Ξ^E contains all possible values of ξ, including various extreme scenarios.

10.2.3.2 Ambiguity Set in Regular Cases

Let S_N indicate the index set of the sampled data in regular situations. $\xi_n^N, n \in S_N$ are the sampled data in regular situations. The empirical probability distribution $\mathbb{P}^N$ and the ambiguity set $\mathcal{B}(\mathbb{P}^N, \varepsilon^N)$ are defined similarly to the extreme case, but the definition of the support set Ξ^N is as follows:

$$\Xi^N := \left\{ \sum_{n \in S_N} y_n \xi_n^N \;\middle|\; \begin{array}{l} y_n \geq 0, \forall n \in S_N \\ \sum_{n \in S_N} y_n = 1 \end{array} \right\} . \tag{10.14}$$

Ξ^N is the convex hull of the regular situation sampling data set $\{\xi_n^N \mid n \in S_N\}$, which is to limit the range of Ξ^N so that it only includes regular situations.

10.2.4 Sizing Model

The Shortfall risk is used to measure the risk of load shedding. According to the definition of Shortfall risk in Definition 2.41, the Shortfall risk of load shedding under the probability distribution $\mathbb{P}$ is

$$\mathrm{SR}^{\mathbb{P}}_{l,\lambda}(g^E(x,\xi)) = \inf\{y \in \mathbb{R} \mid \mathbb{E}^{\mathbb{P}}[l(g^E(x,\xi) - y)] \leq \lambda\} . \tag{10.15}$$

where λ is the acceptable load-shedding amount in extreme situations.

Because the load shedding Shortfall risk is related to the probability distribution, it is required that the load shedding risk is sufficiently small for all probability distributions in the ambiguity set, thus the constraint for the load shedding distributionally robust Shortfall risk is as follows:

$$\sup_{\mathbb{P}\in\mathcal{B}(\mathbb{P}^E,\varepsilon^E)} \mathrm{SR}^{\mathbb{P}}_{l,\lambda}(g^E(x,\xi)) \leq R^E , \tag{10.16}$$

where R^E is the upper limit of load shedding risk, and function l is selected as $l(y) = \max\{y, 0\}$, to measure the loss when load shedding occurs. Equation (10.16) sets an upper limit for the risk of load shedding under the worst-case probability distribution in extreme situations. Similarly, the worst-case expectation of carbon emissions in regular situations is:

$$\sup_{\mathbb{P}\in\mathcal{B}(\mathbb{P}^N,\varepsilon^N)} \mathbb{E}^{\mathbb{P}}[g^N(x,\xi)] . \tag{10.17}$$

The total investment cost of renewable energy generation equipment is:

$$f(x) = \sum_{i\in S_B}(C_i^R x_i^R + C^{Sp} x_i^{Sp} + C^{Se} x_i^{Se}) + \sum_{(i,j)\in S_L} C_{ij}^L n_{ij}^L , \tag{10.18}$$

where C_i^R, C^{Sp}, C^{Se}, and C_{ij}^L are the unit capacity cost of renewable energy generation at bus i, the unit power capacity cost of ES, the unit energy capacity cost of ES, and the cost of each transmission line from bus i to j, respectively.

Considering the limitations of natural resources and construction locations, the capacity variables should all be bounded, so we have the following constraints:

$$\begin{cases} 0 \leq x_i^R \leq X_i^{Ru} , \ 0 \leq x_i^{Sp} \leq X_i^{Spu} , \ 0 \leq x_i^{Se} \leq X_i^{Seu} , \ \forall i \in S_B , \\ 0 \leq n_{ij}^L \leq N_{ij}^{Lu} , \ n_{ij}^L \in \mathbb{Z} , \ \forall (i,j) \in S_L , \end{cases} \tag{10.19}$$

where X_i^{Ru}, X_i^{Spu}, X_i^{Seu}, and N_{ij}^{Lu} represent the upper limit of renewable energy generation capacity at bus i, the upper limit of ES power capacity, the upper limit of ES energy capacity, and the upper limit of the number of new transmission lines from bus i to bus j, respectively.

Finally, we establish the sizing problem as follows:

$$\begin{aligned}
&\min_{x} \left\{ f(x), \sup_{\mathbb{P}\in\mathcal{B}(\mathbb{P}^N,\varepsilon^N)} \mathbb{E}^{\mathbb{P}}[g^N(x,\xi)] \right\} \\
&\text{s.t.} \quad \sup_{\mathbb{P}\in\mathcal{B}(\mathbb{P}^E,\varepsilon^E)} \mathrm{SR}^{\mathbb{P}}_{l,\lambda}(g^E(x,\xi)) \le R^E \ , \ (10.19) \ .
\end{aligned} \tag{10.20}$$

The sizing problem in (10.20) minimizes the total investment cost and the worst-case expectation of carbon emissions in regular situations under the distributionally robust risk constraint of load shedding in extreme situations. After all equipment has been built, the total investment cost is fixed, so the first objective function is deterministic. The operation of the system is affected by the uncertainty of renewable energy generation and load and is considered in the optimal value functions $g^N(x,\xi)$ and $g^E(x,\xi)$ defined in operation optimization. The sizing problem in (10.20) is a bi-objective distributionally robust mixed-integer nonlinear programming problem, so it cannot be solved directly.

10.3 Solution Method

In order to solve the sizing problem in (10.20), we first linearize the nonlinear constraints in it, then use the method based on the Lipschitz constant to transform the distributionally robust Shortfall risk and the worst-case expectation, and finally, the solution algorithm is given using the ϵ-constraint method.

10.3.1 Transforming Nonlinear Constraints

The nonlinear constraints in the sizing model come from two aspects: the product term $n^L_{ij} p^L_{ij,t}$ in the node power balance equation in (10.7) and the complementary constraints of ES charging and discharging power in (10.3e). In this section, we linearize $n^L_{ij} p^L_{ij,t}$ through Boolean auxiliary variables and the big-M method, and relax the constraints in (10.3e).

10.3.1.1 Linearizing the Product Term in the Node Power Balance Equation

Noting that $0 \le n^L_{ij} \le N^{Lu}_{ij}$ and $n^L_{ij} \in \mathbb{Z}$, the variable n^L_{ij} can be decomposed as follows through the Boolean auxiliary variable $z^L_{ij,m}$:

$$\begin{cases} n_{ij}^L = \sum_{m=0}^{M^L} 2^m z_{ij,m}^L \ , \\ z_{ij,m}^L \in \{0, 1\} \ , \ \forall (i, j) \in S_L \ , \ m = 0, 1, \ldots, M^L \ . \end{cases} \tag{10.21}$$

This can be understood as representing n_{ij}^L in binary, and M^L is an integer selected based on N_{ij}^{Lu}, satisfying $N_{ij}^{Lu} < 2^{M^L+1}$. Further, we have:

$$n_{ij}^L p_{ij,t}^L = \sum_{m=0}^{M^L} 2^m p_{ij,t,m}^L \ , \tag{10.22}$$

and the right side is linear with respect to $p_{ij,t,m}^L$. The nonlinear relationship $p_{ij,t,m}^L = z_{ij,m}^L p_{ij,t}^L$ is equivalent to the following linear constraint group:

$$\begin{cases} -M^P z_{ij,m}^L \leq p_{ij,t,m}^L \leq M^P z_{ij,m}^L \ , \\ -M^P (1 - z_{ij,m}^L) \leq p_{ij,t,m}^L - p_{ij,t}^L \leq M^P (1 - z_{ij,m}^L) \ , \end{cases} \tag{10.23}$$

where M^P satisfies $M^P \geq S_{ij}^L$, is a sufficiently large constant. When $z_{ij,m}^L = 1$, according to the second row of (10.23), $p_{ij,t,m}^L = p_{ij,t}^L$ and the first-row constraint does not work; when $z_{ij,m}^L = 0$, according to the first row, $p_{ij,t,m}^L = 0$ and the second-row constraint does not work. Therefore, $p_{ij,t,m}^L = z_{ij,m}^L p_{ij,t}^L$ is equivalent to (10.23).

Therefore, after adding the constraints of (10.21) and (10.23), the product term $n_{ij}^L p_{ij,t}^L$ in formula (10.7) can be equivalently replaced with $\sum_{m=0}^{M^L} 2^m p_{ij,t,m}^L$, that is, the node power balance equation is transformed into the linear constraint:

$$\begin{aligned} &\sum_{(k,i) \in S_L} \left(N_{ki}^L p_{ki,t}^L + \sum_{m=0}^{M^L} 2^m p_{ki,t,m}^L \right) + p_{i,t}^G + p_{i,t}^R + p_{i,t}^{Sd} - p_{i,t}^A \\ &= \sum_{(i,j) \in S_L} \left(N_{ij}^L p_{ij,t}^L + \sum_{m=0}^{M^L} 2^m p_{ij,t,m}^L \right) + p_{i,t}^{Sc} + \xi_{i,t}^D - p_{i,t}^D \ , \ \forall i \in S_B \ , \ t \in S_T \ . \end{aligned} \tag{10.24}$$

Since the binary representation of nonnegative integers is unique, after fixing M^L, n_{ij}^L and $(z_{ij,m}^L; m = 0, 1, \ldots, M^L)$ uniquely determine each other, so there is no harm in adding $z_{ij,m}^L$ to the capacity variable x. According to (10.19), the value range of x is:

$$X := \left\{ \begin{array}{l} x = (x_i^R, x_i^{Sp}, x_i^{Se}, n_{ij}^L, z_{ij,m}^L; i \in S_B, (i,j) \in S_L, m = 0, 1, \ldots, M^L) : \\ 0 \le x_i^R \le X_i^{Ru}, 0 \le x_i^{Sp} \le X_i^{Spu}, 0 \le x_i^{Se} \le X_i^{Seu}, \forall i \in S_B, \\ 0 \le n_{ij}^L \le N_{ij}^{Lu}, n_{ij}^L = \sum_{m=0}^{M^L} 2^m z_{ij,m}^L, \forall (i,j) \in S_L, \\ z_{ij,m}^L \in \{0, 1\}, \forall (i,j) \in S_L, m = 0, 1, \ldots, M^L \end{array} \right\}. \tag{10.25}$$

According to the above analysis, replacing the node power balance constraint equivalently with the form in (10.24) in the two operational optimization models, we obtain $g^E(x, \xi)$ and $g^N(x, \xi)$ as follows:

$$\begin{aligned} g^E(x, \xi) = \min & \sum_{t \in S_T} \sum_{i \in S_B} p_{i,t}^D \Delta_t \\ \text{s.t. } & (10.1)\text{–}(10.6)\ ,\ (10.23)\ ,\ (10.24)\ . \end{aligned} \tag{10.26}$$

$$\begin{aligned} g^N(x, \xi) = \min & \sum_{t \in S_T} \sum_{i \in S_B} c_{i,t}^G \\ \text{s.t. } & p_{i,t}^D = 0\ ,\ \forall i \in S_B\ ,\ t \in S_T\ , \\ & (10.1)\text{–}(10.5)\ ,\ (10.23)\ ,\ (10.24)\ . \end{aligned} \tag{10.27}$$

The nonlinearity in the optimization problems in (10.26) and (10.27) only exists in the complementary constraint of ES charging and discharging power.

10.3.1.2 Relaxation of Complementary Constraints on Energy Storage Charging and Discharging Power

If we remove the constraint of (10.3e) in (10.26) and (10.27), they become linear programming (LP) problems.

$$g_l^E(x, \xi) := \min \sum_{t \in S_T} \sum_{i \in S_B} p_{i,t}^D \Delta_t \tag{10.28a}$$

$$\text{s.t. } (10.1)\text{–}(10.3\text{d})\ ,\ (10.4)\text{–}(10.6)\ , \tag{10.28b}$$

$$(10.23)\ ,\ (10.24)\ . \tag{10.28c}$$

$$g_l^N(x, \xi) := \min \sum_{t \in S_T} \sum_{i \in S_B} c_{i,t}^G \tag{10.29a}$$

$$\text{s.t. } p_{i,t}^D = 0\ ,\ \forall i \in S_B\ ,\ t \in S_T\ ,\ (10.1)\text{–}(10.3\text{d})\ , \tag{10.29b}$$

$$(10.4\text{a})\text{–}(10.5)\ ,\ (10.23)\ ,\ (10.24)\ . \tag{10.29c}$$

Proposition 10.1 *For any $x \in X$ and $\xi \in \Xi^E$, we have $g_l^E(x,\xi) = g^E(x,\xi)$.*

Proof The LP problem defining $g_l^E(x,\xi)$ is obtained from (10.26) by removing the constraint of (10.3e), so $g_l^E(x,\xi) \le g^E(x,\xi)$. We only need to prove $g_l^E(x,\xi) \ge g^E(x,\xi)$. According to (10.6), there are $p_{i,t}^D \ge 0, \forall i \in S_B, t \in S_T$, so $0 \le g_l^E(x,\xi) \le g^E(x,\xi)$. If $g_l^E(x,\xi) = +\infty$, $g^E(x,\xi) = g_l^E(x,\xi)$ holds. Therefore, we only need to consider the case where $g_l^E(x,\xi)$ is finite. Assume that one optimal solution of (10.28) is $(\hat{p}_{i,t}^G, \hat{c}_{i,t}^G, \hat{p}_{i,t}^{Sc}, \hat{p}_{i,t}^{Sd}, \hat{e}_{i,t}^S, \hat{p}_{ij,t}^L, \hat{p}_{i,t}^N, \hat{p}_{i,t}^A, \hat{p}_{i,t}^D, \hat{p}_{ij,t,m}^L)$. We will start from this solution and construct a feasible solution for (10.26) while keeping the objective function value equal to $g_l^E(x,\xi)$.

If $(\hat{p}_{i,t}^G, \hat{c}_{i,t}^G, \hat{p}_{i,t}^{Sc}, \hat{p}_{i,t}^{Sd}, \hat{e}_{i,t}^S, \hat{p}_{ij,t}^L, \hat{\theta}_{i,t}^N, \hat{p}_{i,t}^A, \hat{p}_{i,t}^D, \hat{p}_{ij,t,m}^L)$ is not feasible in (10.26), then only (10.3e) does not hold. There exist t_0 and i_0 such that $\hat{p}_{i_0,t_0}^{Sc}\hat{p}_{i_0,t_0}^{Sd} \ne 0$, so $\hat{p}_{i_0,t_0}^{Sc} > 0$ and $\hat{p}_{i_0,t_0}^{Sd} > 0$. For $\varepsilon > 0$, construct a new solution

$$(\tilde{p}_{i,t}^G, \tilde{c}_{i,t}^G, \tilde{p}_{i,t}^{Sc}, \tilde{p}_{i,t}^{Sd}, \tilde{e}_{i,t}^S, \tilde{p}_{ij,t}^L, \tilde{\theta}_{i,t}^N, \tilde{p}_{i,t}^A, \tilde{p}_{i,t}^D, \tilde{p}_{ij,t,m}^L)\ ,$$

as follows: Let

$$\begin{aligned}
&\tilde{p}_{i,t}^G = \hat{p}_{i,t}^G, \tilde{c}_{i,t}^G = \hat{c}_{i,t}^G, \tilde{e}_{i,t}^S = \hat{e}_{i,t}^S, \tilde{\theta}_{i,t}^N = \hat{\theta}_{i,t}^N, \tilde{p}_{i,t}^D = \hat{p}_{i,t}^D, \forall i \in S_B, t \in S_T\ ,\\
&\tilde{p}_{ij,t}^L = \hat{p}_{ij,t}^L, \tilde{p}_{ij,t,m}^L = \hat{p}_{ij,t,m}^L, \forall i \in S_B, (i,j) \in S_L, t \in S_T, m = 0, 1, \ldots, M^L\ ,\\
&\tilde{p}_{i_0,t_0}^{Sc} = \hat{p}_{i_0,t_0}^{Sc} - \varepsilon, \tilde{p}_{i_0,t_0}^{Sd} = \hat{p}_{i_0,t_0}^{Sd} - \eta_c\eta_d\varepsilon, \tilde{p}_{i_0,t_0}^A = \hat{p}_{i_0,t_0}^A + (1-\eta_c\eta_d)\varepsilon\ ,\\
&\tilde{p}_{i,t}^{Sc} = \hat{p}_{i,t}^{Sc}, \tilde{p}_{i,t}^{Sd} = \hat{p}_{i,t}^{Sd}, \tilde{p}_{i,t}^A = \hat{p}_{i,t}^A, \forall (i,t) \ne (i_0,t_0)\ .
\end{aligned}$$

Let $\varepsilon = \min\{\hat{p}_{i_0,t_0}^{Sc}, \hat{p}_{i_0,t_0}^{Sd}/(\eta_c\eta_d)\}$ and substitute the new solution into (10.26) to verify whether the constraint is valid. It is easy to know that all constraints hold except for (10.3e). Moreover, $\tilde{p}_{i_0,t_0}^{Sc}\tilde{p}_{i_0,t_0}^{Sd} = 0$, i.e., the ES at the bus i_0 does not charge and discharge simultaneously in time period t_0 and the charging and discharging power values do not change for $(i,t) \ne (i_0,t_0)$.

Repeat the above steps for all time periods and all buses until (10.3e) is satisfied, then we get a feasible solution of (10.26), and the objective function value at this point is still equal to $g_l^E(x,\xi)$. $g^E(x,\xi)$ is not greater than the objective function value at any feasible solution, so $g_l^E(x,\xi) \ge g^E(x,\xi)$. Therefore, the proposition conclusion is established. □

The conclusion and proof method of Proposition 10.1 are similar to Proposition 7.1. Proposition 10.1 asserts that when calculating the minimum amount of load shedding, the constraint of ES not charging and discharging at the same time is redundant. The proof method starts from the optimal solution of the relaxed problem and constructs a feasible solution for the unrelaxed problem by simultaneously reducing the charging power and discharging power, keeping the objective function value unchanged. From a physical point of view, allowing ES to charge and discharge at the same time can only reduce the available power of the load due to efficiency reasons, and cannot help reduce load shedding. Similarly to the situation of load shedding, when calculating the minimum carbon emissions, the constraint of ES not charging and discharging at the same time is also redundant, that is:

Proposition 10.2 *For any $x \in X$ and $\xi \in \Xi^N$, we have $g_l^N(x, \xi) = g^N(x, \xi)$.*

The proof of Proposition 10.2 is similar to Proposition 10.1. According to Propositions 10.1 and 10.2, load shedding $g^E(x, \xi)$ and carbon emissions $g^N(x, \xi)$ can be calculated separately through Eqs. (10.28) and (10.29), so both of these functions are the optimal values of LP problems.

10.3.2 Transformation of the Distributionally Robust Shortfall Risk Constraint on Load Shedding

Proposition 10.3 *For any $x \in X$ and $\xi \in \Xi^E$, $g^E(x, \xi)$ is finite.*

Proof According to (10.6), $p_{i,t}^E \geq 0, \forall i \in S_B, t \in S_T$, so $g^E(x, \xi) \geq 0$. Thus, there is a lower bound. We only need to find a feasible solution of (10.28) to show that $g^E(x, \xi)$ is finite.

Construct a solution $(\tilde{p}_{i,t}^G, \tilde{c}_{i,t}^G, \tilde{p}_{i,t}^{Sc}, \tilde{p}_{i,t}^{Sd}, \tilde{e}_{i,t}^S, \tilde{p}_{ij,t}^L, \tilde{\theta}_{i,t}^N, \tilde{p}_{i,t}^A, \tilde{p}_{i,t}^D, \tilde{p}_{ij,t,m}^L)$ as follows: Let

$$\begin{aligned} &\tilde{p}_{i,t}^G = \tilde{c}_{i,t}^G = \tilde{p}_{i,t}^{Sc} = \tilde{p}_{i,t}^{Sd} = \tilde{\theta}_{i,t}^N = 0 \ , \ \forall i \in S_B \ , \ t \in S_T \ , \\ &\tilde{p}_{ij,t}^L = \tilde{p}_{ij,t,m}^L = 0 \ , \ \forall (i, j) \in S_L \ , \ t \in S_T \ , \ m = 0, 1, \ldots, M^L \ , \\ &\tilde{e}_{i,t}^S = \tilde{e}_{i,0}^S \ , \ \tilde{p}_{i,t}^A = x_i^R \xi_{i,t}^R \ , \ \tilde{p}_{i,t}^D = \xi_{i,t}^D \ , \ \forall i \in S_B \ , \ t \in S_T \ . \end{aligned}$$

It is easy to verify that this is a feasible solution, so:

$$0 \leq g^E(x, \xi) \leq \sum_{t \in S_T} \sum_{i \in S_B} \xi_{i,t}^D \Delta_t < +\infty \ . \tag{10.30}$$

Therefore, the proposition conclusion is proven. □

The conclusion and proof process of Proposition 10.3 are similar to Proposition 7.2. From a physical point of view, Proposition 10.3 proves that load shedding is finite by demonstrating that the amount of load shedding is nonnegative and does not exceed the total load. The meaning of the constructed feasible solution is: conventional thermal power units do not generate electricity, renewable energy generation is all curtailed, ES does not charge or discharge, there is no power flow, and no load is supplied, so, in this case, the load shedding is equal to the total load.

10.3.2.1 Calculate the Lipschitz Constant of the Load Shedding Function

Write the given load shedding function $g^E(x, \xi)$ in (10.28) in the following compact form:

$$
\begin{aligned}
g^E(x, \xi) := \min_{\nu}\ & (d^E)^\top \nu \\
\text{s.t. } & A^E \nu \geq \begin{pmatrix} x^\top B_1^E \xi \\ \vdots \\ x^\top B_m^E \xi \end{pmatrix} + E^E x + F^E \xi + h^E ,
\end{aligned}
$$

where d^E, A^E, E^E, F^E, h^E and B_1^E, B_2^E, ..., B_m^E are constant matrices or vectors. Next, define the auxiliary variable $\gamma := (\gamma_{i,t} := x_i^R \xi_{i,t}^R;\, i \in S_B, t \in S_T)$. Based on $0 \leq x_i^R \leq X_i^{Ru}, 0 \leq \xi_{i,t}^R \leq 1, \forall i \in S_B, t \in S_T$, the value range Γ of γ is given as follows:

$$
\Gamma := \left\{ \gamma = (\gamma_{i,t} := x_i^R \xi_{i,t}^R;\, i \in S_B, t \in S_T) \;\middle|\; \begin{array}{l} 0 \leq \gamma_{i,t} \leq X_i^{Ru} \\ \forall i \in S_B, t \in S_T \end{array} \right\} , \tag{10.31}
$$

so

$$
\begin{aligned}
g^E(x, \xi) = \min_{\nu}\ & (d^E)^\top \nu \\
\text{s.t. } & A^E \nu \geq G^E \gamma + E^E x + F^E \xi + h^E ,
\end{aligned}
$$

where G^E is a constant matrix. Solve the following MILP problem to get an upper bound $(\mu^{Eu})_k$ for the maximum value of the dual variable:

$$
\begin{aligned}
\max_{x,\xi,\gamma,\mu,\nu,z}\ & (\mu^{Eu})_k \\
\text{s.t. } & x \in X\ ,\ \xi \in \Xi\ ,\ \gamma \in \Gamma\ ,\ (A^E)^\top \mu = d^E\ ,\ (\mu^{Eu})_k \geq \mu_k\ ,\ (\mu^{Eu})_k \geq -\mu_k\ , \\
& (z)_l \in \{0, 1\}\ ,\ \forall l\ , \\
& 0 \leq \mu \leq M_0(1 - z)\ , \\
& 0 \leq A^E \nu - G^E \gamma - E^E x - F^E \xi - h^E \leq M_0 z\ ,
\end{aligned} \tag{10.32}
$$

where 1 is an all-1 vector, and M_0 is a sufficiently large constant. According to the meaning of the dual variables, $(\mu)_k$ reflects the sensitivity of $g^E(x, \xi)$ to the change of the k-th component on the right-hand side of the constraint, because at the optimal solution,

$$
\begin{aligned}
g^E(x,\xi) &= (\gamma + E^E x + F^E \xi + h^E)^\top \mu \\
&= \left(\begin{pmatrix} x^\top B_1^E \xi \\ \vdots \\ x^\top B_m^E \xi \end{pmatrix} + E^E x + F^E \xi + h^E \right)^\top \mu \\
&= \mu^\top \left(\begin{pmatrix} x^\top B_1^E \\ \vdots \\ x^\top B_m^E \end{pmatrix} + F^E \right) \xi + \mu^\top (E^E x + h^E) \, .
\end{aligned}
$$

Thus, the absolute value of the change sensitivity of $g^E(x,\xi)$ with respect to $(\xi)_l$ is no more than

$$
L_l^E(x) := \sum_k (\mu^{Eu})_k \cdot \left| \left(\begin{pmatrix} x^\top B_1^E \\ \vdots \\ x^\top B_m^E \end{pmatrix} + F^E \right)_{k,l} \right| . \tag{10.33}
$$

Let $L^E(x) := \max_l L_l^E(x)$, then $L^E(x)$ is the desired Lipschitz constant of $g^E(x,\xi)$ as a function with respect to ξ, which is given by the following LP problem:

$$
\begin{aligned}
L^E(x) := \min_{q^E, \tau_{k,l}^E} \; & q^E & \text{(10.34a)} \\
\text{s.t.} \; & \sum_k (\mu^{Eu})_k \cdot \tau_{k,l}^E \le q^E \, , \; \forall l \, , & \text{(10.34b)} \\
& \left(\begin{pmatrix} x^\top B_1^E \\ \vdots \\ x^\top B_m^E \end{pmatrix} + F^E \right)_{k,l} \le \tau_{k,l}^E \, , \; \forall k,l \, , & \text{(10.34c)} \\
& -\left(\begin{pmatrix} x^\top B_1^E \\ \vdots \\ x^\top B_m^E \end{pmatrix} + F^E \right)_{k,l} \le \tau_{k,l}^E \, , \; \forall k,l \, . & \text{(10.34d)}
\end{aligned}
$$

10.3.2.2 Convert the Distributionally Robust Shortfall Risk Constraint of Load Shedding Into a Linear Model

According to Sect. 3.4,

$$
\begin{aligned}
& \sup_{\mathbb{P} \in \mathcal{B}(\mathbb{P}^E, \varepsilon^E)} \mathrm{SR}_{l,\lambda}^{\mathbb{P}}(g^E(x,\xi)) \\
= & \begin{cases} \min\limits_{r \in \mathbb{R}} r \\ \text{s.t.} \quad \sup\limits_{\mathbb{P} \in \mathcal{B}(\mathbb{P}^E, \varepsilon^E)} \mathbb{E}^{\mathbb{P}}[l(g^E(x,\xi) - r)] \le \lambda \, , \end{cases}
\end{aligned}
$$

so Eq. (10.16) is equivalent to:

$$\sup_{\mathbb{P}\in\mathcal{B}(\mathbb{P}^E,\varepsilon^E)} \mathbb{E}^{\mathbb{P}}[l(g^E(x,\xi)-R^E)] \le \lambda \,. \tag{10.35}$$

Note that the function $l(y)=\max\{y,0\}$, so $L^E(x)$ is also the Lipschitz constant of $l(g^E(x,\xi)-R^E)$ as a function of ξ, i.e.,

$$\begin{aligned} & |l(g^E(x,\xi)-R^E)-l(g^E(x,\xi')-R^E)| \\ \le\ & |g^E(x,\xi)-g^E(x,\xi')| \\ \le\ & L^E(x)\|\xi-\xi'\|_1 \,, \ \forall \xi,\xi' \in \Xi \,. \end{aligned}$$

According to Theorem 6.10,

$$\begin{aligned} & \sup_{\mathbb{P}\in\mathcal{B}(\mathbb{P}^E,\varepsilon^E)} \mathbb{E}^{\mathbb{P}}[l(g^E(x,\xi)-R^E)] \\ \le\ & \varepsilon^E L^E(x) + \frac{1}{|S_E|}\sum_{n\in S_E} l(g^E(x,\xi_n^E)-R^E) \,. \end{aligned}$$

Therefore, a conservative approximation of (10.35) is:

$$\varepsilon^E L^E(x) + \frac{1}{|S_E|}\sum_{n\in S_E} l(g^E(x,\xi_n^E)-R^E) \le \lambda \,. \tag{10.36}$$

Substitute the formula for $L^E(x)$ in (10.34) and introduce the auxiliary variable s_n^E to represent $l(g^E(x,\xi_n^E)-R^E)$, and then a conservative linear model for the distributionally robust Shortfall risk constraint of load shedding is obtained as follows:

$$\begin{cases} \varepsilon^E q^E + \dfrac{1}{|S_E|}\displaystyle\sum_{n\in S_E} s_n^E \le \lambda \,, \ \text{(10.34b)–(10.34d)} \,, \\ s_n^E \ge 0 \,, \ s_n^E \ge \displaystyle\sum_{t\in S_T}\sum_{i\in S_B} p_{i,t,n}^D \Delta_t - R^E \,, \ \forall n \in S_E \,, \\ \text{(10.28b)} \,, \ \text{(10.28c)} \,, \ \forall n \in S_E \,. \end{cases} \tag{10.37}$$

10.3.3 Transform the Worst-Case Expectation of Carbon Emissions

Equation (10.29) can be written in the following compact form:

$$g^N(x,\xi) := \min_{\nu} \ (d^N)^\top \nu$$
$$\text{s.t. } A^N \nu \geq \begin{pmatrix} x^\top B_1^N \xi \\ \vdots \\ x^\top B_m^N \xi \end{pmatrix} + E^N x + F^N \xi + h^N \ . \quad (10.38)$$

Different from the situation of load shedding, Eq. (10.29) does not always have feasible solutions under any $x \in X$ and $\xi \in \Xi^N$. Because load shedding is not allowed in regular situations, there are certain requirements for the capacity of generation, transmission, and ES. For this reason, the value range X^N of capacity variable x in regular situations is defined as follows:

$$X^N := \left\{ x \in X \ \middle| \ (10.29\text{b}) \ , \ (10.29\text{c}) \ , \ \forall n \in S_N \right\} \ . \quad (10.39)$$

X^N includes x in X that makes the operation optimization problem of (10.29) feasible for any $\xi \in \{\xi_n^N \mid n \in S_N\}$.

Proposition 10.4 *For any $x \in X^N$ and $\xi \in \Xi^N$, $g^N(x,\xi)$ is finite.*

Proof According to (10.1a), (10.1c), and the lower limit P_i^{Gl} of the power generation of thermal power units is nonnegative according to physical meaning, we have $c_{i,t}^G \geq 0, \forall i \in S_B, t \in S_T$. Therefore, $g^N(x,\xi) = \sum_{t\in S_T} \sum_{i \in S_B} c_{i,t}^G \geq 0$. It has been proven that $g^N(x,\xi)$ has a lower bound. We just need to prove that for any $x_0 \in X^N$ and $\xi_0 \in \Xi^N$, Eq. (10.29) has a feasible solution.

According to the definition of Ξ^N in (10.14), there exist $y_n, n \in S_N$ such that $y_n \geq 0, \forall n \in S_N, \sum_{n\in S_N} y_n = 1$ and $\xi_0 = \sum_{n \in S_N} y_n \xi_n^N$. According to (10.39) and $x_0 \in X^N$, for any ξ_n^N, $n \in S_N$, there exists a feasible solution ν_n of (10.29). We notice that Eq. (10.29) is an LP problem, ξ only appears in the constraints' right-hand term, and the right-hand term is affine in ξ. It is easy to verify that $\sum_{n\in S_N} y_n \nu_n$ is a feasible solution of (10.29) under ξ_0. Thus, the conclusion holds. □

10.3.3.1 Calculate the Lipschitz Constant of the Carbon Emission Function

When $x \in X^N$ is fixed, $g^N(x,\xi)$ is a convex function with respect to ξ. According to the relationship between Lipschitz continuity and local Lipschitz continuity, the maximum local Lipschitz constant at each point in $\{\xi_n^N \mid n \in S_N\}$ is the Lipschitz constant over Ξ^N. Therefore, for each single-point set $\{\xi_n^N\}, n \in S_N$, we calculate the Lipschitz constant:

$$g^N(x, \xi_n^N) = \begin{cases} \min_{\nu} \ (d^N)^\top \nu \\ \text{s.t. } A^N \nu \geq \begin{pmatrix} x^\top B_1^N \xi_n^N \\ \vdots \\ x^\top B_m^N \xi_n^N \end{pmatrix} + E^N x + F^N \xi_n^N + h^N \ . \end{cases} \tag{10.40}$$

Since ξ_n^N is fixed, the auxiliary variable γ is no longer needed. Solve the following MILP to get an upper bound $(\mu_n^{Nu})_k$:

$$\max_{x,\mu,\nu,z} \ (\mu_n^{Nu})_k \tag{10.41a}$$

$$\text{s.t. } x \in X^N \ , \ (A^N)^\top \mu = d^N \ , \ (\mu_n^{Nu})_k \geq \mu_k \ , \ (\mu_n^{Nu})_k \geq -\mu_k \ , \tag{10.41b}$$

$$(z)_l \in \{0, 1\} \ , \ \forall l \ , \tag{10.41c}$$

$$0 \leq \mu \leq M_0(1 - z) \ , \tag{10.41d}$$

$$0 \leq A^N \nu - \begin{pmatrix} x^\top B_1^N \xi_n^N \\ \vdots \\ x^\top B_m^N \xi_n^N \end{pmatrix} - E^N x - F^N \xi_n^N - h^N \leq M_0 z \ . \tag{10.41e}$$

For the single-point set $\{\xi_n^N\}$, the local Lipschitz constant is $\max_l L_{l,n}^N(x)$, where $L_{l,n}^N(x)$ is as follows:

$$L_{l,n}^N(x) := \sum_k (\mu_n^{Nu})_k \cdot \left| \left(\begin{pmatrix} x^\top B_1^N \\ \vdots \\ x^\top B_m^N \end{pmatrix} + F^N \right)_{k,l} \right| \ . \tag{10.42}$$

Let $L^N(x) := \max_{n \in S_N} \max_l L_{l,n}^N(x)$, then $L^N(x)$ is the Lipschitz constant of $g^N(x, \xi)$ as a function of ξ when $x \in X^N$ and $\xi \in \Xi^N$. Furthermore, $L^N(x)$ is given by the following LP problem:

$$L^N(x) = \min_{q^N, \tau_{k,l}^N} \ q^N \tag{10.43a}$$

$$\text{s.t. } \sum_k (\mu_n^{Nu})_k \cdot \tau_{k,l}^N \leq q^N \ , \ \forall n \in S_N, l \ , \tag{10.43b}$$

$$\left(\begin{pmatrix} x^\top B_1^N \\ \vdots \\ x^\top B_m^N \end{pmatrix} + F^N \right)_{k,l} \le \tau_{k,l}^N \ , \ \forall k, l \ , \tag{10.43c}$$

$$-\left(\begin{pmatrix} x^\top B_1^N \\ \vdots \\ x^\top B_m^N \end{pmatrix} + F^N \right)_{k,l} \le \tau_{k,l}^N \ , \ \forall k, l \ . \tag{10.43d}$$

10.3.3.2 Convert the Worst-Case Expectation of Carbon Emissions into a Linear Model

According to Theorem 6.10, we have:

$$\begin{aligned} &\sup_{\mathbb{P} \in \mathcal{B}(\mathbb{P}^N, \varepsilon^N)} \mathbb{E}^{\mathbb{P}}[g^N(x, \xi)] \\ &\le \varepsilon^N L^N(x) + \frac{1}{|S_N|} \sum_{n \in S_N} g^N(x, \xi_n^N) \\ &= \begin{cases} \min \varepsilon^N L^N(x) + \frac{1}{|S_N|} \sum_{n \in S_N} \sum_{t \in S_T} \sum_{i \in S_B} c_{i,t,n}^G \\ \text{s.t. } (10.29\text{b}) \ , \ (10.29\text{c}) \ , \ \forall n \in S_N \ . \end{cases} \end{aligned}$$

Substituting the expression of $L^N(x)$ from (10.43) results in a linear model.

10.3.4 Sizing Model Solution Algorithm

The part involving the ambiguity set in the sizing model has been removed in the previous sections, and the DRO problem was transformed into a traditional bi-objective mathematical optimization problem through sample average approximation (SAA). A common solution method for multi-objective optimization is the weighted sum method. However, this method is not suitable for the sizing model proposed in this chapter. First, due to the existence of Boolean variables, the problem is nonconvex, so some Pareto solutions may not be found through the weighted sum method according to Sect. 2.6.2. Second, the two objective functions, i.e., investment cost and expected carbon emissions, have different units, so it is difficult to select suitable coefficients to add them together, and the selection of coefficients has a significant impact on the result. Therefore, the ϵ-constraint method is used to generate the Pareto frontier of the bi-objective optimization problem. For a given total investment cost budget of F, we solve the following MILP problem:

$$\min \varepsilon^N q^N + \frac{1}{|S_N|} \sum_{n \in S_N} \sum_{t \in S_T} \sum_{i \in S_B} c^G_{i,t,n}$$
$$\text{s.t. } f(x) \le F \ , \ x \in X \ ,$$
$$(10.29\text{b}) \ , \ (10.29\text{c}) \ , \ \forall n \in S_N \ ,$$
$$(10.37) \ , \ (10.43\text{b})\text{–}(10.43\text{d}) \ . \tag{10.44}$$

The optimal solution of (10.44) is denoted as $x(F)$, and the worst-case expectation of carbon emissions is denoted as $c(F)$. The point pair $(F, c(F))$ is on the Pareto frontier of the sizing problem in (10.20). By solving (10.44) under different F, a set of Pareto solutions is obtained. The above solution method is summarized in Algorithm 10.1.

Algorithm 10.1 Calculate the Pareto frontier of the sizing problem in (10.20)

Input: Equation (10.20)'s parameters, the range of investment cost budget $[F^l, F^u]$, and the step size parameter $\delta > 0$.

Output: Pareto solutions $x(F)$ and the points $(F, c(F))$ on the Pareto frontier.

1: Initiation: Solve the MILP problem in (10.32) and obtain μ^{Eu}. Solve the MILP problem in (10.41) and obtain $\mu_n^{Nu}, n \in S_N$. Let $F \leftarrow F_l$.

2: Solve the MILP problem in (10.44) and obtain $x(F)$ and $c(F)$.

3: If $F^u - F < \delta$, terminate; otherwise, update $F \leftarrow F + \delta$ and turn to Step 2.

10.4 Case Study

10.4.1 Settings

The example is based on the IEEE 30-node system in the MATPOWER toolbox (Zimmerman et al. 2011), and its topology map is shown in Fig. 10.1. In the planning scenario, only the conventional thermal power units at bus 1 and bus 2 are still running, with parameters $P^{Gl} = 32$ MW, $P^{Gu} = 80$ MW, $F^G = 0.875$ kgCO$_2$/kWh (Kang et al. 2015). Other conventional thermal power units are out of operation. PV power stations can be built at bus 10 and bus 27. A wind farm can be built at bus 13. The parameters of the renewable energy sites are in Table 10.1. The unit capacity costs of PV power stations and wind farms are set according to the experience data in Qinghai Province. Transmission line parameters including X^L, S^L, and N^L are obtained from the MATPOWER data file, the parameters of the expandable transmission lines are listed in Table 10.2. ESs can be built at bus 8 and bus 21. The related parameters are in Table 10.3. Other parameters of the proposed method include $\varepsilon^E = 0.01$, $\varepsilon^N = 0.005$, $R^E = 0$, $T = 24$, and $\Delta_t = 1$ h. The acceptable load shedding amount λ is set to 1% of the average load in extreme cases, which equals 51.1 MWh.

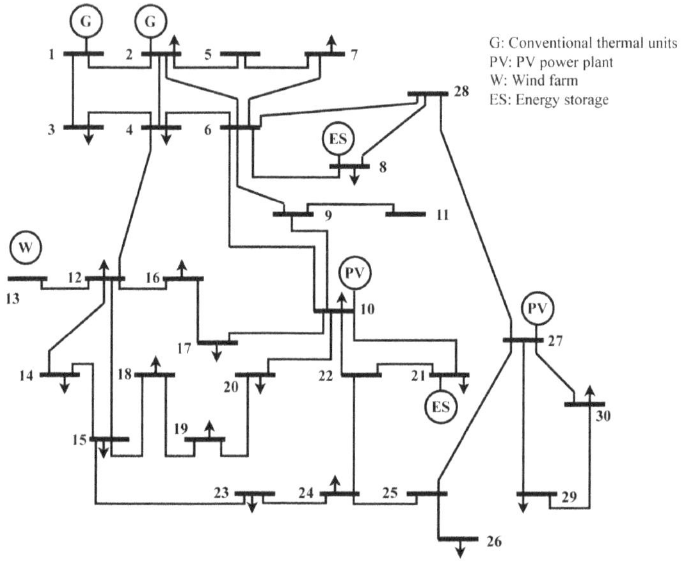

Fig. 10.1 Topology diagram of the low-carbon power grid (Xie et al. 2023)

Table 10.1 Parameters of the renewable power plants in the low-carbon power grid (Xie et al. 2023)

Bus	Type	Maximum capacity X^{Ru} (MW)	Unit capacity cost C^R (CNY/MW)
10	PV	250	3.5×10^6
13	Wind	300	5.5×10^6
27	PV	300	3.5×10^6

Table 10.2 Parameters of the expandable transmission lines in the low-carbon power grid (Xie et al. 2023)

Start bus	End bus	Unit cost C^L (CNY)	Maximum number N^{Lu}
6	8	6.4×10^7	3
21	22	6.4×10^7	3
15	23	3.2×10^7	3
22	24	3.2×10^7	3
23	24	3.2×10^7	3
24	25	3.2×10^7	3
25	27	3.2×10^7	3

Table 10.3 Parameters of the ESs in the low-carbon power grid (Xie et al. 2023)

Parameter	Value	Parameter	Value
Charging efficiency η^{Sc}	0.95	Unit power capacity cost C^{Sp}	1.0×10^6 CNY/MW
Discharging efficiency η^{Sd}	0.95	Unit energy capacity cost C^{Se}	1.2×10^6 CNY/MWh
SoC lower bound ω^{Sl}	0.10	Maximum power capacity X^{Spu}	200 MW
SoC upper bound ω^{Su}	0.90	Maximum energy capacity X^{Seu}	1,000 MWh

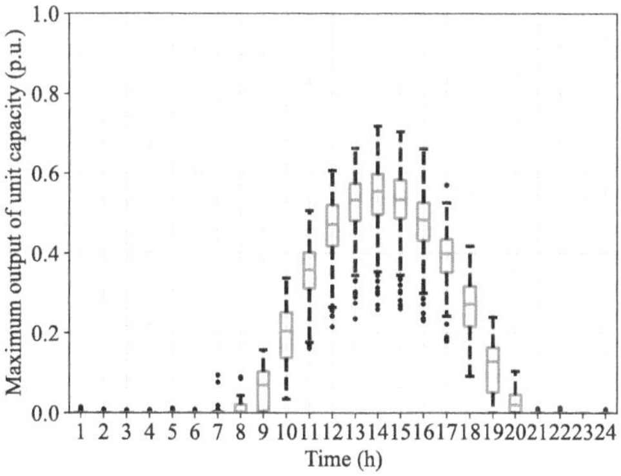

Fig. 10.2 Boxplot of PV power generation data at bus 10 (the *long box* represents the 50% confidence interval, the *line inside the box* represents the median, and the *points outside the box* and the *dashed line* represent outliers) (Xie et al. 2023)

The unit capacity renewable power generation ξ^R data is generated according to the historical data of Qinghai Province and is shown in Figs. 10.2, 10.3, and 10.4. The load demand ξ^D data is generated according to the load data in MATPOWER. The variability of the total load power for each time period is shown in Fig. 10.5, and the average load at each bus is plotted in Fig. 10.6. A total of 366 sets of data were generated, each of which includes the unit capacity power generation ξ^R of 3 renewable power plants and the load power ξ^D of 30 buses for T time periods. Based on the total renewable generation and load demand, we selected 91 sets of extreme data and the remaining 275 sets represent regular cases. The data was then further split into planning and testing data. In the basic example, a total of 12 sets of extreme data and 48 sets of normal data were used. The example tests were run on a laptop with an Intel Core i5-5200U processor and 8 GB of memory. The proposed method is implemented in the MATLAB environment, with the MILP model established by the YALMIP toolbox (Lofberg 2004) and solved by Gurobi 9.1 (Gurobi Optimization 2023).

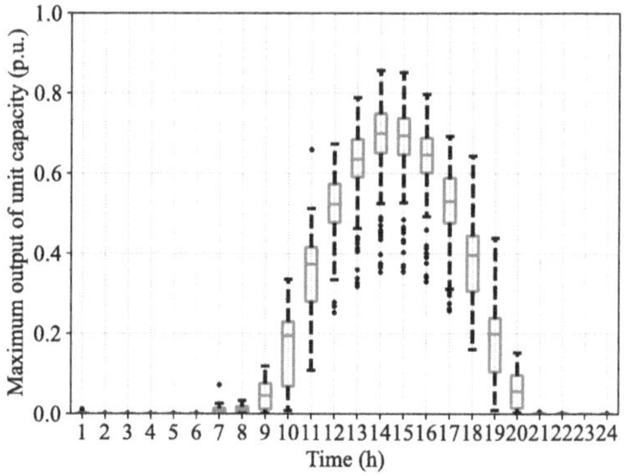

Fig. 10.3 Boxplot of PV power generation data at bus 27 (Xie et al. 2023)

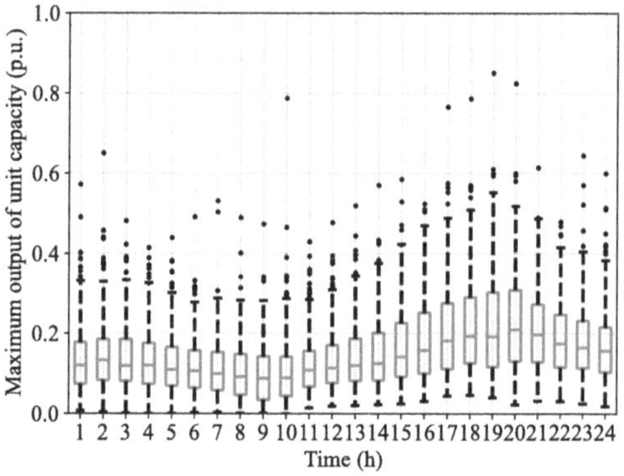

Fig. 10.4 Boxplot of wind power generation data at bus 13 (Xie et al. 2023)

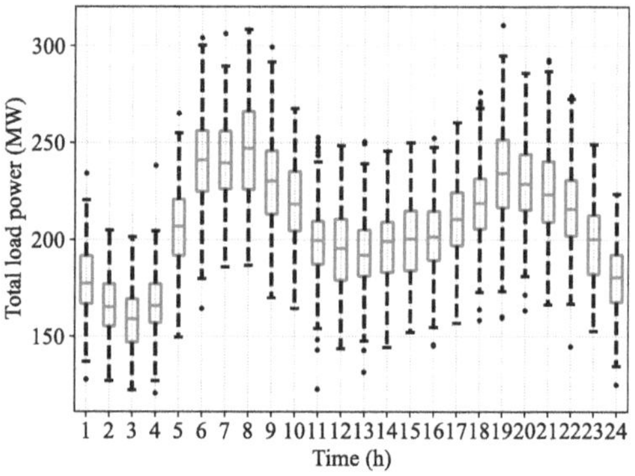

Fig. 10.5 Boxplot of total load power in each time period (Xie et al. 2023)

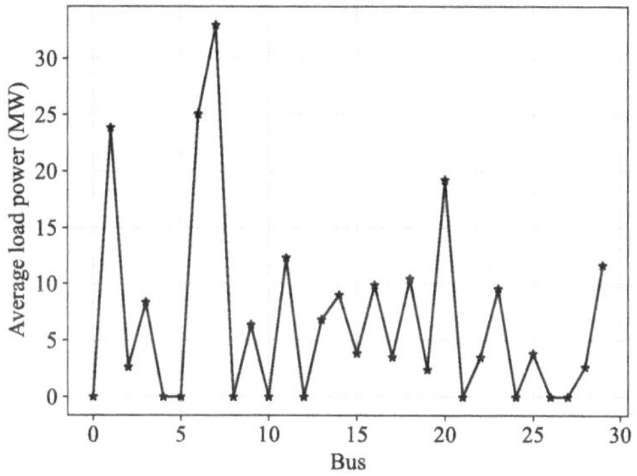

Fig. 10.6 Average load power at each bus (Xie et al. 2023)

10.4.2 Other Methods Used for Comparative Analysis

The method proposed in this chapter is denoted as DRO-C. In addition to the proposed method, three other methods are considered for comparison:

The first method is stochastic programming (SP), denoted as SP;

The second method is robust optimization (RO), denoted as RO;

The third method is the DRO based on the Wasserstein distance, considering only discrete probability distributions, denoted as DRO-D.

The following provides a detailed introduction to each of these three methods.

10.4.2.1 Stochastic Programming Method

Assuming that the empirical probability distributions $\mathbb{P}^E$ and $\mathbb{P}^N$ are accurate, the sizing model for the SP method is as follows:

$$\begin{aligned}&\min_{x}\ \left\{f(x), \mathbb{E}^{\mathbb{P}^N}[g^N(x,\xi)]\right\}\\&\text{s.t. } \mathrm{SR}_{l,\lambda}^{\mathbb{P}^E}(g^E(x,\xi)) \leq R^E\ ,\ x \in X\ .\end{aligned} \tag{10.45}$$

Because $\mathbb{P}^E \in \mathcal{B}(\mathbb{P}^E, \varepsilon^E)$ and $\mathbb{P}^N \in \mathcal{B}(\mathbb{P}^N, \varepsilon^N)$, the SP method is more optimistic than the proposed method. The nonlinear constraints are converted into linear constraints using the method in Sect. 10.3.1, and then the sizing problem is transformed into a traditional bi-objective MILP problem using SAA, and the Pareto frontier is computed using the ϵ-constraint method. The MILP problem to be solved for the given budget F is as follows:

$$\min \frac{1}{|S_N|} \sum_{n \in S_N} \sum_{t \in S_T} \sum_{i \in S_B} c^G_{i,t,n} \tag{10.46a}$$

$$\text{s.t. } f(x) \leq F \ , \ x \in X \ , \ \frac{1}{|S_E|} \sum_{n \in S_E} s^E_n \leq \lambda \ , \tag{10.46b}$$

$$(10.29\text{b}), \ (10.29\text{c}), \ \forall n \in S_N \ , \tag{10.46c}$$

$$s^E_n \geq 0 \ , \ s^E_n \geq \sum_{t \in S_T} \sum_{i \in S_B} p^D_{i,t,n} \Delta_t - R^E \ , \ \forall n \in S_E \ , \tag{10.46d}$$

$$(10.28\text{b}), \ (10.28\text{c}), \ \forall n \in S_E \ . \tag{10.46e}$$

10.4.2.2 Robust Optimization Method

Optimizing the objective function under the worst-case sampling scenario, the sizing model for the RO method is as follows:

$$\min_x \left\{ f(x), \ \sup_{n \in S_N} g^N(x, \xi^N_n) \right\} \tag{10.47a}$$

$$\text{s.t. } \sup_{n \in S_E} g^E(x, \xi^E_n) \leq \lambda \ , \ x \in X \ , \tag{10.47b}$$

in which the total investment cost $f(x)$ depends on the capacity allocation policy and is irrelevant to the uncertainty. It minimizes the worst-case carbon emissions in the second objective function and restricts load shedding in extreme cases in the constraints. By applying the ϵ-constraint method, the Pareto frontier of the sizing model in the RO method can be solved through the following MILP:

$$\begin{aligned}
&\min s \\
&\text{s.t. } f(x) \leq F \ , \ x \in X \ , \\
&\quad s \geq \sum_{t \in S_T} \sum_{i \in S_B} c^G_{i,t,n} \ , \ (10.29\text{b}) \ , \ (10.29\text{c}) \ , \ \forall n \in S_N \ , \\
&\quad \sum_{t \in S_T} \sum_{i \in S_B} p^D_{i,t,n} \Delta_t \leq \lambda \ , \ (10.28\text{b}) \ , \ (10.28\text{c}) \ , \ \forall n \in S_E \ .
\end{aligned}$$

10.4.2.3 Distributionally Robust Optimization Method Based on Wasserstein Distance Considering Only Discrete Probability Distributions

The ambiguity set used in this method is different from the proposed method. It only contains discrete probability distributions that have the same supporting set as the empirical probability distribution, i.e.,

$$\mathcal{B}^d(\mathbb{P}^E, \varepsilon^{Ed}) := \left\{ \mathbb{P} := \sum_{n \in S_E} q_n \mathbf{1}_{\xi_n^E} \;\middle|\; \begin{array}{r} q_n \geq 0, \forall n \in S_E, \\ \sum_{n \in S_E} q_n = 1, \\ d_W(\mathbb{P}, \mathbb{P}^E) \leq \varepsilon^{Ed} \end{array} \right\} . \quad (10.48)$$

The ambiguity set of probability distributions for regular cases $\mathcal{B}^d(\mathbb{P}^N, \varepsilon^{Nd})$ is defined in a similar manner. Based on these two ambiguity sets of probability distributions, the sizing model for the DRO-D method is as follows:

$$\begin{aligned} &\min_x \left\{ f(x), \sup_{\mathbb{P} \in \mathcal{B}^d(\mathbb{P}^N, \varepsilon^{Nd})} \mathbb{E}^{\mathbb{P}}[g^N(x, \xi)] \right\} \\ &\text{s.t.} \sup_{\mathbb{P} \in \mathcal{B}^d(\mathbb{P}^E, \varepsilon^{Ed})} \mathrm{SR}_{l,\lambda}^{\mathbb{P}}(g^E(x, \xi)) \leq R^E \ , \ x \in X \ . \end{aligned} \quad (10.49)$$

When $\varepsilon^{Ed} = \varepsilon^E$, $\mathbb{P}^E \in \mathcal{B}^d(\mathbb{P}^E, \varepsilon^{Ed}) \subsetneq \mathcal{B}(\mathbb{P}^E, \varepsilon^E)$, because $\mathcal{B}(\mathbb{P}^E, \varepsilon^E)$ contains probability distributions in forms other than discrete probability distribution. Therefore, the DRO-D model is more conservative than the SP model, but more optimistic than the proposed method. By the method in Sect. 6.3, Eq. (10.49) is equivalent to a conventional bi-objective MILP problem, which can be further solved by the ϵ-constraint method to find the Pareto frontier. The MILP problem to be solved at a fixed budget F is as follows:

$$\min \varepsilon^{Nd} \rho^N + \sum_{n \in S_N} \frac{\gamma_n^N}{|S_N|} + \tau^N \quad (10.50a)$$

$$\text{s.t. } \rho^N \|\xi_m^N - \xi_n^N\|_1 + \mu_m^N + \gamma_n^N \geq 0 \ , \ \forall m, n \in S_N \ , \quad (10.50b)$$

$$- \mu_n^N + \tau^N \geq \sum_{t \in S_T} \sum_{i \in S_B} c_{i,t,n}^G \ , \ \forall n \in S_N \ , \quad (10.50c)$$

$$(10.29b) \ , \ (10.29c) \ , \ \forall n \in S_N \ , \quad (10.50d)$$

$$\rho^E \|\xi_m^E - \xi_n^E\|_1 + \mu_m^E + \gamma_n^E \geq 0 \ , \ \forall m, n \in S_E \ , \quad (10.50e)$$

$$- \mu_n^E + \tau^E \geq 0 \ , \ -\mu_n^E + \tau^E \geq \sum_{t \in S_T} \sum_{i \in S_B} p_{i,t,n}^D \Delta_t - R^E \ , \ \forall n \in S_E \ , \quad (10.50f)$$

$$(10.28b) \ , \ (10.28c) \ , \ \forall n \in S_E \ , \quad (10.50g)$$

$$f(x) \leq F \ , \ x \in X \ , \quad (10.50h)$$

$$\rho^N \geq 0 \ , \ \rho^E \geq 0 \ , \ \varepsilon^{Ed} \rho^E + \sum_{n \in S_E} \frac{\gamma_n^E}{|S_E|} + \tau^E \leq \lambda \ . \quad (10.50i)$$

10.4.3 Results and Comparison

Each of the four methods requires solving a MILP problem with about 350,000 constraints and 150,000 variables (including 14 boolean variables) to compute a point on the Pareto frontier, and the computation time is around 1 h. Since it is used for planning purposes, computation time is not the primary consideration. If ε^{Ed} and ε^{Nd} are set to $\varepsilon^{Ed} = \varepsilon^{E} = 0.01$ and $\varepsilon^{Nd} = \varepsilon^{N} = 0.005$, then the results of SP and DRO-D are very close, so we set $\varepsilon^{Ed} = 0.2$ and $\varepsilon^{Nd} = 0.1$. Figure 10.7 depicts the Pareto frontier for the four methods. As total investment costs increase, carbon emissions decrease. The SP and DRO-C models differ primarily when the investment cost is less than 4×10^9 CNY. In other cases, the sensitivity of daily carbon emissions to investment cost is roughly -3×10^{-4} $kgCO_2$/CNY, implying that in this example, every reduction of 300 $kgCO_2$ in daily carbon emissions requires an additional investment of 10^6 CNY. From top to bottom, the Pareto frontiers belong to the RO, DRO-C, DRO-D, and SP methods, respectively. From left to right, the left endpoints of the Pareto frontier belong to the SP, DRO-D, DRO-C, and RO methods, respectively, reflecting the size relationship of the feasible domains of various methods. Therefore, SP is the most optimistic method, and the results of DRO-D are still very close to SP. RO is the most conservative method, while DRO-C is in between. These observations verify the theoretical analysis of the four methods presented earlier.

Setting the daily carbon emissions to no more than 2.4×10^6 $kgCO_2$, the sizing results of different methods are compared in Table 10.4, with Shortfall risk being tested in the test extreme scenario data, and average carbon emissions being tested in the test regular scenario data. The penetration rate of renewable energy, i.e., the proportion of renewable energy output to total load, is about 50%. The results show that the DRO-C method's scheme meets the load shedding limit in the test data, i.e., the load shedding risk does not exceed 0, and the daily carbon emissions do not exceed 2.4×10^6 $kgCO_2$. In contrast, SP is overly optimistic in both load shedding and carbon emissions, because it does not take into account the inexactness of the empirical probability distribution. The empirical probability distribution is generated

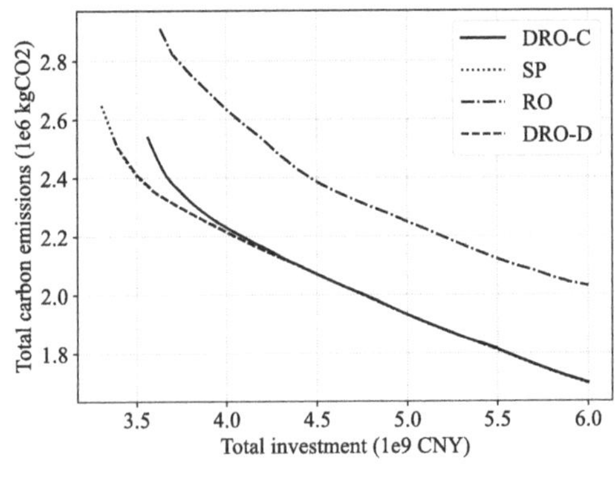

Fig. 10.7 Pareto frontier of carbon emissions and total investment costs of the low-carbon power grid (Xie et al. 2023)

Table 10.4 Sizing results of the low-carbon power grid under different methods (Xie et al. 2023)

Method	DRO-C	SP	RO
Total investment cost (CNY)	3.67×10^9	3.62×10^9	4.46×10^9
Renewable capacity ($x_{10}^R, x_{13}^R, x_{27}^R$) (MW)	(250, 127, 163)	(247, 198, 135)	(250, 240, 205)
ES power capacity (x_8^{Sp}, x_{21}^{Sp}) (MW)	(109, 34.3)	(81.1, 34.9)	(84.4, 43.2)
ES energy capacity (x_8^{Se}, x_{21}^{Se}) (MWh)	(753, 160)	(548, 187)	(709, 266)
New transmission line	(6, 8) × 3	(6, 8) × 2	(6, 8) × 2
	(24, 25)	(24, 25)	(24, 25) × 2
	(25, 27) × 2	(25, 27)	(25, 27) × 2
Test load shedding risk	−0.29	0.24	−0.48
Test daily emissions (kgCO_2)	2.37×10^6	2.47×10^6	2.13×10^6

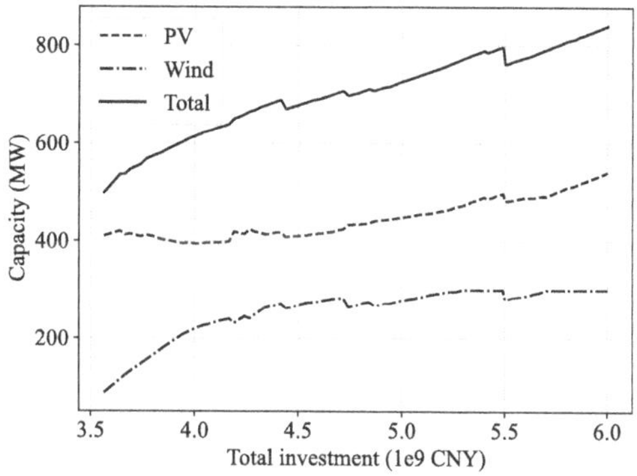

Fig. 10.8 Renewable energy generation capacities in the Pareto solutions of the low-carbon power grid (Xie et al. 2023)

from planning data. SP's different performance in planning and testing data reflects both the inexactness of the empirical probability distribution and indicates that the results given by the SP method are less robust and more sensitive to the distribution. Although the results of RO meet the requirements for load shedding and carbon emissions, its investment cost is 22% higher than the proposed DRO-C method. Therefore, among these methods, the DRO-C method provides the optimal balance between robustness and optimality in this case.

The Pareto solutions of the proposed method are shown in Figs. 10.8, 10.9, 10.10, and 10.11, including how the capacities and costs of each part change when the budget varies. The results show that as the budget increases or the carbon emission cap decreases, the capacity of renewable power generation and ES tends to increase. However, because the variables for transmission grid extension are discrete, there are some stepwise phenomena. Roughly speaking, in this case, for each reduction in daily carbon emissions of 100 kgCO_2, approximately 40 kW of additional installed capacity for renewable power generation is needed.

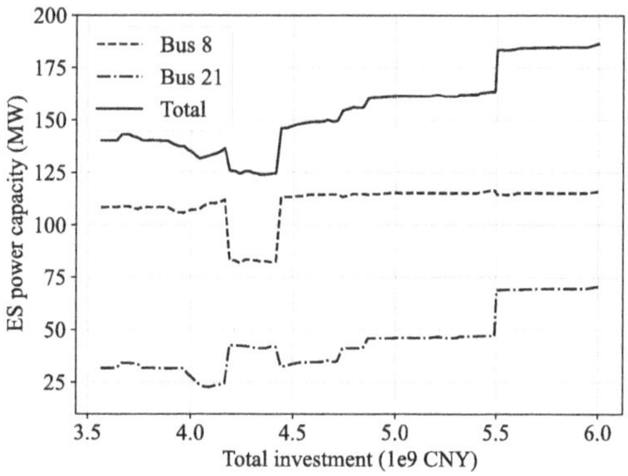

Fig. 10.9 ES power capacities in the Pareto solutions of the low-carbon power grid (Xie et al. 2023)

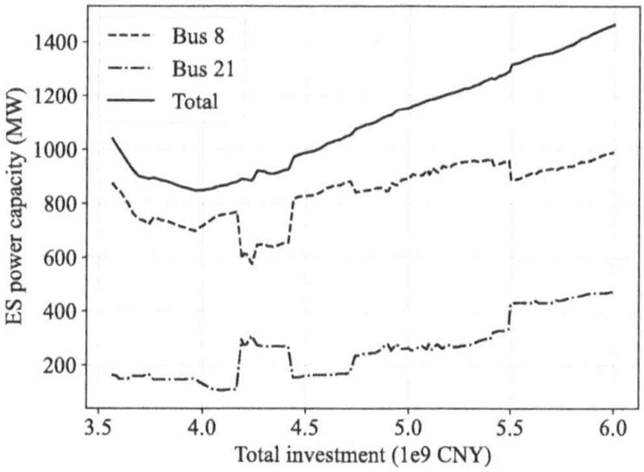

Fig. 10.10 ES energy capacities in the Pareto solutions of the low-carbon power grid (Xie et al. 2023)

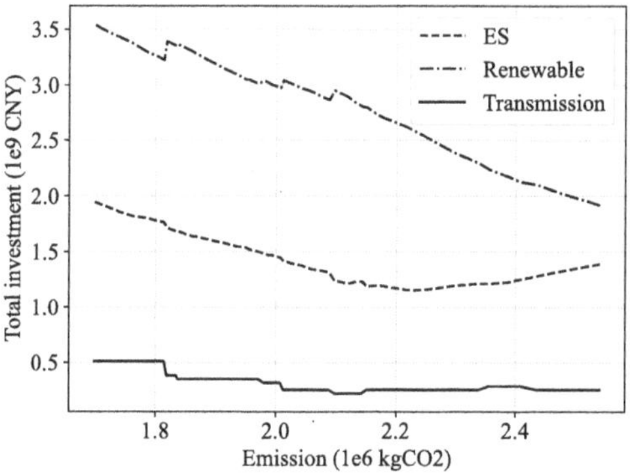

Fig. 10.11 Composition of investment costs in the Pareto solutions of the low-carbon power grid (Xie et al. 2023)

10.4.4 Parameter Sensitivity Analysis

This section investigates the influence of certain parameters on sizing results, including the acceptable load-shedding amount, conventional thermal power unit capacity, unit cost of ES, and load demand. In this section, the daily carbon emissions cap is set to 2.4×10^6 $kgCO_2$, at which point the penetration of renewables is about 50% or higher.

Table 10.5 shows the sizing results under different acceptable load shedding amounts λ. The value of λ affects the size of the feasible domain. When λ increases, the feasible domain expands, so the total ES capacity decreases, the need for transmission line expansion decreases, and the total investment cost decreases. The total renewable generation capacity does not change significantly because the load demand and conventional fossil fuel power unit capacity remain unchanged.

The capacities of the two conventional thermal power units are changed at the same time, and the results are shown in Table 10.6. As the capacity of the conventional thermal power units increases, the regulation ability strengthens, and they can bear more peak-shaving pressure. Therefore, the optimal solution reduces the capacity of renewable power generation, ES, and grid expansion. Hence, the total investment cost also decreases.

In order to study the impact of the unit capacity cost of ES on the results, C^{Sp} and C^{Se} are multiplied by a constant coefficient at the same time. The results under different coefficients are shown in Table 10.7. As ES becomes cheaper, the total capacity of renewable power generation remains basically unchanged, the capacity of ES shows a growing trend, and grid expansion shows a slight downward trend, indicating that ES and grid extension have a certain complementarity. The total investment cost significantly decreases as the cost of ES decreases.

Table 10.5 Sizing results of the low-carbon power grid under different acceptable load shedding values (Xie et al. 2023)

Acceptable load shedding λ (MWh)	51.1	76.7	102.2	127.8
Total investment cost (CNY)	3.67×10^9	3.54×10^9	3.45×10^9	3.44×10^9
Renewable capacity $(x_{10}^R, x_{13}^R, x_{27}^R)$ (MW)	(250, 127, 163)	(250, 155, 142)	(250, 216, 103)	(250, 221, 99.6)
ES power capacity (x_8^{Sp}, x_{21}^{Sp}) (MW)	(109, 34.3)	(86.3, 41.2)	(85.1, 30.1)	(84.6, 30.3)
ES energy capacity (x_8^{Se}, x_{21}^{Se}) (MWh)	(753, 160)	(584, 247)	(526, 124)	(516, 118)
New transmission line	(6, 8) × 3 (24, 25) (25, 27) × 2	(6, 8) × 2 (24, 25) (25, 27)	(6, 8) × 2	(6, 8) × 2

Table 10.6 Sizing results of the low-carbon power grid under different conventional thermal power unit capacities (Xie et al. 2023)

Thermal unit capacity (MW)	150	160	170	180
Total investment cost (CNY)	4.35×10^9	3.67×10^9	3.44×10^9	3.43×10^9
Renewable capacity ($x_{10}^R, x_{13}^R, x_{27}^R$) (MW)	(250, 72.4, 283)	(250, 127, 163)	(250, 212, 104)	(225, 231, 129)
ES power capacity (x_8^{Sp}, x_{21}^{Sp}) (MW)	(109, 45.3)	(109, 34.3)	(81.2, 27.6)	(51.9, 42.5)
ES energy capacity (x_8^{Se}, x_{21}^{Se}) (MWh)	(927, 336)	(753, 160)	(537, 131)	(319, 266)
New transmission line	(6, 8) × 3 (22, 24) (24, 25) × 3 (25, 27) × 3	(6, 8) × 3 (24, 25) (25, 27) × 2	(6, 8) × 2	(6, 8) (24, 25) (25, 27)

Table 10.7 Sizing results of the low-carbon power grid under different ES unit capacity costs (Xie et al. 2023)

Multiple of ES unit capacity cost	2.0	1.0	0.50	0.25
Total investment cost (CNY)	4.91×10^9	3.67×10^9	3.05×10^9	2.72×10^9
Renewable capacity ($x_{10}^R, x_{13}^R, x_{27}^R$) (MW)	(242, 134, 159)	(250, 127, 163)	(250, 131, 159)	(250, 133, 152)
ES power capacity (x_8^{Sp}, x_{21}^{Sp}) (MW)	(108, 39.7)	(109, 34.3)	(113, 31.8)	(116, 35.2)
ES energy capacity (x_8^{Se}, x_{21}^{Se}) (MWh)	(674, 225)	(753, 160)	(806, 150)	(821, 150)
New transmission line	(6, 8) × 3 (24, 25) × 2 (25, 27) × 2	(6, 8) × 3 (24, 25) (25, 27) × 2	(6, 8) × 3 (24, 25) (25, 27)	(6, 8) × 3 (24, 25) (25, 27)

The load data is multiplied by a coefficient, and the impact of different load levels on the results is compared in Table 10.8. As the carbon emission cap is fixed, to supply the growing load, a larger capacity of renewable energy generation should be allocated, more transmission lines should be added, and ES should be further expanded for peak shaving.

Table 10.8 Sizing results of the low-carbon power grid under different ES unit capacity costs (Xie et al. 2023)

Load multiple	1.05	1.00	0.95	0.90
Total investment cost (CNY)	4.91×10^9	3.67×10^9	3.07×10^9	2.51×10^9
Renewable capacity ($x_{10}^R, x_{13}^R, x_{27}^R$) (MW)	(250, 120, 267)	(250, 127, 163)	(222, 127, 130)	(213, 127, 87.5)
ES power capacity (x_8^{Sp}, x_{21}^{Sp}) (MW)	(108, 56.4)	(109, 34.3)	(79.6, 33.9)	(56.6, 34.6)
ES energy capacity (x_8^{Se}, x_{21}^{Se}) (MWh)	(723, 421)	(753, 160)	(534, 160)	(340, 167)
New transmission line	(6, 8) × 3 (21, 22) (22, 24) (24, 25) × 3 (25, 27) × 3	(6, 8) × 3 (24, 25) (25, 27) × 2	(6, 8) × 2 (24, 25) (25, 27)	(6, 8)

10.4.5 Adopt Renewable Energy Generation and Energy Storage to Replace Conventional Thermal Power Units

The amount of new renewable energy generation and ES needed is calculated through case studies in the situation where some conventional thermal power units are retired without an increase in the load level. At this time, grid extension is not allowed, and the total investment cost is minimized under carbon emission constraints, where the carbon emission cap is proportional to the total capacity of the still-operating conventional thermal power units. To simulate the process of conventional thermal power unit retirement, we start from the original case of six conventional thermal power units, i.e., 80 MW at bus 1, 80 MW at bus 2, 50 MW at bus 22, 55 MW at bus 27, 30 MW at bus 23, and 40 MW at bus 13. For these six conventional thermal power units to meet the load shedding risk requirements, the load data is multiplied by 0.87 and then kept constant. The retirement of conventional thermal power units follows the order of buses 27, 22, 13, and 23. The sizing results for renewable energy generation and ES are shown in Table 10.9. To retire a 30 MW conventional thermal power unit, approximately 100 MW of renewable power generation needs to be allocated. When the remaining capacity of the conventional thermal power units is small, the demand for ES capacity grows more rapidly.

10.4.6 Analysis of Energy Storage Site Selection

This section will extend the sizing model and its solutions to the renewable-transmission-ES coordination planning model considering ES site selection, explor-

Table 10.9 The sizing results of renewable and ES capacities when conventional thermal power units are phased out (Xie et al. 2023)

Conventional thermal power unit capacity (MW)	335	280	190	160
Total investment cost (CNY)	0	7.27×10^8	3.08×10^9	1.04×10^{10}
Renewable capacity (MW)	0	174	542	682
ES power capacity (MW)	0	7.3	59.1	88.4
ES energy capacity (MWh)	0	19.3	470	754
Daily emission ($kgCO_2$)	3.79×10^6	3.17×10^6	2.15×10^6	1.81×10^6
Renewable penetration rate (%)	0	16	43	52

ing the impact of ES site selection on investment costs and carbon emissions through examples.

10.4.6.1 Coordinated Renewable-Transmission-Storage Planning Model Considering Energy Storage Site Selection

Let z_i^S be a Boolean variable indicating whether ES is built at bus i. $z_i^S = 1$ indicates that ES is built at bus i, and $z_i^S = 0$ indicates no ES is built. Therefore, the ES capacity at bus i should satisfy:

$$0 \le x_i^{Sp} \le X_i^{Spu} z_i^S \ , \ 0 \le x_i^{Se} \le X_i^{Seu} z_i^S \ . \tag{10.51}$$

Include z_i^S in the planning variable x, and consider (10.51) and $z_i^S \in \{0, 1\}$ within the range of planning variable values, we get

$$X := \left\{ \begin{array}{l} x = (x_i^R, z_i^S, x_i^{Sp}, x_i^{Se}, n_{ij}^L, z_{ij,m}^L; i \in S_B, (i, j) \in S_L, m = 0, 1, \ldots, M^L) : \\ 0 \le x_i^R \le X_i^{Ru}, z_i^S \in \{0, 1\}, 0 \le x_i^{Sp} \le X_i^{Spu} z_i^S, 0 \le x_i^{Se} \le X_i^{Seu} z_i^S, \forall i \in S_B, \\ 0 \le n_{ij}^L \le N_{ij}^{Lu}, n_{ij}^L = \sum_{m=0}^{M^L} 2^m z_{ij,m}^L, \forall (i, j) \in S_L, \\ z_{ij,m}^L \in \{0, 1\}, \forall (i, j) \in S_L, m = 0, 1, \ldots, M^L \end{array} \right\} . \tag{10.52}$$

There may be fixed investment costs at the ES location, denoted as $C_i^S, i \in S_B$, so the total investment cost is:

$$\begin{aligned} f(x) &:= \sum_{i \in S_B} (C_i^R x_i^R + C_i^S z_i^S + C^{Sp} x_i^{Sp} + C^{Se} x_i^{Se}) + \sum_{(i,j) \in S_L} C_{ij}^L n_{ij}^L \\ &= \sum_{i \in S_B} (C_i^R x_i^R + C_i^S z_i^S + C^{Sp} x_i^{Sp} + C^{Se} x_i^{Se}) + \sum_{(i,j) \in S_L} C_{ij}^L \sum_{m=0}^{M^L} 2^m z_{ij,m}^L . \end{aligned} \tag{10.53}$$

Substitute X and $f(x)$ from (10.52) and (10.53) into the sizing problem, and we get the renewable-transmission-ES coordinated planning model considering ES location as follows:

$$\begin{aligned} &\min_x \left\{ f(x), \sup_{\mathbb{P} \in \mathcal{B}(\mathbb{P}^N, \varepsilon^N)} \mathbb{E}^{\mathbb{P}}[g^N(x, \xi)] \right\} \\ &\text{s.t.} \quad \sup_{\mathbb{P} \in \mathcal{B}(\mathbb{P}^E, \varepsilon^E)} \text{SR}_{l,\lambda}^{\mathbb{P}}(g^E(x, \xi)) \le R^E \ , \ x \in X \ . \end{aligned}$$

Since only the definition and range of x have been changed, and $f(x)$ is still a linear function, the solution in Sect. 10.3 remains valid. When a carbon emission cap or budget cap is given, the planning model is transformed into a MILP problem for solution.

10.4.6.2 Case Study Results and Analysis

To explore the impact of ES location on the planning results, 4 scenarios are set. The basic parameters are consistent with the previous examples. In addition, all transmission lines can be expanded by 3 lines, with costs proportional to capacity. The carbon emission cap is set at 2.4×10^6 $kgCO_2$, minimizing total investment costs. The fixed investment cost of ES is set to 0. In Scenario 1, the optional locations for ES are the same as in the basic example, which means ES can be configured at bus 8 and bus 21. In Scenario 2, ES can be configured at all buses. In Scenario 3, ES can be configured on the power generation side, that is, at buses 1, 2, 10, 13, 27. In Scenario 4, ES can be configured on the load side, at load buses 2–4, 7, 8, 10, 12, 14-21, 23, 24, 26, 29, 30.

Table 10.10 compares the results of the 4 scenarios. Scenario 1 has the highest total investment cost and the largest number of new transmission lines, because ES can only be installed at two bus locations, and the energy regulation process in the power grid is limited by transmission capacity. Scenario 2 has the lowest total investment cost, as all bus locations can accommodate ES, resulting in the largest feasible area. The optimal solution of Scenario 2 deploys ES in 11 out of the total 30 bus locations. The computational time is the longest, owing to the largest variable value range

Table 10.10 Comparison of ES site selection in the low-carbon power grid (Xie et al. 2023)

Scenario	1	2	3	4
Configurable ES Locations	Buses 8 and 21	All buses	Generation side	Load side
Total investment cost (CNY)	3.67×10^9	3.35×10^9	3.41×10^9	3.40×10^9
Renewable capacity (MW)	541	527	528	532
The number of buses with ES	2	11	4	9
ES power capacity (MW)	143	154	151	151
ES energy capacity (MWh)	913	917	898	914
Number of new transmission lines	6	0	1	1
Number of constraints	7.4e5	1.03e6	7.7e5	9.3e5
Number of variables	2.4e5	3.7e5	2.6e5	3.2e5
Number of Boolean variables	123	123	123	123
Computation time (s)	13,259	32,408	25,960	29,657

in Scenario 2, but it is still acceptable for planning problems. Further, it can be solved using the MILP problem acceleration algorithm, especially the acceleration algorithm for scenario-based stochastic planning, to further reduce the computation time (Li and Grossmann 2021). The total investment costs of Scenario 3 and Scenario 4 are close, higher than Scenario 2 but significantly lower than Scenario 1, and both scenarios require only one new transmission line. In comparison to Scenario 4, Scenario 3 can achieve similar results by configuring ES at only 4 bus locations on the generation side, offering more advantages when the fixed investment cost of ES is high. In conclusion, the proposed renewable-transmission-storage coordination planning model considering ES location selection is effective.

10.5 Conclusion

This chapter proposes an optimal sizing method for renewable generation, transmission, and ES in a low-carbon power grid based on DRO. It takes into account the inexactness of empirical probability distribution via the Wasserstein distance-based probabilistic distribution ambiguity set. Under the distributionally robust risk constraint of load shedding in extreme cases, the bi-objective sizing model minimizes the total investment cost and worst-case expectation of carbon emissions in normal situations. In the proposed solution method, Boolean variables and the big-M method are used to transform nonlinear constraints, proving that the complementary constraints of storage charging-discharging power can be removed. Distributionally robust risk and worst-case expectation are transformed into linear models based on the Lipschitz constant method. Using the ϵ-constraint method, the Pareto frontier is calculated.

The example is based on the IEEE-30 node system and renewable power generation resource data are from Qinghai Province. The proposed method is compared with SP, RO, and DRO based on Wasserstein distance that only considers discrete probability distributions. The results show that the proposed method can achieve a good balance between robustness and optimality. The sensitivity of daily carbon emissions to total investment costs is approximately -3×10^{-4} kgCO_2/CNY. To reduce daily carbon emissions by 100 kgCO_2, it is roughly necessary to allocate 40 kW of renewable power generation. The ES capacity is affected by a variety of factors. When the proportion of conventional thermal power units in the system is high, the demand for ES capacity is small; otherwise, the demand for ES capacity rapidly increases with the reduction of the capacity of conventional thermal power units.

References

Chaube A, Chapman A, Minami A et al (2021) The role of current and emerging technologies in meeting Japan's mid- to long-term carbon reduction goals. Appl Energy 304:117669

Dong W, Chen X, Yang Q (2022) Data-driven scenario generation of renewable energy production based on controllable generative adversarial networks with interpretability. Appl Energy 308:118387

Gurobi Optimization, LLC (2023) Gurobi optimizer reference manual. https://www.gurobi.com/documentation/current/refman/index.html [2023-12-20]

Kang C, Zhou T, Chen Q et al (2015) Carbon emission flow from generation to demand: a network-based model. IEEE Trans Smart Grid 6(5):2386–2394

Li C, Grossmann IE (2021) A review of stochastic programming methods for optimization of process systems under uncertainty. Front Chem Engin 2

Lofberg J (2004) YALMIP: a toolbox for modeling and optimization in MATLAB. In: 2004 IEEE international conference on robotics and automation, pp 284–289

Weng Y, Cai W, Wang C (2021) Evaluating the use of BECCS and afforestation under China's carbon neutral target for 2060. Appl Energy 299:117263

Xie R, Wei W, Li M et al (2023) Sizing capacities of renewable generation, transmission, and energy storage for low-carbon power systems: a distributionally robust optimization approach. Energy 263:125653

Zimmerman RD, Murillo-Sánchez CE, Thomas RJ (2011) MATPOWER: steady-state operations, planning, and analysis tools for power systems research and education. IEEE Trans Power Syst 26(1):12–19

Chapter 11
Sizing Wind Farm and Energy Storage Considering Wake Effect

Abstract The wake effect within a wind farm can precipitate wind speed deficits, subsequently leading to a decline in the power generation of downstream wind turbines. This chapter proposes a bi-objective distributionally robust optimization (DRO) model, which aims to determine the capacities of wind power generation and energy storage while considering the wake effect. To encapsulate the uncertainties of wind power and demand, an ambiguity set, grounded in the Wasserstein distance, is established. It is worth noting that wind power uncertainty is directly impacted by the wind power generation capacity, which is determined in the first stage. Consequently, the proposed model typifies a DRO problem with endogenous uncertainty, also referred to as decision-dependent uncertainty. To solve the proposed model, a method utilizing the approximation of stochastic programming through Lipschitz moduli is developed. This method effectively transforms the DRO model into a linear programming problem. Following this, an iterative algorithm is formulated, which incorporates methods for evaluating the Lipschitz moduli.

This chapter introduces the work in Xie et al. (2023).

11.1 Introduction

The wake effect within a wind farm is a phenomenon that results in a reduction of gross energy production due to alterations in wind speed, instigated by the interaction of wind turbines (WTs) with each other. The wake created by upstream WTs diminishes the kinetic energy of the wind, subsequently reducing the wind speed at the downstream WTs. Given that the wake effect can potentially curtail power output by approximately 20% as per Gao et al. (2016), it plays a significant role and is an indispensable factor in the planning and operation of wind farms.

The motivation for the proposed method is rooted in several key factors. First, the real-time flexibility of the power system is instrumental in accommodating wind power, making it crucial to consider the interplay between wind farms and the power system. Second, during the planning phase, the empirical probability distribution may not be accurate, thus necessitating the application of the distributionally robust

R. Xie and W. Wei, *Distributionally Robust Optimization and its Applications in Power System Energy Storage Sizing*,
https://doi.org/10.1007/978-981-97-2566-3_11

optimization (DRO) technique. Third, energy storage (ES) represents a significant flexibility resource for wind-integrated power systems, supporting both operational reliability and wind power accommodation. Therefore, it should be considered concurrently in the proposed method.

This chapter presents a novel distributionally robust method for sizing wind farms and ES, taking into account the wake effect. The wake effect introduces a decision-dependent factor, where uncertain wind power depends not only on random wind speed but also on the sizing decision. This variability gives rise to a DRO problem with endogenous uncertainty, also known as decision-dependent uncertainty (DDU). This form of uncertainty fundamentally differentiates the problem from traditional DRO problems that do not involve DDU, as the decision impacts both the ambiguity set and the worst-case distribution. DRO with DDU has been studied in Luo and Mehrotra (2020), which focuses on ambiguity sets containing only discrete distributions. However, given that wind and load uncertainties possess continuous probability distributions, we need a solution method tailored for the DRO problem with DDU and Wasserstein-distance-based ambiguity sets consisting of general distributions. Therefore, the key contributions of this chapter include:

First, this chapter presents a DRO model with DDU specifically designed for the sizing of wind power generation and energy storage systems (ESSs) in power systems. Unlike traditional models, this model optimizes the worst-case expected fuel cost under normal operating conditions, while also constraining the worst-case expected load shedding under extreme conditions. Unique to this model is the consideration of the wake effect within a wind farm, achieved by simulating the available wind power and approximating it with a piecewise linear model. Owing to the wake effect, the Wasserstein-distance-based ambiguity set of the uncertainties becomes dependent on the sizing decisions, thereby introducing a decision-dependent characteristic to the model.

Second, to address the proposed DRO model with DDU built on the Wasserstein distance, we derive a stochastic programming (SP) approximation of the model, aided by the use of Lipschitz moduli. Subsequently, we establish an iterative algorithm, integrated with methods for calculating the upper bounds of the Lipschitz moduli, to resolve the resulting SP problem. This algorithm is innovative in its provision of a systematic method for solving the Wasserstein-distance-based DRO model that considers both continuous and discrete probability distributions, while it is capable of handling DDUs. This approach provides a robust and comprehensive solution for tackling complex DRO problems in wind farm planning.

The remainder of this chapter is structured as follows. Section 11.2 introduces a linear approximation of the available wind power function that incorporates the wake effect. Section 11.3 proposes a distributionally robust wind–storage sizing model that also considers the wake effect, with a corresponding solution method developed in Sect. 11.4. Section 11.5 presents a series of case studies to illustrate the practical application of the model. Finally, Sect. 11.6 concludes the chapter.

11.2 Considering Wake Effect in the Linearized Function of Available Wind Power

In this section, we explore the available wind power within a wind farm, taking into account the impact of the wake effect. Initially, we utilize the Jensen wake model to simulate the wake effect produced by a single WT. Subsequent simulations are conducted to estimate the available wind power in a wind farm, taking into consideration varying total WT capacities and real-time wind conditions. This establishes a relationship represented through a nonlinear function. For the sake of computational simplicity, we then further refine this nonlinear function into a linearized form.

11.2.1 Determining Available Wind Power Through Simulation

Figure 11.1 provides a visual representation of the wake effect observed between WTs. The wind is depicted as originating from the left-hand side, with the wind speed at WT 1 identified as v_1^W. As WT 1 harnesses kinetic energy from the wind, the speed on its right side decreases below v_1^W. This phenomenon is known as the wake effect of WT 1. Similarly, at WT 2, the wind speed from the left side is v_2^W, and on its right side, the speed is further diminished due to the wake effect of WT 2. WT 3 follows a similar pattern. Consequently, $v_1^W > v_2^W > v_3^W$, indicating that the wake effect contributes to a decrease in wind speed and power output at the WTs located downstream.

The Jensen wake model, a popular analytical tool for analyzing wake effects, is widely used (Serrano González et al. 2014). Assuming the wind direction aligns parallel to the line connecting two WTs, the Jensen wake model enables the calculation of the wind speed at the downstream WT as follows (Feijoo and Cidras 2000):

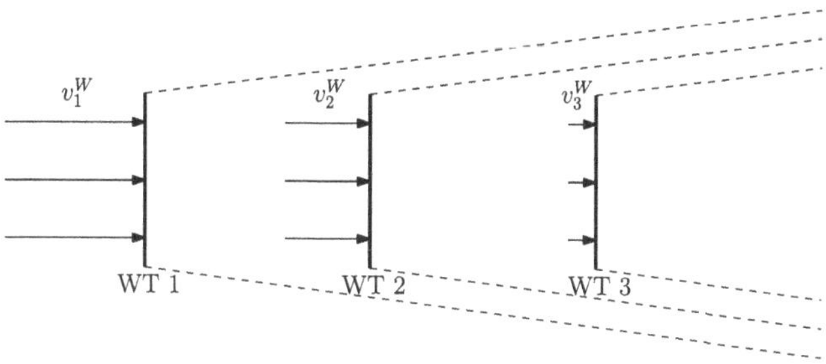

Fig. 11.1 Illustration of the wake effect between wind turbines (Xie et al. 2023)

$$v_2^W = v_1^W \left(1 - \left(1 - \sqrt{1 - C^T}\right)\left(\frac{D^R}{D^R + 2K^W d^W}\right)^2\right), \tag{11.1}$$

$$K^W = 0.5/\log\left(H^H/Z^R\right). \tag{11.2}$$

In the above equations, v_1^W and v_2^W represent the wind speed at the upstream and downstream WTs, respectively, while d^W indicates the distance separating the two WTs. The thrust factor (C^T), the turbine rotor diameter (D^R), the hub height (H^H), and the roughness length (Z^R) are all parameters. K^W stands for the wake decay constant. As demonstrated in (11.1), the proximity between the two WTs directly influences the strength of the wake and the resulting wind speed deficit. The closer the WTs are to one another, the greater the intensity of the wake effect and the more significant the reduction in wind speed.

For demonstrative purposes, let us consider the wind farm layout depicted in Fig. 11.2: The wind farm is set to be constructed within a pre-determined rectangular area. All WTs are of the same type, share identical parameters, and are evenly spaced. The wind direction is parallel to the rows of WTs. The distance between each adjacent row is fixed and sufficient to negate any significant wake effect between them. A similar assumption can be found in Feijoo and Cidras (2000). However, other wind farm layouts and wake models can be applied in simulations, and the subsequent process from Sects. 11.2.2 to 11.4 can still proceed. Assuming that the length of a row is R^P and there is more than one WT in a row ($n^W > 1$), the distance between adjacent WTs can be calculated as $d^W = R^P/(n^W - 1)$. The wind speed at the i-th WT is then determined as follows:

$$v_i^W = v_1^W \left(1 - \left(1 - \sqrt{1 - C^T}\right)\left(\frac{D^R}{D^R + 2K^W d^W}\right)^2\right)^{i-1}. \tag{11.3}$$

The available wind power of a single WT can be computed utilizing the subsequent function (Villanueva et al. 2011).

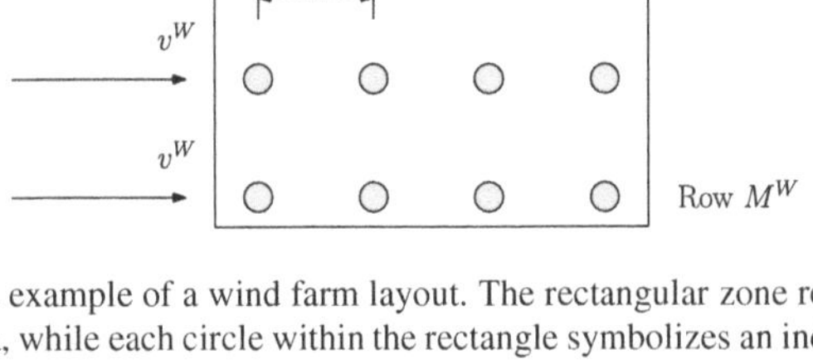

Fig. 11.2 Here is an example of a wind farm layout. The rectangular zone represents the entirety of the wind farm area, while each circle within the rectangle symbolizes an individual wind turbine (Xie et al. 2023)

$$p^W(v) := \begin{cases} 0\,, & \text{if } v \le V^C \text{ or } v > V^F\,, \\ P^R \frac{v^2-(V^C)^2}{(V^R)^2-(V^C)^2}\,, & \text{if } V^C < v \le V^R\,, \\ P^R\,, & \text{if } V^R < v \le V^F\,. \end{cases} \tag{11.4}$$

In this equation, p^W represents the available wind power, and v is the wind speed at the given WT. The parameters include the cut-in speed (V^C), cut-out speed (V^F), rated speed (V^R), and rated power (P^R). By aggregating the available power from all WTs, we can determine the total available wind power of the wind farm as $f^W := M^W \sum_{i=1}^{n^W} p^W(v_i^W)$, where M^W signifies the number of rows. Given the number of WTs in a row, n^W, and the initial wind speed, $\nu^W := v_1^W$ at the wind farm, we can uniquely define the total capacity of WTs, $x^W = P^R M^W n^W$, and the total available wind power f^W. Therefore, we can characterize the variation of available wind power in relation to x^W and ν^W through simulations.

For clarity, let us denote the upper bounds of n^W and ν^W as $\bar{n}^W$ and $\bar{\nu}^W$, respectively. We will use $N^V + 1$ wind speed values for discretization. For $n^W = 1, 2, \ldots, \bar{n}^W$ and $\nu^W = 0, \bar{\nu}^W/N^V, 2\bar{\nu}^W/N^V, \ldots, \bar{\nu}^W$, we calculate the corresponding values of f^W and x^W. This process results in tuples (x_l^W, ν_l^W, f_l^W) for $l = 1, 2, \ldots, \bar{n}^W(N^V + 1)$. The fluctuation of f^W in relation to x^W and v^W is depicted in Fig. 11.3. When ν^W is held constant and x^W ranges from 0 to $\bar{x}^W := P^R M^W \bar{n}^W$, the value of f^W initially rises proportionally. However, its rate of growth diminishes over time. In fact, f^W may even decrease when x^W becomes excessively large due to the severity of the wake effect. This phenomenon emphasizes the importance of considering the wake effect when determining the size of the wind farm.

For the sake of simplicity, we posit that all WTs within a wind farm are of the same type when simulating available wind power. However, this model can be easily modified to accommodate wind farms with heterogeneous WTs. This allows for the calculation of the available wind power of the wind farm under varying total capacities and initial wind speeds. It is important to note that the method proposed from Sects. 11.2.2 to 11.4 remains applicable to scenarios involving wind and ES sizing.

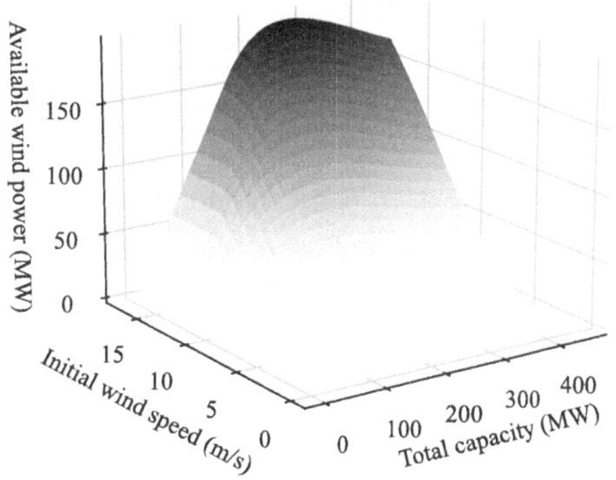

Fig. 11.3 The available wind power, denoted as f^W, is a function of two key parameters: the total capacity, represented as x^W, and the initial wind speed, symbolized as ν^W (Xie et al. 2023)

The accuracy of sizing decisions can be enhanced if layout design optimization is concurrently considered in the sizing problem, as suggested in Ma et al. (2022). However, wind farm layout optimization presents itself as a complex combinatorial problem, making the discovery of a global optimum challenging. Therefore, merging the layout design phase into the sizing phase becomes complex, leading us to adopt certain simplifications to enhance computational efficiency. One potential approach to incorporate layout design into the proposed distributionally robust wind farm sizing method is by integrating the layout design within the available wind power simulation. In this way, the benefits brought about by the layout design can be considered in the sizing model via the available wind power function.

11.2.2 Linearization of Available Wind Power Function

Given that the relationship between f^W and x^W, v^W is notably nonlinear, we aim to simplify computation by establishing a piecewise linear approximation for the available wind power function. Initially, to eradicate the nonlinear term in (11.4), we introduce an auxiliary variable, denoted as ξ^W.

$$\xi^W(v^W) := \begin{cases} 0\,, & \text{if } v^W \leq V^C\,, \\ \frac{(v^W)^2-(V^C)^2}{(V^R)^2-(V^C)^2}\,, & \text{if } v^W > V^C\,. \end{cases} \tag{11.5}$$

Subsequently, the available wind power, f^W, can be interpreted as a function of x^W and ξ^W, as illustrated in Fig. 11.4. The actual wind power output can vary between 0 and f^W, thus the feasible region for wind power lies below the plane in Fig. 11.4 and above zero. This can be effectively approximated by its convex hull. The Multi-Parametric Toolbox (MPT) 3.0 (Herceg et al. 2013), can be employed to identify the convex hull of a set of points, either represented by its set of extreme points or a linear inequality set describing the facets. We utilize MPT to determine the convex hull of the feasible region for wind power, resulting in a polyhedron, as described below.

$$\mathcal{P}^W := \left\{ (x^W, \xi^W, p^W) \;\middle|\; \begin{array}{l} p^W \geq 0, x^W \leq \bar{x}^W, \xi^W \leq \bar{\xi}^W \\ p^W \leq a_k^{(1)} x^W + a_k^{(2)} \xi^W + a_k^{(3)}, \forall k \in \mathcal{K}_0^a \end{array} \right\}. \tag{11.6}$$

In this context, $a_k^{(1)}$, $a_k^{(2)}$, and $a_k^{(3)}$ serve as parameters within the inequalities. The term $\min_{k\in\mathcal{K}_0^a}\{a_k^{(1)}x^W + a_k^{(2)}\xi^W + a_k^{(3)}\}$ provides a piecewise linear approximation of the available wind power function.

The formula (11.6) might comprise a multitude of inequalities. To enhance computational efficiency, we propose Algorithm 11.1 to select a representative subset of inequalities that still maintains an acceptable level of precision. Algorithm 11.1 begins with a cuboid that encompasses the original polyhedron and incrementally

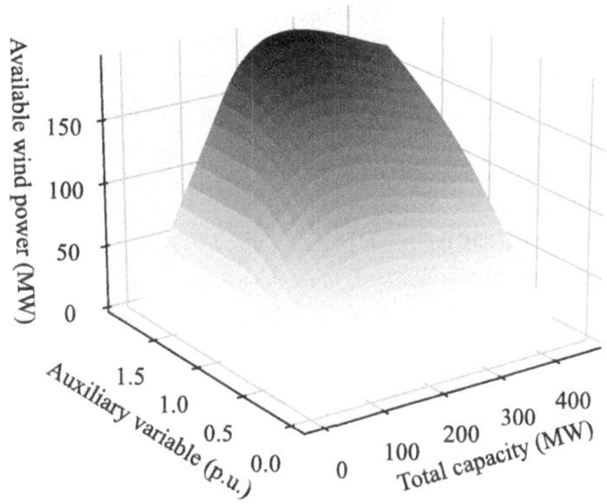

Fig. 11.4 Use the auxiliary variable ξ^W to represent the available wind power function (Xie et al. 2023)

adds the most critical remaining inequality in each iteration until the approximation error falls below a designated threshold. Algorithm 11.1 is guaranteed to converge due to the finite number of inequalities involved.

Algorithm 11.1 Selection of representative inequalities

Input: Error tolerance $\varepsilon > 0$, $\bar{x}^W$, $\bar{\xi}^W$, and coefficients $(a_k^{(1)}, a_k^{(2)}, a_k^{(3)})$, $k \in \mathcal{K}_0^a$ in the inequalities.

Output: An inequality index set $\mathcal{K}^a \subset \mathcal{K}_0^a$ and the corresponding approximate polyhedron $\mathcal{P}$.

1: Initiation: Let $\mathcal{K}^a \leftarrow \varnothing$.

2: Let $\mathcal{P}$ be

$$\left\{ (x^W, \xi^W, p^W) \;\middle|\; \begin{array}{l} 0 \le x^W \le \bar{x}^W, 0 \le \xi^W \le \bar{\xi}^W \\ 0 \le p^W \le \bar{p}^W \\ p^W \le a_k^{(1)} x^W + a_k^{(2)} \xi^W + a_k^{(3)}, \forall k \in \mathcal{K}^a \end{array} \right\}.$$

3: Find the vertices (x_l^W, ξ_l^W, p_l^W), $l \in \mathcal{I}$ of $\mathcal{P}$ using MPT, where $\mathcal{I}$ is the index set.

4: Compute the maximum error of approximation:

$$\max_{l \in \mathcal{I}} \left\{ \max_{k \in \mathcal{K}_0^a \setminus \mathcal{K}^a} \left\{ p_l^W - (a_k^{(1)} x_l^W + a_k^{(2)} \xi_l^W + a_k^{(3)}) \right\} \right\}.$$

The internal max function calculates the approximation error at the point (x_l^W, ξ_l^W). If the error does not exceed ε, the process is terminated. However, if it does, and assuming $k_0 \in \mathcal{K}_0^a \setminus \mathcal{K}^a$ achieves the aforementioned maximum, then $\mathcal{K}^a$ is updated to $\mathcal{K}^a \leftarrow \mathcal{K}^a \cup \{k_0\}$. After this update, the process returns to Step 2.

11.3 Sizing Model

In this section, we initially suggest models designed to assess operational costs in relation to the sizing strategies. Following this, we construct ambiguity sets. These sets then serve as the foundation upon which we establish the distributionally robust sizing model.

11.3.1 Operation Optimization

Firstly, we take into account the operational stage, factoring in both the wind power generation and ESS capacities along with associated uncertainties. We develop two distinct models designed to evaluate operational costs under two conditions. The first model comes into play under normal conditions, where the primary objective is to minimize total fuel cost without permitting load shedding. The second model is designed for extreme conditions, where the focus shifts to minimizing total load shedding.

11.3.1.1 Power System Model

Let $S^T := \{1, 2, \ldots, T\}$, S^B, and S^L represent the set of periods, buses, and transmission lines, respectively. We symbolize the period length as Δ_t. The constraints governing the operation of the power system are outlined as follows.

$$f_{i,t}^G = \max_{k \in \mathcal{K}^b} \left\{ b_{i,k}^{(1)} p_{i,t}^G \Delta_t + b_{i,k}^{(2)} \right\} \ , \ \forall i \in S^B \ , \ t \in S^T \ , \tag{11.7a}$$

$$\underline{P}_i^G \le p_{i,t}^G \le \overline{P}_i^G \ , \ \forall i \in S^B \ , \ t \in S^T \ , \tag{11.7b}$$

$$\underline{R}_i^G \Delta_t \le p_{i,t}^G - p_{i,t-1}^G \le \overline{R}_i^G \Delta_t \ , \ \forall i \in S^B \ , \ t \in S^T \ , \tag{11.7c}$$

$$p_{i,T}^G = p_{i,0}^G \ , \tag{11.7d}$$

$$0 \le p_{i,t}^{SC} \le x_i^{SP} \ , \ 0 \le p_{i,t}^{SD} \le x_i^{SP} \ , \ \forall i \in S^B, t \in S^T \ , \tag{11.7e}$$

$$e_{i,t}^S = e_{i,t-1}^S + \left(\eta^{SC} p_{i,t}^{SC} - \frac{p_{i,t}^{SD}}{\eta^{SD}} \right) \Delta_t \ , \ \forall i \in S^B \ , \ t \in S^T \ , \tag{11.7f}$$

$$e_{i,T}^S = e_{i,0}^S \ , \tag{11.7g}$$

$$\underline{\omega}^S x_i^{SE} \le e_{i,t}^S \le \overline{\omega}^S x_i^{SE} \ , \ \forall i \in S^B \ , \ t \in S^T \ , \tag{11.7h}$$

$$p_{ij,t}^L = \frac{\theta_{i,t}^B - \theta_{j,t}^B}{X_{ij}^L}, -S_{ij}^L \le p_{ij,t}^L \le S_{ij}^L \ , \ \forall (i,j) \in S^L \ , \ t \in S^T \ , \tag{11.7i}$$

$$0 \le p_{i,t}^D \le \zeta_{i,t}^D, \ p_{i,t}^C \ge 0 \ , \ \forall i \in S^B \ , \ t \in S^T \ , \tag{11.7j}$$

$$\sum_{(k,i)\in S^L} p_{ki,t}^L + p_{i,t}^G + \zeta_{i,t}^W + p_{i,t}^{SD} - p_{i,t}^C = \sum_{(i,j)\in S^L} p_{ij,t}^L + p_{i,t}^{SC} + \zeta_{i,t}^D - p_{i,t}^D \ , \ \forall i \in S^B \ , \ t \in S^T \ . \tag{11.7k}$$

The model for traditional generators is comprised of constraints (11.7a)–(11.7d). Here, the power generation and fuel cost of the traditional generator at bus i during period t are represented as $p_{i,t}^G$ and $f_{i,t}^G$ respectively. As per constraint (11.7a), the fuel cost is depicted as the apex of a series of linear functions, where $b_{i,k}^{(1)}, b_{i,k}^{(2)}, k \in \mathcal{K}^b$ are immutable coefficients. Constraint (11.7b) defines the power generation range as $[\underline{P}_i^G, \overline{P}_i^G]$. The ramp rates' lower and upper bounds, denoted by $\underline{R}_i^G$ and $\overline{R}_i^G$, are encapsulated within constraints (11.7c) and (11.7d). The latter constraint guarantees uninterrupted operation.

The ESS model is composed of constraints (11.7e)–(11.7h). Here, $p_{i,t}^{SC}$ and $p_{i,t}^{SD}$ symbolize the charging and discharging power of the ESS at bus i during period t, while $e_{i,t}^S$ denotes the energy stored at the conclusion of period t. The power and energy capacities of the ESS at bus i are represented by x_i^{SP} and x_i^{SE}, respectively. Constraint (11.7e) restricts the charging/discharging power. The dynamics of stored energy are governed by constraints (11.7f) and (11.7g), factoring in charging and discharging efficiencies η^{SC} and η^{SD}. Lastly, constraint (11.7h) establishes the ESSs' state-of-charge (SoC) lower limit $\underline{\omega}^S$ and upper limit $\overline{\omega}^S$. The complementary constraint $p_{i,t}^{SC} p_{i,t}^{SD} = 0$, which prohibits simultaneous charging and discharging, is intentionally excluded in accordance with Propositions 10.1 and 10.2. This is done under the understanding that the optimization of fuel cost or load-shedding values will not be influenced by this exclusion.

The DC power flow model employed in (11.7i) is where $p_{ij,t}^L$ signifies the active power flow from node i to node j at time t. The reactance and capacity of the line are denoted as X_{ij}^L and S_{ij}^L, respectively, while the voltage angle is represented by $\theta_{i,t}^B$. The load demand, load shedding, and power curtailment at bus i during period t are symbolized by $\zeta_{i,t}^D$, $p_{i,t}^D$, and $p_{i,t}^C$, respectively, with their bounds specified in (11.7j). The nodal power balance equation is encapsulated in constraint (11.7k), in which $\zeta_{i,t}^W$ characterizes the wind power generation at bus i during period t.

11.3.1.2 Operation Optimization in Normal Conditions

Under normal conditions, the objective is to reduce the total fuel cost associated with traditional generators.

$$g^N(x,\zeta) := \min_{f^G, p^G, p^{SC}, p^{SD}, e^S, p^L, \theta^B, p^C} \sum_{t\in S^T} \sum_{i\in S^B} f_{i,t}^G \tag{11.8a}$$

$$\text{s.t. } f_{i,t}^G \geq b_{i,k}^{(1)} p_{i,t}^G \Delta_t + b_{i,k}^{(2)} \ , \ \forall k \in \mathcal{K}^b \ , \ i \in S^B \ , \ t \in S^T \ , \tag{11.8b}$$

$$p_{i,t}^D = 0 \ , \ \forall i \in S^B \ , \ t \in S^T \ , \ \text{(11.7b)–(11.7k)} \ . \tag{11.8c}$$

The objective (11.8a) is designed to minimize the total fuel cost. The fuel cost model depicted in (11.7a) is converted into equivalent inequalities in (11.8b), facilitated by the minimization process in the objective. The remaining constraints in the power system model are incorporated, and load shedding is set to zero in (11.8c), given that load shedding is not permissible under normal circumstances. We denote the capacity variable as $x := (x_i^W, x_i^{SP}, x_i^{SE}; i \in S^B)$ and the uncertainty realization as $\zeta := (\zeta_{i,t}^W, \zeta_{i,t}^D; i \in S^B, t \in S^T)$. As (x, ζ) varies, the optimal objective value in (11.8a) adjusts accordingly, represented as $g^N(x, \zeta)$.

11.3.1.3 Operation Optimization in Extreme Conditions

In extreme conditions, the aim is to minimize the total amount of load shedding.

$$g^E(x, \zeta) := \min_{p^D, p^G, p^{SC}, p^{SD}, e^S, p^L, \theta^B, p^C} \sum_{t \in S^T} \sum_{i \in S^B} p_{i,t}^D \Delta_t \tag{11.9a}$$

$$\text{s.t.} \quad (11.7b)\text{–}(11.7k) \; . \tag{11.9b}$$

The total load shedding is minimized as per (11.9a), adhering to the power system constraints outlined in (11.9b). Fuel cost is not a primary consideration during extreme conditions. Similarly, the optimal objective value is symbolized by $g^E(x, \zeta)$, a function of x and ζ. This representation indicates the total load shedding under the given parameters (x, ζ).

11.3.2 Decision-Dependent Ambiguity Sets Based on Wasserstein Distance

To model uncertainties, we employ ambiguity sets. These are formed by a collection of distributions centered around the empirical distribution, with the distances between them measured using the Wasserstein distance.

11.3.2.1 Empirical Distributions

Assuming we possess some empirical data for wind and load demand in extreme conditions, denoted as $(\hat{\xi}_{i,t,n}^W, \hat{\zeta}_{i,t,n}^D; i \in S^B, t \in S^T), n \in S^E$, where S^E represents the index set of extreme scenarios. The available wind power can then be calculated based on this data.

$$\hat{\zeta}_{i,t,n}^W(x_i^W) = \min \left\{ \min_{k \in \mathcal{K}^a} \{a_{i,k}^{(1)} x_i^W + a_{i,k}^{(2)} \hat{\xi}_{i,t,n}^W + a_{i,k}^{(3)}\}, \hat{\xi}_{i,t,n}^W x_i^W \right\} , \tag{11.10}$$

Here, $\min_{k\in\mathcal{K}^a}\{a_{i,k}^{(1)}x_i^W + a_{i,k}^{(2)}\hat{\xi}_{i,t,n}^W + a_{i,k}^{(3)}\}$ represents the linearized approximation provided by Algorithm 11.1. Moreover, $\hat{\xi}_{i,t,n}^W x_i^W$ provides a theoretical upper limit for the available wind power, excluding consideration of the wake effect and cut-out wind speed, as illustrated in (11.4) and (11.5). The term $\hat{\zeta}_{i,t,n}^W(x_i^W)$ underlines that the available wind power is dependent on the decision taken. Consequently, the data for scenario $n \in S^E$ are defined as $\hat{\zeta}_n(x) := (\hat{\zeta}_{i,t,n}^W(x_i^W), \hat{\zeta}_{i,t,n}^D; i \in S^B, t \in S^T)$. Based on these data, the empirical probability distribution in extreme conditions, $\mathbb{P}_0^E(x)$, can be determined.

$$\mathbb{P}_0^E(x) := \frac{1}{|S^E|}\sum_{n\in S^E}\delta_{\hat{\zeta}_n(x)} = \frac{1}{|S^E|}\sum_{n\in S^E}\left(\times_{i\in S^B}\delta_{\hat{\zeta}_{i,n}^W(x_i^W)}\right)\times\delta_{\hat{\zeta}_n^D}\,.$$

Here, $|S^E|$ signifies the number of scenarios within S^E. The symbol δ represents the Dirac distribution with a unit mass at the subscript point, with $\times$ indicating the product operation of distributions. Consequently, each available wind power sample $\hat{\zeta}_n(x)$ is assigned a probability of $1/|S^E|$ under the empirical distribution $\mathbb{P}_0^E(x)$. Further, let us denote the index set of normal scenarios as S^N. In a similar fashion, the corresponding empirical distribution $\mathbb{P}_0^N(x)$ can be established.

11.3.2.2 Sample Space

For the wind farm located at bus i, the sample space $\Xi_i^{WE}(x_i^W)$ of the wind power output ζ_i^W in extreme conditions is constructed as follows.

$$\Xi_i^{WE}(x_i^W) := \left\{\zeta_i^W \;\middle|\; \begin{array}{l} \zeta_{i,t}^W \le a_{i,k}^{(1)}x_i^W + a_{i,k}^{(2)}\overline{\xi}_i^W + a_{i,k}^{(3)}, \forall k \in \mathcal{K}^a \\ 0 \le \zeta_{i,t}^W \le x_i^W \end{array}\right\}.$$

The wind power $\zeta_{i,t}^W$ must not surpass the limits determined in Algorithm 11.1, nor should it exceed the capacity x_i^W. Once again, we denote the sample space of wind power as $\Xi_i^{WE}(x_i^W)$ to underscore the fact that it is dependent on decision-making factors.

In extreme conditions, the sample space of the load demand is denoted by $\Xi^{DE} := \{\zeta^D \mid 0 \le \zeta_{i,t}^D \le \overline{\zeta}_i^D\}$. Here, $\overline{\zeta}_i^D$ represents the maximum conceivable load demand at bus i.

In normal conditions, the sample space is defined by the convex hull of the corresponding scenarios.

$$\Xi_i^{WN}(x_i^W) := \left\{\zeta_i^W = \sum_{n\in S^N}\lambda_n\hat{\zeta}_{i,n}^W(x_i^W) \;\middle|\; \lambda_n \ge 0, \sum_{n\in S^N}\lambda_n = 1\right\}.$$

In this context, $\lambda_n, n \in S^N$ serve as the coefficients of the convex combination, and all such combinations of $\hat{\zeta}_{i,n}^W(x_i^W), n \in S^N$ are encompassed within $\Xi_i^{WN}(x_i^W)$. The sample space Ξ^{DN} of load demand under normal conditions follows a similar pattern. When compared to normal conditions, the sample space of uncertain variables during extreme conditions is expanded to account for potential extreme scenarios. Therefore, $\Xi_i^{WN}(x_i^W)$ is a subset of $\Xi_i^{WE}(x_i^W)$ and Ξ^{DN} is a subset of Ξ^{DE}.

11.3.2.3 Ambiguity Set

We suggest the following ambiguity set $\mathcal{B}^E(\varepsilon^E, x)$ for the uncertain variable ζ under extreme conditions.

$$
\begin{aligned}
\mathcal{B}^E(\varepsilon^E, x) := \Bigg\{ & \frac{1}{|S^E|} \sum_{n \in S^E} \left(\times_{i \in S^B} \mathbb{P}_{i,n}^W \right) \times \mathbb{P}_n^D \;\Bigg|\\
& \mathbb{P}_{i,n}^W \in \mathcal{M}_1(\Xi_i^{WE}(x_i^W)) \;,\; \Pi_i^W = \frac{1}{|S^E|} \sum_{n \in S^E} \mathbb{P}_{i,n}^W \times \delta_{\hat{\zeta}_{i,n}^W(x_i^W)} \;,\\
& \int_{(\Xi_i^{WE}(x_i^W))^2} \|\zeta_i^W - \tilde{\zeta}_i^W\|_1 \Pi_i^W(\mathrm{d}\zeta_i^W, \mathrm{d}\tilde{\zeta}_i^W) \le \varepsilon_i^{WE} x_i^W \;,\; \forall i \in S^B \;,\\
& \mathbb{P}_n^D \in \mathcal{M}_1(\Xi^{DE}) \;,\; \Pi^D = \frac{1}{|S^E|} \sum_{n \in S^E} \mathbb{P}_n^D \times \delta_{\hat{\zeta}_n^D} \;,\\
& \int_{(\Xi^{DE})^2} \|\zeta^D - \tilde{\zeta}^D\|_1 \Pi^D(\mathrm{d}\zeta^D, \mathrm{d}\tilde{\zeta}^D) \le \varepsilon^{DE} \Bigg\} \;.
\end{aligned}
\tag{11.11}
$$

The distributions within $\mathcal{B}^E(\varepsilon^E, x)$ are composed of the product distributions of $\mathbb{P}_{i,n}^W$ and $\mathbb{P}_n^D$, averaged. Here, $\mathcal{M}_1(\cdot)$ represents the set of all distributions supported on the range enclosed within the brackets that have finite first-order moments. Π_i^W is a joint distribution with marginals $\frac{1}{|S^E|}\sum_{n \in S^E} \mathbb{P}_{i,n}^W$ and the empirical distribution $\mathbb{P}_{i,0}^{WE} := \frac{1}{|S^E|}\sum_{n \in S^E} \delta_{\hat{\zeta}_{i,n}^W(x_i^W)}$. The Wasserstein distance between these distributions is constrained to be no greater than $\varepsilon_i^{WE} x_i^W$, where ε_i^{WE} is a parameter that determines the size of the ambiguity set. Since a larger capacity can lead to more significant fluctuations in wind power, we utilize the product $\varepsilon_i^{WE} x_i^W$ as the bound. The constraints on the load demand distribution $\mathbb{P}_n^D$ follow a similar bound, denoted as ε^{DE}. Let $\varepsilon^E := (\varepsilon_i^{WE}, \varepsilon^{DE}; i \in S^E)$. The ambiguity set then encompasses $\mathbb{P}_0^E(x)$ for all nonnegative values of ε^E and reduces to a singleton when $\varepsilon^E = 0$, the zero vector. Moreover, the joint distribution Π_i^W is necessitated to be the average of the product distribution of $\mathbb{P}_{i,n}^W$ and the Dirac distribution at $\tilde{\zeta}_{i,n}^W(x_i^W)$. In this manner, $\mathbb{P}_{i,n}^W$ is associated with the scenario $n \in S^E$ and, moreover, $\mathbb{P}_{k,n}^W, \forall k \in S^B$ and $\mathbb{P}_n^D$, so that the correlation between the wind farms and loads can be taken into account. Since the structure of Π_i^W is restricted, the constraints for the wind power distribution in (11.11) differ from the pure Wasserstein distance constraint:

$$d^W\left(\frac{1}{|S^E|}\sum_{n\in S^E}\mathbb{P}^W_{i,n},\mathbb{P}^{WE}_{i,0}\right)\le\varepsilon^{WE}_i x^W_i \ .$$

In the latter case, $\mathbb{P}^W_{i,n}$ cannot be associated with the scenario data, thereby neglecting the correlation between the loads and wind farms.

The ambiguity set $\mathcal{B}^N(\varepsilon^N, x)$ under normal conditions can be defined in a similar manner. This is achieved by substituting S^E, Ξ^{WE}_i, Ξ^{DE}, ε^{WE}_i, and ε^{DE} with S^N, Ξ^{WN}_i, Ξ^{DN}, ε^{WN}_i, and ε^{DN}, respectively. Both of these ambiguity sets are dependent on decisions made.

Certain literature on Wasserstein-distance-based DRO employs ambiguity sets that only incorporate discrete distributions supported on the sample data sets, as seen in Luo and Mehrotra (2020). However, such an approach may not be appropriate for wind power and load demand uncertainties because these can change continuously. In contrast, the Wasserstein-distance-based ambiguity sets constructed in this chapter accommodate general distributions.

11.3.3 Sizing Model

Suppose $\overline{x}^W_i$, $\overline{x}^{SP}_i$, and $\overline{x}^{SE}_i$ represent the upper bounds of x^W_i, x^{SP}_i, and x^{SE}_i, respectively. The feasible set for the capacity variable x would be defined as follows.

$$\mathcal{X}:=\left\{x\ \middle|\ \begin{array}{l}0\le x^W_i\le\overline{x}^W_i, 0\le x^{SP}_i\le\overline{x}^{SP}_i, 0\le x^{SE}_i\le\overline{x}^{SE}_i, \forall i\in S^B\\ g^N(x,\hat{\zeta}_n)<+\infty, \forall n\in S^N\end{array}\right\}.$$

In this context, $g^N(x,\hat{\zeta}_n)<+\infty, \forall n\in S^N$ implies that the operation problem (11.8) is feasible for all normal scenarios. In other words, there will be no need for load shedding under normal conditions. As $g^N(x,\zeta)$ is the optimal value of a linear programming (LP) problem, it stands as a convex function on its domain according to Sect. 2.5.4.

The sizing model is

$$\min_{x\in\mathcal{X}}\left\{\sum_{i\in S^B}(C^W_i x^W_i + C^{SP}x^{SP}_i + C^{SE}x^{SE}_i), I^F\right\} \tag{11.12a}$$

$$\text{s.t.}\quad \sup_{\mathbb{P}\in\mathcal{B}^E(\varepsilon^E,x)}\mathbb{E}^{\mathbb{P}}[g^E(x,\zeta)]\le\overline{g}^E , \tag{11.12b}$$

$$I^F = \sup_{\mathbb{P}\in\mathcal{B}^N(\varepsilon^N,x)}\mathbb{E}^{\mathbb{P}}[g^N(x,\zeta)] . \tag{11.12c}$$

In (11.12a), the first objective is to minimize the total investment cost, and the second objective function, as defined in (11.12c), represents the worst-case fuel cost expectation when the distribution $\mathbb{P}$ varies within the ambiguity set $\mathcal{B}^N(\varepsilon^N, x)$ for normal conditions. Constraint (11.12b) stipulates an upper bound $\overline{g}^E$ for the worst-case load shedding expectation under extreme conditions. It is worth noting that the worst-case distributions differ between normal and extreme conditions. In (11.12), the wind farm capacities are approximated by continuous variables. If the obtained capacity is not divisible by the capacity of a single WT, we round the obtained capacity to the nearest integer multiples of the WT capacity to ensure the number of WTs is an integer. For large-scale wind farms, this simplification should not significantly affect the optimality, as will be examined in Sect. 11.5.

The complexity of solving the sizing model (11.12) stems from three primary factors: 1. The model includes worst-case expectations; 2. The ambiguity sets within the model are decision-dependent; 3. The functions $g^E(x, \zeta)$ and $g^N(x, \zeta)$ are delineated as the optimal values of LP problems. To address these challenges, we will be advancing a solution approach in the subsequent section.

The endogenous uncertainty inherent in the sizing problem primarily originates from the wake effect observed in wind farms. As discussed in Sect. 11.2, the accessible wind power, denoted as f^W, of a wind farm can be considered a function of the total capacity x^W, and the auxiliary variable ξ^W which mirrors the initial wind speed. If the wake effect were to be disregarded, then $f^W = \xi^W x^W$ holds true when $0 \le \xi^W \le 1$ and $0 \le x^W \le \bar{x}^W$. Under these circumstances, the uncertainty of wind power can be fully epitomized by ξ^W, which can be viewed as the accessible wind power per unit capacity as identified in Chap. 10. Nevertheless, the sizing problem accounting for the wake effect is characterized by endogenous uncertainty. The manifestation of endogenous uncertainty implies that the uncertain variable ranges, the ambiguity sets, and the DRO model all become dependent on the decision-making process.

11.4 Solution Method

Initially, we convert the DRO sizing model, as presented in (11.12), into an SP problem. Following this transformation, an iterative algorithm is proposed.

11.4.1 Stochastic Programming Approximation

Initially, we employ Lipschitz constants to convert the worst-case expectations into expectations relative to the empirical distributions, supplemented by compensation terms. This transformation process is captured in the following proposition.

Proposition 11.1 (Conservative approximation of the worst-case expectation) *For any $x \in \mathcal{X}$ and $\varepsilon^E \ge 0$,*

$$\sup_{\mathbb{P}\in\mathcal{B}^E(\varepsilon^E,x)} \mathbb{E}^{\mathbb{P}}[g^E(x,\zeta)]$$
$$\le \sum_{i\in S^B} \varepsilon_i^{WE} x_i^W L_i^{WE}(x) + \varepsilon^{DE} L^{DE}(x) + \mathbb{E}^{\mathbb{P}_0^E(x)}[g^E(x,\zeta)] \,. \tag{11.13}$$

Here, $L_i^{WE}(x)$ signifies the Lipschitz modulus of $g^E(x,\zeta)$, considered as a function of ζ_i^W, and $L^{DE}(x)$ represents the Lipschitz modulus of $g^E(x,\zeta)$ when viewed as a function of ζ^D. For instance,

$$L_i^{WE}(x) := \inf\left\{ L \;\middle|\; \begin{array}{l} \left|g^E(x,\zeta) - g^E(x,\tilde{\zeta})\right| \le L\left\|\zeta_i^W - \tilde{\xi}_i^W\right\|_1 \\ \forall \zeta_i^W, \tilde{\xi}_i^W \in \Xi_i^{WE}(x_i^W), \text{ s.t. } \zeta^D = \tilde{\zeta}^D \in \Xi^{DE} \\ \zeta_j^W = \tilde{\zeta}_j^W \in \Xi_j^{WE}(x_j^W), \forall j \ne i \end{array} \right\} .$$

The proof of Proposition 11.1 is analogous to the proof of Theorem 6.10 by reference Esfahani and Kuhn (2018). The full content of the proof can be found in Xie et al. (2023). The worst-case expectation under normal conditions exhibits a parallel attribute, with Lipschitz moduli $L_i^{WN}(x)$ and $L^{DN}(x)$.

Employing the sample average approximation (SAA) technique, $\mathbb{E}^{\mathbb{P}_0^E(x)}[g^E(x,\zeta)] = \sum_{n\in S^E} g^E(x,\hat{\zeta}_n(x))/|S^E|$. Consequently, the succeeding SP problem serves as an approximation of the sizing model as indicated in (11.12).

$$\min \left\{ \sum_{i\in S^B} (C_i^W x_i^W + C^{SP} x_i^{SP} + C^{SE} x_i^{SE}), I^F \right\} \tag{11.14a}$$
$$\text{s.t. } 0 \le x_i^W \le \overline{x}_i^W \,,\; 0 \le x_i^{SP} \le \overline{x}_i^{SP} \,,\; 0 \le x_i^{SE} \le \overline{x}_i^{SE} \,,\; \forall i \in S^B \,, \tag{11.14b}$$
$$\sum_{i\in S^B} \varepsilon_i^{WE} x_i^W L_i^{WE}(x) + \varepsilon^{DE} L^{DE}(x) + \frac{1}{|S^E|}\sum_{n\in S^E} g_n^E \le \overline{g}^E \,, \tag{11.14c}$$
$$g_n^E \ge \sum_{t\in S^T}\sum_{i\in S^B} p_{i,t,n}^D \Delta_t \,,\; (11.9b) \,,\; \forall n \in S^E \,, \tag{11.14d}$$
$$I^F = \sum_{i\in S^B} \varepsilon_i^{WN} x_i^W L_i^{WN}(x) + \varepsilon^{DN} L^{DN}(x) + \frac{1}{|S^N|}\sum_{n\in S^N} g_n^N \,, \tag{11.14e}$$
$$g_n^N \ge \sum_{t\in S^T}\sum_{i\in S^B} f_{i,t,n}^G \,,\; (11.8b) \,,\; (11.8c) \,,\; \forall n \in S^N \,, \tag{11.14f}$$
$$\zeta_{i,t,n}^W \le a_{i,k}^{(1)} x_i^W + a_{i,k}^{(2)} \hat{\xi}_{i,t,n}^W + a_{i,k}^{(3)} \,,\; \forall k \in \mathcal{K}^a \,,\; n \in S^E \cup S^N \,, \tag{11.14g}$$
$$0 \le \zeta_{i,t,n}^W \le \hat{\xi}_{i,t,n}^W x_i^W \,,\; \forall i \in S^B \,,\; n \in S^E \cup S^N \,. \tag{11.14h}$$

Here, (11.14b) and (11.14f) encompass the requirement $x \in \mathcal{X}$; g_n^E and g_n^N are auxiliary variables signifying $g^E(x,\zeta_n(x))$ and $g^N(x,\zeta_n(x))$, respectively.

11.4.2 Solution Algorithm

To solve the SP problem as presented in (11.14), we must address two crucial aspects: 1. Determining the Lipschitz moduli $L_i^{WE}(x)$, $L^{DE}(x)$, $L_i^{WN}(x)$, and $L^{DN}(x)$; 2. Approaching the bi-objective optimization problem.

Addressing the first issue, although accurately calculating the Lipschitz moduli can be challenging, it is feasible for us to find their upper bounds. This is based on the fact that the functions $g^E(x, \zeta)$ and $g^N(x, \zeta)$ are defined by LP problems, and their dual variables can mirror their changing rates in ζ. This approach has been introduced in Chap. 10, and can also be found in Xie et al. (2023). The upper bounds are denoted by $\tilde{L}_i^{WE}(x)$, $\tilde{L}^{DE}(x)$, $\tilde{L}_i^{WN}(x)$, and $\tilde{L}^{DN}(x)$, respectively.

Calculating $\tilde{L}_i^{WE}(x)$ and $\tilde{L}^{DE}(x)$ involves solving a Karush-Kuhn-Tucker (KKT) condition embedded optimization while calculating $\tilde{L}_i^{WN}(x)$ and $\tilde{L}^{DN}(x)$ only requires solving some LP problems. To accelerate the computation, we use uniform upper bounds for $L_i^{WE}(x)$ and $L^{DE}(x)$, i.e., $\tilde{L}_i^{WE} \geq L_i^{WE}(x)$, $\tilde{L}^{DE} \geq L^{DE}(x)$, $\forall x \in \mathcal{X}$; and x-dependent upper bounds for $\tilde{L}_i^{WN}(x)$ and $\tilde{L}^{DN}(x)$, i.e., $\tilde{L}_i^{WN}(x) \geq L_i^{WN}(x)$ and $\tilde{L}^{DN}(x) \geq L^{DN}(x)$.

In relation to the second issue, we employ the ϵ-constraint method as a means to manage the two objectives. Specifically, the investment cost objective is substituted with a budget constraint, and the revised model focuses solely on minimizing the expectation of the fuel cost, as outlined below.

$$
\begin{aligned}
&\min \ I^F \\
&\text{s.t.} \ \ (11.14\text{b})\text{–}(11.14\text{h}) \ , \\
&\qquad \sum_{i \in S^B} (C_i^W x_i^W + C^{SP} x_i^{SP} + C^{SE} x_i^{SE}) \leq \overline{C} \ .
\end{aligned} \tag{11.15}
$$

Here, $\overline{C}$ represents the budget. With a fluctuating budget, the Pareto frontier of the bi-objective sizing model can be calculated. The uniform upper bounds $\tilde{L}_i^{WE}$ and $\tilde{L}^{DE}$ are computed in advance and utilized in the solution process of (11.15), as demonstrated in Algorithm 11.2. This algorithm finds an approximate solution iteratively: in each iteration k, the $\tilde{L}_i^{WN}(x_k)$ and $\tilde{L}^{DN}(x_k)$ from the current iteration are used to approximate the upper bounds of Lipschitz moduli in the next iteration by setting $L_i^{WN} \leftarrow \tilde{L}_i^{WN}(x_k), \forall i \in S^B$, $L^{DN} \leftarrow \tilde{L}^{DN}(x_k)$.

Algorithm 11.2 Solving problem (11.15)

Input: The parameters of problem (11.12) include the budget $\overline{C}$, an error tolerance $\varepsilon > 0$, and uniform upper bounds $\tilde{L}_i^{WE}$ and $\tilde{L}^{DE}$.

Output: Solution x_k and the worst-case expectation of fuel cost I^F.

1: Initiation: Let $L_i^{WN} \leftarrow 0, \forall i \in S^B$, $L^{DN} \leftarrow 0$, $k \leftarrow 0$, and $x_k \leftarrow 0$.

2: Solve the LP problem

$$\begin{aligned}
\min \ & \sum_{i \in S^B} \varepsilon_i^{WN} x_i^W L_i^{WN} + \varepsilon^{DN} L^{DN} + \frac{1}{|S^N|} \sum_{n \in S^N} g_n^N \\
\text{s.t.} \ & \sum_{i \in S^B} (C_i^W x_i^W + C^{SP} x_i^{SP} + C^{SE} x_i^{SE}) \leq \overline{C} , \\
& \sum_{i \in S^B} \varepsilon_i^{WE} x_i^W \tilde{L}_i^{WE} + \varepsilon^{DE} \tilde{L}^{DE} + \frac{1}{|S^E|} \sum_{n \in S^E} g_n^E \leq \overline{g}^E , \\
& (11.14b) , \ (11.14d) , \ (11.14f)\text{–}(11.14h) .
\end{aligned}$$

If the solution is infeasible, the process is terminated. Otherwise, $k \leftarrow k + 1$; let x_k represent the optimal solution and I^F denote the optimal value.

3: If $k > 0$ and $\|x_k - x_{k-1}\| \leq \varepsilon$, output the results and terminate the process. Otherwise, let $L_i^{WN} \leftarrow \tilde{L}_i^{WN}(x_k), \forall i \in S^B$, $L^{DN} \leftarrow \tilde{L}^{DN}(x_k)$. Subsequently, return to Step 2.

11.5 Case Studies

This section employs a modified IEEE 30-bus system to evaluate the proposed method. We examine the impact of the wake effect on the sizing results, compare the proposed method with traditional SP and robust optimization (RO) methods, and conduct a sensitivity analysis. All experiments are performed on a laptop equipped with an Intel i7-12700H processor and 16 GB of RAM. The LP and mixed-integer linear programming (MILP) problems are solved using Gurobi 9.5 within the MATLAB environment.

11.5.1 Benchmark

The benchmark case study is rooted in a modified version of the IEEE 30-bus system. Some key parameters are listed in Table 11.1. There are potential sites for wind farms at buses 13 and 27, with the WT parameters sourced from Yang et al. (2019). The proposed wind farms are planned on square areas with 25 and 20 rows, respectively. The wind speed data is generated based on the meteorological data from Qinghai province, China (Staffell and Pfenninger 2016). Potential sites for ESSs are identified

Table 11.1 Parameters (Xie et al. 2023)

Parameter	Value	Parameter	Value
T	24	Δ_t	1 h
η^{SC}	0.95	η^{SD}	0.95
$\underline{\omega}^S$	0.10	$\overline{\omega}^S$	0.90
C^{SP}	1.0×10^6 CNY/MW	C^{SE}	1.2×10^6 CNY/MWh
C^W	5.5×10^6 CNY/MW	$\overline{g}^E$	220 MWh

at buses 13, 23, and 27. All possess capacity boundaries with $\overline{x}^{SP} = 1$ GW and $\overline{x}^{SE} = 5$ GWh. The load demand data is generated using normal distributions with specified bounds.

In total, 365 sets of data are generated. Upon analyzing the load demand and wind conditions, 89 sets are selected to represent extreme conditions, while the remaining 272 sets represent normal conditions. In the benchmark case, $|S^E| = 24$ and $|S^N| = 72$ data sets are utilized for sizing. During the out-of-sample tests, 65 and 200 sets of data are used for validation, which represent the distributions of extreme and normal conditions, respectively. The parameters ε^E and ε^N, which control the size of the ambiguity sets, are set as follows.

$$\varepsilon_i^{WE} = \varepsilon_0/|S^E|^{1/T}\ ,\ i = 13, 27\ ,\ \varepsilon^{DE} = 20\varepsilon_0/|S^E|^{1/(20T)}\ ,$$
$$\varepsilon_i^{WN} = \varepsilon_0/|S^N|^{1/T}\ ,\ i = 13, 27\ ,\ \varepsilon^{DN} = 20\varepsilon_0/|S^N|^{1/(20T)}\ .$$

Here, 20 represents the number of buses with load, and $\varepsilon_0 = 0.05$ is an auxiliary scalar parameter used to control ε^E and ε^N.

To illustrate the endogenous uncertainty, we plot the available wind power sample points in Fig. 11.5. Here, points sharing the same horizontal coordinate indicate the distribution and probability density function (PDF) of the available wind power under a specific wind power capacity. It is evident that the distributions change as capacity fluctuates. The total computation time for the linearized available wind power function is less than 1 s. The errors between the available wind power model, as per (11.10), and the simulated value average about 5%, primarily resulting from the convex hull approximation.

The sizing results, corresponding with different investment budgets, are depicted in Fig. 11.6. The minimum investment required is 2.83×10^8 CNY to equip ESSs and ensure the reliable supply of load demand. As the investment budget expands, the capacities of wind power generation increase because renewable power can aid in reducing fuel costs. When the investment budget reaches 32×10^8 CNY, the capacity of wind power generation accounts for approximately 30% of the total generation capacity. The total capacity of ESSs does not vary significantly when the penetration of wind power is relatively low but begins to grow with the increase of wind power generation due to the flexibility requirement in the power system. We compare the results before and after the obtained capacity is rounded to integer multiples of the

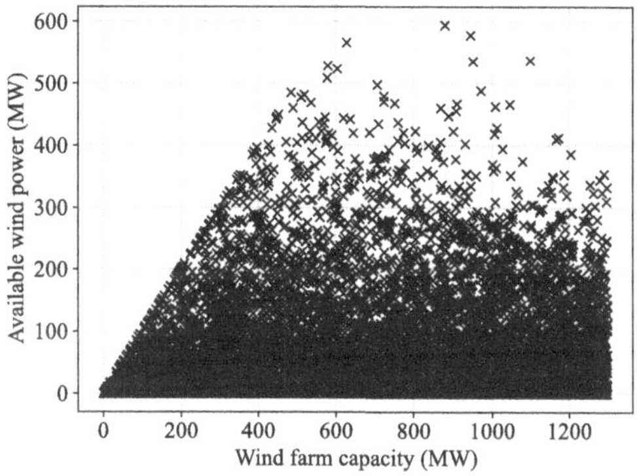

Fig. 11.5 Available wind power samples of the wind farm at bus 13 (Xie et al. 2023)

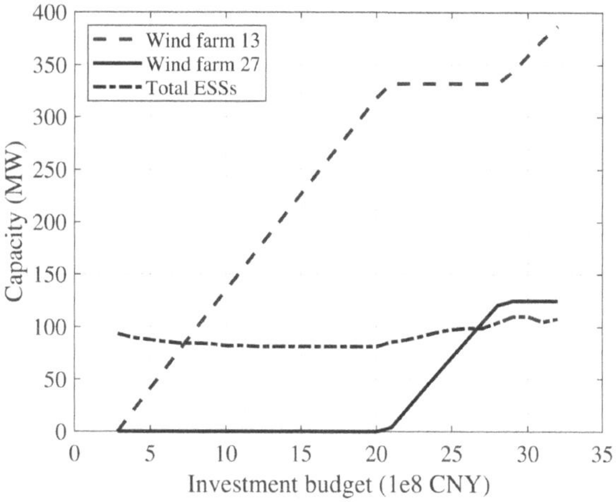

Fig. 11.6 Sizing results (Xie et al. 2023)

single WT capacity, and the error is below 0.5%. Consequently, we can optimize with continuous capacity variables and then round the results without a significant loss of optimality.

11.5.2 Comparison of Different Methods

To underscore the influence of the wake effect and the benefits of the DRO technique, we juxtapose the proposed method with traditional methods. For the sake of comparison, all the methods utilize the same data.

- SP1: The traditional SP method, which does not take the wake effect into account, is one such method used for comparison. In this method, the empirical distributions are treated as the actual distributions. The wake effect is not considered, so the available wind power in (11.10) becomes $\hat{\zeta}^W_{i,t,n}(x^W_i) = \hat{\xi}^W_{i,t,n} x^W_i$. The sizing model

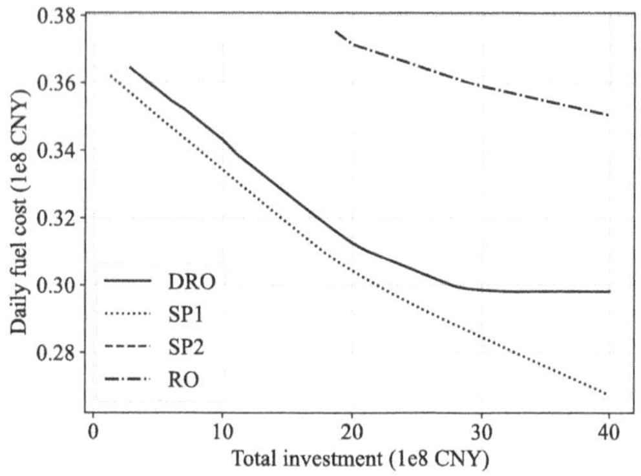

Fig. 11.7 Sizing result comparison (Xie et al. 2023)

can be equivalently transformed into an LP problem by following a solution method similar to the one described in Sect. 11.4.

- SP2: Traditional SP method while considering the wake effect. This method is similar to the first SP method, but the available wind power is still calculated according to (11.10).
- RO: RO method, which takes the wake effect into account. This method optimizes over the worst-case scenarios in $\tilde{\Xi}^E(x) := \{\hat{\zeta}_n(x) \mid n \in S^E\}$ and $\tilde{\Xi}^N(x) := \{\hat{\zeta}_n(x) \mid n \in S^N\}$. The sizing model is

$$\min_{x \in X} \left\{ \sum_{i \in S^B} (C_i^W x_i^W + C^{SP} x_i^{SP} + C^{SE} x_i^{SE}),\ \sup_{n \in S^N} g^N(x, \hat{\zeta}_n(x)) \right\}$$
$$\text{s.t.}\ \sup_{n \in S^E} g^E(x, \hat{\zeta}_n(x)) \le \overline{g}^E .$$

The Pareto frontiers are illustrated in Fig. 11.7. Initially, we analyze the influence of the wake effect. When the investment budget is under 17×10^8 CNY, the Pareto frontiers for SP1 and SP2 nearly overlap since the capacities for wind power generation are minimal, rendering the wake effect negligible as seen in Fig. 11.5. However, as the investment budget increases, the disparity between the Pareto frontiers of SP1 and SP2 becomes more pronounced. SP1's projected fuel cost significantly undercuts that of SP2, indicating that overlooking the wake effect can lead to an overly optimistic fuel cost estimation. Consequently, it is crucial to take into account the wake effect in future power systems that plan to incorporate large-scale wind power.

Next, we draw a comparison between SP2, RO, and DRO. All three models consider the wake effect but employ distinct optimization tactics to address uncertainties. The Pareto frontier of DRO is positioned above SP2 and below RO, and its minimum feasible investment budget also falls between those of SP2 and RO. This demonstrates that DRO's outcomes, while slightly more expensive than SP2, are significantly less conservative than those of RO.

Table 11.2 Comparison of sizing results (Xie et al. 2023)

Method	DRO	SP1	SP2	RO
(x_{13}^W, x_{27}^W) (MW)	(359, 125)	(497, 41)	(383, 125)	(99, 125)
$(x_{13}^{SP}, x_{23}^{SP}, x_{27}^{SP})$ (MW)	(55, 0, 55)	(0, 4, 15)	(40, 0, 10)	(196, 145, 266)
$(x_{13}^{SE}, x_{23}^{SE}, x_{27}^{SE})$ (MWh)	(119, 0, 73)	(0, 5, 19)	(127, 0, 13)	(258, 191, 521)
Estimate g^E (MWh)	220	199	129	0
Test g^E (MWh)	73	223	133	1
Estimate g^N (10^8 CNY)	0.2986	0.2847	0.2954	0.3589
Test g^N (10^8 CNY)	0.2959	0.3045	0.2957	0.3326
Number of variables	218, 361	218, 361	218, 361	218, 337
Number of constraints	570, 907	487, 963	570, 907	553, 673

To further analyze the differences between these methods, we detail the sizing results under the investment budget of 30×10^8 CNY in Table 11.2. Here, the estimate g^E and g^N are derived from their corresponding optimization methods, while the test g^E and g^N represent average values from out-of-sample tests. Regarding the estimate fuel cost g^N, we observe the progression SP1 $<$ SP2 $<$ DRO $<$ RO, transitioning from the most optimistic to the most conservative methods. Correspondingly, the total capacity of ESSs escalates, while that of wind power generation diminishes. This is attributable to the fact that while low-cost wind power generation introduces uncertainties into the system, ESSs can offer flexibility.

The estimate load shedding g^E remains within the upper limit of 220 MWh across all methods, yet the test g^E of SP1 exceeds this bound. The test g^E of DRO is 73 MWh, which, although higher than that of RO, is significantly lower than those of SP1 and SP2. The test g^N of SP1 and SP2 surpass their estimate values, attributable to the fact that the distribution represented by the test data differs from the empirical distribution, a variance overlooked by the SP-based methods. In contrast, the test g^N under DRO falls below the estimate value, as this method already factors in the potential inexactness of the empirical distribution and the test distribution is more favorable than the worst-case distribution. Regarding the test g^N, the progression is SP2 $\approx$ DRO $<$ SP1 $<$ RO. Given that the DRO method outperforms SP2 in terms of load shedding and is more cost-effective than RO, DRO emerges as the most balanced solution, striking an optimal compromise between optimality and robustness.

All four methods exhibit similar LP solution times, averaging around 50 s. This can be attributed to the fact that the LP problems they resolve share similar patterns and problem scales, even though variances in models result in different performance outcomes. Nonetheless, the computational efficiency of all four methods is acceptable for planning purposes.

11.5.3 Parameter Sensitivity Analysis

In the sensitivity analysis, we delve into the impacts of the scalar parameter ε_0 with respect to the ambiguity set size, the maximum acceptable load shedding expectation $\overline{g}^E$ in extreme conditions, and the unit cost of ESS. In this subsection, we set an upper limit of 0.31×10^8 CNY for the estimate g^N and aim to minimize the investment cost.

Table 11.3 illustrates the results under different ε_0 values. A larger ε_0 results in broader ambiguity sets and subsequently more conservative results, as evidenced by the increasing investment cost and the decreasing test g^E and g^N. It is worth noting that the test g^N surpasses the upper bound when $\varepsilon_0 = 0.01$ and 0.02, suggesting that the ambiguity sets are not sufficiently extensive to encompass the distribution represented by the test data. Therefore, it is vital to select appropriate parameters in order to construct an ambiguity set of an optimal size, ensuring that the sizing results are robust yet not excessively conservative.

Table 11.4 presents the results under varying $\overline{g}^E$ values. As $\overline{g}^E$ increases, the investment cost decreases. The capacities of ESSs exhibit significant changes, while that of wind power generation remains nearly the same. This is due to the fact that larger ESSs contribute to improved reliability. It is important to note that the test g^E consistently falls below $\overline{g}^E$ and the test g^N remains within the bounds, which further substantiates the efficacy of the proposed method.

Table 11.3 Results under different ambiguity set size parameter ε_0 (Xie et al. 2023)

ε_0	0.01	0.02	0.05	0.08
(x_{13}^W, x_{27}^W) (MW)	(321, 0)	(326, 0)	(332, 9)	(332, 26)
$(x_{13}^{SP}, x_{23}^{SP}, x_{27}^{SP})$ (MW)	(0, 0, 31)	(0, 0, 44)	(13, 0, 74)	(5, 0, 161)
$(x_{13}^{SE}, x_{23}^{SE}, x_{27}^{SE})$ (MWh)	(0, 0, 41)	(0, 0, 60)	(17, 0, 124)	(6, 0, 280)
Investment (10^8 CNY)	18.45	19.09	21.26	24.78
Test g^E (MWh)	208	184	110	37
Test g^N (10^8 CNY)	0.3110	0.3104	0.3087	0.3068

Table 11.4 Results under different maximum acceptable load shedding expectation $\overline{g}^E$ (Xie et al. 2023)

$\overline{g}^E$ (MWh)	150	220	290	360
(x_{13}^W, x_{27}^W) (MW)	(332, 8)	(332, 9)	(332, 9)	(332, 9)
$(x_{13}^{SP}, x_{23}^{SP}, x_{27}^{SP})$ (MW)	(7, 0, 147)	(13, 0, 74)	(7, 0, 38)	(0, 4, 15)
$(x_{13}^{SE}, x_{23}^{SE}, x_{27}^{SE})$ (MWh)	(9, 0, 257)	(17, 0, 124)	(9, 0, 49)	(0, 5, 20)
Investment (10^8 CNY)	23.44	21.26	19.88	19.24
Test g^E (MWh)	42	110	183	228
Test g^N (10^8 CNY)	0.3087	0.3087	0.3087	0.3085

Table 11.5 Results under different ESS unit cost parameter κ (Xie et al. 2023)

κ	0.6	0.8	1.0	1.2
(x_{13}^{W}, x_{27}^{W}) (MW)	(332, 8)	(332, 8)	(332, 9)	(332, 9)
$(x_{13}^{SP}, x_{23}^{SP}, x_{27}^{SP})$ (MW)	(14, 0, 73)	(13, 0, 74)	(13, 0, 74)	(12, 0, 75)
$(x_{13}^{SE}, x_{23}^{SE}, x_{27}^{SE})$ (MWh)	(18, 0, 122)	(17, 0, 123)	(17, 0, 124)	(16, 0, 125)
Investment (10^8 CNY)	20.24	20.75	21.26	21.77

The impact of the ESS unit cost is examined by multiplying C^{SP} and C^{SE} by a constant κ. The results under different κ values are outlined in Table 11.5. As κ increases, the investment cost also grows, while the capacities of wind power generation and ESSs remain relatively stable. This suggests that ESSs and wind power serve distinct functions within the system and cannot easily replace each other, even when the unit cost varies.

11.6 Conclusion

This chapter explores the sizing problem of wind power generation and ES in power systems while considering the wake effect. This problem is framed as a DRO model with DDU. We develop a solution methodology that consists of an SP approximation based on Lipschitz moduli, evaluations of the Lipschitz constants, and an iterative algorithm. Simulations conducted on the modified IEEE 30-bus system demonstrate the effectiveness of the proposed method and lead to the following insights: 1. The impact of the wake effect becomes more prominent as the wind power capacity increases within a confined area. 2. The DRO model surpasses the traditional SP and RO models in achieving a balance between optimality and robustness.

References

Esfahani PM, Kuhn D (2018) Data-driven distributionally robust optimization using the Wasserstein metric: performance guarantees and tractable reformulations. Math. Program. 171:115–166

Feijoo A, Cidras J (2000) Modeling of wind farms in the load flow analysis. IEEE Trans. Power Syst. 15(1):110–115

Gao X, Yang H, Lu L (2016) Optimization of wind turbine layout position in a wind farm using a newly-developed two-dimensional wake model. Appl. Energy 174:192–200

Herceg M, Kvasnica M, Jones C et al (2013) Multi-parametric toolbox 3.0. In: Proceeding of the European control conference. Zürich, Switzerland, pp 502–510

Luo F, Mehrotra S (2020) Distributionally robust optimization with decision dependent ambiguity sets. Optim. Lett. 14:2565–2594

Ma Y, Xie K, Zhao Y et al (2022) Bi-objective layout optimization for multiple wind farms considering sequential fluctuation of wind power using uniform design. CSEE J. Power Energy Syst. 8(6):1623–1635

Serrano González J, Burgos Payán M, Santos JMR et al (2014) A review and recent developments in the optimal wind-turbine micro-siting problem. Renew. Sustain. Energy Rev. 30:133–144

Staffell I, Pfenninger S (2016) Using bias-corrected reanalysis to simulate current and future wind power output. Energy 114:1224–1239

Villanueva D, Pazos JL, Feijoo A (2011) Probabilistic load flow including wind power generation. IEEE Trans. Power Syst. 26(3):1659–1667

Xie R, Wei W, Chen Y (2023) Sizing grid-connected wind power generation and energy storage with wake effect and endogenous uncertainty: a distributionally robust method. arXiv:2212.14665. Accessed 03 Jan 2024 [2024-01-03]

Yang K, Kwak G, Cho K et al (2019) Wind farm layout optimization for wake effect uniformity. Energy 183:983–995

Chapter 12
Planning Stand-Alone Electric Vehicle Charging Stations on Highways

Abstract This chapter puts forward a comprehensive two-stage method for determining the locations and capacities of stand-alone electric vehicle charging stations on highways. In the first stage, we determine the locations where individual vehicles necessitate charging services via a Monte Carlo simulation, which utilizes traffic demand and battery data. Subsequently, we propose an integer programming model to pinpoint the optimal locations for charging stations to guarantee that every vehicle has access to at least one charging station without depleting its battery. In the second stage, a distributionally robust optimization (DRO) model based on Kullback-Leibler divergence is constructed to optimize the capacities of renewable generation and energy storage units within each charging station. Two reformulations of the DRO model are proposed. The first reformulation, based on value-at-risk, results in a more accurate mixed-integer linear programming problem. The second reformulation, based on conditional value-at-risk, provides a more tractable, albeit conservative, linear programming approximation. This chapter introduces the research presented in Xie et al. (2018).

12.1 Introduction

The extensive adoption of renewable energy sources and the surge in electric vehicles (EVs) are widely recognized as effective strategies to lessen our dependence on fossil fuels and promote sustainable development. To stimulate the prevalence of EVs and enhance the driving experience, the construction of sufficient charging facilities is of paramount importance.

In this chapter, we explore the optimal planning of stand-alone, renewable-powered charging stations within a highway network, independent of a power grid. This need often emerges in areas of natural beauty, like Qinghai Lake in northwest China, or remote regions where a power grid is unavailable, but renewable resources are abundant. Such a framework holds promise for the future, as it is entirely green, producing no carbon emissions or air pollution.

R. Xie and W. Wei, *Distributionally Robust Optimization and its Applications in Power System Energy Storage Sizing*,
https://doi.org/10.1007/978-981-97-2566-3_12

The issue under investigation is characterized by three features. Firstly, under normal conditions, the highway network experiences less congestion compared to an urban traffic system. This allows vehicles to consistently travel at the maximum permitted speed, with energy consumption primarily dependent on the distance traveled. Second, renewable generation is volatile and must be coordinated with energy storage (ES) units. A Monte Carlo simulation method that incorporates traffic data is used for planning off-grid EV charging stations in Dong et al. (2016), but this approach neglects the uncertainty of renewable power output. Third, although the stochastic nature of renewable resources can be modeled by probability distributions, obtaining an exact model is challenging without sufficient historical data, particularly at the planning stage.

To counteract the volatility of renewable generation, the charging facilities in a stand-alone fast charging station must operate in harmony with ES units. This makes the optimal sizing of storage units a critical issue for realizing transportation electrification and sustainable mobility. An effective coordination and sizing strategy will ensure efficient energy utilization, reduce operational costs, and deliver reliable service to EV users.

This chapter tackles the issue of planning stand-alone photovoltaic (PV) powered charging stations within a highway network. Considering the forecasted traffic demands within the system, the first stage determines the optimal locations for charging stations and their respective demands. This is achieved by solving a mixed-integer linear programming (MILP) problem and implementing a Monte Carlo simulation. Furthermore, this stage also factors in driving patterns and the correlation of demands. The capacities of PV panels and storage units at each station are optimized in the second stage by formulating a distributionally robust optimization (DRO) problem. A reference probability distribution can be easily derived from basic information, although it may lack precision. We measure the difference between two distributions using the Kullback-Leibler (KL) divergence, ensuring that the KL divergence from any potential distribution to the reference distribution does not exceed a specified value. We demonstrate that the DRO model can be transformed into an MILP based on value-at-risk (VaR) and the sampling average approximation (SAA), and into a linear programming (LP) problem based on conditional value-at-risk (CVaR). The latter is more conservative but substantially more tractable.

The charging station placement problem and its MILP model are set forth in Sect. 12.2. The DRO formulation for capacity planning of a single charging station and MILP/LP-based solution methods are elaborated in Sect. 12.3. Case studies are presented in Sect. 12.4. Finally, conclusions are drawn in Sect. 12.5.

12.2 The First Stage: Determining the Placement of Charging Stations

In this initial stage, the Monte Carlo simulation method is implemented to derive the distribution of charging demand, utilizing traffic demand data. Following this, the optimal locations for the charging stations are ascertained using an MILP model. The charging demand for each station is then determined by conducting a post-siting charging demand simulation.

12.2.1 Electric Vehicle Model

The performance of EVs significantly influences charging demand. We, therefore, establish an EV model that incorporates elements such as battery capacity, state-of-charge (SoC), and energy consumption per unit distance, based on the study cited as Dong et al. (2016). Additional information can be inferred from existing data. In the Monte Carlo simulation, each vehicle's parameters are generated in accordance with this statistical EV model.

12.2.1.1 Battery Capacity

The battery capacities of EVs are random variables, subject to variability due to different vehicle types and degradation over time. As per the study (Dong et al. 2016), the battery capacity Cap of a specific type of EV can be generated from either a gamma distribution or a normal distribution, using the provided parameters. This process is repeated until the result falls within the defined lower and upper bounds.

12.2.1.2 State-of-Charge and Travel Range

We operate under the assumption that, post-charging, the SoC of the battery is evenly distributed between 80 and 90% (SoC_i). A driver is presumed to consider recharging the vehicle once the battery SoC drops to between 15 and 30% (SoC_c). The energy consumption is assumed to be a linear function of the traveled distance, according to Qian et al. (2011). Consequently, the maximum travel range (Ran_{mc}) is proportionate to the battery capacity (Cap). From this, we can derive the actual travel range with SoC from SoC_i to SoC_c (Ran_{ac}) and the travel range with SoC_c (Ran_{sc}). These can be calculated using (12.1) and (12.2).

$$Ran_{ac} = \eta_t \cdot (SoC_i - SoC_c) \cdot Ran_{mc} , \tag{12.1}$$

$$Ran_{sc} = \eta_t \cdot SoC_c \cdot Ran_{mc} . \tag{12.2}$$

The parameter η_t in these equations represents the efficiency factor, corresponding to the energy loss during the process of acceleration and deceleration.

12.2.1.3 Velocity and Charging Time

On uncongested highways, vehicles are able to travel at the maximum speed limit. This information is utilized in the charging demand simulation, wherein the passage of time is considered. It is assumed that the charging time t_c is directly proportional to the increase in battery SoC, and inversely proportional to the rated power p of the charger (Perez et al. 2017). This relationship is illustrated in (12.3):

$$t_c = \frac{(SoC_i - SoC_c) \cdot Cap}{\eta_c \cdot p} . \tag{12.3}$$

In this equation, η_c represents the charging efficiency parameter.

12.2.2 *Data of Traffic Demand*

12.2.2.1 Number of Electric Vehicles

Let B_t denote the number of EVs entering the highway system within a single day. The distribution of B_t can be derived from demand forecasts.

12.2.2.2 Origin-Destination Pairs

The highway system encompasses N_e entry and exit points. We define an $N_e \times N_e$ origin-destination (OD) matrix, denoted as Λ. Each element, Λ_{ij}, represents the fraction of vehicles traveling from node i to node j, with the total sum of Λ_{ij} equating to 1 across all i and j. Under conditions of no congestion, travelers consistently opt for the shortest and fastest path.

12.2.2.3 Other Parameters

The departure time of a vehicle refers to the moment it commences its journey from the origin. The hourly distribution of departure times forms part of the traffic demand data, which can be acquired from forecasts or calibrated through statistical analysis

(Dong et al. 2016; Mu et al. 2014). We assume that within each hour, the departure times are uniformly distributed.

The precision of the aforementioned information will not significantly hamper the proposed method's effectiveness. This is because uncertainties in the probability distributions of PV generation and demand are addressed in the second stage through a DRO model.

12.2.3 Monte Carlo Simulation

Monte Carlo simulation (MCS) is used to determine the spatial distribution of charging points. For each vehicle, when the battery's SoC decreases to a pre-set threshold, a virtual charging point is marked and the SoC is replenished. The MCS for a single vehicle proceeds as follows: 1. Generate the vehicle's battery capacity, departure time, initial SoC, and SoC thresholds for initiating and completing charging based on the assumed distribution. 2. Identify the shortest path according to the vehicle's origin and destination. 3. Mark all charging points along its path, and treat destinations as charging points. More details about the MCS can be found in the study (Dong et al. 2016).

12.2.4 Charging Station Placement Optimization

Let us denote S_p as the set of all charging points for EVs, containing N_p elements, and S_c as the set of pre-determined candidate sites for charging stations, containing N_c elements. These candidate sites are selected based on geographical conditions and construction-related factors. For any given $i \in S_c$ and $j \in S_p$, we can say that candidate site i covers charging point j if the EV, which generates j, can reach i with the remaining battery energy. Consequently, we can establish an $N_c \times N_p$ matrix, denoted as Γ, where

$$\Gamma_{ij} = \begin{cases} 1, \text{ if } i \text{ covers } j \,, \\ 0, \text{ otherwise} \,. \end{cases} \tag{12.4}$$

The construction strategy is represented by a vector of binary variables, $z = [z_1, z_2, \ldots, z_{N_c}]^T$, where $z_i = 1$ signifies the construction of a charging station at site $i \in S_c$, and $z_i = 0$ denotes the absence of one. In order to meet the charging demands while minimizing the number of charging stations, the location planning problem can be formulated as follows:

$$\min \sum_{i \in S_c} z_i \tag{12.5a}$$

$$\text{s.t.} \quad \sum_{i \in S_c : \Gamma_{ij}=1} z_i \geq 1 \,, \; \forall j \in S_p \,, \tag{12.5b}$$

$$z_i \in \{0, 1\} \,, \; \forall i \in S_c \,. \tag{12.5c}$$

The expression (12.5b) guarantees that all charging points will be covered.

12.2.5 Simulation of Charging Demand at Charging Stations

Once the locations of the charging stations are established, the temporal distributions of their charging demands can be derived from the MCS. We make the assumption that a vehicle will be recharged at a station if its SoC falls below 15–30%, or if it is unable to reach the next charging station or its final destination without recharging. The elapse of time is also a consideration within this process. Figure 12.1 illustrates the simulation procedure for a single vehicle. It operates under the assumption that the vehicle's speed while traveling is equal to the maximum permissible speed. Following the MCS, we obtain the demand profiles for each charging station.

We have the capability to incorporate various types of chargers and simulate queuing times using the MCS. However, this would necessitate the simultaneous simulation of all vehicles and charging stations. The flexibility of the MCS also allows us to integrate more practical considerations, thereby enhancing the accuracy and reliability of our predictive models.

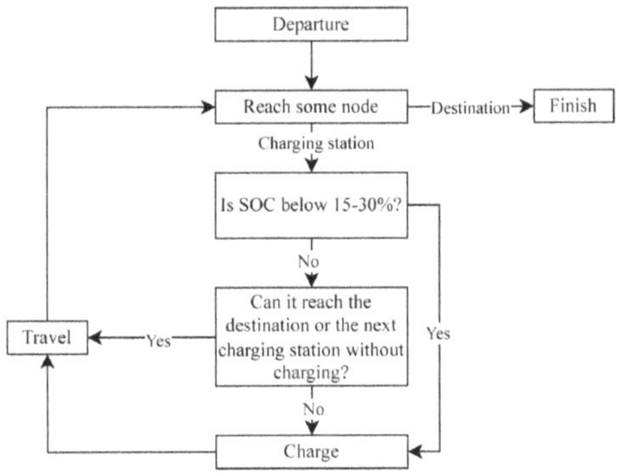

Fig. 12.1 Monte Carlo simulation for one vehicle (Xie et al. 2018)

12.3 The Second Stage: Capacity Determination

We are preparing to construct stand-alone PV-powered charging stations at the locations determined in Sect. 12.2. The challenge lies in determining the capacities of the PV panels and the ES unit. Our goal is to minimize investment costs while satisfying as much of the charging demand as possible. This must be achieved while considering the random weather conditions and potential inaccuracies in load distribution forecasts.

12.3.1 Deterministic Optimization Problem

For the sake of simplicity, we will use data sampled from a typical day, divided into 24 periods, to construct our deterministic model. We will operate under the assumption that the capacity planning problem for each charging station is independent. This takes into account the spatial and temporal correlation of loads, which has already been considered in the MCS.

The energy flow within a fast charging station is depicted in Fig. 12.2. Here, the capacity of PV generation is represented by x_g. In period t, the actual generation is $\xi_g(t)x_g$, where $\xi_g(t) \in [0, 1]$ represents an uncertain factor; the demand is represented by $\xi_c(t)$. The generated energy can either be used to serve the load, represented by $p_1(t)$, or charge the storage unit, represented by $p_2(t)$. Concurrently, the storage unit, with a capacity of x_s, can supply a portion of the load with power $p_3(t)$.

Let us denote $\overline{w}$ and $\underline{w}$ as the maximum and minimum SoC of the storage unit, respectively. Let η_1 represent the charging efficiency and η_2 the discharging efficiency. w_0 is the initial SoC, Δ_t is the duration of period t, and T is the number of periods, with $S_t = \{1, 2, \ldots, T\}$. The constraints for the ES operation can be formulated as:

$$E_s(t) = w_0 x_s + \sum_{i=1}^{t} \left(\eta_1 p_2(i) - \frac{p_3(i)}{\eta_2} \right) \Delta_t \ , \ \forall t \in S_t \ , \tag{12.6}$$

$$\underline{w} x_s \leq E_s(t) \leq \overline{w} x_s \ , \ \forall t \in S_t \ , \tag{12.7}$$

$$p_2(t) \leq p_2^m x_s \ , \ p_3(t) \leq p_3^m x_s \ , \ \forall t \in S_t \ . \tag{12.8}$$

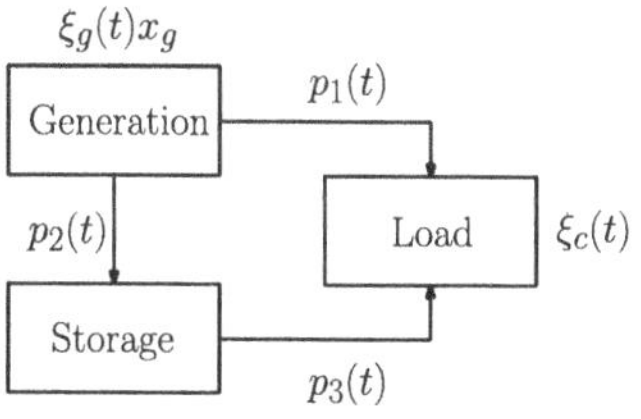

Fig. 12.2 Energy flow relationships in a charging station (Xie et al. 2018)

Equation (12.6) portrays the dynamics of the SoC; (12.7) confines the range of SoC based on the installed capacity x_s; (12.8) establishes the maximum charging and discharging rates, which are proportional to the capacity x_s. Given that the efficiency coefficients η_1 and η_2 are strictly less than 1 due to unavoidable energy dissipation, simultaneous charging and discharging does not constitute an optimal dispatch strategy. Therefore, the complementarity condition $p_2(t)p_3(t) = 0, \forall t$ is naturally upheld according to Proposition 7.1 and is consequently omitted from the formulations.

We are going to assume that the investment expenses for PV panels and ES units correlate directly with their capacities. This correlation is expressed through constant coefficients C_g for PV panels and C_s for ES units. With this in mind, we can construct the following deterministic model for capacity planning:

$$\min C_g x_g + C_s x_s \tag{12.9a}$$

$$\text{s.t. } x_g \geq 0 \ , \ \ x_s \geq 0 \ , \tag{12.9b}$$

$$p_1(t), p_2(t), p_3(t) \geq 0 \ , \ \ \forall t \in S_t \ , \tag{12.9c}$$

$$\xi_g(t) x_g \geq p_1(t) + p_2(t) \ , \ \ \forall t \in S_t \ , \tag{12.9d}$$

$$\xi_c(t) \leq p_1(t) + p_3(t) \ , \ \ \forall t \in S_t \ , \tag{12.9e}$$

$$E_s(t) = w_0 x_s + \sum_{i=1}^{t} \left(\eta_1 p_2(i) - \frac{p_3(i)}{\eta_2} \right) \Delta_t \ , \ \ \forall t \in S_t \ , \tag{12.9f}$$

$$\underline{w} x_s \leq E_s(t) \leq \overline{w} x_s \ , \ \ \forall t \in S_t \ , \ \ E_s(T) = w_0 x_s \ , \tag{12.9g}$$

$$p_2(t) \leq x_g \ , \ \ p_2(t) \leq p_2^m x_s \ , \ \ p_3(t) \leq p_3^m x_s \ , \ \ \forall t \in S_t \ . \tag{12.9h}$$

The objective is to minimize the overall investment cost. The constraints (12.9b) and (12.9c) establish non-negativity for the variables. The constraints (12.9d) and (12.9e) represent adequacy for both generation and load supply. At the planning stage, it is important to note that we do not enforce strict power balancing. Excessive power can be managed through curtailment. The ES model is defined by constraints (12.9f)–(12.9h). These constraints also ensure identical SoC values at the beginning and end of each period to complete a daily cycle.

12.3.2 Distributionally Robust Optimization Problem

In the DRO model, we take into account all probability distributions in close proximity to the reference probability distribution $\mathbb{P}_0$. To measure the distance from $\mathbb{P}_0$ to another probability distribution $\mathbb{P}$, we utilize KL divergence, as outlined in Definition 5.4. The ambiguity set, based on KL divergence, is articulated as follows:

$$D = \left\{ \mathbb{P} \in \mathcal{M} \ \middle| \ D_{\text{KL}}(\mathbb{P} \parallel \mathbb{P}_0) \leq d_{\text{KL}} \right\} \ . \tag{12.10}$$

In this context, $\mathcal{M}$ signifies the set encompassing all potential probability distributions, while d_{KL} stands for the tolerance level of the KL divergence.

To simplify notation, define vectors $x = [x_g, x_s]^T$, $\xi_g = [\xi_g(1), \ldots, \xi_g(T)]^T$, $\xi_c = [\xi_c(1), \ldots, \xi_c(T)]^T$, $\xi = (\xi_g; \xi_c)$, $p_i = [p_i(1), \ldots, p_i(T)]^T$, $i = 1, 2, 3$, and $p = [p_1; p_2; p_3]$. Additionally, we define the feasible region of ξ for a specific x as follows: $C(x) = \{\xi \mid \exists p, \text{s.t. (12.9c)–(12.9h)}\}$.

For fixed x and ξ, a failure state occurs when the generation and storage unit cannot fulfill the charging request during a given period, implying that $\xi \notin C(x)$. Given the uncertainty of PV output and charging demand, it becomes necessary to contemplate the probability of such a failure event. If the exact probability distribution of these uncertain factors is available, and if α represents the acceptable probability of failure, then a chance constraint can be applied:

$$\Pr[\xi \in C(x)] \geq 1 - \alpha \ .$$

This allows us to maintain the quality of service, where Pr represents the probability. However, acquiring a probability distribution with high accuracy can be challenging during the planning stage due to a lack of operating experience and historical data. To address this issue, we consider the set of inexact probability distributions defined by (12.10). The reference distribution in this set is comparatively easier to obtain. Additionally, we can manipulate the variation range of the probability distribution via the parameter d_{KL}. As a result, the probability of failure also becomes uncertain, depending on the probability distribution in set D. Therefore, it is wise to ensure service quality even in the worst-case scenario, leading to a robust chance constraint.

$$\inf_{\mathbb{P} \in D} \Pr[\xi \in C(x)] \geq 1 - \alpha \ . \tag{12.11}$$

This implies that the quality of the service is reliant on the probability distribution, and it consistently maintains a value not less than $1 - \alpha$.

Theorem 5.14 establishes that if set D is formed using the KL divergence in (12.10), then the robust chance constraint (12.11) is equivalent to:

$$\Pr{}^0[\xi \in C(x)] \geq 1 - \alpha_{1+} \ , \tag{12.12}$$

where $\Pr^0$ implies that the probability is assessed with respect to the reference distribution $\mathbb{P}_0$ and

$$\alpha_{1+} = \max\{\alpha_1, 0\} \ , \ \alpha_1 = 1 - \inf_{y \in (0,1)} \left\{ \frac{\exp(-d_{\mathrm{KL}}) y^{1-\alpha} - 1}{y - 1} \right\} \ .$$

Furthermore, the ratio $(\exp(-d_{\mathrm{KL}}) y^{1-\alpha} - 1)/(y - 1)$ exhibits convexity over y within the interval of $(0, 1)$. Consequently, α_1 can be readily calculated using a bisection line search.

At this point, it becomes straightforward to introduce the capacity planning model for a single charging station.

$$\min\ C_g x_g + C_s x_s \tag{12.13a}$$
$$\text{s.t. } x_g \geq 0\ ,\ \ x_s \geq 0\ , \tag{12.13b}$$
$$\Pr{}^0[\xi \in C(x)] \geq 1 - \alpha_{1+}\ . \tag{12.13c}$$

This represents a conventional chance-constrained program linked with the reference distribution, where the inaccuracies of probability distributions have been adjusted using a corrective parameter, α_{1+}. However, the chance constraint continues to pose a significant challenge in solving (12.13).

12.3.3 Equivalent Formulation Based on Value-at-Risk

In this subsection, we utilize the concept of VaR to derive an equivalent formulation of (12.13). This serves as the foundational principle for our two models designed to address and solve (12.13).

12.3.3.1 Loss Function

For given x and ξ, a failure is triggered if no (p_1, p_2, p_3) exists that satisfies conditions (12.9c)–(12.9h). To measure the severity of failure under scenario ξ for a specific capacity planning strategy x, we introduce the following loss function, denoted as $g(x, \xi)$.

$$g(x, \xi) = \min\ g \tag{12.14a}$$
$$\text{s.t. } Ap \leq b(\xi) - B(\xi)x\ , \tag{12.14b}$$
$$\xi_c(t) - p_1(t) - p_3(t) \leq g\ ,\ \ \forall t \in S_t\ . \tag{12.14c}$$

The first inequality consolidates constraints (12.9c), (12.9d), and (12.9f)–(12.9h); A represents a constant matrix; $B(\xi)$ and $b(\xi)$ emphasize the coefficients affected by uncertainty. It is evident that program (12.14) is always feasible, as $p = 0$ and $g = \max_t\{\xi_c(t)\}$ consistently provide a feasible solution. Moreover, given that $0 \leq p_1(t) \leq \xi_g(t)x_g$ and $0 \leq p_3(t) \leq p_3^m x_s$, program (12.14) is bounded from below. As a result, the function $g(x, \xi)$ is well-defined. With the loss function defined, the failure event corresponds to $g(x, \xi) > 0$, and constraint (12.12) can be reformulated as:

$$\Pr{}^0[g(x, \xi) \leq 0] \geq 1 - \alpha_{1+}\ . \tag{12.15}$$

12.3.3.2 Value-at-Risk Representation of the Chance Constraint

According to the definition of VaR in Definition 2.39, the chance constraint (12.15) is equivalent to:

$$(1-\alpha_{1+})\text{-VaR}(g(x,\xi)) \leq 0\,. \tag{12.16}$$

In simpler terms, the ratio of scenarios from all the uncertain data ξ resulting in $g(x,\xi) > 0$ should not surpass α_{1+}. This aligns consistently with (12.15).

12.3.3.3 An Equivalent Mixed-Integer Linear Programming Model

Consider $\xi_1, \xi_2, \ldots, \xi_q$ as q scenarios sampled from the random variable ξ. We utilize q binary variables $z_c(1), z_c(2), \ldots, z_c(q)$ to denote potential failures: $z_c(k) = 1$ signifies that a failure might occur in ξ_k. Let M represent the upper limit of charging demands, thus $g(x,\xi) \leq M$ is always valid. Now, consider the following inequality:

$$g(x,\xi_k) \leq M z_c(k)\,. \tag{12.17}$$

When $z_c(k) = 0$, $g(x,\xi_k)$ is not permitted to exceed 0, which mandates that $\xi_k \in C(x)$ must be true; when $z_c(k) = 1$, there is essentially no constraint for $g(x,\xi_k)$, signifying that a failure event could potentially occur in scenario ξ_k. It is important to note that constraint (12.17) is equivalent to:

$$\begin{aligned}
&Ap^k \leq b(\xi^k) - B(\xi^k)x\,,\\
&\xi_c^k(t) - p_1^k(t) - p_3^k(t) \leq M z_c(k)\,,\ \forall t\,.
\end{aligned}$$

We are now in a position to introduce the VaR and SAA-based MILP model.

$$\min\ C_g x_g + C_s x_s \tag{12.18a}$$

$$\text{s.t. } x_g \geq 0\,,\ x_s \geq 0\,, \tag{12.18b}$$

$$Ap^k \leq b(\xi^k) - B(\xi^k)x\,,\ \forall k\,, \tag{12.18c}$$

$$\xi_c^k(t) - p_1^k(t) - p_3^k(t) \leq M z_c(k)\,,\ \forall t\,,\ \forall k\,, \tag{12.18d}$$

$$\frac{1}{q}\sum_{k=1}^{q} z_c(k) \leq \alpha_{1+}\,,\ z_c(k) \in \{0,1\}\,,\ \forall k\,. \tag{12.18e}$$

In the MILP model (12.18), failure events can occur in a maximum of $q\alpha_{1+}$ out of q scenarios, thereby ensuring the service quality as defined in (12.11).

12.3.4 Conservative Formulation Based on Conditional Value-at-Risk

The number of binary variables in the MILP model (12.18) equals the number of sample scenarios, which may hinder the efficient resolution of the MILP model (12.18). To mitigate this challenge, we devise an LP approximation for the chance-constrained program (12.13) leveraging the properties of CVaR as revealed in Lemma 2.8 and Theorem 2.6.

According to Lemma 2.8, β-VaR $\leq$ β-CVaR is always valid. Therefore, a conservative approximation of the constraint (12.15) is:

$$(1-\alpha_{1+})\text{-CVaR}(g(x,\xi)) \leq 0 \,. \tag{12.19}$$

This inequality serves as a sufficient condition for both (12.15) and (12.16). This conservative substitution aligns with the principles of robust optimization (RO) and is acceptable within a planning problem context. Moving forward, we will restructure (12.19) into a format that is compatible with solvers.

As stated in Theorem 2.6, the left-hand side of (12.19) equates to the optimal value of the following minimization problem:

$$\min_{\gamma} \left\{ \gamma + \frac{1}{\alpha_{1+}} \int_{\xi \in \mathbb{R}^K} \max\{g(x,\xi)-\gamma, 0\} \mathrm{d}\mathbb{P}_0(\xi) \right\} \,. \tag{12.20}$$

By implementing SAA, the integral within (12.20) translates to discrete summations over the sampled scenarios, yielding the following result:

$$\min_{\gamma} \left\{ \gamma + \frac{1}{q\alpha_{1+}} \sum_{k=1}^{q} \max\{g(x,\xi_k)-\gamma, 0\} \right\} \,. \tag{12.21}$$

By incorporating the auxiliary variable s_k, the feasible region outlined by (12.19) can be represented by

$$\begin{aligned}
&\exists \gamma \in \mathbb{R}, s_k \in \mathbb{R}^+, \text{s.t. } Ap^k \leq b(\xi^k) - B(\xi^k)x \;, \;\; \forall k \;, \\
&\xi_c^k(t) - p_1^k(t) - p_3^k(t) - \gamma \leq s_k \;, \;\; \forall k \;, \;\; \forall t \;, \\
&\gamma + \frac{1}{q\alpha_{1+}} \sum_{k=1}^{q} s_k \leq 0 \,.
\end{aligned}$$

Finally, the model based on CVaR and SAA, which utilizes LP, can be expressed as follows:

$$\min\ C_g x_g + Cb_s x_s \tag{12.22a}$$

$$\text{s.t. } x_g \geq 0\ ,\ x_s \geq 0\ ,\ Ap^k \leq b(\xi^k) - B(\xi^k)x\ ,\ \forall k\ , \tag{12.22b}$$

$$\xi_c^k(t) - p_1^k(t) - p_3^k(t) - \gamma \leq s_k\ ,\ \forall t\ ,\ \forall k\ , \tag{12.22c}$$

$$\gamma \in \mathbb{R}\ ,\ \gamma + \frac{1}{q\alpha_{1+}} \sum_{k=1}^{q} s_k \leq 0\ ,\ s_k \geq 0\ ,\ \forall k\ . \tag{12.22d}$$

By implementing the planning strategies $\{x_g, x_s\}$ proposed by (12.22), the probability of failure events is typically less than α, even under the worst-case probability distribution in D.

12.4 Case Study

12.4.1 Settings

We are considering a straight highway scenario. The highway has ten entrances and exits, with geographical information provided in Table 12.1. The speed limit is set to 120 km/h. The scenario takes into account 4 types of EVs, with their respective battery capacity distributions listed in Table 12.2. We assume an energy consumption rate of 1/6 kWh/km and the rated power of chargers is 60 kW. B_t is generated by a normal distribution with $\mu = 10{,}000$, $\sigma = 2{,}000$, an upper limit of 18,000, and a lower limit of 2,000. The initial battery SoC of EVs is evenly distributed between 30 and 80%. The OD matrix and departure time distribution are displayed in Figs. 12.3 and 12.4, respectively (Dong et al. 2016). Solar radiation data have been obtained from the National Solar Radiation Data Base (Sengupta et al. 2018). All other relevant parameters are summarized in Table 12.3. The system uses Lithium-ion batteries (Hu et al. 2017). Both the MILP and LP models are developed with the YALMIP toolbox (Lofberg 2004) within the MATLAB environment and are solved using Gurobi 7.5.

Table 12.1 Parameters of entrances and exits (Xie et al. 2018)

No.	Location (km)	No.	Location (km)
1	0	6	160
2	30	7	210
3	50	8	240
4	90	9	270
5	130	10	300

Table 12.2 Battery settings (Xie et al. 2018)

EV type	I	II	III	IV
Proportion (%)	10	50	20	20
Distribution	Gamma	Gamma	Normal	Normal
Parameter	$\alpha = 7.5, \beta = 4.7$	$\alpha = 4.5, \beta = 6.3$	$\mu = 23.0, \sigma = 9.5$	$\mu = 85.3, \sigma = 28.1$
Upper bound (kWh)	65.0	72.0	40.0	120.0
Lower bound (kWh)	15.0	10.0	9.6	51.2

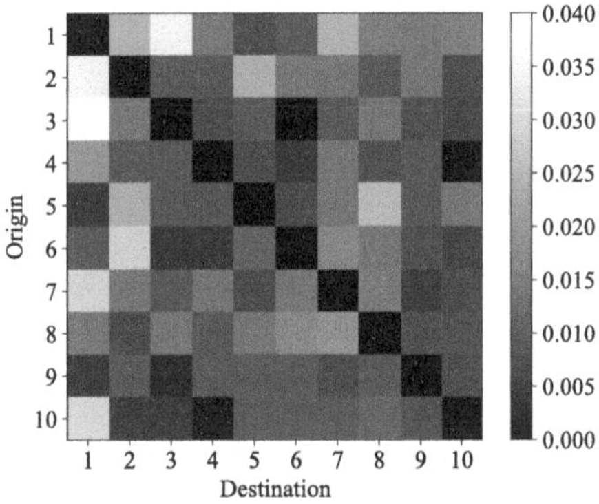

Fig. 12.3 Heatmap of the OD matrix (Xie et al. 2018)

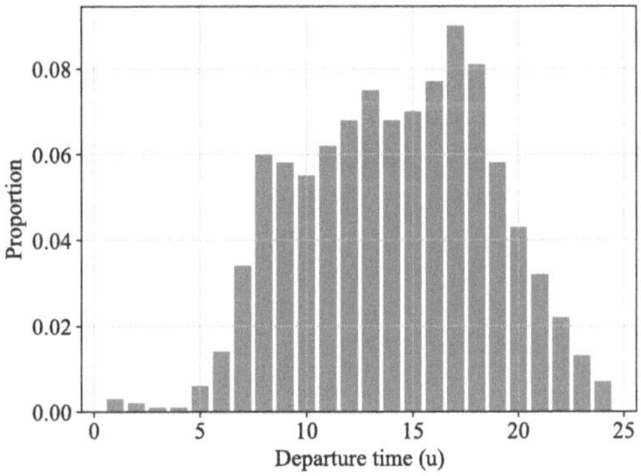

Fig. 12.4 Histogram of EV departure time (Xie et al. 2018)

12.4.2 Results

Initially, we presume that charging stations can be placed every 0.2 km along the highway, resulting in 1,501 potential sites. We utilize the MILP model (12.5) for the placement of charging stations, as we have ample locational options for the installation of charging facilities. For the comparative study, we assume that charging

Table 12.3 Other parameters (Xie et al. 2018)

Parameter	Value	Parameter	Value
C_g	\$1,770/kW	C_s	\$360/kWh (Hu et al. 2017)
p_2^m	0.2	p_3^m	0.2
η_1	90%	η_2	90%
$\overline{w}$	90%	$\underline{w}$	10%
Δ_t	1 h	T	24
α	5%	d_{KL}	0.01
w_0	0.3		

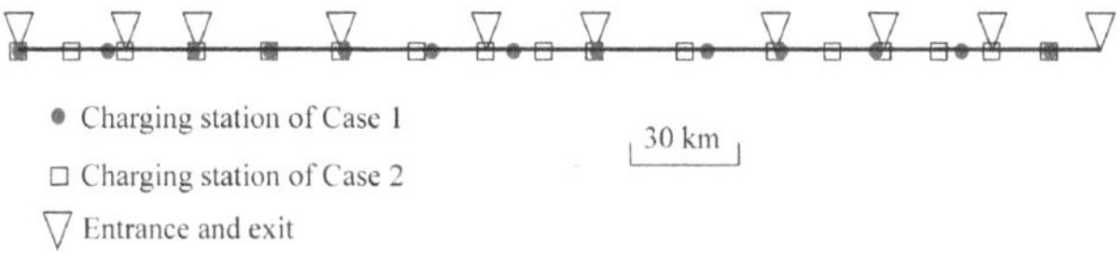

Fig. 12.5 Siting results of charging stations (Xie et al. 2018)

stations are positioned at every entrance/exit node, as well as at midpoints between two adjacent nodes. These two arrangements are referred to as Case 1 and Case 2, respectively. In each case, we examine two models for capacity planning: the MILP model (12.18) and the LP model (12.22). These models will be referred to as MILP and LP, respectively, hereafter.

Figure 12.5 portrays the placement outcomes in both cases, demonstrating that Case 2 requires more stations than Case 1 to fulfill all the charging demands along the highway, given that the candidate locations in Case 2 are more restrictive. The charging stations in Case 1 are not uniformly distributed, despite the potential for arbitrary placement, due to the spatial distribution of traffic demands. The capacity planning results in both Case 1 and Case 2 are summarized in Tables 12.4 and 12.5, where stations are indexed in sequence from 0 km to 300 km. In Case 1, charging stations situated in the middle have larger installed capacities, as most vehicles have to traverse this area. In contrast, in Case 2, the capacities of stations 8 and 9 in the middle are significantly smaller than those of the nearby stations, since they are in close proximity to each other and can therefore share charging demands. It can also be observed that the LP model offers planning strategies with larger component capacities than the MILP model does. This is due to the conservative approximation of the chance constraint (12.15) using the CVaR inequality (12.19), which leads to improved reliability and increased cost.

Table 12.6 presents several other comparative elements in the two cases. The charging points coverage rate signifies the percentage of charging points that can be serviced by the planned charging stations. This is tested on another set of approximately 300,000 charging points generated by MCS using the same statistical EV model and traffic demand data. The planning strategy in Case 1 can cater to all charg-

Table 12.4 Sizing results in Case 1 (Xie et al. 2018)

Station No.	PV capacity (MW)		Storage capacity (MWh)	
	LP	MILP	LP	MILP
1	2,614	2,452	7,216	4,534
2	4,631	4,144	12,365	12,249
3	6,337	5,406	16,433	15,668
4	6,656	5,849	18,824	18,577
5	6,763	6,026	19,774	17,622
6	6,662	5,667	20,088	19,861
7	6,932	6,137	20,473	18,551
8	7,636	7,031	22,468	20,522
9	6,966	6,023	19,462	16,310
10	6,287	5,597	17,479	15,691
11	4,861	4,520	13,988	12,062
12	3,464	3,133	9,584	9,550
13	1,455	1,348	4,457	4,052

Table 12.5 Sizing results in Case 2 (Xie et al. 2018)

Station No.	PV capacity (MW)		Storage capacity (MWh)	
	LP	MILP	LP	MILP
1	1,888	1,729	5,102	5,053
2	2,057	1,924	7,313	6,423
3	3,009	2,776	10,826	10,096
4	6,072	5,416	19,778	19,721
5	5,325	4,978	16,204	15,884
6	5,311	5,016	16,383	14,169
7	6,839	6,210	18,228	17,312
8	3,319	3,173	9,752	9,886
9	1,827	1,694	5,463	4,902
10	6,675	6,569	19,262	17,801
11	7,172	6,661	18,759	17,616
12	6,084	5,456	15,806	15,882
13	5,893	5,690	18,188	17,394
14	5,243	5,197	17,100	15,708
15	2,970	2,738	11,409	10,264
16	1,705	1,461	6,872	5,420
17	940	809	2,577	2,668

Table 12.6 Comparison of Case 1 and Case 2 (Xie et al. 2018)

	Case 1	Case 2
Number of stations	13	17
Charging points coverage (%)	100.0	99.7
Investment cost (LP)	$\$1.99 \times 10^8$	$\$2.07 \times 10^8$
Investment cost (MILP)	$\$1.80 \times 10^8$	$\$1.94 \times 10^8$

ing requests, while the planning scheme of Case 2 fails to cover 0.3% of charging points. For both the LP and the MILP models of capacity planning, the total investment cost in Case 1 is lower, because the sites are better suited and fewer charging stations will be constructed. In both cases, the schemes provided by the LP model incur higher costs.

12.4.3 Parameter Sensitivity Analysis

We delve into the influence of two specific parameters on the efficacy of planning strategies: the number of samples, denoted as q, and the KL-divergence threshold, represented by d_{KL}. This exploration involves a comparison of three key factors: investment cost, failure rates (illustrated by the percentage of supply inadequacy), and computation time. To facilitate these tests, we acquire additional PV generation data from Sengupta et al. (2018), from which we generate 10,000 samples. This robust sample size allows for an accurate representation of the true distribution of PV output.

Initially, we address the capacity planning issues of 13 charging stations in Case 1 by utilizing varying values of the sample number q. The results are presented in Table 12.7, where "maximum" denotes the highest value, "mean" represents the average value across the 13 stations, and computation time is displayed as an average value. It is apparent that the MILP model yields lower costs. While the cost within the LP model shows a decrease with an increase in q, the MILP model does not display a similar monotonic trend. The computation time for the MILP model sees an exponential increase with q. However, the LP model consistently resolves within less than a minute.

Table 12.7 also presents the failure rates. For this test, we set $d_{\mathrm{KL}} = 0.01$ and $\alpha = 5.0\%$, which implies $\alpha_{1+} \approx 2.5\%$. It is evident that the failure rates suggested by the MILP model typically exceed 2.5%, although they consistently fall below the designated value of 5.0%. When $q = 100$, the failure probability surpasses the acceptable $\alpha = 5.0\%$, largely due to the sample size being insufficient to adequately reflect the dispersion of real data. However, when the sample size is sufficient, the MILP model's failure rate is closer to the preset value of 5.0%. This is because the VaR condition (12.16) effectively encapsulates the chance constraint (12.15). In

Table 12.7 Testing results under different sample numbers (Xie et al. 2018)

Sample number			100	200	400	800
LP	Total investment (10^8\$)		2.022	1.991	1.959	1.915
	Failure rate	Maximum	0.019	0.018	0.019	0.021
		Mean	0.0099	0.0102	0.0122	0.0155
		Severe	0.027	0.024	0.027	0.026
	Average computation time (s)		3.970	6.731	13.489	30.638
MILP	Total investment (10^8\$)		1.751	1.795	1.801	1.784
	Failure rate	Maximum	0.073	0.042	0.038	0.040
		Mean	0.0387	0.0326	0.0289	0.0315
		Severe	0.043	0.043	0.046	0.048
	Average computation time (s)		7.081	14.287	52.419	860.227

contrast, the LP model, due to the conservatism introduced by the CVaR inequality (12.19), consistently yields lower failure rates that are always below 5%.

To assess the robustness of the DRO strategy under more challenging conditions, we create a discrete distribution within set D for each q. This distribution mirrors the range of the reference distribution. However, the probabilities associated with failure events are increased, while those linked to remaining scenarios are decreased until the KL divergence reaches its threshold, that is, $D_{\mathrm{KL}}(\mathbb{P} \parallel \mathbb{P}_0) = d_{\mathrm{KL}}$. The probability of failure under such adverse distribution is then computed and presented in Table 12.7. The results indicate that service quality is effectively maintained, thus exemplifying the robustness of our proposed method. However, when $q = 100$, the KL divergence from the actual distribution to the reference distribution surpasses $d_{\mathrm{KL}} = 0.01$. As a result, the maximum test failure rate in the MILP case can still exceed 5.0%.

We examine the effect of probability distribution accuracy on capacity planning outcomes in Case 1 by varying the value of d_{KL} from 0 to 0.05. For this series of tests, we employ 200 scenarios in SAA and compute the failure rates under the corresponding planning outcomes using 10,000 samples from the test set. The results are depicted in Table 12.8. When $d_{\mathrm{KL}} = 0$ and $\alpha_{1+} = \alpha$, the DRO model simplifies into a traditional chance-constrained program. This model results in a strategy with the lowest cost; nevertheless, if the probability distribution of PV output deviates from the reference, the service quality becomes unreliable. As d_{KL} increases, α_{1+} correspondingly decreases, leading to more conservative planning strategies that have higher costs and lower failure chances. When the imprecision of the probability distribution is considered, both capacity planning models deliver satisfactory strategies from a reliability standpoint. However, the larger the d_{KL} chosen, the higher the cost incurred. A balance must therefore be struck between reliability and economy, which ultimately depends on the decision-maker's risk tolerance.

Table 12.8 Testing results under different KL divergence thresholds (Xie et al. 2018)

KL divergence threshold			0.000	0.005	0.010	0.020	0.050
α_{1+}			0.050	0.031	0.025	0.018	0.008
LP	Total investment (10^8\$)		1.880	1.959	1.991	2.041	2.129
	Failure rate	Maximum	0.025	0.017	0.018	0.012	0.010
		Mean	0.0195	0.0122	0.0102	0.0069	0.0048
MILP	Total investment (10^8\$)		1.691	1.749	1.795	1.824	1.955
	Failure rate	Maximum	0.069	0.055	0.042	0.038	0.024
		Mean	0.0530	0.0402	0.0326	0.0279	0.0128

12.4.4 Impact of Forecast Errors

The statistical performance of the planning strategy is contingent upon the actual distribution of traffic demand and PV generation data. We examine the impact of errors in data distributions on reliability (or supply adequacy). The planning strategy for Case 1, as depicted in Fig. 12.5, is evaluated. We scrutinize inexact data encompassing the OD matrix Λ, daily EV number B_t, and PV generation output ξ_g. Given that the coverage of charging points does not significantly vary under these circumstances, our primary focus is on the performance of capacity planning. For these tests, model parameters are selected as $\alpha = 0.05$, $d_{\text{KL}} = 0.01$, and $q = 200$.

Firstly, we investigate the effects of OD matrix forecast errors. Assume the real OD matrix, denoted as Λ', differs from the one Λ used during planning. We posit that the traffic flow percentage at the first five entrances/exits is overestimated. In other words, Λ' is derived from Λ by scaling each element in the first five rows and columns by a constant θ ($0 < \theta < 1$). Following this, Λ' is normalized, meaning the entries of Λ' are increased proportionally so that the sum of all elements totals 1. This implies that the overall traffic demand in the highway transportation system remains constant. Table 12.9 displays the failure rates under planning strategies suggested by both the MILP and LP models. As θ decreases (which corresponds to an increase in forecast error), the failure rates rise. The CVaR-based LP formulation maintains a maximum failure rate below 5% when $\theta \geq 0.7$. However, for the VaR-based MILP formulation, the maximum failure rates rapidly surpass the acceptable limit of 5% with just a 0.1 reduction in θ.

Subsequently, we assess the impact of the total number of EVs. In line with the benchmark case, B_t is assumed to adhere to a normal distribution with a mean (μ) of 10,000 and a standard deviation (σ) of 2,000. We adjust the distribution of the EV number by modifying the value of μ. The results, summarized in Table 12.10, suggest that for the planning strategy proposed by the MILP model, reliability can be assured in the worst-case scenario when μ does not exceed 10,000. Conversely,

Table 12.9 Failure rates under different OD matrix forecast errors (Xie et al. 2018)

θ		0.50	0.60	0.75	1.00
LP	Maximum	0.081	0.059	0.031	0.018
	Mean	0.0224	0.0178	0.0118	0.0102
MILP	Maximum	0.133	0.095	0.066	0.042
	Mean	0.045	0.040	0.034	0.033

Table 12.10 Failure rates under different daily EV number forecast errors (Xie et al. 2018)

μ		9,000	10,000	11,000	12,000
LP	Maximum	0.004	0.018	0.024	0.037
	Mean	0.0018	0.0102	0.016	0.0289
MILP	Maximum	0.019	0.042	0.073	0.094
	Mean	0.0103	0.0326	0.0465	0.0705

Table 12.11 Failure rates under different PV generation forecast errors (Xie et al. 2018)

σ_1		0.0	0.1	0.2	0.3
LP	Maximum	0.018	0.018	0.043	0.063
	Mean	0.0072	0.0136	0.0374	0.0559
MILP	Maximum	0.042	0.043	0.078	0.095
	Mean	0.0249	0.0353	0.0629	0.0860

for the planning strategy suggested by the LP model, the failure probability remains under the threshold even when μ increases to 12,000 or higher.

To simulate the volatility of PV generation, we multiply the historical data of PV output by a random variable that follows a normal distribution with a mean of 1 and a standard deviation of σ_1, and then test the failure rates. The results are illustrated in Table 12.11. As previously discussed, the proposed model has the inherent ability to account for ambiguous distribution. When $\sigma_1 = 0$, the PV output aligns with the historical data, and the volatility of PV output exerts minimal impact on reliability when $\sigma_1 \leq 0.1$. Under the planning strategy proposed by the MILP model, failures exceed 5% if $\sigma_1 \geq 0.1$. For the planning strategy offered by the LP model, failures surpass 5% when $\sigma_1 \geq 0.2$. This shows that the LP model is more robust facing data uncertainty.

12.5 Conclusion

This chapter introduces a two-stage approach for addressing the planning issue of standalone renewable-powered charging stations. The transportation system is modeled and the DRO model accounts for the inexactness associated with the probability distributions of traffic demand and solar power generation. Based on risk theory, MILP and LP reformulations are developed, striking a balance between model accuracy and computational efficiency.

The primary insights from the case studies are as follows: Firstly, the solution derived from the traditional chance-constrained program fails to maintain a desired level of supply reliability facing distributional uncertainty. However, the DRO approach proposed in this chapter can handle numerous uncertainties in a data-driven manner, including traffic demand, daily EV numbers, and PV generation, without requiring exact probability distributions of uncertain factors. Secondly, the MILP and LP reformulations each have their own set of benefits and drawbacks. However, the latter is recommended due to its higher tractability and acceptable level of conservatism for a planning problem. Lastly, the threshold of KL divergence in the ambiguity set should be carefully determined to balance reliability and cost. Generally, the more data we possess, the smaller the threshold will be, as we have higher confidence in the accuracy of the reference distribution.

References

Dong X, Mu Y, Jia H et al (2016) Planning of fast EV charging stations on a round freeway. IEEE Trans Sustain Energy 7(4):1452–1461

Hu X, Zou C, Zhang C et al (2017) Technological developments in batteries: a survey of principal roles, types, and management needs. IEEE Power Energ Mag 15(5):20–31

Lofberg J (2004) YALMIP: a toolbox for modeling and optimization in MATLAB. In: 2004 IEEE international conference on robotics and automation, pp 284–289

Mu Y, Wu J, Jenkins N et al (2014) A spatial-temporal model for grid impact analysis of plug-in electric vehicles. Appl Energy 114:456–465

Perez HE, Hu X, Dey S et al (2017) Optimal charging of Li-ion batteries with coupled electro-thermal-aging dynamics. IEEE Trans Veh Technol 66(9):7761–7770

Qian K, Zhou C, Allan M et al (2011) Modeling of load demand due to EV battery charging in distribution systems. IEEE Trans Power Syst 26(2):802–810

Sengupta M, Xie Y, Lopez A et al (2018) The national solar radiation data base (NSRDB). Renew Sustain Energy Rev 89:51–60

Xie R, Wei W, Khodayar ME et al (2018) Planning fully renewable powered charging stations on highways: a data-driven robust optimization approach. IEEE Trans Transp Electrification 4(3):817–830

Chapter 13
Energy Hub Sizing in Multi-carrier Energy Networks

Abstract The strategic use of natural gas, electric power, and heat in a cascading manner can yield synergistic effects, potentially leading to the development of integrated energy systems. In these infrastructures, the energy hub serves as the critical link among various energy systems, performing essential functions of energy production, conversion, and storage. The capacity of the energy hub significantly impacts the degree of integration among these systems and the flexibility of the overall system's operations. This chapter introduces a data-driven two-stage distributionally robust model for energy hub capacity planning using an ambiguity set by the Kullback-Leibler divergence. The model aims to minimize the overall expenditure by reducing the construction cost and the expected life-cycle operating cost under the worst-case distribution confined within the ambiguity set. Leveraging duality theory and sampling average approximation, the proposed model is converted into an equivalent convex program with a nonlinear objective and linear constraints. This is then solved using an outer approximation algorithm, which only requires the solution of a linear programming problem. This chapter introduces the work in Cao et al. (2020).

13.1 Introduction

The growing use of renewable energy resources like wind and solar power significantly curtails carbon dioxide emissions from the electricity sector. However, the unpredictable nature of wind farms and photovoltaic (PV) panels' output necessitates large backup capacity and flexible resources to offset real-time imbalances, posing considerable challenges to power system operations. Gas-fired units offer a solution, given their ability to adapt quickly to changes in renewable output. Additionally, heating systems, with their substantial thermal inertia, can function as energy storage (ES) (Heinen et al. 2017; Good and Mancarella 2019). Moreover, the combined dispatch of natural gas, heat, and power generation proves to be more efficient than independent operations (Li et al. 2013; Fang et al. 2018). In essence, integrating multi-carrier energy can enhance overall energy efficiency through cascaded energy usage and improve system flexibility by capitalizing on the quick-response and

R. Xie and W. Wei, *Distributionally Robust Optimization and its Applications in Power System Energy Storage Sizing*,
https://doi.org/10.1007/978-981-97-2566-3_13

storage capabilities of gas and heating systems. Given these advantages, energy system integration has gained significant attention in recent years, leading to the emergence of multi-carrier energy systems (Krause et al. 2011; Kienzle et al. 2011). In such integrated infrastructures, the connecting facility, commonly referred to as the energy hub (Geidl and Andersson 2007; Geidl et al. 2007), plays a pivotal role in energy production, conversion, and storage.

This chapter presents a two-stage distributionally robust stochastic programming (DR–SP) model designed to determine the optimal sizing of facilities within a distribution-level energy hub.

First, a DR–SP approach is introduced to tackle the problem of energy hub capacity planning within multi-carrier energy networks. The model considers uncertainties in renewable generation and load, represented by ambiguous probability distributions centered around an empirical probability distribution derived from moderate historical data. The Kullback-Leibler (KL) divergence quantifies the distance between these two probability distributions. The model's objective is to minimize the sum of investment and anticipated operating costs throughout the service period under the worst-case distribution confined within the ambiguity set. Supply reliability under extreme conditions is modeled using robust chance constraints. Compared to traditional stochastic programming (SP) and robust optimization (RO) approaches, the proposed method requires moderate uncertainty information and offers robust planning strategies with an adjustable level of conservatism, determined by manipulating the KL divergence parameter within the ambiguity set.

Secondly, we introduce practical reformulations of the proposed model, originally nonconvex and nonlinear due to the complex min-max-min structure of the objective function and the robust chance constraints. These constraints are linearized utilizing convex approximation and the sampling average approximation (SAA) method. By leveraging dual theory, the objective function is transformed into a strictly minimization-oriented one. We establish its convexity and construct an outer approximation (OA) algorithm for efficient problem resolution using linear programming (LP) solvers. The effectiveness of our proposed methodology is substantiated through various case studies.

The chapter is structured as follows: Sect. 13.2 introduces the mathematical formulation of the DR–SP model, which is used for capacity planning of the energy hub. Section 13.3 outlines the development of our solution strategy. To demonstrate the efficacy of the proposed model and algorithm, case studies are presented in Sect. 13.4. The chapter concludes with a summary and final thoughts in Sect. 13.5.

13.2 Mathematical Formulation

13.2.1 Energy Hub Model

Figure 13.1 illustrates the topology and energy flow variables of the energy hub. The energy inputs consist of electricity and natural gas, while the outputs are electricity and heat. The energy hub is composed of several components: a combined heat and power (CHP) unit, heat pump (HP), electricity storage unit (ESU), and thermal storage unit (TSU). Unlike residential energy hubs that connect their outputs to end users, the hub depicted in Fig. 13.1 connects to energy systems on both sides. The operation constraints and energy flows within the energy hub can be modeled as follows:

$$p_{i,t}^{out} = p_{i,t}^{CHP} + p_{i,t}^{d} - p_{i,t}^{c} , \tag{13.1a}$$
$$h_{i,t}^{out} = h_{i,t}^{CHP} + h_{i,t}^{HP} + h_{i,t}^{d} - h_{i,t}^{c} , \tag{13.1b}$$
$$g_{i,t}^{in} = p_{i,t}^{CHP}/\eta_i^{gp} + h_{i,t}^{CHP}/\eta_i^{gh} , \tag{13.1c}$$
$$h_{i,t}^{HP} = COP_i \cdot p_{i,t}^{in} , \tag{13.1d}$$
$$W_{i,t+1}^{E} = W_{i,t}^{E}(1-\mu_i^E) + (p_{i,t}^{c}\eta_{i,c}^{E} - p_{i,t}^{d}/\eta_{i,d}^{E})\Delta t , \tag{13.1e}$$
$$W_{i,t+1}^{T} = W_{i,t}^{T}(1-\mu_i^T) + (h_{i,t}^{c}\eta_{i,c}^{T} - h_{i,t}^{d}/\eta_{i,d}^{T})\Delta t . \tag{13.1f}$$

The equations in (13.1a) and (13.1b) outline the conditions for balancing electrical and thermal power. The input-output relations of the CHP unit and HP are defined by (13.1c) and (13.1d). Equations (13.1e) and (13.1f) describe the charging dynamics of both the ESU and TSU. The model naturally satisfies the complementarity of charging and discharging, as stipulated in (13.1), since simultaneous charging and discharging is not the optimal strategy due to the principle of dissipativity. A comprehensive analysis is available in Li et al. (2016). The imposition of strict complementarity in model (13.1) can be achieved via binary variables, ensuring the solution method discussed later is not compromised.

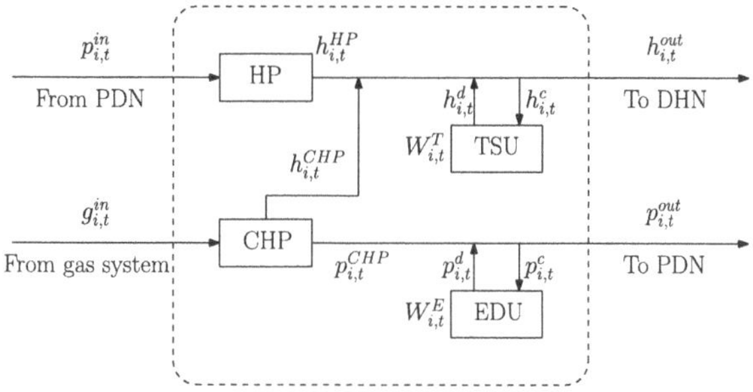

Fig. 13.1 Topology and energy flow of energy hub. DHN and PDN are the abbreviations of district heating network and power distribution network, respectively (Cao et al. 2020)

13.2.2 Network Model

The topology of the power distribution network (PDN) is inherently radial, with power flow within the PDN able to be established in a recursive manner. This is based on the linear branch flow model (Baran and Wu 1989).

$$p_{j,t} + P_{ij,t} = \sum_{k \in S_j^{B,d}} P_{jk,t} + p_{j,t}^d \ , \tag{13.2a}$$

$$q_{j,t} + Q_{ij,t} = \sum_{k \in S_j^{B,d}} Q_{jk,t} + q_{j,t}^d \ , \tag{13.2b}$$

$$V_{j,t} = V_{i,t} - (r_{ij} P_{ij,t} + x_{ij} Q_{ij,t}) / V_0 \ . \tag{13.2c}$$

Equations (13.2a) and (13.2b) denote the conditions for balancing active and reactive power. The total active power injection at bus j, referred to as $p_{j,t}$, encompasses the active output from local generators ($p_{j,t}^{TU}$), wind farms ($p_{j,t}^{w}$), and the energy hub ($p_{j,t}^{out}$). Equation (13.2c) represents the forward voltage drop along a distribution line. Although model (13.2) does not account for network losses, it does consider reactive power and bus voltage. Therefore, it proves more appropriate for distribution systems than the widely recognized DC power flow model, where reactive power is neglected and bus voltage magnitudes are assumed to be constant.

A district heating network (DHN) consists of symmetrically aligned supply and return pipelines. These pipelines facilitate the injection or withdrawal of heat at each source or load node through a heat exchanger that connects the supply and return aspects. As noted in reference Liu et al. (2016), the physical model of the DHN is dictated by both hydraulic and thermal conditions. This chapter employs the constant-flow variable temperature model, wherein mass flow rates, represented as $\dot{m}$, maintain a steady value within the modeled hydraulic conditions of the DHN. The thermal flow model in the DHN is

$$h_{i,t}^S = c_p \dot{m}_{i,t}^S (\tau_{i,t}^S - \tau_{i,t}^R) \ , \tag{13.3a}$$

$$h_{i,t}^L = c_p \dot{m}_{i,t}^L (\tau_{i,t}^S - \tau_{i,t}^R) \ , \tag{13.3b}$$

$$\tau_{b,t}^o = (\tau_{b,t}^i - T_t^a) \exp\left(-\frac{\lambda_b L_b}{c_p \dot{m}_{b,t}}\right) + T_t^a \ , \tag{13.3c}$$

$$\sum_{b \in S_i^{P,e}} (\tau_{b,t}^o \dot{m}_{b,t}) = \tau_{i,t}^n \sum_{b \in S_i^{P,e}} \dot{m}_{b,t} \ , \tag{13.3d}$$

$$\tau_{b,t}^i = \tau_{i,t}^n \ , \ \forall b \in S_i^{P,s} \ . \tag{13.3e}$$

Equations (13.3a) and (13.3b) describe the thermal energy exchange occurring at both heat source and heat load nodes. The temperature fluctuations along the supply/return pipelines are outlined in (13.3c). Equation (13.3d) illustrates the temperature of the mixed fluid at the confluence nodes, while Eq. (13.3e) specifies the temperature of mass flows departing from a confluence node. For a more comprehensive understanding of the DHN, kindly refer to Liu et al. (2016).

Gas flows within a pipeline network adhere to the principles of partial differential equations, as referenced in Wang et al. (2018). The phenomena of gas-electricity coupling predominantly occur at the transmission level. However, this chapter shifts the focus towards heat-power integration at the distribution level, investigating the interconnection between PDN and DHN within urban environments. This is due to the economic impracticality of delivering heat over extended distances. As a result, the model for the upstream-level natural gas transportation network is not considered in this study. It is worth noting that transients within a gas distribution network typically exhibit a higher speed. The primary operational constraint for an energy hub is the maximum rate of gas delivery. This can be approximately determined by implementing time-variable upper limits on the inflow.

13.2.3 Traditional Planning Model

Within the deterministic formulation, both renewable output and load demands are precisely known. The process unfolds in two stages. In the first stage, the capacities of energy hub components are decided. In the second stage, operational constraints are taken into account over the course of three typical days, representative of spring/autumn, summer, and winter seasons. The objective function aims to minimize the aggregate of construction costs and life-cycle operational expenses. The following assumptions are made in the planning problem:

First, the planning of the energy hub is orchestrated under the guidance of a government agency. The goal is to minimize both the initial investment cost associated with the hub and the cumulative operational cost of the integrated energy system over a ten-year span. Given that we consider a distribution-level energy hub, the operational cost is regarded as comparable to the system's overall cost. Consequently, these two costs can be combined and jointly minimized.

Second, the connectivity architecture of the energy hub is predetermined and unchangeable. Potential areas of investment encompass the CHP unit, HP, ESU, and TSU. The model can easily incorporate electric boilers and diverse battery types. However, for simplicity and clarity, we have chosen one representative facility for each functionality within the model. The operational cost is primarily composed of the fuel expenses of local generators in the PDN and the CHP unit located within the energy hub.

Third, the cost of the heat pump is overlooked for two principal reasons: On the one hand, the heat pump utilizes electricity, the production cost of which has already been accounted for. On the other hand, any existing costs associated with the heat pump are paid to the PDN by the energy hub. This is essentially a domestic financial matter within the integrated energy system and hence, does not factor into the objective function.

Therefore, the deterministic model of energy hub planning is as follows:

$$\begin{aligned} \min\ & f_C + N_d \cdot f_O \\ \text{s.t. } & \text{Cons-PF}\ ,\ \text{Cons-TF}\ , \\ & \text{Cons-EH}\ ,\ \text{Cons-BD}\ . \end{aligned} \tag{13.4}$$

In this formulation, Cons-PF refers to the linearized branch flow equations indicated by (13.2). Cons-TF is shorthand for the thermal flow constraints expressed in (13.3). Cons-EH encompasses the operational conditions of the energy hub, as detailed in (13.1). Lastly, Cons-BD brings together all the lower- and upper-bound constraints of the decision variables. The calculation of the construction cost is as follows:

$$f_C = I^{CHP} C^{CHP} + I^{HP} C^{HP} + I^E C^E + I^T C^T\ . \tag{13.5}$$

This calculation encompasses the investment costs associated with the CHP unit, heat pumps, ESU, and TSU within the energy hub. Following this, the daily operation cost function is as follows:

$$f_O = \sum_t \left(\sum_i \omega_g g_{i,t}^{in} + \sum_j F(p_{j,t}^{TU}) \right)\ . \tag{13.6}$$

This includes the fuel costs for the CHP units and local generators. The convex quadratic function $F(p_{j,t}^{TU})$ can be approximated by a piecewise linear function using established methods, such as the one stated in Wu (2011). As a result, we can safely assume the objective function to be linear without any loss of generality. N_d represents the number of service days, which is 3,650 in this study. To maintain brevity, a typical day is selected to illustrate the model. However, in practical implementation, we incorporate three representative days from spring/autumn, summer, and winter, assigned weights of 0.5, 0.25, and 0.25, respectively, to compute the daily operation cost f_O. In this particular context, we do not take into account demand growth and the net present value of operation cost, given that a 10-year planning horizon is relatively moderate. However, these factors can be readily integrated by adjusting certain parameters.

The boundaries of the operating variables for the energy hub are influenced by the capacities of the CHP unit C^{CHP}, the heat pump C^{HP}, the ESU C^E, and the TSU C^T. The relationships between these variables are as follows:

$$(p_{i,t}^{CHP}/\eta_i^{gp} + h_{i,t}^{CHP}/\eta_i^{gh}) \le C^{CHP}\ , \tag{13.7a}$$

$$p_{i,t}^{CHP}/\eta_i^{gp} \ge R_g^p C^{CHP}\ ,\ h_{i,t}^{CHP}/\eta_i^{gp} \le R_g^h C^{CHP}\ , \tag{13.7b}$$

$$h_{i,t}^{HP} \le C^{HP}\ , \tag{13.7c}$$

$$W_{i,t}^E \le C^E\ ,\ p_{i,t}^d \le R_d^E C^E\ ,\ p_{i,t}^c \le R_c^E C^E\ , \tag{13.7d}$$

$$W_{i,t}^C \le C^T\ ,\ h_{i,t}^d \le R_d^T C^T\ ,\ h_{i,t}^c \le R_c^T C^T\ . \tag{13.7e}$$

These relationships are encompassed within Cons-BD. The initial two equations indicate the polyhedral operation region of the CHP unit, taking into account the maximum fuel import rate as per (13.7a), as well as the minimum electric output and maximum thermal output detailed in (13.7b), where R_g^h and R_g^p are constants. Equation (13.7c) mirrors the capacity of the HP. Equations (13.7d) and (13.7e) confine the charging/discharging rates of storage units based on their respective capacities, with R_d^E, R_c^E, R_d^T and R_c^T constants.

Denote the first-stage decision variables by the vector x, which includes the facility capacities C^{CHP}, C^{HP}, C^E, and C^T. The vector ξ symbolizes the uncertain parameters, which encompass the output of wind farms $p_{j,t}^w$ and system loads $p_{j,t}^d$ and $h_{i,t}^L$. The vector y consists of the second-stage decision variables, incorporating elements of both the power and thermal flow models. With these notations clearly defined, we can express the deterministic model for energy hub planning as an LP problem. This can be succinctly written in matrix form as follows:

$$\begin{aligned} &\min C^\top x + Q(x, \xi) \\ &\text{s.t. } x \in X \; . \end{aligned} \tag{13.8}$$

The feasible set of first-stage decisions is represented by $X = \{x \mid 0 \le x \le x^u\}$. In the objective function, the first term $c^\top x$ corresponds to the construction cost f_C as defined in (13.5). Meanwhile, the second term $Q(x, \xi)$ signifies the optimal operating cost linked with parameter ξ under a specific x. The optimal operating cost can be expressed as follows:

$$\begin{aligned} Q(x, \xi) = &\min_y \; p^\top y \\ &\text{s.t. } Ax + By + C\xi + d \le 0 \; . \end{aligned} \tag{13.9}$$

The objective function here corresponds to the operational cost as outlined in (13.6). The constraints involved, encapsulate those in Cons-PF, Cons-TF, Cons-EH, and Cons-BD, except those included in X.

In the absence of uncertainty, problem (13.8) simplifies to an LP problem, which can be solved effortlessly. To understand why the planning problem is structured in a two-stage format, we need to examine how uncertainty is addressed. To actively take into account the uncertain nature of parameter ξ, both SP and RO models can be established based on the deterministic formulation (13.8). Assuming that we have a set of representative samples like $\xi^1, \xi^2, \ldots, \xi^n$. For the scenario-based SP, given the knowledge of probabilities $\pi_1, \pi_2, \ldots, \pi_n$ associated with each sample, the SP model can be established as follows:

$$\begin{aligned} &\min c^\top x + \sum_n \pi_n p^\top y^n \\ &\text{s.t. } x \in X \; , \; Ax + By^n + C\xi^n + d \le 0 \; , \; \forall n \; . \end{aligned} \tag{13.10}$$

Conversely, if there is no available information regarding the probabilities, we can turn to the RO model. This alternative approach offers a solution when dealing with uncertainty and is structured as follows:

$$\begin{aligned} &\min_{x} c^{\top} x + \max_{\xi^n} \left\{ \min_{y^n} \ p^{\top} y^n \right\} \\ &\text{s.t. } x \in X \ , \ \ Ax + By^n + C\xi^n + d \le 0 \ , \ \forall n \ . \end{aligned} \tag{13.11}$$

From (13.10) and (13.11), we can make the following observations:

Firstly, the second-stage decision y^n hinges on the value of the first-stage decision x and the uncertain data ξ^n. This suggests that once the capacity planning strategies for the energy hub have been implemented at the first stage, they become immutable during the operational periods. The second stage simulates the system's daily operations once the uncertain data (such as renewable generation and load demand) can be accurately predicted. This implies that each unit's output can adjust to the actual value of uncertain data, thereby minimizing operational costs. In essence, this approach integrates planning and operation into a comprehensive model by considering detailed operational constraints and data uncertainty during the planning stage. This holistic model provides an effective framework for decision-making in the context of energy hub planning.

Secondly, while the SP and RO models share the same constraints, they differ in their objective functions. The former integrates an expected cost in the second stage, while the latter accounts for the worst-case scenario. The SP model demands a more detailed understanding of the uncertainty. In practical scenarios, we might not have exact values for π_n, but we can still extract useful information from the available finite historical data. This may include an empirical distribution and an assessment of how far the true distribution deviates from the empirical one. Conversely, if we employ the RO model (13.11), we inherently discard all distribution information and may end up with overly conservative planning strategies. This predicament highlights the need for a new computationally manageable methodology that effectively utilizes ambiguous information.

13.2.4 Uncertainty Model

The histogram is the most commonly used empirical distribution. For instance, suppose we have a total of M samples to fit into N bins, with $M_1, M_2, \ldots, M_N$ samples in each respective bin. In this case, the representative scenario in each bin is the expected value of ξ within that bin. The corresponding probability is $\pi_i = M_i/M$, where $i = 1, \ldots, N$. Thus, the distribution of P_0 is represented by the set $\{\pi_1, \ldots, \pi_N\}$.

We take into account all potential probability distributions that are sufficiently close to P_0. More precisely, we consider all elements in the following set:

$$W = \{P \mid D_{\mathrm{KL}}(P||P_0) \leq d_{\mathrm{KL}}\} , \tag{13.12}$$

where d_{KL} is a threshold parameter determining the size of the ambiguity set and reflecting the confidence level. $D_{\mathrm{KL}}(P||P_0)$ is the KL divergence from P to P_0 defined in Definition 5.4.

13.2.5 Proposed Planning Model

Using the compact form (13.8) and the ambiguity set W defined in (13.12), we can formulate the DR–SP model for energy hub planning as follows:

$$\min c^{\top} x + \sup_{P \in W} \mathbb{E}^{P}[Q(x, \xi)] \tag{13.13a}$$

$$\text{s.t. } x \in X , \tag{13.13b}$$

$$\inf_{P' \in W'} \Pr\{D_{loss}(\xi) \leq 0\} \geq 1 - \alpha . \tag{13.13c}$$

The model features $\mathbb{E}^{P}[Q(x, \xi)]$, which stands for the expected value of the operation cost function $Q(x, \xi)$ under the assumption that the uncertain parameter ξ adheres to distribution P. The constraint, as outlined in (13.13c), mandates that during extreme conditions, all loads must be catered to, with the minimal probability set at $1 - \alpha$. Consequently, this implies that the likelihood of load shedding must not exceed α. This requirement is coined as a robust chance constraint due to its consideration of the worst-case distribution. $D_{loss}(\xi)$ symbolizes the minimal unserved load during the extreme conditions, contingent upon the uncertain data ξ.

$$\begin{aligned} D_{loss}(\xi) = \min\ & g \\ \text{s.t. }\ & A'x + B'y + C'\xi + d' \leq 0 , \\ & \sum_{k \in S_j^{b,d}} P_{jk,t} + p_{i,t}^{d} - P_{ij,t} - p_{j,t} \leq g , \\ & c_p \dot{m}_{i,t}^{L} (\tau_{i,t}^{S} - \tau_{i,t}^{R}) - h_{i,t}^{L} \leq g . \end{aligned} \tag{13.14}$$

In the equation, the coefficient matrices A', B', C', and d' correspond to matrices A, B, C, and d, respectively, albeit excluding the conditions of nodal energy balancing. The last two inequalities represent relaxed nodal energy balancing conditions. In this context, load shedding is measured by a slack variable g, while ξ encompasses all uncertain parameters, including elements such as $p_{j,t}^{w}$, $p_{j,t}^{d}$, and $h_{i,t}^{L}$. It is important to note that in (13.14), there is no requirement for g to be nonnegative. A negative g will not affect planning or operational strategies because its function is solely to evaluate the probability of load shedding, and it will not be used for any operational purpose.

The objective function, as presented in (13.13a), takes into account the expectation $\mathbb{E}^P[Q(x, \xi)]$ in the worst-case distribution under normal circumstances. Considering these elements, the planning model (13.13) combines the benefits of both SP and RO: it does not require an exact probability distribution, and the optimal strategy remains resilient to any perturbations in the probability distribution of uncertain parameters. In fact, due to the storage capabilities of ESU and TSU, along with the flexibility provided by multi-energy generation and conversion, the system also demonstrates robustness to changes in the parameter ξ.

It is important to clarify that the worst-case distributions in the objective function (13.13a) for regular days and the robust chance constraint (13.13c) for extreme days are distinct. Consequently, we can establish separate ambiguity sets for (13.13a) and (13.13c). To differentiate between the two, we utilize P' and W' in the reliability constraint (13.13c), setting them apart from the terms used in the objective function (13.13a).

13.3 Solution Method

In the formulation presented in (13.13), the maximum expectation in (13.13a) combined with the minimum probability evaluation over the ambiguity set in (13.13c) creates a scenario that prevents the equation from being solved directly. In this section, we will develop tractable reformulations to overcome this issue.

13.3.1 Reformulate the Robust Chance Constraint

The robust chance constraint (13.13c) presents a challenge due to its dependency on an infimum evaluation over an infinite set. If we assume that constraint (13.13c) is met, then the probability calculated under the reference distribution P_0 must exceed $1 - \alpha$. The modifier's value, in this case, depends on the divergence measure and the confidence level. When KL divergence is used to determine the distance between probability distributions, Theorems 5.13 and 5.14 indicate that the constraint (13.13c) equates to a conventional chance constraint as follows:

$$\mathrm{Pr}_0\{D_{loss} \leq 0\} \geq 1 - \alpha_+ \, . \tag{13.15}$$

Here, Pr_0 symbolizes the probability evaluated under the reference distribution P_0. The value of α_+ can be computed by the following method:

$$\alpha_+ = \max\left\{0, 1 - \inf_{z \in (0,1)} \left\{\frac{\exp(-d_{\mathrm{KL}})z^{1-\alpha} - 1}{z - 1}\right\}\right\} . \tag{13.16}$$

The univariate function $h(z) = (\exp(-d_{\mathrm{KL}})z^{1-\alpha} - 1)/(z - 1)$ exhibits convexity in z over the open interval $(0, 1)$. As a result, its minimum can be efficiently computed using traditional optimization techniques such as the golden section search method or the first-order optimality condition, which necessitates that $\partial h(z)/\partial z = 0$. Moreover, it is significant to emphasize that $\alpha_+ < \alpha$. This suggests that Eq. (13.15) provides a more conservative approach compared to a conventional chance constraint assessed at P_0.

Nonetheless, it is key to note that the chance constraint (13.15) remains non-convex according to Nemirovski and Shapiro (2007). A feasible approach to overcome this challenge lies in finding a conservative yet convex approximation. Evidently, (13.15) is equivalent to:

$$\mathbb{E}^{P_0}[1_{\mathbb{R}_+}(D_{loss})] = \mathrm{Pr}_0\{D_{loss} > 0\} \leq \alpha_+ \,. \tag{13.17}$$

In this context, $\mathbb{E}^{P_0}(\cdot)$ represents the expectation tied to the empirical distribution P_0, while $1_{\mathbb{R}_+}(x)$ serves as the indicator function of $\mathbb{R}_+$. At this point, our task is to identify a convex function $\psi(x)$ that offers an overestimation of $1_{\mathbb{R}_+}(x)$. This step is crucial to guarantee the following approximation is conservative:

$$\mathbb{E}^{P_0}[1_{\mathbb{R}_+}(D_{loss})] \leq \mathbb{E}^{P_0}[\psi(D_{loss})] \leq \alpha_+ \,. \tag{13.18}$$

If we impose the conditions that 1. $\psi(x)$ is nondecreasing and 2. $\psi(0) = 1$, then $\psi(x)$ effectively provides an overestimation of $1_{\mathbb{R}_+}(x)$. In the context of this chapter, $\psi(D_{loss})$ has been chosen in the following:

$$\psi(D_{loss}) = \max\{0, D_{loss}/\beta + 1\} \,. \tag{13.19}$$

In the above equation, $\beta > 0$ is a constant. However, the parameter β will be fine-tuned in the final problem to ensure that $\psi(D_{loss})$ delivers a satisfactory approximation.

Assume that $\xi^1, \xi^2, \ldots, \xi^K$ are representative scenarios with corresponding probabilities $\pi_1, \pi_2, \ldots, \pi_K$. In such a case, the inequality $\mathbb{E}^{P_0}[\psi(D_{loss})] \leq \alpha_+$ translates to

$$\sum_k \pi_k \max\{0, D_{loss}(\xi^k)/\beta + 1\} \leq \alpha_+ \,.$$

By multiplying both sides of the equation by β and introducing an auxiliary variable ϕ_k, the aforementioned inequality can be linearized as follows.

$$\begin{aligned} & D_{loss}(\xi^k) + \beta \leq \phi_k \,, \;\; \phi_k \geq 0 \,, \;\; \forall k \,, \\ & \sum_k \pi_k \phi_k \leq \beta\alpha_+ \,, \;\; \beta > 0 \,. \end{aligned} \tag{13.20}$$

The value of $D_{loss}(\xi_k)$ is derived from problem (13.14). In the final problem, the need for the minimization operator is eliminated, and β also becomes a decision variable. Given that we have utilized α_+, a more cautious threshold than α, we can ensure the reliability requirement during extreme days, even if the probabilities $\pi_1, \ldots, \pi_K$ are not entirely accurate.

13.3.2 Reformulate the Objective Function

Given a specific planning strategy x, the problem that seeks to find the worst-case expectation can be formulated as follows:

$$\sup_{P \in W} \mathbb{E}^P[Q(x, \xi)] . \tag{13.21}$$

Problem (13.21) involves optimizing a probability distribution, making it an infinite-dimensional optimization problem. As per Theorem 5.12, the worst-case expectation in (13.21) can be equated to the following expression:

$$\min_{\lambda \geq 0} \lambda \log \left\{ \sum_n \pi_n \left[\exp(Q(x, \xi^n)/\lambda) \right] \right\} + \lambda d_{\text{KL}} . \tag{13.22}$$

In this case, the decision variable is defined as a nonnegative scalar λ. Let us define $\theta_n = Q(x, \xi_n)$.

Proposition 13.1 *The following function is convex in λ and θ:*

$$H(\theta, \lambda) = \lambda \log \left(\sum_n \pi_n \exp(\theta_n/\lambda) \right) . \tag{13.23}$$

Proof First of all, the following function is convex

$$h_1(\theta) = \log \left(\sum_{n=1}^{N} \exp(\theta_n) \right) ,$$

according to Example 3.14 in Boyd and Vandenberghe (2004). Secondly, the function $h_2(\theta)$ with

$$h_2(\theta) = h_1(\theta + b) \ , \ b = [\log \pi_1, \ldots, \log \pi_N]^\top ,$$

is convex because the mapping $\theta \mapsto \theta + b$ is affine. Therefore,

$$h_2(\theta) = \log\left(\sum_{n=1}^{N} \pi_n \exp(\theta_n)\right),$$

is convex. Because $H(\theta, \lambda) = \lambda h_2(\theta/\lambda)$ is the perspective of $h_2(\theta)$, it is convex in θ and λ (Boyd and Vandenberghe 2004). □

While θ_n is treated as a constant in problem (13.22), it will serve as a decision variable in the final problem. This inherent convexity significantly simplifies the development of the algorithm, aiding in its efficiency and effectiveness.

13.3.3 Solution Algorithm

Drawing from the foregoing discussions, the energy hub planning problem (13.13) can be reformulated. This new formulation now features a convex objective function along with linear constraints, as shown below.

$$\min\ c^\top x + \lambda \log\left(\sum_n \pi_n \exp(\theta_n/\lambda)\right) + \lambda d_{\mathrm{KL}} \tag{13.24a}$$

$$\text{s.t. } x \in X\ ,\ \lambda \geq 0\ , \tag{13.24b}$$

$$\theta_n = p^\top y^n\ ,\ \forall n\ , \tag{13.24c}$$

$$Ax + By^n + C\xi^n + d \leq 0\ ,\ \forall n\ , \tag{13.24d}$$

$$A'x + B'y^k + C'\xi^k + d' \leq 0\ ,\ \forall k\ , \tag{13.24e}$$

$$My^k + N\xi^k \leq g^k\ ,\ \forall k\ , \tag{13.24f}$$

$$g^k + \beta \leq \phi_k\ ,\ \phi_k \geq 0\ ,\ \forall k\ , \tag{13.24g}$$

$$\sum_{k=1}^{K} \pi_k \phi_k \leq \beta\alpha_+\ ,\ \beta > 0\ . \tag{13.24h}$$

In (13.24), we generate scenarios under both normal and extreme conditions. These scenarios come from three typical days and two extreme days respectively, each independently denoted by n and k. The optimal second-stage cost, represented as $Q(x, \xi^n) = p^T y^n$, is referred to as θ_n within (13.24c). Under normal conditions, as indicated in (13.24d), the shedding of load is not permitted. Constraints (13.24e) through (13.24f) detail the minimum unserved nodal demands. Finally, (13.24g)–(13.24h) are representative of the reliability constraints under extreme conditions which are derived from (13.20), where the loss function $D_{loss}(\xi^k) = g^k$.

The objective function (13.24a) is demonstrated to be convex, while constraints (13.24b)–(13.24h) are polyhedral. Given these properties, any local algorithm or solver, if successful, should converge to the global optimum of (13.24). However, based on preliminary testing, general-purpose nonlinear programming solvers such as IPOPT and NLOPT have failed to achieve convergence when attempting to solve problem (13.24). In response to this challenge, we developed an outer approximation (OA) algorithm to iteratively solve (13.24), the basic procedures of which are summarized in Algorithm 13.1. This algorithm commences by linearizing the convex region defined by $H(\theta, \lambda) \leq \sigma$, and then successively generates cutting planes to approximate its boundary with increasing precision. As a result, only LP problems need to be solved. The finite convergence of the OA algorithm for convex programs has been confirmed in Hogan (1973).

Algorithm 13.1 Outer approximation

Input: Parameters of problem (13.24); convergence tolerance $\varepsilon > 0$.
Output: Optimal solution of problem (13.24).

1: Initiation: $m \leftarrow 0$. Solve SP model (13.10) to initialize θ^0. $\lambda^0 \leftarrow 1000$. Calculate $H^m \leftarrow H(\theta^m, \lambda^m)$ and the gradient of $H(\theta, \lambda)$ at (θ^m, λ^m) as follows:

$$g_H^m \leftarrow \left[\left(\frac{\partial H}{\partial \theta} \right)^\top, \frac{\partial H}{\partial \lambda} \right] \Bigg|_{(\theta^m, \lambda^m)}^\top . \tag{13.25}$$

2: Solve the master LP

$$\begin{aligned} &\min c^\top x + \sigma + \lambda d_{\text{KL}} \\ &\text{s.t. (13.24b)–(13.24h)} , \\ &H^v + (g_H^v)^\top \begin{bmatrix} \theta - \theta^v \\ \lambda - \lambda^v \end{bmatrix} \leq \sigma , \\ &v = 0, 1, \ldots, m . \end{aligned} \tag{13.26}$$

Update $m \leftarrow m + 1$.

3: If the change of optimal values is no larger than ε, terminate and output the final optimal solution; otherwise, calculate H^m and g_H^m by (13.25), add the constraint

$$H^m + (g_H^m)^\top \begin{bmatrix} \theta - \theta^m \\ \lambda - \lambda^m \end{bmatrix} \leq \sigma , \tag{13.27}$$

to the master problem (13.26), and go to Step 2.

In order to draw comparisons between the proposed DR–SP model and traditional SP and RO approaches, the supply reliability constraint during extreme days is enhanced. Following the process of convex approximation and SAA, the SP model, denoted by (13.10), can be articulated as follows:

$$\begin{aligned}
&\min c^\top x + \sum_n \pi_n p^\top y^n \\
&\text{s.t. } x \in X, Ax + By^n + C\xi^n + d \le 0 \ , \ \forall n \ , \\
&\quad A'x + B'y^k + C'\xi^k + d' \le 0 \ , \ \forall k \ , \\
&\quad My^k + N\xi^k \le g^k \ , \ \forall k \ , \\
&\quad g^k + \beta \le \phi_k \ , \ \phi_k \ge 0 \ , \ \forall k \ , \\
&\quad \sum_{k=1}^{K} \pi_k \phi_k \le \beta\alpha \ , \ \beta > 0 \ .
\end{aligned} \tag{13.28}$$

The RO model, represented by (13.11), can be modified as follows:

$$\begin{aligned}
&\min_x c^\top x + \max_{\xi^n} \left\{ \min_{y^n} p^\top y^n \right\} \\
&\text{s.t. } x \in X \ , \ Ax + By^n + C\xi^n + d \le 0 \ , \ \forall n \ , \\
&\quad A'x + B'y^k + C'\xi^k + d' \le 0 \ , \ \forall k \ , \\
&\quad My^k + N\xi^k \le 0 \ , \ \forall k \ .
\end{aligned} \tag{13.29}$$

In (13.29), there is a strict prohibition on load shedding even during the most extreme circumstances, adhering to the basic principles of RO. This differentiates it from a traditional RO model, which generally employs an uncertainty set. Instead, (13.29) handles a finite number of scenarios. This approach can be perceived as equivalent to conceptualizing the convex hull of $\{\xi^n\}$ as the uncertainty set.

13.4 Case Studies

The efficacy of the proposed model and algorithm is validated via a test system, which integrates a modified IEEE 33-bus PDN and a 10-node DHN. The system's topology can be referenced in Fig. 13.2. Over a span of ten years, two energy hubs are set for investment. The relevant data concerning the energy hub components can be found in Table 13.1. All experiments were conducted on a laptop equipped with an Intel i5-7300HQ CPU and 8GB of memory. All the optimization models were developed using YALMIP (Lofberg 2004) in MATLAB, and the LP was solved using CPLEX 12.8.

Regarding the uncertain factors, profiles for electrical/thermal demands and wind generation during three typical days (spring/autumn, summer, and winter) are applied in the normal condition constraint (13.24d). Two extreme days, specifically those in summer and winter, are specially addressed in (13.24e)–(13.24f). The forecast data is illustrated in Fig. 13.3. It is assumed that the forecast error adheres to a normal distribution with a zero mean, while the standard deviation is 0.2 times the forecast

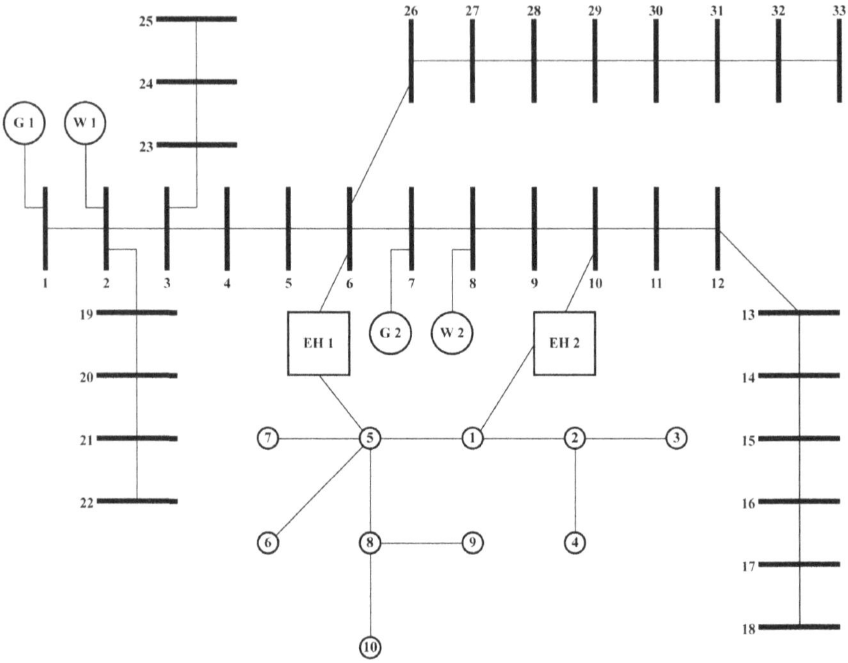

Fig. 13.2 Topology of the test system (Cao et al. 2020)

Table 13.1 Component parameters (Cao et al. 2020)

	Parameters	Construction cost
CHP	$\eta_{gp} = 0.9$, $\eta_{gh} = 0.9$	1,000,000 (\$/MW)
HP	$COP = 3$	1500,000 (\$/MW)
ESU	$\eta_c^E = 0.98$, $\eta_d^E = 0.98$, $\mu^E = 0.01$	200,000 (\$/MWh)
TSU	$\eta_c^H = 0.95$, $\eta_d^H = 0.95$, $\mu^H = 0.01$	150,000 (\$/MWh)

values for wind power and 0.1 times the forecast values for system load. For the SAA implementation, 5,000 scenarios are generated for each representative day. Utilizing the back-forward scenario reduction method in Dupačová et al. (2003), 100 scenarios are selected for both normal and extreme days. A value of $d_{\mathrm{KL}} = 0.0124$ is employed in (13.12). In the robust chance constraint (13.13c), the reliability level is sustained at no less than 95%, implying values of $\alpha = 0.05$ and $\alpha_+ = 0.0229$, as per (13.16).

The proposed DR–SP model (13.24) is evaluated in comparison with the traditional SP model (13.28) and RO model (13.29). An analysis of the planning results generated by these three models is provided in Table 13.2. Evidently, in terms of the optimal value, the DR–SP model is more conservative than the traditional SP model and less conservative than the RO model. This is because the SP model solely considers the reference distribution P_0, while the RO model overlooks the dispersion

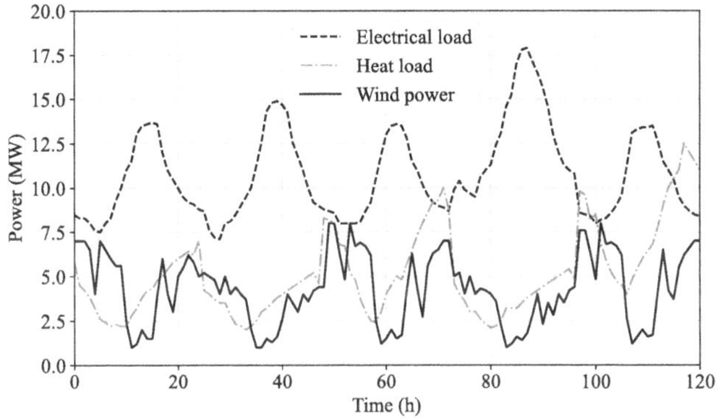

Fig. 13.3 The displayed data represent electrical loads, heat loads, and wind power outputs across various cases. The data, from left to right, correspond to three typical days in the spring/autumn seasons, followed by summer and winter. Additionally, extreme days for both summer and winter are also included (Cao et al. 2020)

Table 13.2 Planning results (Cao et al. 2020)

		DR–SP	SP	RO
EH 1	CHP (MW)	5.8732	5.6406	6.1008
	HP (MW)	2.6416	2.6325	2.5780
	ESU (MWh)	3.8214	4.0569	4.0688
	TSU (MWh)	6.8870	6.7553	6.6172
EH 2	CHP (MW)	5.9455	5.7004	6.1711
	HP (MW)	2.6013	2.5917	2.5393
	ESU (MWh)	3.8665	4.0572	4.0786
	TSU (MWh)	6.8611	6.7520	6.5141
Optimal value (10^7 \$)		8.0282	7.9391	8.6954
Investment cost (10^7 \$)		2.3283	2.2826	2.3547
Operation cost (10^7 \$)		5.6999	5.6565	6.3407
Computation time (s)		847	128	91

effect of uncertain parameters. Interestingly, the RO model shows a preference for investing in CHP units due to their flexibility, even though they result in the highest cost. Contrarily, the DR–SP model, when compared to the SP model, ensures system flexibility under normal conditions and reliability during extreme conditions. However, this comes with a slight increase in expenditure, amounting to an additional 1.12%. Table 13.2 also presents the computation time for each model. Algorithm 13.1, applied to the DR–SP model, converges within 6 iterations, taking a total of 847 s. This duration is considered acceptable within the context of a planning problem.

The primary advantage of the DR–SP model is its optimal solution's insensitivity to probability distribution perturbations. In this experiment, we generated data sets comprising 10,000 scenarios, each with a distribution distinct from the reference

Table 13.3 Failure rates under different KL divergence (Cao et al. 2020)

d_p^0		0.005 (%)	0.010 (%)	0.020 (%)	0.030 (%)	0.050 (%)
DR–SP	Pr_1	2.36	2.92	3.74	4.41	5.52
	Pr_2	2.22	2.75	3.55	4.19	5.27
SP	Pr_1	4.26	5.02	6.15	7.06	8.57
	Pr_2	4.13	4.88	5.99	6.89	8.37

one. We calculated their KL divergence d_p^0 in accordance with Definition 5.4. We then assessed the probability of demand loss events for each data set, with results presented in Table 13.3. Here, Pr_1/Pr_2 signifies the failure probability on extreme summer or winter days. Under the planning strategies generated by the conventional SP model, the failure probabilities on extreme weather days exceed the set value once $d_p^0 > 0.01$. Conversely, under the planning strategies provided by the DR–SP model, the reliability level surpasses 97% even when the actual data follows a different distribution and $d_p^0 < 0.0124$, the value utilized in W. This denotes the distributional robustness of the planning strategies. In fact, the reliability level can be sustained even if d_p^0 escalates to 0.03. This is attributable to our use of the convex function $\psi(x)$ to approximate the indicator function $1_{\mathbb{R}_+}(x)$, culminating in the conservative inequality (13.18). However, a further increase in d_p^0 violates the reliability requirement, as indicated in the final column of Table 13.3.

ESUs are key components in the daily operation of energy hubs. As we look to the future, the unit capacity costs of both ESUs and TSUs are anticipated to continually decrease. In this experiment, we adjusted the investment cost coefficients of the storage units by multiplying their values in Table 13.1 with a scalar, ζ. The results are presented in Table 13.4. From this, it is evident that as investment costs decrease, the storage capacity swiftly increases. Concurrently, there is a significant reduction in the capacity of the CHP unit, while the capacity of the heat pump experiences slight growth due to the increasing size of the TSU. These outcomes suggest that in future integrated systems, the demand for ES may experience rapid growth.

Lastly, we assessed the effect of wind penetration level. For this test, the wind farm output displayed in Fig. 13.2 is multiplied by a constant. The capacity refers to the maximum output throughout the day. Results are presented in Table 13.5. The data reveals that as more wind energy becomes available, excess electricity produced during the night can either be stored in the ESU or converted into heat and stored in the TSU, resulting in a significant increase in the capacity of storage units. Simultaneously, a larger heat pump is utilized to convert electric power into thermal energy, which serves to meet peak heat demand and charge the TSU. As a consequence, the capacity of the CHP unit decreases, given that more energy will be generated by wind turbines, which incur no operational costs. In contrast, dispatching the CHP unit involves non-zero gas fuel costs.

Table 13.4 Results under different storage costs (Cao et al. 2020)

ζ		1.0	0.8	0.6	0.4
EH 1	CHP (MW)	5.8732	5.3492	3.3455	3.0995
	HP (MW)	2.6416	2.7203	2.9550	3.0594
	ESU (MWh)	3.8214	5.8367	14.6498	15.7582
	TSU (MWh)	6.8870	8.1579	13.4595	14.2451
EH 2	CHP (MW)	5.9455	5.4157	3.3896	3.1393
	HP (MW)	2.6013	2.6767	2.9060	3.0067
	ESU (MWh)	3.8665	5.8203	14.5244	15.6491
	TSU (MWh)	6.8611	8.1389	13.4434	14.2272
Optimal value (10^7 \$)		8.0282	7.9438	7.7950	7.5944
Investment cost (10^7 \$)		2.3283	2.3636	2.5397	2.5890
Operation cost (10^7 \$)		5.6999	5.5802	5.2553	5.0054

Table 13.5 Results under different wind penetration levels (Cao et al. 2020)

Wind capacity (MW)		6	8	10	12
EH 1	CHP (MW)	6.4103	5.8732	4.9633	3.6916
	HP (MW)	2.5695	2.6416	2.8034	2.9743
	ESU (MWh)	3.0106	3.8214	6.2506	10.7163
	TSU (MWh)	5.5171	6.8870	8.9294	12.1764
EH 2	CHP (MW)	6.4923	5.9455	5.0227	3.7299
	HP (MW)	2.5327	2.6013	2.7678	2.9206
	ESU (MWh)	3.0770	3.8665	6.2979	10.6220
	TSU (MWh)	5.5207	6.8611	8.9499	12.1630
Optimal value (10^7 \$)		8.8974	8.0282	7.2216	6.4713
Investment cost (10^7 \$)		2.3429	2.3283	2.3534	2.4182
Operation cost (10^7 \$)		6.5545	5.6999	4.8682	4.0531

13.5 Conclusion

This chapter presents an innovative, data-driven, robust stochastic programming model that facilitates the decision-making process regarding component capacity within multi-carrier energy networks. This proposed method displays two unique and advantageous characteristics. Initially, it does not depend on the accurate distributions of uncertain elements, but instead, it integrates their dispersion effect. Secondly, its computational manageability is ensured owing to its convex characteristic and an LP-based algorithm. By efficiently integrating operational constraints and reliability prerequisites into the planning phase, the model is capable of delivering a more comprehensive and insightful guide for decision-makers.

The implications drawn from case studies are as follows: Methodologically, the model's conservatism is acceptable for planning applications. Its reliability exhibits robustness against fluctuations in renewable generation and load probability distributions, a quality referred to as distributional robustness. In essence, the proposed model delivers planning strategies that are both more robust and less conservative than existing methods, with adjustable conservatism to suit the needs of decision-makers. From an investment perspective, the continuous decline in storage construction costs coupled with the rising penetration of renewables suggests that the demand for ES in the integrated energy system will experience substantial growth.

References

Baran M, Wu F (1989) Network reconfiguration in distribution systems for loss reduction and load balancing. IEEE Trans Power Deliv 4(2):1401–1407

Boyd S, Vandenberghe L (2004) Convex optimization. Cambridge University Press, Cambridge

Cao Y, Wei W, Wang J et al (2020) Capacity planning of energy hub in multi-carrier energy networks: a data-driven robust stochastic programming approach. IEEE Trans Sustain Energy 11(1):3–14

Dupačová J, Gröwe-Kuska N, Römisch W (2003) Scenario reduction in stochastic programming. Math Program 95:493–511

Fang J, Zeng Q, Ai X et al (2018) Dynamic optimal energy flow in the integrated natural gas and electrical power systems. IEEE Trans Sustain Energy 9(1):188–198

Geidl M, Andersson G (2007) Optimal power flow of multiple energy carriers. IEEE Trans Power Syst 22(1):145–155

Geidl M, Koeppel G, Favre-Perrod P et al (2007) Energy hubs for the future. IEEE Power Energy Mag 5(1):24–30

Good N, Mancarella P (2019) Flexibility in multi-energy communities with electrical and thermal storage: a stochastic, robust approach for multi-service demand response. IEEE Trans Smart Grid 10(1):503–513

Heinen S, Hewicker C, Jenkins N et al (2017) Unleashing the flexibility of gas: innovating gas systems to meet the electricity system's flexibility requirements. IEEE Power Energy Mag 15(1):16–24

Hogan WW (1973) Application of a general convergence theory for outer approximation algorithms. Math Program 5:151–168

Kienzle F, Ahčin P, Andersson G (2011) Valuing investments in multi-energy conversion, storage, and demand-side management systems under uncertainty. IEEE Trans Sustain Energy 2(2):194–202

Krause T, Andersson G, Fröhlich K et al (2011) Multiple-energy carriers: modeling of production, delivery, and consumption. Proc IEEE 99(1):15–27

Li S, Sui J, Jin H et al (2013) Full chain energy performance for a combined cooling, heating and power system running with methanol and solar energy. Appl Energy 112:673–681

Li Z, Guo Q, Sun H et al (2016) Sufficient conditions for exact relaxation of complementarity constraints for storage-concerned economic dispatch. IEEE Trans Power Syst 31(2):1653–1654

Liu X, Wu J, Jenkins N et al (2016) Combined analysis of electricity and heat networks. Appl Energy 162:1238–1250

Lofberg J (2004) YALMIP: a toolbox for modeling and optimization in MATLAB. In: 2004 IEEE international conference on robotics and automation, pp 284–289

Nemirovski A, Shapiro A (2007) Convex approximations of chance constrained programs. SIAM J Optim 17(4):969–996

Wang C, Wei W, Wang J et al (2018) Equilibrium of interdependent gas and electricity markets with marginal price based bilateral energy trading. IEEE Trans Power Syst 33(5):4854–4867

Wu L (2011) A tighter piecewise linear approximation of quadratic cost curves for unit commitment problems. IEEE Trans Power Syst 26(4):2581–2583

Index

R. Xie and W. Wei, *Distributionally Robust Optimization and its Applications in Power System Energy Storage Sizing*,
https://doi.org/10.1007/978-981-97-2566-3

The manufacturer's authorised representative in the EU is Springer Nature Customer Service Centre GmbH, Europaplatz 3, 69115 Heidelberg, Germany. If you have any concerns regarding our products, please contact ProductSafety@springernature.com

Printed and bound by CPI Group (UK) Ltd, Croydon, CR0 4YY

01/10/2026

02225849-0015